Adaptive Control Systems

ELECTRICAL ENGINEERING AND ELECTRONICS

A Series of Reference Books and Textbooks

1. Rational Fault Analysis, *edited by Richard Saeks and S. R. Liberty*
2. Nonparametric Methods in Communications, *edited by P. Papantoni-Kazakos and Dimitri Kazakos*
3. Interactive Pattern Recognition, *Yi-tzuu Chien*
4. Solid-State Electronics, *Lawrence E. Murr*
5. Electronic, Magnetic, and Thermal Properties of Solid Materials, *Klaus Schröder*
6. Magnetic-Bubble Memory Technology, *Hsu Chang*
7. Transformer and Inductor Design Handbook, *Colonel Wm. T. McLyman*
8. Electromagnetics: Classical and Modern Theory and Applications, *Samuel Seely and Alexander D. Poularikas*
9. One-Dimensional Digital Signal Processing, *Chi-Tsong Chen*
10. Interconnected Dynamical Systems, *Raymond A. DeCarlo and Richard Saeks*
11. Modern Digital Control Systems, *Raymond G. Jacquot*
12. Hybrid Circuit Design and Manufacture, *Roydn D. Jones*
13. Magnetic Core Selection for Transformers and Inductors: A User's Guide to Practice and Specification, *Colonel Wm. T. McLyman*
14. Static and Rotating Electromagnetic Devices, *Richard H. Engelmann*
15. Energy-Efficient Electric Motors: Selection and Application, *John C. Andreas*
16. Electromagnetic Compossibility, *Heinz M. Schlicke*
17. Electronics: Models, Analysis, and Systems, *James G. Gottling*

18. Digital Filter Design Handbook, *Fred J. Taylor*

19. Multivariable Control: An Introduction, *P. K. Sinha*

20. Flexible Circuits: Design and Applications, *Steve Gurley, with contributions by Carl A. Edstrom, Jr., Ray D. Greenway, and William P. Kelly*

21. Circuit Interruption: Theory and Techniques, *Thomas E. Browne, Jr.*

22. Switch Mode Power Conversion: Basic Theory and Design, *K. Kit Sum*

23. Pattern Recognition: Applications to Large Data-Set Problems, *Sing-Tze Bow*

24. Custom-Specific Integrated Circuits: Design and Fabrication, *Stanley L. Hurst*

25. Digital Circuits: Logic and Design, *Ronald C. Emery*

26. Large-Scale Control Systems: Theories and Techniques, *Magdi S. Mahmoud, Mohamed F. Hassan, and Mohamed G. Darwish*

27. Microprocessor Software Project Management, *Eli T. Fathi and Cedric V. W. Armstrong (Sponsored by Ontario Centre for Microelectronics)*

28. Low Frequency Electromagnetic Design, *Michael P. Perry*

29. Multidimensional Systems: Techniques and Applications, *edited by Spyros G. Tzafestas*

30. AC Motors for High-Performance Applications: Analysis and Control, *Sakae Yamamura*

31. Ceramic Materials for Electronics: Processing, Properties, and Applications, *edited by Relva C. Buchanan*

32. Microcomputer Bus Structures and Bus Interface Design, *Arthur L. Dexter*

33. End User's Guide to Innovative Flexible Circuit Packaging, *Jay J. Miniet*

34. Reliability Engineering for Electronic Design, *Norman B. Fuqua*

35. Design Fundamentals for Low-Voltage Distribution and Control, *Frank W. Kussy and Jack L. Warren*

36. Encapsulation of Electronic Devices and Components, *Edward R. Salmon*

37. Protective Relaying: Principles and Applications, *J. Lewis Blackburn*

38. Testing Active and Passive Electronic Components, *Richard F. Powell*

39. Adaptive Control Systems: Techniques and Applications, *V. V. Chalam*

40. Computer-Aided Analysis of Power Electronic Systems, *Venkatachari Rajagopalan*

Additional Volumes in Preparation

Electrical Engineering-Electronics Software

1. Transformer and Inductor Design Software for the IBM PC, *Colonel Wm. T. McLyman*
2. Transformer and Inductor Design Software for the Macintosh, *Colonel Wm. T. McLyman*
3. Digital Filter Design Software for the IBM PC, *Fred J. Taylor and Thanos Stouraitis*

Adaptive Control Systems

TECHNIQUES AND APPLICATIONS

V. V. Chalam

Director
Institute of Technology
Banaras Hindu University
Varanasi, India

MARCEL DEKKER, INC. New York and Basel

Library of Congress Cataloging-in-Publication Data
Chalam, V. V.
Adaptive control.
(Electrical engineering and electronics ; 39)
Includes bibliographies and index.
1. Adaptive control systems. I. Title. II. Series.
TJ217.C47 1987 629.8'36 87-5320
ISBN 0-8247-7650-X

MARCEL DEKKER, INC.
270 Madison Avenue, New York, New York 10016

Current printing (last digit):
10 9 8 7 6 5 4 3 2 1

PRINTED IN THE UNITED STATES OF AMERICA

To

The Kindly Light

Yat karoṣi yad aśnāsi yaj juhoṣi dadāsi yat |
Yat tapasyasi Kaunteya tat kuruṣva mad arpaṇam ||

Whatever you do, whatever you eat, whatever you offer in sacrifice, whatever you gift away, whatever austerity you practise, do it as an offering to Me.

— Bhagavadgītā

Preface

Adaptive control is a field that has come of age. I was fascinated by adaptive control twenty-five years ago while a doctoral student at the Moore School of Electrical Engineering, University of Pennsylvania. Since then I have had sustained interest in this area, as I have been teaching this topic to graduate students and guiding their theses. In recent years I have found, as have many of my co-workers in this discipline, that there is no up-to-date book covering all the aspects of adaptive control in a systematic manner with consistent terminology and compatible notation. This produces difficulties for graduate students and researchers, since valuable material is widely scattered in a large number of papers, reports, proceedings of symposia, and so on, with different authors employing different symbols and terms. Comprehensive and integrated coverage of the subject in one volume is needed. The basic objective of the present work is to present a unified treatment so as to "give the part, Wholeness in a sense."

I do realize that no one can write a "once-and-for-all" book on this fast-evolving subject. The past few years have seen significant advances in adaptive control. One can only hope to provide state-of-the-art coverage of the adaptive control area by identifying certain important issues and trends. In the effort to encompass a broad range of topics not normally found in a single book on adaptive control, many details and certain methodologies must be omitted to keep the book to a reasonable length. However, extensive references and bibliography are given as a source of pertinent literature for further study.

To quote from Don Quixote, "And when all is said, he that publishes a book runs a very great hazard, since nothing can be more impossible than to compose one that may secure the approbation of every reader." All the same, I have dared to compose the book, keeping in mind the words of Lord Krishna in the Bhagavadgītā:

> Sahajam karma Kaunteya sadoṣam api na tyajet |
> Sarvārambhā hi doṣeṇa dhūmenā'gnir ivā' vṛtāḥ ||

This Sanskrit stanza may be translated as: "One's own innate duty ought not to be abandoned, though tainted with blemish. All undertakings indeed are clouded by flaws, as fire is by smoke."

This book is addressed to three primary audiences. It serves as a research monograph for researchers, as a reference work for control engineers, and as an advanced-level textbook for graduate students. In recent years the advent of powerful 8- and 16-bit microprocessors has contributed to a qualitative change in the area of control. Industry has moved enthusiastically in applying computer- and microprocessor-based control systems, resulting in increasingly complex and sophisticated real-time systems. The discipline of adaptive control contains many new ideas and techniques which are novel to practicing engineers. This book is expected to serve as an up-to-date single source of vital information on basic concepts, analytical tools, and a wide variety of application trends, so that potential users may choose, commission, and operate adaptive controllers, wherever relevant.

The reader is required to have a mathematical knowledge of linear and nonlinear control systems, discrete-time control, optimal control, and stochastic processes. Familiarity with microprocessor applications will be very helpful to the reader.

I take great pleasure in acknowledging my deep sense of indebtedness to many who made this effort successful. First of all, my gratitude is due to those countless researchers "standing on whose shoulders" I could get an overview of the entire field. A vast majority of the material in this book is drawn from the technical literature. I wish to thank Springer-Verlag, Elsevier Science Publishers (North-Holland), Pergamon Press, Marcel Dekker, Inc., and the publishers of IEEE Transactions, IEE Proceedings, Automatica, International Journal of Control, Transactions of Institute of Measurement and Control, and others, for permitting me to use the material published in their books and journals. I am also grateful to Drs. F. Davoli, B. Egardt, P. J. Gawthrop, G. C. Goodwin, M. Jamshidi, K. S. Narendra, P. C. Parks, H. Unbehauen, and other authors for giving permission to use published material.

My sincere thanks are due to fellow workers in many countries, interaction with whom generated the motivation to undertake this project. I am also highly obligated to researchers all over the world who readily provided me with reprints of their papers and reports. Pertinent questions raised by my graduate students also contributed to my probing further into the literature and whetted my desire to take up this work.

It is more than a duty to acknowledge with grateful thanks the help received from the authorities of:

1. The Banaras Hindu University, for granting one year's sabbatical leave
2. The Indian Institute of Technology, Delhi, for inviting me as a Visiting Professor in the Curriculum Development Cell of their Department of Electrical Engineering and providing library and other facilities
3. The University Grants Commission, New Delhi, for providing the funds for the preparation of the manuscript
4. The British Council, New Delhi, for arranging short visits to a few universities in the United Kingdom where research is going on in adaptive control.

I am thankful to Mr. Satyanarayan and Mr. C. H. Pandey for their painstaking typing in spite of personal commitments and constraints on their time.

Special mention must be made of the moral support and encouragement from Nāgarāj, Chittaranjan, Ramā, and Srīnāth. I have also to record that what sustained me during the travails of the period of this writing was the music of the great masters.

Without the emotional support, understanding, and patience of my wife, Nāgarathnā, this book would not have seen the light of the day.

Finally, the following words of a Hindu savant have always inspired me: "We usually undertake a project as if it were our own personal wish and it is not easily given to us to know that even behind the so-called personal wish, the pull and push of the Universal is at work. Whether the wave knows it or not, the force behind it is of the Ocean. Blessed is he who is allowed to maintain a consciousness, even occasionally, of this tremendous way of working of the Divine."

V. V. Chalam

Contents

PROLOGUE

I do not know what I may appear to the world, but to myself I seem to have been only like a boy playing on the sea-shore, and diverting myself in now and then finding a smoother pebble or a prettier shell than ordinary, whilst the great ocean of truth lay all undiscovered before me.

Sir Isaac Newton

1

Motivation, Definitions, and Classification

> There is beginning before the beginning,
> as preparation of the wick in the morning
> for the evening Pradeep (lamp).
>
> Rabindranath Tagore

1.1 INTRODUCTION

Adaptive control is an important area of modern control, dealing as it does with the control of systems in the presence of uncertainties, structural perturbations, and environmental variations. There has been a recent upsurge of interest in this field, since adaptive controllers can be implemented simply because of the availability of increasingly versatile digital hardware. Adaptive control techniques have also benefited from the steady and even spectacular reduction in the cost/performance ratio of microelectronic devices in recent years. This has resulted in a wide variety of industrial applications in situations which were not considered easily implementable earlier. Furthermore, many theoretical problems that had baffled researchers have also been solved during the past few years. There have been major advances in the design of adaptive control systems, especially in terms of global stability. Several adaptive control strategies have been successfully applied in diverse practical problems. Adaptive control's time has come!

The growth of interest in adaptive control is shown in Fig. 1.1, which indicates the dramatic increase in interest during the past few years. Adaptive control techniques have great potential, as only these methods can cope with increasingly complex systems in the presence of extreme changes in system parameters and input signals.

In the succeeding chapters a unified presentation is given of the various aspects of adaptive control. Because the field of adaptive systems is of great breadth and depth, it is possible to cover only a selected set of topics. However, a comprehensive overview of the entire field is given,

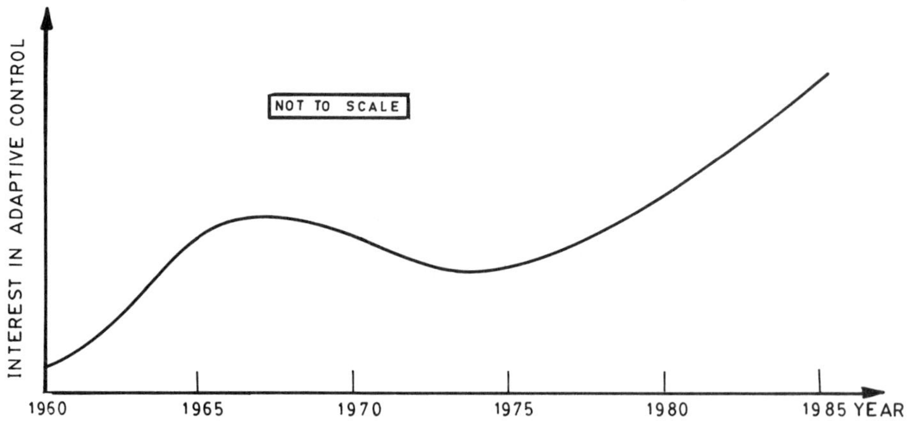

FIG. 1.1 Growth of interest in adaptive control.

including recent trends in theory and applications, so that the reader will readily recognize that this is a discipline of considerable theoretical elegance, challenging practical problems, and significant promise.

The present chapter is organized as follows: Section 1.1 deals with some important landmarks, some early work, and the surveys, bibliographies, and books that have come out on this subject. Section 1.2 considers the important questions regarding why adaptive control is to be used, while at the same time emphasizing the need for probing whether conventional control schemes can be applied to solve the problem involved satisfactorily. Several proposed definitions are included in Section 1.3, even though it is realized that there is, at present, no universally acceptable definition. Essential aspects of adaptive control are discussed in Section 1.4. Section 1.5 considers different classifications of adaptive control schemes and introduces the two main approaches: model-reference adaptive control systems and self-tuning regulators. Concluding remarks are given in Section 1.6.

1.1.1 Some Important Landmarks

Adaptive control was first proposed by Draper and Li [1] as far back as 1951. They dealt with a control system which could optimize the performance of an internal combustion engine in the presence of uncertainties in the performance characteristics. This type of control, which automatically searches for an optimal operating point, is designated as an <u>optimalizing control</u> or sometimes a <u>peak-holding control</u>. Their further work was reported in 1955 [2]. The term adaptive was used in the control literature in 1954, when Tsien [3] described Ashby's [4] model of the human brain. In 1955 Benner and Drenick [5] presented a control system with "adaptive" characteristics.

The next major step in adaptive control was taken by Whitaker et al. [6] in 1958 when they considered adaptive aircraft flight control systems, employing a reference model to obtain error signals between the actual and desired behavior. These error signals were used to modify the controller parameters to attain ideal behavior to the extent possible in spite of uncertain and varying system dynamics. Such systems are referred to as model reference adaptive control systems (MRAC systems). Their further work was reported in Refs. 7-9.

Li and van der Velde [10] in 1960 considered another type of adaptive control system, based on the automatic compensation of parameter uncertainties introduced by a limit cycle in the control loop. Such a system is called a self-oscillating adaptive system.

Petrov et al. [11] introduced another approach in 1963 for adaptive system design based on the invariance of the system trajectory in relation to parameter values when the control input is generated from a switching function and relay. Such systems are termed variable structure systems.

Bellman [12] in the United States and Fel'dbaum [13] in the USSR reported in 1960-1961 on the application of dynamic programming concepts in the design of controllers for systems with probabilistic uncertainties. The interesting dual property of control inputs as identification (parameter estimation) signals and actuation (control) signals led to the term dual control.

Lainiotis [14] introduced the partitioning approach to adaptive control in 1971. Partitioned adaptive controls are shown [15] to have several interesting properties as well as a completely decoupled parallel-processing structure that is computationally attractive and robust.

Astrom and Wittenmark [16] were responsible for the development of another important form of adaptive control, self-tuning regulators (STR), which are relatively easy to implement in view of the availability of microprocessors. This seminal work led to a great upsurge of interest throughout the world in developing new versions of STRs and exploring a variety of applications, even in situations earlier thought to be very difficult. It is probably pertinent to say that applications based on self-tuning regulators have now outnumbered those based on other methods.

Another important breakthrough came in 1974 when Monopoli [17] introduced the augmented error approach to the design of stable MRAC systems. This led to a considerable amount of research in developing adaptive algorithms for globally asymptotically stable systems and resulted in 1980 in several important papers by Narendra et al. [18-20], Morse [21], Goodwin et al. [22], and Egardt [23,24], both for continuous-time and discrete-time adaptive systems. Not only has there been a resolution of the longstanding problem of stability, but also a unification of different approaches to adaptive control and a realization of the equivalence of MRAC systems and STR.

Adaptive control has now reached a considerable degree of maturity with respect to significant theoretical and algorithmic advances, and adaptive control systems have important current and potential applications.

1.1.2 Some Early Work

An important symposium was held as far back as 1959 at the Wright Air Development Center (WADC), Dayton, Ohio, on self-adaptive flight control systems [25]. During this symposium, many interesting ideas were suggested and these formed the basis for later work on model reference adaptive control [MRAC]. Even the concept involving identification followed by adaptive control action was indicated, and this ultimately led to the introduction of self-tuning regulators.

Although there has been a tremendous outpuring of papers on various aspects of adaptive control in recent years, it is interesting to note some of the early papers [26-54] published 25 years ago. Similarly, among the large number of doctoral dissertations all over the world, some of the early ones are represented by Refs. 55-65.

The early adaptive schemes were not implementable with the then available hardware. Further, many of the proposed systems were not properly understood, as theory had not kept pace and there were doubts about the stability and performance of such systems. Designs based on stability concepts using Lyapunov functions were developed and synthesized in the late 1960s. It was only in 1980 that the longstanding problems of global stability were solved.

1.1.3 Surveys, Bibliographies, and Books

The first survey of adaptive control systems was presented as far back as 1958 by Aseltine et al. [66] and the first bibliography was prepared as early as 1959 by Stromer [67]. Cooper et al. [68] give in 1960 a survey of the philosophy and state of the art of adaptive systems. Jacobs [69] reviews self-adjusting systems in 1961. Truxal [70] in 1963 and Zypkin [71] in 1973 survey the adaptive control scene. Other important surveys on specific aspects of adaptive systems are by Åström and Eykoff [72] in 1971 on system identification; Landau [73] in 1972 and [74] in 1974 on model-reference adaptive systems along with an extensive list of references; Lindorff and Caroll [75] in 1973 on adaptive control using the Lyapunov approach; Narendra and Kudva [76] in 1974 on stable adaptive schemes for identification and control; Wittenmark [77] in 1975 on stochastic adaptive control; and Unbehauen and Schmid [78] in 1975 on the status and applications of adaptive control. Asher et al. [79] give in 1976 an exhaustive bibliography of 700 papers divided into several categories. Sternby [80] considers in 1980 extremum adaptive control systems. Parks et al. [81] survey the applications of adaptive control systems in 1980. A survey on theory and applications of adaptive control has been presented in 1983 by Astrom [82]. In 1984 Voronov et al. [83] give an account of the status of adaptive control in the USSR. In 1985

Unbehauen [83a] surveys system identification methods using parameter estimation.

There has been an ever-increasing interest in adaptive control and this has led to the publication of special issues of important technical journals in recent years: IEEE Proceedings in 1976 [84]; IEEE Transactions on Automatic Control in 1977 [85]; Optimal Control Applications and Methods in 1982 [86]; and Automatica in 1984 [87].

Among the early books on adaptive control are those by Bellman [88] and Mishkin and Braun [89] in 1961, followed by Westcott [90] in 1962, and by Caruthers et al. [91]. Other relevant books on certain aspects of adaptive control are given in Refs. 92-101. Among the recent books are those by Saridis [102], who covers the important area of self-organizing control of stochastic systems; Egardt [103] dealing with stability of adaptive controllers; and Landau [104] on model reference adaptive systems with special reference to the application of hyperstability concepts. Three publications edited by Narendra and Monopoli [105], Unbehauen [106], and Harris and Billings [107] are a collection of interesting papers on specific topics presented at symposia and workshops, held during 1979-1981 in the United States, West Germany, and England, respectively. Other related topics are covered by Ioannou and Kokotovic [108], Åström and Wittenmark [109], and Goodwin and Sin [110].

1.2 WHY USE ADAPTIVE CONTROL

Many engineering problems demand automatic control in the presence of uncertainties and unforeseen large changes in system parameters and input signals. These structural perturbations and environmental variations which result in the degradation of performance may be cited in a number of examples:

1. Aircraft control: The dynamic behavior of a supersonic aircraft varies widely as a result of the difference in air density close to ground level and at 30,000 ft altitude. The aerodynamic characteristics and hence the control characteristics vary widely with altitude, speed, and the configuration of the aircraft. The ratio of variation of some parameters lies between 10 and 50.

2. Missile control: In a missile the mass and the center of gravity alter rapidly as the fuel is expended.

3. Process control: In a continuous process chemical plant there may be changes in the parameters when there are changes in the ambient temperature and changes in the rates of inflow or outflow.

4. Electric drives: In a paper mill the moment of inertia of the winding reel changes as the paper is wound on, so the motor torque required to maintain constant tension in the web must also change. Similar problems occur in rolling mills, wiring machines, machine tools, and so on.

Further, with thyristor control of motors there are structural perturbations when the firing angle changes, affecting the parameters of the plant.

5. Navigational course control of ships: In a ship or an oil tanker the dynamic characteristics of the transfer function (rudder angle divided by course angular speed) vary with ship load, speed, the depth of water, and the environment in which it operates (i.e., waves, wind, current, etc.).

6. Metallurgical processes: The parameters of the various processes vary from batch to batch. Starting conditions are not always the same. The material introduced is never exactly the same. The reactor characteristics are also variable in the course of its life.

7. Satellite attitude control: Some satellites are used for observation of celestial phenomena which are highly irregular and short-lived. There is need for very quick reorientation of the satellite so as to increase the available observation time.

In general, in many practical control problems we may come across

1. Changes in the plant transfer function, either in its order or in the values of some parameters due to variations in the environment, the size and properties of raw materials, the plant throughput, the characteristics involving alterations in the coefficients, and/or wear and tear of some important components.
2. Stochastic disturbances.
3. Changes in the nature of inputs.
4. Propagation of disturbances along a chain of unit processes.
5. Nonlinear behavior as in the case of complex chemical or biochemical reactions.
6. Appreciable dead time.
7. Unknown parameters, as when a control system for a new process is commissioned.

In all such situations a conventional controller cannot maintain the performance of the system at acceptable levels. There is, therefore, a need for a special class of control systems which can automatically compensate for these unforeseen variations in parameters and input signals. Early attempts to overcome such problems involved designing an intentionally nonlinear control system with adjustment of, say, the loop gain. Considerable analytical effort is needed to specify the optimum parameter settings. Such a fixed compensator can be successful over a limited range of operating conditions. Outside this range the compensation is inadequate and there is degradation of the performance.

Practical control systems are usually very complex and their dynamics are not known in detail. The extreme changes in the parameters and/or inputs or random fluctuations occurring within the system are often not measurable. In such situations the control engineer has to employ adaptive control systems which automatically sense and correct themselves whenever drastic disturbances and/or severe changes in the parameters occur. It is also possible to introduce a simple learning mechanism in the adaptive part of the system. Any complexity should be added to the system only if improved performance is mandatory and the costs are justifiable. An adaptive control system has to be designed so as to guarantee stability and also robustness in the presence of disturbances. The designer has also to keep in mind the need for high speed of the adaptation algorithm to compensate for parameter variations and the need to use only a small amount of computer time and capacity.

One should explore alternative solutions before considering adaptive control. There are situations in which the variations in parameters are not significant and the degree of improvement demanded is not such as to justify the increased cost and complexity of adaptive control. In such cases conventional methods may be adequate as a constant-gain feedback may be able to cope with the variations in the system dynamics. Such instances are referred to by Åström [82,111,112], Horowitz [113], Parks et al. [81], and Kidd [114].

In an interesting study on when adaptive control is useful, Jacobs [115] has reported on three different studies:

1. A Monte Carlo study [116] in which the performance of two adaptive algorithms are compared with each other. Examples are provided of (a) problems where adaptive control is no better than classical control, (b) problems where adaptive control is better, (c) problems where one adaptive control is better than another, and (d) problems where neither classical nor adaptive control is effective, if the process is uncontrollable.
2. A laboratory study [117] indicating that adaptive control is not required for a water-level control problem.
3. An industrial case study [118] indicating improvement in performance of adaptive control of pH effluent.

We must note that adaptive control techniques are much more complex and expensive than conventional control schemes, in spite of the availability of inexpensive microcomputers for implementation. Before considering adaptive control for any application, one should examine whether the extra cost involved is justifiable in terms of significant improvement in performance.

1.3 DEFINITIONS

Several definitions have been proposed for adaptive control. Several terms such as "adaptive," "learning," "self-organizing," "self-tuning," "optimalizing," and "self-optimizing" have been introduced by different workers in this field. Some terms have given rise to controversies and misunderstanding.

None of the definitions are general enough to include all the work done by researchers in this area. There is no universally acceptable definition at present. A precise definition is somewhat difficult because of several forms of uncertainties present in a system and because of different methodologies involved to tackle the situation. However, in recent years there has been a greater realization of the equivalence of different approaches and a unified view is steadily emerging.

A few of the well-known definitions are given below. A knowledge of these enables the reader to be aware of the extensive scope of this important discipline.

Self-Organizing Control

A control process is called self-organizing [119] if reduction of the a priori uncertainties pertaining to the effective control of the process is accomplished through information acquired from subsequent observations of the accessible inputs and outputs as the control process evolves.

Adaptation

Adaptation [101] is considered to be a process of modifying the parameters of the structure of the system and the control actions. The current information is used to obtain a definite (usually optimal) state of the system when the operating conditions are uncertain and time varying.

Adaptive Systems

An adaptive system [48] continuously searches for the optimum within its allowed class of possibilities by an orderly trial-and-error process and can give performance vastly superior to that of a fixed system.

Adaptive Control

An adaptive control system [120] must provide continuous information about the present state of the plant, that is, it must identify the process; it must compare present system performance with the desired or optimum performance and make a decision to adapt the system so as to tend toward optimum performance; and finally it must initiate a proper modification so as to drive the system toward the optimum. These three functions are inherent in an adaptive system.

Adaptive Control Processes

In adaptive control processes we can learn about the nature of the unknown elements as the process proceeds [88]. The multistage aspects of the process must be made to compensate for the initial lack of information. The mathematical theory developed to treat control processes of this type is called the theory of adaptive control processes.

Dual Control

Under dual control we consider those optimal systems with incomplete information about the plant and with active accumulation of information during the control process. Fel'dbaum introduced the word dual, as he noted that control when applied to a system has two effects: in addition to its effect on the state, the present control might affect the future state uncertainty. Furthermore, the control signal has a twofold purpose: (1) to investigate the characteristics of the process, and (2) to implement the regulation and to direct the process to the required state. These two actions are combined in dual control.

Learning

Learning [101] in a system involves a process of forcing the system to have a particular response to a specific input signal (action) by repeating the input signals and then correcting the system externally.

Self-Learning

Self-learning [101] is learning without external corrections, that is, without punishments or rewards. Any additional information regarding the correctness or incorrectness of the system's reaction is not given.

Learning Systems

A learning system [122] is one that learns the unknown information about a process and uses it as an experience for future decisions or controls, so that the performance of the system will be gradually improved.

Alternatively, according to Saridis et al. [119], a learning system is one in which information about the unknown features of a process or its environment is acquired by the system and the obtained experience is used for future estimation, recognition, classification, decision, or control to improve the performance of the system.

Learning Control

Learning controllers [123] are capable of estimating the unknown information during the system's operation and then determining an optimal control such that the performance of the system will be gradually improved.

Alternatively, according to Saridis et al. [119], a learning system is called a learning control system if the acquired information is used to control a process with unknown features. Learning control may be performed off-line (in which case the controller is trainable) or on-line.

Adaptive and Learning Control

The possibility of controlling plants under incomplete information is based on the application of adaptation and learning in automatic systems

[101] which reduces initial uncertainty by using information obtained during the process of control.

The preceding sample of definitions indicates that there is no agreement among researchers. It may also be pointed out that the names adaptive system, adaptive control, learning systems, or learning control have been used rather loosely by practicing engineers to describe systems with additional feedback loops. Åström [82] states that a meaningful definition of adaptive control which makes it possible to look at a regulator and decide whether it is adaptive or not is still missing. However, there appears to be a consensus [79] that adaptive control means the control of uncertain systems, with a controller structure that includes a subsystem for the on-line estimation of unknown parameter values or system structure and a subsystem for the generation of suitable control inputs based on estimated parameter values or system structure.

There is a clear need to incorporate some of the ideas in the above definitions into the formulation of a new noncontroversial definition. This is no easy task in the constantly evolving discipline of adaptive control.

1.4 ESSENTIAL ASPECTS OF ADAPTIVE CONTROL

The basic functions common to most adaptive control systems are the following:

1. Identification of unknown parameters or measurement of an index of performance (IP)
2. Decision on the control strategy
3. On-line modification of the parameters of the controller or the input signal

Depending on how these functions are brought about, we have different types of adaptive controllers. To outline the essential aspects of adaptive control, the following "definition"]74] may be considered:

> An adaptive system measures a certain index of performance (IP) using the inputs, the states, and the outputs of the adjustable system. From the comparison of the measured IP values and a set of given ones, the adaptation mechanism modifies the parameters of the adjustable system or generates an auxiliary input in order to maintain the IP values close to the set of given ones.

This "definition" is illustrated in Fig. 1.2 [74]. There are many practical difficulties in implementing the three blocks in Fig. 1.2, as the systems are generally rather complicated. Furthermore, it is not always

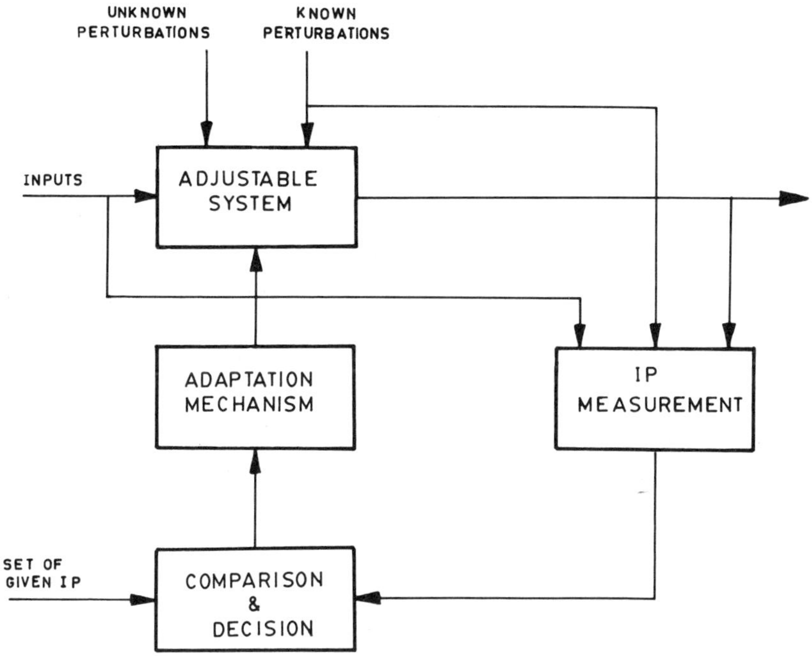

FIG. 1.2 Essential components of an adaptive control system. (From Ref. 74.)

possible for an adaptive system to be split into three distinct blocks as shown in Fig. 1.2.

The "adjustable system" refers to a system whose performance can be adjusted either by modifying its parameters (i.e., its internal structure) or by modifying its input signals. "IP measurement" may be done either directly or indirectly, as, for example, by "identification" of the dynamic parameters of the system. Hence identification, or parameter estimation, is a vital aspect of an adaptive control system. There are several methods by which this can be accomplished. A brief account of recursive parameter estimation is given in Section 5.3. The "comparison and decision" block "compares" the set of given IP and the measured IP and then "decides" whether the measured values are within acceptable limits. If not, this block impels the "adaptation mechanism" to act appropriately to modify system performance, either by altering the parameters of the adjustable system, in which case the system will be called a parameter-adaptive control system or by changing the input signal, in which case the system is said to be a signal-synthesis adaptive control system.

1.5 CLASSIFICATION

1.5.1 Classification in General

Adaptive control systems may be classified by many criteria, leading to several classifications and categories. Specifically, they may be classified [74,123] according to:

1. The adaptation mechanism: parameter-adaptive or signal-synthesis adaptive systems
2. The operating conditions: deterministic, stochastic, or learning systems
3. The nature of the basic mathematical equations used: linear, distributed parameter, continuous-time, discrete-time systems, or hybrid systems in which there is discrete-time adaptation of continuous-time systems
4. The index of performance specified: static (e.g., efficiency), dynamic (e.g., shape of response), parametric (e.g., damping factor), or functional of state variables and inputs (e.g., quadratic integral criterion)
5. The need of test signals to measure the index of performance correctly and quickly
6. The nature of the "comparison-decision block": (a) subtractor, when the given IP is specified uniquely and can be represented by a certain signal; (b) extremal, when a certain IP is either maximized or minimized; or (c) belonging to a certain domain of values, when a parameter (say, damping factor) is desired to be maintained within a specified range of values
7. The nature of uncertainty: parametric or structural uncertainty and time-invariant or time-varying uncertainties
8. The nature of the constraints imposed either on the desired system or on the nature of the adaptation process.

1.5.2 Actively Adaptive or Passively Adaptive Systems

Adaptive control methods are also classified into two categories [124], actively adaptive or passively adaptive, according to how the available information is being utilized in the on-line calculation of the control. An important element of adaptive control is the learning of drifting parameters. As the process unfolds, additional information becomes available either accidentally through past control actions or as a result of active probing, which itself is a possible control policy.

An actively adaptive controller utilizes, in addition to the available real-time information, the knowledge that future observations will be made and regulates its adaptation or learning accordingly. This is done by anticipating how future estimation will be beneficial to the control objective.

A passively adaptive controller, on the contrary, while utilizing the available real-time measurements, does not take into account the fact that future measurements will be made.

An actively adaptive control gives a better performance but it is also more complicated and computationally more demanding than passively adaptive control. The reader is referred to the work of Tse and Bar-Shalom [124,125] for further insight into the nature of actively adaptive controls.

1.5.3 Model Reference Adaptive Control

There are two principal approaches to adaptive control, namely, model reference adaptive control (MRAC), to be considered briefly now, and self-tuning regulators, dealt with briefly in Section 1.5.4.

MRAC systems are important since they lead to relatively easy-to-implement systems with a fairly high speed of adaptation as there is no need for identification of the plant dynamic performance. They can be used in a variety of situations. MRAC was originally developed by Whitaker and his co-workers at MIT in 1958 [6] for designing adaptive autopilots.

In MRAC the aim is to make the output of an unknown plant approach asymptotically the output of a given reference model, which is part of the control system, as shown in Fig. 1.3.

In the MRAC system in Fig. 1.3 the given set of IP appearing in Fig. 1.2 is replaced by one dynamic-type IP, which becomes the reference IP. To generate this reference IP, a reference model is used, and this is excited by the same input as the adjustable system. In this case the comparison between the given IP and the measured IP is obtained directly by

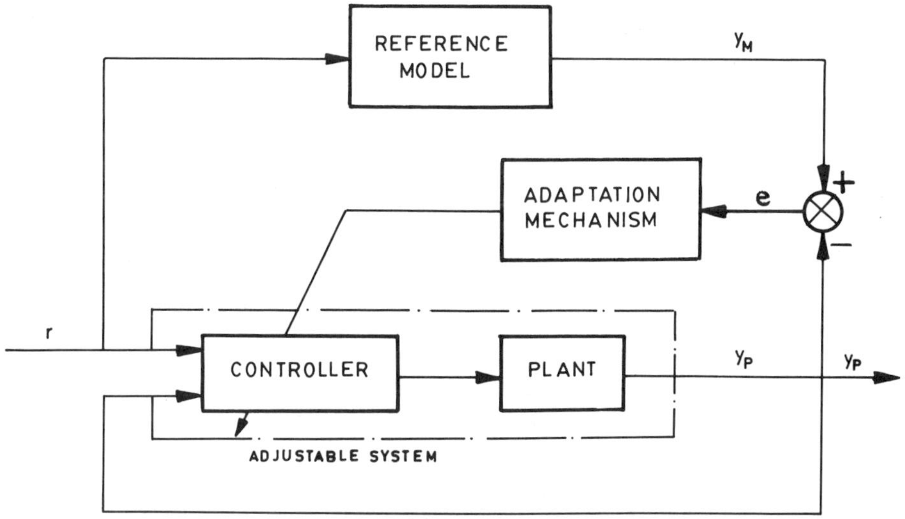

FIG. 1.3 Model reference adaptive control system (parameter adaptation).

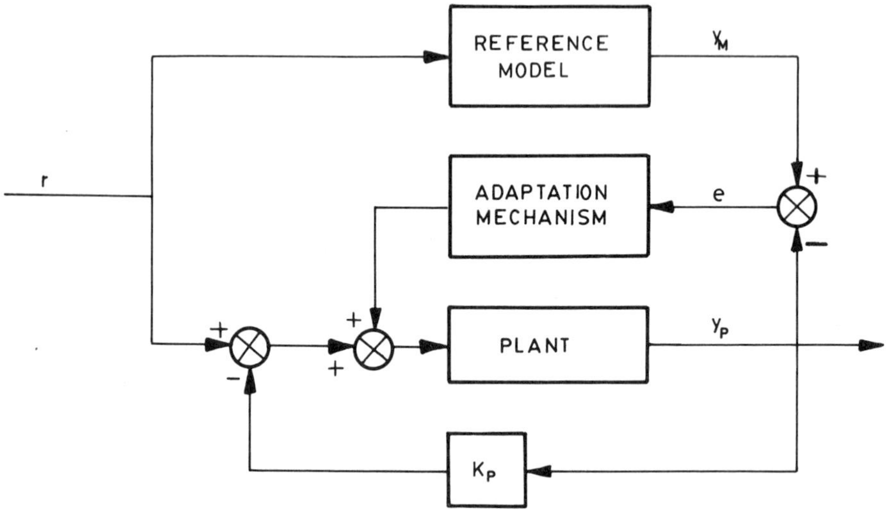

FIG 1.4 Model reference adaptive control system (signal synthesis adaptation).

comparing the output of the reference model and that of the adjustable system. This "error" is used by the adaptation mechanism to modify the controller parameters, so as to minimize the "error." This is a parameter-adaptive system. One can also arrange the adaptation mechanism to generate an auxiliary input signal to minimize the difference between the output of the reference model and that of the adjustable system. In this case the system will be a signal-synthesis MRAC (Fig. 1.4). The main problem is to determine the adaptation mechanism such that not only is the error brought to zero but also the resulting system is stable. This is a difficult problem and considerable research effort has gone on to design MRAC from the point of view of stability; if stability is not guaranteed, the adaptive system is not of practical utility.

MRAC systems can also be further classified. The MRAC shown in Fig. 1.3 is called "direct" because the controller parameters are directly updated to minimize the error between the model and plant outputs. There are also MRAC systems where the controller parameters are updated indirectly and such systems are called "indirect." In indirect control the plant parameters are estimated and the control parameters are adjusted on the basis of these estimates. It may, however, be mentioned at this stage that it has been demonstrated [126] that for a particular parameterization of the plant, both direct and indirect control lead to identical equations.

Further, MRAC can be used with continuous-time as well as discrete-time systems, whereas STR involve only discrete-time systems. MRAC systems are mostly applied in deterministic systems, whereas there has been a considerable progress in the development of stochastic STRs. There has been significant research going on regarding MRAC systems in the presence of stochastic disturbances. Moreover, recent research has demonstrated the equivalence of MRAC and STR methods. This will be dealt with in a later section.

1.5.4 Self-Tuning Regulators

The self-tuning regulator is another important form of an adaptive system, originally introduced by Åström and Wittenmark in 1973 [16]. In STR a design procedure for known plant parameters is first chosen. This is applied to the unknown plant using recursively estimated values of the parameters. Such a system is shown in Fig. 1.5. In the figure, the block "regulator design" represents the on-line solution for a system with known parameters. This is called the underlying design problem. To evaluate adaptive control schemes it is helpful to find this underlying design problem as this will give the characteristics of the system under the ideal conditions when the parameters are known exactly. In Fig. 1.5 perturbation signals are not shown. However, these are necessary to obtain good estimates.

STR have become very popular in recent years because of their versatility and the ease with which they can be implemented with microprocessors. There are many industrial applications of STR. The self-tuning adaptive controllers have the advantage of being able to adapt for any (especially unmeasurable) disturbances if designed well.

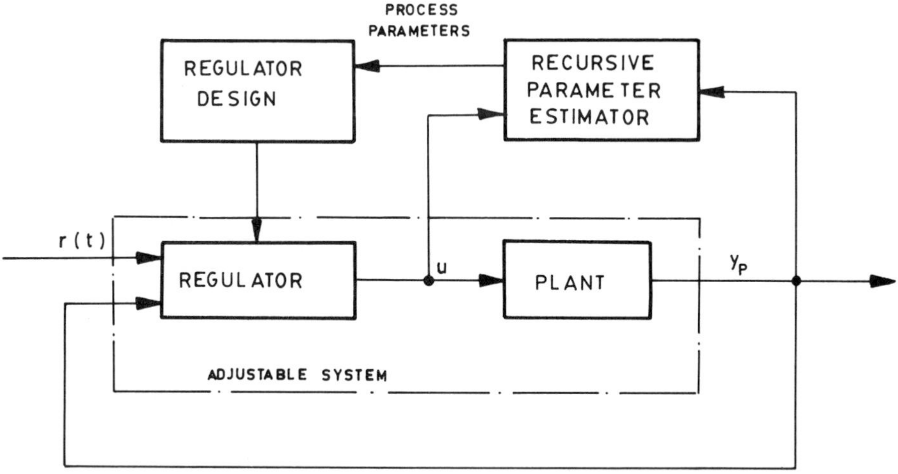

FIG. 1.5 Self-tuning regulator.

The STR shown in Fig. 1.5 is called an explicit STR. It is based on estimation of an explicit process model. Sometimes, the process may be reparameterized in terms of the regulator parameters. This results in simpler algorithms as the design calculations are avoided. Such an STR is called an implicit STR, since it is based on estimation of an implicit process model.

The direct MRAC can be related to the implicit STR and the indirect MRAC to the explicit STR.

1.6 CONCLUDING REMARKS

The purpose of the chapter has been to introduce the area of adaptive control and clarify why and when adaptive control is used. Some important landmarks and early work are discussed. Although there is no standardized definition for adaptive control, several proposed definitions of various terms occurring in this field have been given to give an idea of the scope and objectives of adaptive control. After discussing the essential aspects of adaptive control, we have considered classification of these adaptive control systems, in particular, as model reference adaptive control systems and self-tuning regulators, which are discussed in detail in Chapters 2 through 6. It must be borne in mind that in addition to these two major approaches, there are other adaptive control strategies, which form the subject matter of Chapter 7.

REFERENCES

1. Draper, C. S., and Li, Y. T., Principles of optimalizing control systems and an application to the internal combustion engine, ASME Publication (1951).
2. Draper, C. S., and Li, Y. T., Principles of optimalizing control systems and an application to the internal combustion engine, Report No. ARG-TI, Aerophysics Research Group, Aeronautical Engineering Dept., MIT (1955).
3. Tsien, H. S., Engineering Cybernetics, McGraw-Hill, New York, 1954.
4. Ashby, W. R., Design for a Brain, John Wiley, New York, 1960.
5. Benner, A. H., and Drennick, R. F., An adaptive servosystem, IRE Conv. Rec., Pt. 4, 8-14 (1955).
6. Whitaker, H. P., Yamron, J., and Kezer, A., Design of model reference adaptive control systems for aircraft, Report No. R-164, Instrumentation Lab., MIT (1958).
7. Whitaker, H. P., Massachusetts Institute of Technology presentation, Proceedings of Self-Adaptive Flight Control Systems Symp., WADC Tech. Report 59-49, pp. 16-78 (1959).

8. Whitaker, H. P., Design capabilities of model reference adaptive systems, Proc. Nat. Electro. Conf., 18: 241-249 (1962).
9. Osburn, P. V., Whitaker, H. P., and Kezer, A., New developments in the design of model reference adaptive control systems, ISA Paper No. 61-39 (1961).
10. Li, Y. T., and van der Velde, W. E., The philosophy of nonlinear adaptive control, Proc. IFAC Congress, Moscow (1960).
11. Petrov, B. N., Ulanov, G. M., and Emel'yanov, S. V., Invariantnost i optimizacia sistem avtomalicheskogo regulirovania s zsestoko i peremennoi stukturoi, IFAC Congress, Basel (1963).
12. Bellman, R., Adaptive Control Processes: A Guided Tour, Princeton University Press, Princeton, N.J., 1961.
13. Fel'dbaum, A. A., Dual control theory I-IV, Auto. Rem. Contr., 21: 874-880; 21: 1033-1039; 22: 1-12; 22: 109-121 (1960-1961).
14. Lainiotis, D. G., Upadhyay, T. N., and Deshpande, J. G., Optimal adaptive control of linear systems, Proc. IEEE Conf. Decision & Contr., 247-257 (1971).
15. Lainiotis, D. G., Partitioning: A unifying framework for adaptive systems, Pt. I: Estimation; Pt. II: Control, Proc. IEEE, 64: 1126-1143, 1182-1198 (1976).
16. Åström, K. J., and Wittenmark, B., On self-tuning regulators, Automatica, 9: 185-199 (1973).
17. Monopoli, R. V., Model reference adaptive control with an augmented error signal, IEEE Trans. Auto. Contr., AC-19: 474-484 (1974).
18. Narendra, K. S., Lin, Y. H., and Valavani, L. S., Stable adaptive controller design, Pt. II: Proof of stability, IEEE Trans. Auto. Contr., AC-25: 440-448 (1980).
19. Narendra, K. S., and Lin, Y. H., Stable discrete adaptive control, IEEE Trans. Auto. Contr., AC-25: 456-461 (1980).
20. Lin, Y. H., and Narendra, K. S., A new error model for adaptive systems, IEEE Trans. Auto. Contr., AC-25: 585-587 (1980).
21. Morse, A. S., Global stability of parameter-adaptive control systems, IEEE Trans. Auto. Contr., AC-25: 433-439 (1980).
22. Goodwin, G. C., Ramadge, P. J., and Caines, P. E., Discrete-time multivariable adaptive control, IEEE Trans. Auto. Contr., AC-25: 449-456 (1980).
23. Egardt, B., Stability analysis of continuous-time adaptive control systems, SIAM J. Contr. Optimiz., 18: 540-558 (1980).
24. Egardt, B., Stability analysis of discrete-data adaptive control schemes, IEEE Trans. Auto. Contr., AC-25: 710-716 (1980).
25. Gregory, P. C. (ed.), Proceedings of the Self-Adaptive Flight Control Systems Symposium, WADC Report No. 59-49 (1959).
26. Gentfe, W. K., Optimalizing control—design of a fully automatic control system for a turbojet aircraft, Proc. WESCON, Pt. 4 (1957).
27. Drennich, R. F., and Shahbender, R. A., Adaptive servomechanisms, AIEE Trans., Applications in Industry, 282-291 (1957).

28. Anderson, G. W., Aseltine, J. A., Mancini, A. R., and Sarture, C. W., A self-adjusting system for optimum dynamic performance, IRE Nat. Conv. Rec., Pt. 4 (1958).
29. Kalman, R. E., Design of a self-optimizing control system, Trans. ASME, 80: 468-478 (1958).
30. Taylor, C. F., Problems of nonlinearity in adaptive or self-optimizing systems, Proc. IRE (1958).
31. Anderson, G. W., Buland, R. W., and Cooper, G. R., The Aeronutronic self-optimizing automatic control system, Proceedings of Self-Adaptive Flight Control Symp., WADC Report No. 59-49 (1959).
32. Anderson, G. W., Buland, R. W., and Cooper, G. R., Use of cross-correlation in an adaptive control system, Proc. Nat. Elec. Conf., 15 (1959).
33. Gibson, J. E., and McVey, E. S., Multidimensional adaptive control, Proc. Nat. Elec. Conf., 15: 12-26 (1959).
34. Haddad, R. A., and Mishkin, E., On the identification and command problems in computer-controlled adaptive systems, Proc. WESCON (1959).
35. Margolis, M., and Leondes, C. T., Parameter tracking servo for adaptive control, IRE Trans. Auto. Contr., AC-4: 100-111 (1959).
36. Merriam, C. W., III, Use of a mathematical error criterion in the design of adaptive control systems, AIEE Trans., 78, Pt. II: 506-512 (1959).
37. Marx, M. F., Recent adaptive control work at the General Electric Co., Proceedings of Self-Adaptive Flight Control Symp., WADC Report No. 59-49 (1959).
38. Schuck, O. H., Honeywell's history and philosophy in adaptive control field, Proceedings of Self-Adaptive Flight Control Symp., WADC Report No. 59-49 (1959).
39. Staffin, R., Executive-controlled adaptive systems, AIEE Trans., Applications in Industry, 78, Pt. II: 523 (1959).
40. Widrow, B., Adaptive sampled-data systems—A statistical theory of adaptation, WESCON Conv. Rec., Pt. 4, 74-85 (1959).
41. Gibson, J. E., Mechanizing the adaptive principle, Control Eng., 7: 109 (1960).
42. Braun, L., Jr., Mishkin, E., and Truxal, J. G., On the approximate identification of process dynamics in computer-controlled adaptive systems, Proc. IFAC Congress, Moscow (1960).
43. Caldwell, E. G., Applications of adaptive systems, NORAIR Report (1961).
44. Margolis, M., and Leondes, C. T., On a theory of adaptive control systems: The method of learning model, Proc. IFAC Congress, Moscow (1960).
45. Margolis, M., and Leondes, C. T., On the philosophy of adaptive control systems, Report No. TN-59-1199, Dept. of Eng., UCLA, Los Angeles (1960).

46. Smith, K. C., Adaptive control through sinusoidal response, IRE Trans. Auto. Contr., AC-5 (1960).
47. Tou, J. T., Joseph, P., and Lewis, J., A study of digital adaptive control systems, Report No. 1, NONR-110(8), Office of Naval Research, Washington, D.C. (1960).
48. Widrow, B., Adaptive sampled-data systems, Proc. IFAC Congress, Moscow (1960).
49. Mathias, R. A., and Van Nice, R. I., Adaptive control systems: A survey, Electro Technology, 66: 116-125 (1960).
50. Gibson, J. E., and Meditch, J. S., A class of predictive adaptive controls, Report, Contr. & Instr. Syst. Lab., Purdue Univ. (1961).
51. Gibson, J. E., Self-optimizing or adaptive control systems, Auto. Rem. Contr., 2: 586-595 (1961).
52. McGrath, R. J., Rajaraman, V., and Rideout, V. C., A parametric perturbation adaptive control system, IRE Trans. Auto. Contr., AC-6: 154 (1961).
53. Tutt, G. E., and Waymeyer, W. K., Model feedback applied to flexible booster control, IRE Trans. Auto. Contr., AC-6: 135-142 (1961).
54. Weygandt, C. N., and Puri, N. N., Transfer function tracking and adaptive control systems, IRE Trans. Auto. Contr., AC-6: 162-166 (1961).
55. Merriam, C. W., III, Synthesis of Adaptive Controls, Doctoral thesis, MIT (1958).
56. Staffin, R., Executive-controlled adaptive systems, Doctoral thesis, Polytech. Inst. of Brooklyn (1958).
57. Braun, L., Jr., On adaptive control systems, Doctoral thesis, Polytech. Inst. of Brooklyn (1959).
58. Margolis, M., On the theory of process adaptive control system: The learning model approach, Doctoral thesis, Univ. of Calif., Los Angeles (1959).
59. Lichtenberger, W. W., The identification of linear processes by means of correlating filters, Doctoral thesis, Univ. of Illinois (1960).
60. Osburn, P. V., Adaptive control systems: An analytical design method for model reference adaptive control systems, Doctoral rhesis, MIT (1960).
61. Puri, N. N., Tracking and adaptive control systems, Doctoral thesis, Univ. of Pennsylvania (1960).
62. Schiewe, A. J., A three dimensional adaptive control system, Doctoral thesis, Purdue Univ. (1960).
63. Donalson, D. D., The theory and stability analysis of a model referenced parameter tracking technique for adaptive automatic control system, Doctoral thesis, Univ. of Calif., Los Angeles (1961).
64. Eveleigh, V. W., A comparison of two approaches to extrema searching adaptive systems, Doctoral thesis, Purdue Univ. (1961).
65. Meditch, J. S., A class of predictive adaptive controls, Doctoral thesis, Purdue Univ. (1961).

66. Aseltine, J. A., Mancini, A. R., and Sarture, C. W., A survey of adaptive control systems, IRE Trans. Auto. Contr., AC-6: 102-108 (1958).
67. Stromer, P. R., Adaptive self-optimizing control systems—A bibliography, IRE Trans. Auto. Contr., AC-5: 65-68 (1959).
68. Cooper, G. R., Gibson, J. E., Eveleigh, V. W., Lindenlaub, J. C., Meditch, J. S., and Raible, R. H., A survey of the philosophy and state-of-the-art of adaptive systems, Report No. TR-EE-60-1, USAF, WADD, Purdue Univ. (1960).
69. Jacobs, O. L. R., A review of self-adjusting systems in automatic control, J. Electron. Contr., 10: 311-322 (1961).
70. Truxal, J. G., Adaptive control—A survey, Proc. IFAC Congress, Basel, 386-392 (1963).
71. Zypkin, J., Adaptive system theory—Today and tomorrow, Proc. IFAC Symposium, Ischia, Italy, 47-67 (1973).
72. Åström, K. J., and Eykoff, P., System identification—A survey, Automatica, 7: 123-162 (1971).
73. Landau, I. D., Model reference adaptive systems—A survey: What is possible and why? ASME Trans., Dyn. Syst. Meas. Control., 94: 119-132 (1972).
74. Landau, I. D., A survey of model reference adaptive techniques (Theory and applications), Automatica, 10: 353-379 (1974).
75. Lindorff, D. P., and Caroll, R. L., Survey of adaptive control using Lyapunov design, Int. J. Contr., 18: 897-914 (1973).
76. Narendra, K. S., and Kudva, P., Stable adaptive schemes for system identification and control, Pts. I & II, IEEE Trans. Syst., Man, & Cybern., SMC-4: 542-560 (1974).
77. Wittenmark, B., Stochastic adaptive control methods—A survey, Int. J. Contr., 21: 705-730 (1975).
78. Unbehauen, H., and Schmid, Ch., Status and application of adaptive control systems, Auto. Contr. Theory Appl., 3: 1-12 (1976).
79. Asher, R. B., Andrisani, D., and Dorato, P., Bibliography on adaptive control systems, Proc. IEEE, 64: 1226-1240 (1975).
80. Sternby, J., Extremum control systems—An area for adaptive control, Proceedings of JACC (1980).
81. Parks, P. C., Schaufalberger, W., Schmid, Ch., and Unbehauen, H., Applications of adaptive control systems, in Methods and Applications of Adaptive Control (H. Unbehauen, ed.), Springer-Verlag, Berlin, pp. 161-198, 1980.
82. Åström, K. J., Theory and applications of adaptive control—A survey, Automatica, 19: 471-486 (1983).
83. Voronov, A. A., and Rutkovsky, V. Yu., State-of-the-art and prospects of adaptive systems, Automatica, 20: 547-557 (1984).
83a. Unbehauen, H., System identification methods using parameter estimation—A survey, Proceedings 2nd Int. Symp. on System Analysis and Simulation, Berlin, D.D.R. (1985).

84. IEEE Proceedings, Special issue on adaptive systems, 64: 1121-1264 (1976).
85. IEEE Transactions Auto. Contr., Mini-issue on NASA's advanced control law program, AC-22: 752-806 (1977).
86. Opt. Contr. Appl. & Methods, Special issue on self-tuning control, 3 (1982).
87. Automatica, Special issue on adaptive control, 20 (1984).
Bellman, R., Adaptive Control Processes: A Guided Tour, Princeton University Press, Princeton, N.J., 1961.
89. Mishkin, E., and Braun, L., Adaptive Control Systems, McGraw-Hill, New York, 1961.
90. Westcott, J. H., An Exposition of Adaptive Control, Pergamon Press, Oxford, 1962.
91. Caruthers, F. P., and Levenstein, H., Adaptive Control Systems, Pergamon Press, Oxford, 1963.
92. Krasavskii, A. A., Dynamics of Continuous Self-Organizing Systems (in Russian), Fizmatgiz, Moscow, 1963.
93. Hammond, P. H. (ed.), Theory of Self-Adaptive Control Systems, Plenum Press, New York, 1966.
94. Kulikovski, R., Optimal and Adaptive Processes in Automatic Control Systems (in Russian), Nauka, Moscow, 1966.
95. Sawaragi, Y., Sunahara, Y., and Nakamizo, T., Statistical Decision Theory in Adaptive Control Systems, Academic Press, New York, 1966.
96. Sworder, D., Optimal Adaptive Control Systems, Academic Press, New York, 1966.
97. Eveleigh, V. W., Adaptive Control and Optimization Techniques, McGraw-Hill, New York, 1967.
98. Yakowitz, S. J., Mathematics of Adaptive Control Processes, American Elsevier, New York, 1969.
99. Davies, W. D. T., System Identification for Self-Adaptive Control, Wiley-Interscience, New York, 1970.
100. Mendel, J. M., and Fu, K. S., Adaptive, Learning, and Pattern Recognition Systems: Theory and Applications, Academic Press, New York, 1970.
101. Tsypkin, Ya. Z., Adaptation and Learning in Automatic Systems, Academic Press, New York, 1971.
102. Saridis, G. N., Self-Organizing Control of Stochastic Systems, Marcel Dekker, New York, 1977.
103. Egardt, B., Stability of Adaptive Controllers, Springer-Verlag, Berlin, 1979.
104. Landau, I. D., Adaptive Control: The Model Reference Approach, Marcel Dekker, New York, 1979.
105. Narendra, K. S., and Monopoli, R. V. (eds.), Applications of Adaptive Control, Academic Press, New York, 1980.

106. Unbehauen, H. (ed.), Methods and Applications in Adaptive Control, Springer-Verlag, Berlin, 1980.
107. Harris, C. J., and Billings, S. A. (eds.), Self-Tuning and Adaptive Control: Theory and Applications, Peter Peregrinus, London, 1981.
108. Ioannou, P. A., and Kokotovic, P. V., Adaptive Systems with Reduced Models, Springer-Verlag, Berlin, 1983.
109. Åstrom, K. J., and Wittenmark, B., Computer-Controlled Systems: Theory and Design, Prentice-Hall, Englewood Cliffs, N.J., 1984.
110. Goodwin, G. C., and Sin, K. S., Adaptive Filtering Prediction and Control, Prentice-Hall, Englewood Cliffs, N.J., 1984.
111. Åström, K. J., Design principles for self-tuning regulators, in Methods and Applications in Adaptive Control (H. Unbehauen, ed.), Springer-Verlag, Berlin, pp. 1-20, 1980.
112. Aström, K. J., Why use adaptive techniques for steering large tankers ? Int. J. Contr., 32: 689-708 (1980).
113. Horowitz, I. M., Synthesis of Feedback Systems, Academic Press, New York, 1963.
114. Kidd, P. T., Comments on "Self-tuning and stable adaptive control of a batch polymerization reactor," Automatica, 20: 481-482 (1984).
115. Jacobs, O. L. R., When is adaptive control useful? Proceedings of IMA Conf. on Control Theory, Sheffield (1980).
116. Jacobs, O. L. R., and Saratchandran, P., Comparison of adaptive controllers, Automatica, 16: 89-97 (1980).
117. Jacobs, O. L. R., On nonlinear adaptive control, Eng. Lab. Report No. 1323, Oxford Univ. (1980).
118. Jacobs, O. L. R., Hewkins, P. F., and White, C., On-line computer control of pH in an industrial process, IEE Proc., 127, Pt. D: 161-168 (1980).
119. Saridis, G. N., Mendel, J. M., and Nikolic, Z. Z., Report on definitions of self-organizing control processes and learning systems, IEEE Control Systems Society Newsletter (1973).
120. Gibson, J. E., Nonlinear Automatic Control, McGraw-Hill, New York (1962).
121. Fel'dbaum, A. A., Optimal Control Systems, Academic Press, New York, 1965.
122. Fu, K. S., Learning control systems—Review and outlook, IEEE Trans. Auto. Contr., AC-15: 210-221 (1970).
123. Lainiotis, D. G., Scanning the special issue on adaptive systems, IEEE Proc., 64: 1123-1125 (1976).
124. Tse, E., and Bar-Shalom, Y., An actively adaptive control for discrete-time systems with random parameters, IEEE Trans. Auto. Contr., AC-18: 109-117 (1973).
125. Tse, E., and Bar-Shalom, Y., Actively adaptive control for nonlinear stochastic systems, IEEE Proc., 64: 1172-1181 (1976).
126. Narendra, K. S., and Valavani, L. S., Direct and indirect adaptive control, Automatica, 15: 653-664 (1979).

BIBLIOGRAPHY

Amourous, M., Babary, J. P., and El-Jai, A., Adaptive open loop control for a class of distributed parameter systems, Automatica, 14: 603-607 (1978).

Automatica, Special section, System identification tutorials, 16: 501-587 (1980).

Automatica, Special issue, Identification and system parameter estimation, 17: 1-252 (1981).

Billings, S. A., and Young, P. C. (eds.), Identification and System Parameter Estimation, 2 vols., Pergamin Press, Oxford, 1985.

Currie, M. G., Theory and applications of adaptive systems, Doctoral thesis, Dept. of Engineering, UCLA (1968).

Deshpande, J. G., Upathdyay, T. N., and Lainiotis, D. G., Adaptive control of linear stochastic system, Automatica, 9: 107-115 (1973).

Dion, J. M., Etude des systemes adaptifs a amortissement constant, Doctoral thesis, Inst. Polytechnique de Grenoble, France (1977).

Dyurki, I., and Kocish, Y., Experimental analog computer investigation of a simple adaptive system, Autom. Rem. Contr., 29: 349-352 (1968).

Edwards, J. B., and Owens, D. H., First order type models for multivariable process control, IEE Proc., 124: 1083-1088 (1977).

Elliott, H., Non-MRA model matching, in Methods and Applications in Adaptive Control (H. Unbehauen, ed.), Springer-Verlag, Berlin, pp. 122-130, 1980.

Elliott, H., and Wolovich, W. A., Parameter adaptive identification and control, IEEE Trans. Auto. Contr., AC-24: 592-599 (1979).

Falconer, D. D., Adaptive filter theory and applications, in Analysis and Optimization of Systems, Springer-Verlag, Berlin, 1980.

Florentin, J. J., Optimal probing, adaptive control of a simple Bayesian system, Int. J. Contr., 13: 165 (1971).

Franklin, G. F., and Powell, T. D., Digital Control of Dynamic Systems, Addison-Wesley, Reading, Mass., 1980.

Garland, B., Adaptive and optimal control of time-delay systems, Doctoral thesis, University of Bath, U.K. (1978).

Hang, C. C., A new form of stable adaptive observer, IEEE Trans. Auto. Contr., AC-21: 544-547 (1976).

Hirsch, J. J., and Pelti, P., Real-time identification using adaptive discrete models, Proceedings 3rd IFAC Symp. on Sensitivity, Adaptivity, and Optimality, Ischia, pp. 290-297 (1973).

Isermann, R., Digital Control Systems, Springer-Verlag, Berlin, 1981.

Isermann, R. (ed.), System Identification Tutorials, Pergamin Press, Oxford, 1981.

Jacobs, O. L. R., Two uses of the term "adaptive" in automatic control, IEEE Trans. Auto. Contr., AC-9: 574-575 (1964).

Johnson, C. R., Jr., Adaptive parameter matrix and output vector estimation via an equation error formulation, IEEE Trans. Syst., Man and Cybernetics, SMC-9: 392-397 (1979).

Johnson, C. R., Jr., A convergence proof for a hyperstable adaptive recursive filter harf, IEEE Trans. Inf. Th., IT-25: 745-749 (1979).

Kailath, T., Linear Systems, Prentice-Hall, Englewood Cliffs, N.J., 1980.

Kalman, R. E., Engler, T. S., and Bucy, R. S., Fundamental study of adaptive control systems, Tech. Report No. ASD-TR-61-27, Vol. I, Wright Patterson Air Force Base, Ohio (1962).

Kazakov, I. E., Dynamics of self-adaptive systems with external continuous adjustment of the compensating network in the presence of random distirubances, Autom. Rem. Contr., 21: 1040 (1961).

Kazakov, I. E., On a statistical theory of continuous self-adaptive systems, Energetika i Avtomatika, Acad. of Science, USSR, Moscow (1962).

Kazakov, I. E., and Evlanov, On a theory of self-adaptive systems with the gradient method, Proc. 2nd IFAC Congress (1963).

Kirschbaum, H. S., The application of Markov chains to discrete extremum-seeking adaptive systems, in Adaptive Control Systems (F. P. Caruthers and H. Levenstein, eds.), Pergamon Press, Oxford, pp. 179-202, 1963.

Kostynk, V. I., Extremal control without search oscillations using self-adaptive model of the plant, Avtomatika, 10 (1965).

Kostynk, V. I., A relationship between the self-adaptive systems without search, differential systems, and the systems based on sensitivity theory, Avtomatika, 11 (1966).

Kraft, L. G., Controllable form state variables obtained from an arbitrarily fast adaptive observer, Proceedings IEEE Conf. Decision & Control (1976)

Kreisselmeier, G., Algebraic separation in realizing a linear state feedback control law by means of an adaptive observer, IEEE Trans. Auto. Contr., AC-25: 238-242 (1979).

Kreisselmeier, G., Adaptive control via adaptive observation and asymptotic feedback matrix synthesis, IEEE Trans. Auto. Contr., AC-26: 717-722 (1980).

Krutova, I. V., and Rutkovskii, V. Yu, A self-adaptive system with a model reference—I & II, Engineering Cybernetics (USSR) (1964).

Kudva, P., Stable adaptive systems, Doctoral thesis, Yale University, New Haven, Conn. (1975).

Kudva, P., and Narendra, K. S., An identification procedure for discrete multivariable system, IEEE Trans. Auto. Congr., 19: 549-552 (1974).

Kunze, E. G., Adaptive control by a sensitivity method without need for on-line identification, in Methods and Applications in Adaptive

Control (H. Unbehauen, ed.), Springer-Verlag, Berlin, pp. 291-300, 1980.
Lainiotis, D. G., Optimal adaptive estimation: Structure and parameter adaptation, IEEE Trans. Auto. Contr., AC-16: 160-170 (1971).
Lainiotis, D. G., Upadhyay, T. N., and Deshpande, J. G., Optimal adaptive control: A nonlinear separation theorem, Int. J. Contr., 15: 877-888 (1972).
Landau, I. D., and Unbehauen, H., The development of adaptive systems in Germany and France, ASME Trans., J. Dyn. Syst. Meas. Contr., 96: 405-413 (1974).
Li, Y. T., Performance indices of adaptive control systems, in Adaptive Control Systems (F. B. Caruthers and H. Levenstein, eds.), Pergamon Press, Oxford, pp. 11-20, 1963.
Ljung, L., and Anderson, B. D. O., Adaptive control, where are we? Automatica, 20: 499-500 (1984).
Maletinsky, V., On-line parameter estimation of continuous processes, Proc. IFAC Congress, Boston, Paper No. 11.6 (1975).
Maletinsky, V., and Schanfelberger, W., Suboptimum adaptive control, Proceedings IFAC Congress, Zurich, pp. 129-143 (1974).
Marshall, J. E., Identification strategies for time-delay systems, in Methods and Applications in Adaptive Control (H. Unbehauen, ed.), Springer-Verlag, Berlin, pp. 141-150, 1980.
McCorkell, C., and Swamick, B. H., The development of computer algorithms for multivariable adaptive control, Int. J. Contr., 22: 837-850 (1975).
Mehra, R. K., and Lainiotis, D. G. (eds.), System Identification: Advances and Case Studies, Academic Press, New York, 1976.
Mendel, J. M., Application of artificial intelligence techniques to a space-craft control problem, NASA Report No. CR-755 (1967).
Mendel, J. M., and Zapalac, J. J., The application of techniques of artificial intelligence to control system design, in Advances in Control Systems, Vol. 6 (C. T. Leondes, ed.), Academic Press, New York, 1968.
Mosca, E. (ed.), Proceedings Workshop on Adaptive Control, Inst. of Systems and Information, University of Florence, Italy (1982).
Nagumo, J., and Noda, K., A learning method for system identification, IEEE Trans. Auto. Contr., AC-12: 282-287 (1967).
Narendra, K. S., and McBride, L. E., Multi-parameter self-optimizing systems using correlation techniques, IEEE Trans. Auto. Contr., AC-9: 31-38 (1964).
Nuyan, S., and Carroll, R. L., Minimal order arbitrarily fast adaptive observers and identifiers, IEEE Trans. Auto. Contr., AC-24: 510-513 (1979).
Pearson, A. E., On adaptive optimal control of nonlinear processes, J. Basic Eng., 86: No. 1 (1964).

Pearson, A. E., and Sarachik, P. E., On the formulation of adaptive optimal control problems, J. Basic Eng., 87: No. 1 (1965).

Perlis, H. J., A self-adaptive system with auxiliary adjustment prediction, IEEE Trans. Auto. Contr., AC-10: No. 3 (1965).

Porter, B., and Tatnall, L., Performance characteristics of an adaptive hydraulic servomechanism, Int. J. Contr., 11: 741-757 (1970).

Propoi, A. I., and Tsypkin, Ya. Z., On adaptive synthesis of optimal systems, Dakl. Akad. Nauk (USSR), 175: No. 6 (1967).

Puri, N. N., and Weygandt, C. N., Transfer function tracking of a linear time-varying system by means of an arbitrary simple lag network, Proceedings JACC, pp. 200-203 (1963).

Puri, N. N., and Weygandt, C. N., Transfer function tracking for a linear time-varying system by means of orthogonal networks, IEEE Trans. Appl. Ind., 82: 234-238 (1963).

Ramadge, P. J., and Goodwin, G. C., Design of restricted complexity adaptive control systems, IEEE Trans. Auto. Contr., AC-24: 584-588 (1979).

Rosen, R. (ed.), Anticipatory Systems, Pergamon Press, Oxford, 1985.

Ruban, A. I., Design of algorithms of adaptive control with identification, Autom. Rem. Contr., 44: 1346-1355 (1984).

Saratchandran, P., On the performance of an adaptive controller under different estimators, Proc. JACC, San Francisco (1980).

Saridis, G. N., Stochastic approximation methods for identification and control—A survey, IEEE Trans. Auto. Contr., AC-19: 798-809 (1974).

Saridis, G. N., and Dao, T. K., A learning approach to the parameter adaptive self-organizing control problem, Automatica, 9: 589-598 (1972).

Saridis, G. N., and Lobbia, R. N., Parameter identification and control of linear discrete-time systems, IEEE Trans. Auto. Contr., AC-17: 52-60 (1972).

Sarture, C. W., and Aseltine, J. A., Performance criteria in adaptive control, in Adaptive Control Systems (F. P. Caruthers and H. Levenstein, eds.), Pergamon Press, Oxford, pp. 1-10, 1963.

Schmid, Ch., CAD of adaptive systems, Preprints IFAC Symp. on Computer-Aided Design of Control Systems, Zurich (1979).

Schmid, Ch., KEDDC—A computer-aided analysis and design package for control systems, Proc. Amer. Contr. Conf., Arlington, pp. 211-212 (1982).

Schmid, Ch., and Unbehauen, H., Identification and CAD of adaptive systems using the KEDDC package, Proceedings 5th IFAC Symp. on Identification & System Parameter Estimation, Darmstadt, F.R.G., pp. 1087-1094 (1979).

Sefl, O., Principy adaptivity v technicke kybernetice, in Kybernetika a Jeji Veuzitv, Prague, 1965.

Sefl, O., Adaptivni systemy, in Sounru Praci a Automatizace, Prague, 1965.

Shinners, S. M., Optimal and adaptive control systems, Electro-Technology, 74: 63-80 (1964).

Sinha, N. K., Gupta, M. M., and Nikiforuk, P. N., Recent advances in adaptive control, J. Cybernetics, 6: 79-100 (1976).

Sklansky, J., Learning systems for automatic control, IEEE Trans. Auto. Contr., AC-11: 6-19 (1966).

Smyth, R. K., and Nahi, N. E., On the stability and design of dither adaptive systems, Proc. IFAC Congress (1963).

Solov'ev, I. G., Modifications of regularized adaptive control algorithms, Autom. Rem. Contr., 44: 1204-1210 (1984).

Smith, F. W., Contactor control by adaptive pattern recognition techniques, Tech. Report No. 6762-1, Stanford Electronics Labs., Stanford, Calif. (1964).

Stear, E. B., and Gregory, P. C., Capabilities and limitations of some adaptive techniques, Proc. Nat. Aerospace Elec. Conf., pp. 644-660 (1962).

Stein, G., and Saridis, G. N., A parameter adaptive control technique, Automatica, 5: 731-739 (1969).

Sternby, J., Topics in dual control, Report No. TFRT-1012, Dept. of Auto. Contr., Lund Inst. of Tech., Lund, Sweden (1977).

Sternby, J., Performance limits in adaptive control, IEEE Trans. Auto. Contr., AC-24: 645-647 (1979).

Taylor, W. K., Self-organizing control systems using pattern recognition, Proceedings 2nd IFAC Congress (1963).

Thathachar, M. A. L., and Gajendran, F., A structural principle for adaptive control and identification, Tech. Report No. 44/45, Dept. of Elec. Eng., Indian Inst. of Science, Bangalore, India (1977).

Thathachar, M. A. L., and Gajendran, F., A structural principle for a class of adaptive systems, Int. J. Contr., 31: 519-538 (1980).

Trulsson, E., and Ljung, L., Direct minimization methods for adaptive control of non-minimum phase systems, Proc. IFAC Symp. Identification & System Parameter Estimation, Washington, D.C., pp. 631-635 (1982).

Truxal, J. G., Trends in adaptive control, Proc. Nat. Electronics Conf. (1960).

Tse, E., and Athans, M., Adaptive stochastic control for a class of linear systems, IEEE Trans. Auto. Contr., AC-17: 38-51 (1972).

Tse, E., Further comments on "Adaptive stochastic control for a class of linear systems," IEEE Trans. Auto. Contr., AC-18: 324-326 (1973).

Tsypkin, Ya. Z., Adaptation, training and self-organization in automatic systems, Autom. Rem. Contr., 27: 16-51 (1966).

Tsypkin, Ya. Z., Foundations of the Theory of Learning Systems, Academic Press, New York, 1973.

Unbehauen, H., A laboratory experiment for teaching system identification and adaptive control, Proc. IFAC Symp. on Identification & Syst. Par. Estn., University of York, U.K. (1985).

Unbehauen, H., Gohring, B., and Bauer, B., *Parameterschatzverfahren zur Systemidentifikation*, R. Oldenbourg-Verlag, Munich, F.R.G., 1974.

van Nine, R. I., and Mathias, R. A., Extremal adaptation in the presence of noise and drift, in *Adaptive Control Systems* (F. P. Caruthers and Levenstein, eds.), Pergamon Press, Oxford, pp. 123-142 (1963).

Weygandt, C. N., and Puri, N. N., Multivariable adaptive control systems, IEEE Trans. Appl. Ind., **82**: 89-94 (1963).

Widrow, B., A statistical theory of adaptation, in *Adaptive Control Systems* (F. P. Caruthers and H. Levenstein, eds.), Pergamon Press, Oxford, pp. 97-122 (1963).

Wieslander, J., and Wittenmark, B., An approach to adaptive control using real time identification, Automatica, **7**: 211-217 (1971).

Yadykin, I. B., Controller adaptability and two-level algorithms of adjustment of adaptive control system parameters, Autom. Rem. Contr., **44**: 631-641 (1983).

Yuan, J. S-C., Adaptive decoupling control of linear multivariable systems, Doctoral thesis, University of Toronto, Canada (1976).

Zemlyakov, S. D., and Yu Rutskovskii, V., Optimizing self-adjusting control systems with a model, Autom. Rem. Contr., **27**: 407-414 (1966).

Zemlyakov, S. D., and Yu Rutskovskii, V., Generalized adaptation algorithms for a class of search-less self-adjusting systems with a model, Autom. Rem. Contr., **28**: 935-940 (1967).

Zinober, A. S. I., Self-adaptive near-optimal control of diffusion equation, IEE Proc., **127**, Pt. D: 290-295 (1980).

2

Model Reference Adaptive Systems

2.1 INTRODUCTION

In recent years model reference adaptive control (MRAC) has become a very efficient and systematic method for controlling plants with unknown (or partially known) parameters. As mentioned in Section 1.5.3, in this procedure the objectives of control are specified in terms of a reference model and the design problem involves the determination, from all available on-line data, of the value of either the adjustable parameters or the control input such that the error between the plant and model outputs approaches zero asymptotically.

This chapter discusses different configurations of MRAC systems in Section 2.2. Section 2.3 deals with the classification of MRAC systems, while the mathematical description of these systems is given in Section 2.4. An equivalent representation of an MRAC system as a nonlinear time-varying system is considered in Section 2.5 and this will be of importance while designing MRAC systems by the hyperstability approach, discussed later in Section 3.4. The relevant equations are formulated for the direct and indirect MRAC systems in Section 2.6. Concluding remarks are given in Section 2.7.

2.2 DIFFERENT CONFIGURATIONS OF MODEL REFERENCE ADAPTIVE SYSTEMS

Some typical configurations of model reference adaptive systems (MRAS) are considered [1,2] here.

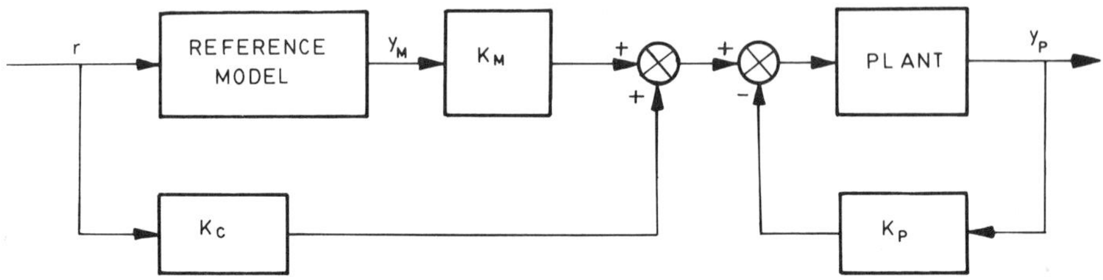

FIG. 2.1 LMFC system. (From Ref. 2.)

2.2.1 Linear Model-Following Control System

Linear model-following control (LMFC) systems (Fig. 2.1) are useful when the parameters of the plant and of the model are known and a quadratic performance index is to be minimized. This figure may be redrawn (Fig. 2.2) so that the role played by the reference model is more evident.

2.2.2 Adaptive Model-Following Control System

Adaptive model-following control (AMFC) systems (Fig. 2.3) are used when the parameters of the plant are unknown or important variations occur. The adaptation mechanism in Fig. 2.3 modifies the parameters K_C and K_P to assure "good" model following. Such a scheme is also called a model

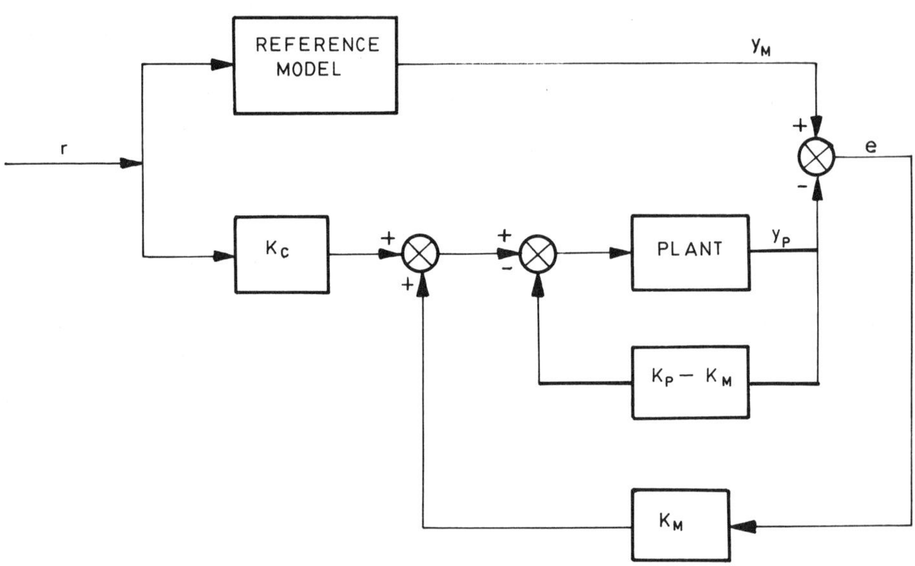

FIG. 2.2 LMFS system. (From Ref. 2.)

reference adaptive control system. This type of MRAC system is applied in adaptive autopilots, in-flight simulators, adaptive speed controllers for electric drives, and so on, in which the reference models specify the desired behavior.

2.2.3 Signal Synthesis MRAC System

Figure 2.4 shows another configuration of an MRAC system in which the K_c and K_p are constants and are designed by assuming that the controlled plant behaves always as the reference model. When parameter variations occur, the adaptation mechanism synthesizes an auxiliary input signal to assure that the plant behaves exactly as the reference model. Such a configuration is used when only part of the controlled plant is subject to parameter variations or when pure integration is inherent, as in the position control system shown in Fig. 2.5. Other applications include positioning an optical telescope [3] and rudder control [4].

The signal synthesis adaptation involves the adjustment of only the input variable, whereas the parameter adaptation requires real-time adjustment of a few plant parameters. Although the latter is effective in many cases, it has some disadvantages, namely, the need for matchable parameters, perfect model following, and direct adjustability of the plant parameters. A design procedure for a signal synthesis model reference adaptive controller has been given by Raju and Riad [4a].

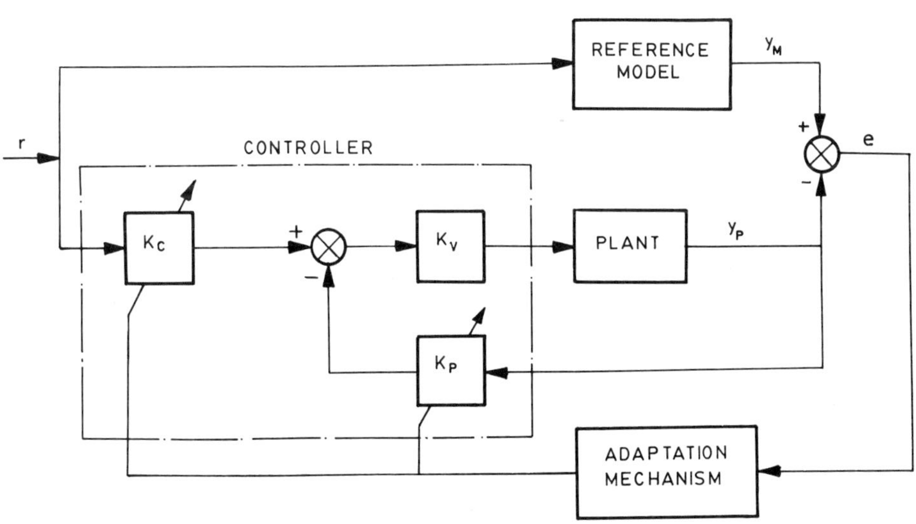

FIG. 2.3 AMFC system (From Ref. 2.)

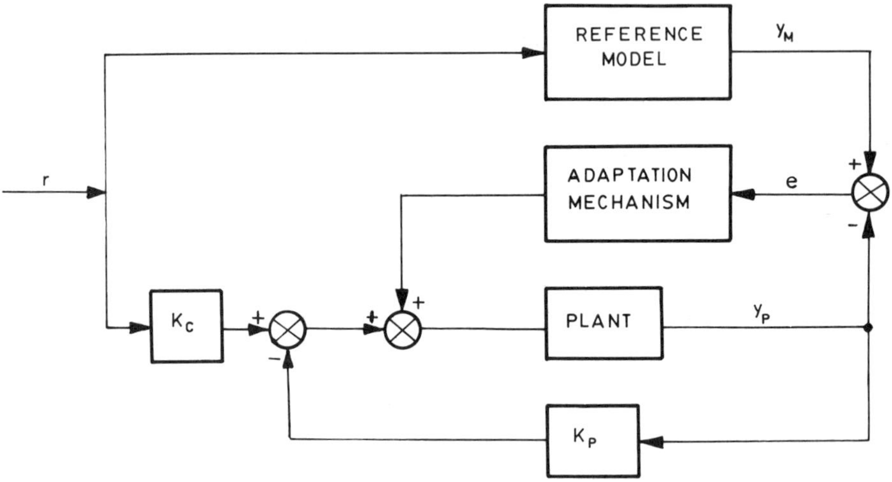

FIG. 2.4 Signal synthesis MRAC system. (From Ref. 2.)

2.2.4 Adaptive State Observer

In Fig. 2.6 an adjustable model having the same mathematical structure as the plant is the observer. The adaptation mechanism enables the adjustable model to track the parameters and the states of the plant. In the presence of noise we can obtain an estimate of the state of the plant, with an appropriate choice of the adaptation mechanism. Some nonlinear identification problems also may be solved by the MRAS [5, 6]. Such a method is also called an output error method.

2.2.5 Series-Parallel MRAS

A series-parallel placement of the adjustable system is shown in Fig. 2.7. To avoid the use of pure differentiators in the adjustable system, asymptotically stable low-pass filters called state variable filters are used in both parts. This method is also called the equation error method [7, 8] or the generalized error method [5]. This scheme can also be used for control, in which case the reference model is split into parts: one in series with the adjustable system and the other in parallel with the adjustable system.

2.2.6 Adaptive State Regulator with Adjustable Model

This scheme (Fig. 2.8) can be used for adaptive state variable regulation when the parameters of the plant are unknown and the states are not accessible. Adaptation depends on the nature of the input that acts on the plant. Such configurations are used when the reference signal r is almost constant (or slowly varying) and large variations of plant parameters occur. Such a

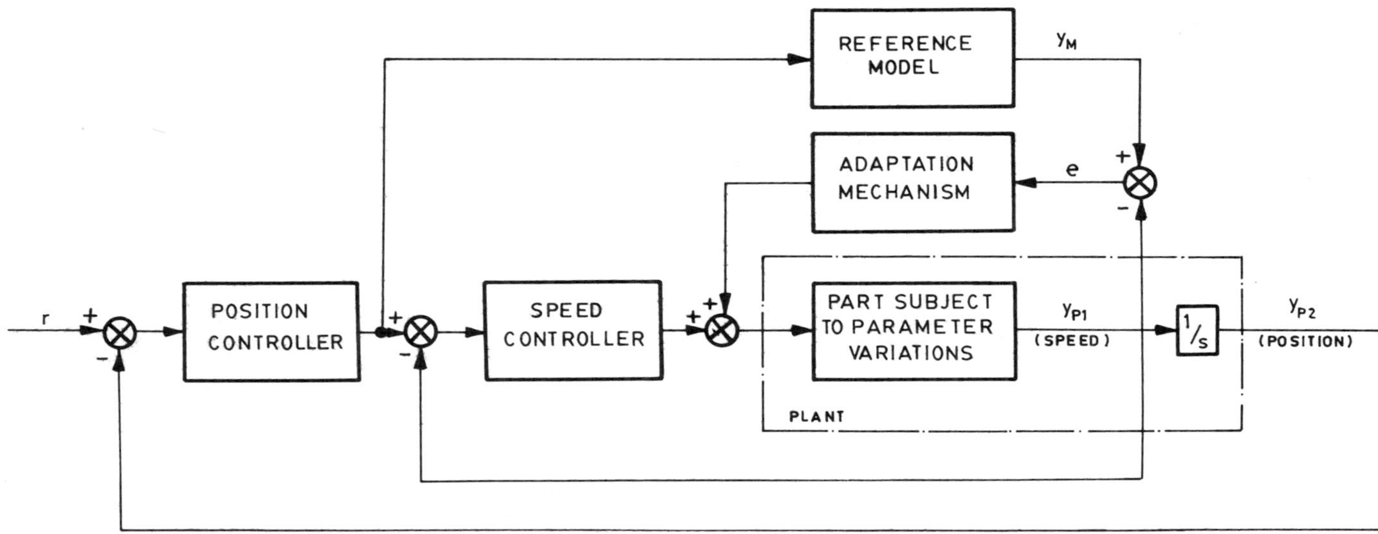

FIG. 2.5 Adaptive position control system. (From Ref. 2.)

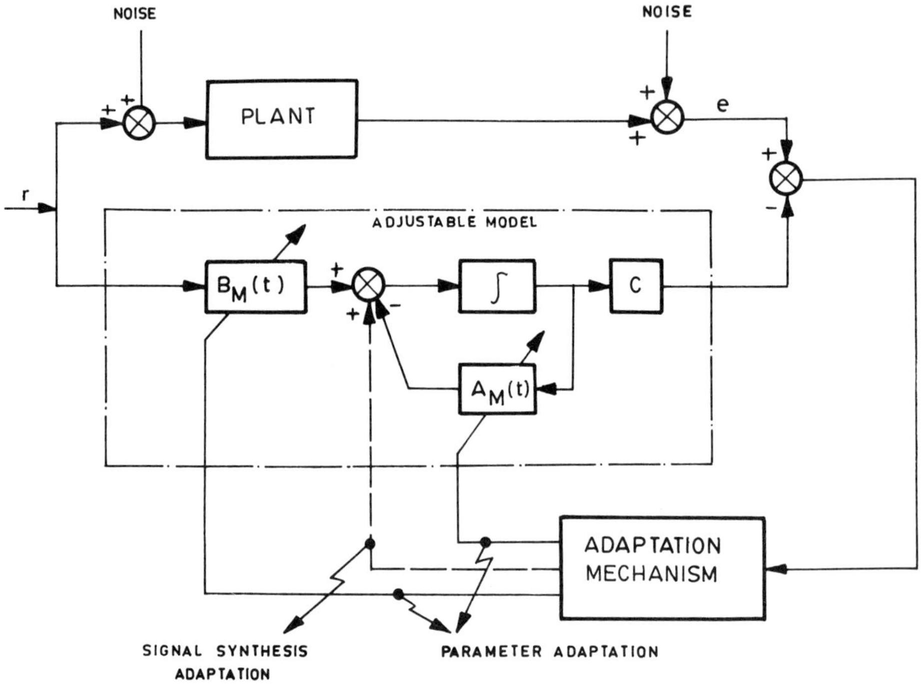

FIG. 2.6 Adaptive state observer. (From Ref. 2.)

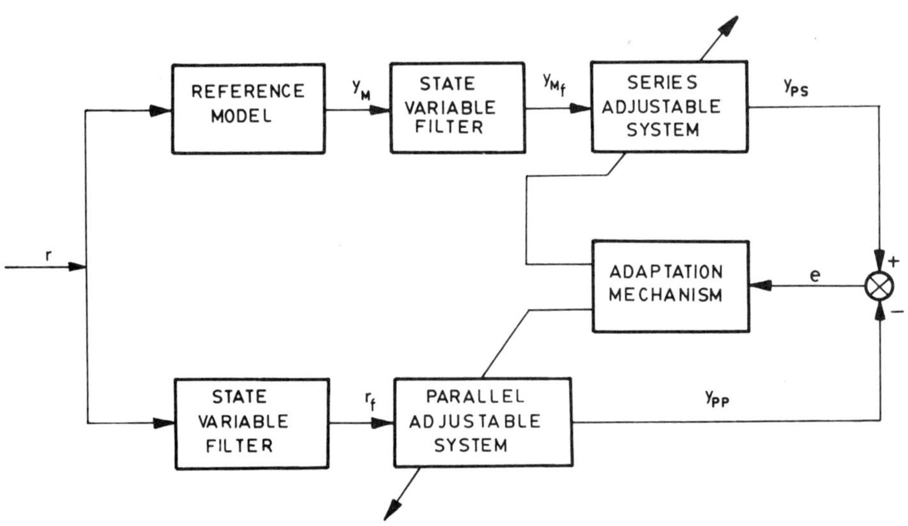

FIG. 2.7 Series-parallel MRAS. (From Ref. 2.)

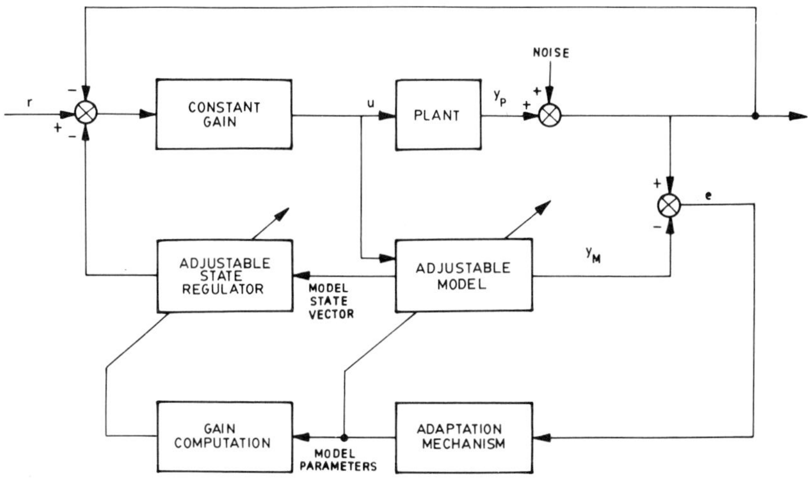

FIG. 2.8 Adaptive state regulator with adjustable. (From Ref. 2.)

scheme is useful when the desired dynamic characteristics of the control system are constrained by the values of the plant parameters.

2.2.7 Two-Level MRAC System

In this configuration (Fig. 2.9) the AMFC system is superimposed on the adaptive state regulator, so as to improve the performance of the latter.

2.2.8 Extremal MRAC System

It was mentioned in Section 1.5.1 that extremal adaptive control results when a certain IP is either maximized or minimized. One example is the adjustment of the blade angles in a water turbine to achieve maximum power. In recent years extremal adaptive control has been used by Keviczky and Haber [9], Bamberger and Isermann [10], and Sternby [11], especially because it is now possible to implement rather complex algorithms in microcomputers which have become relatively inexpensive [12].

In an extremal adaptive system, in the absence of disturbances, the steady-state relationship between the input and output should be a function with an extremum. The objective of control is to stay as close to this extremum as possible despite noise, drift, and so on.

An example of an extremal adaptive system incorporating MRAC is shown in Fig. 2.10. An adjustable model is used to identify the parameters of the plant nonlinear characteristic approximated by a parabola and from this information to compute the input that will drive the plant to the extremum.

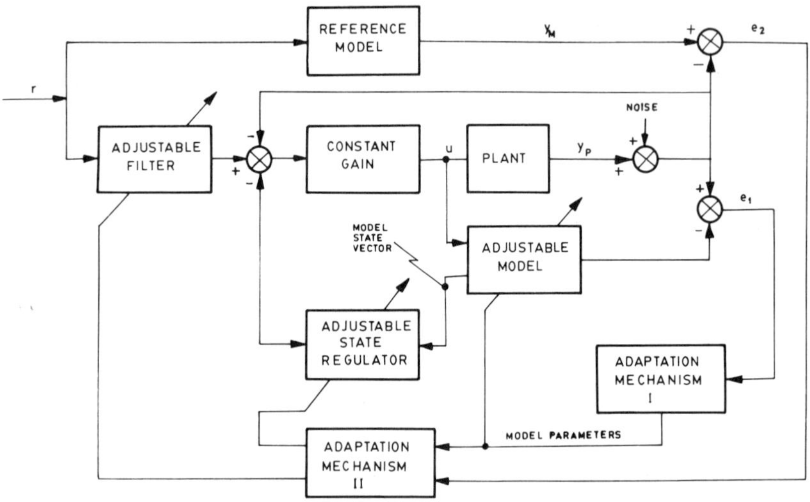

FIG. 2.9 Two-level MRAS. (From Ref. 2.)

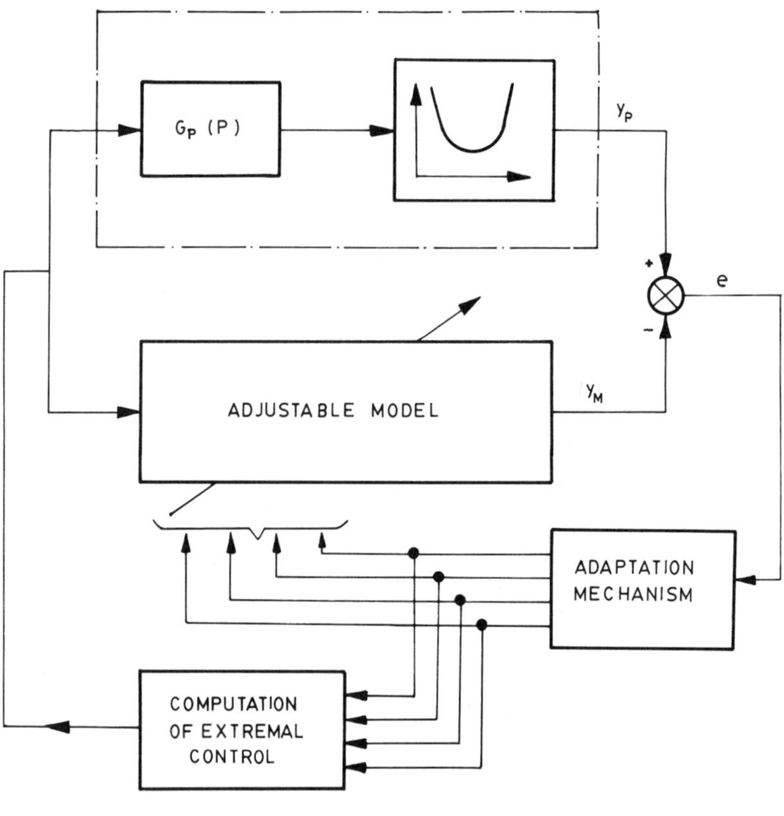

FIG. 2.10 Extremal MRAC system. (From Ref. 2.)

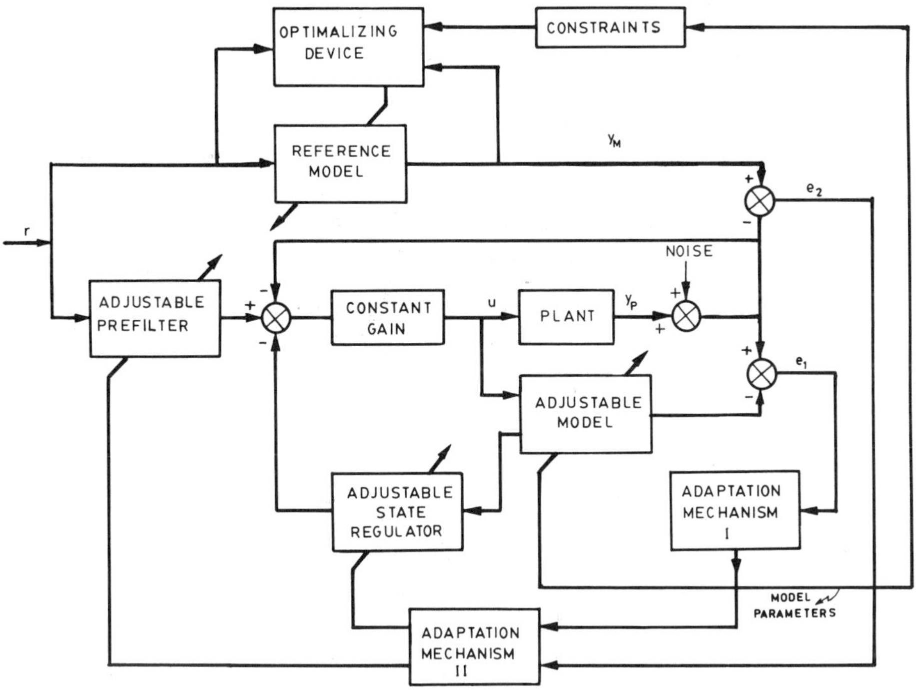

FIG. 2.11 Self-optimizing MRAS. (From Ref. 2.)

2.2.9 Self-Optimizing MRAS

Assuming high-speed adaptation, the reference model can be a slowly time-varying one whose parameters are optimized by an optimalizing device so that a performance index is either minimized or maximized. This results in a self-optimizing MRAS (Fig. 2.11).

2.3 CLASSIFICATION OF MRAS

2.3.1 Classification with Respect to Structure

There are many types of MRAS, which may be classified as follows [2]:

1. Parallel MRAS: Figure 2.12 illustrates a parallel MRAS, which is the most commonly used scheme.
2. Series MRAS: A series MRAS is shown in Fig. 2.13.
3. Series-parallel MRAS: There can be two alternative structures for series-parallel MRAS (Figs. 2.14 and 2.15).

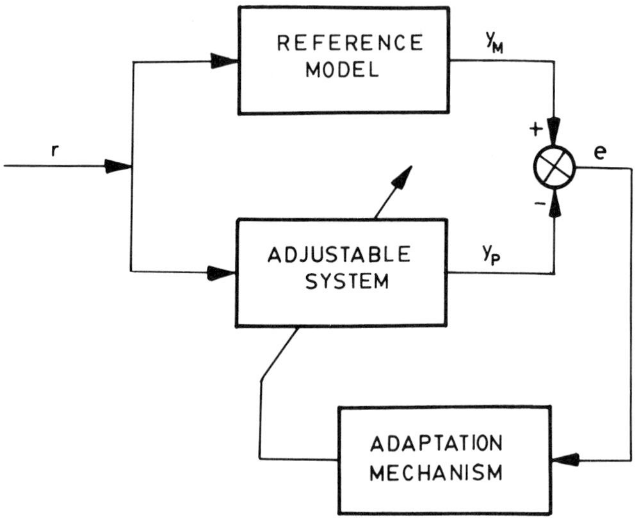

FIG. 2.12 Parallel MRAS. (From Ref. 2.)

It is usual to consider these three structures from separate points of view. However, a unified approach to their analysis and synthesis is possible. Figures 2.7, 2.8, and 2.9 illustrate the implementation of the series-parallel structures. Similar implementation is possible for series MRAS also.

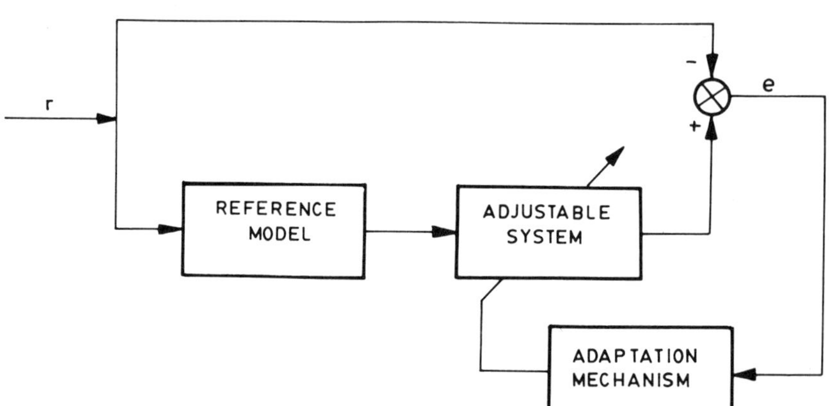

FIG. 2.13 Series MRAS. (From Ref. 2.)

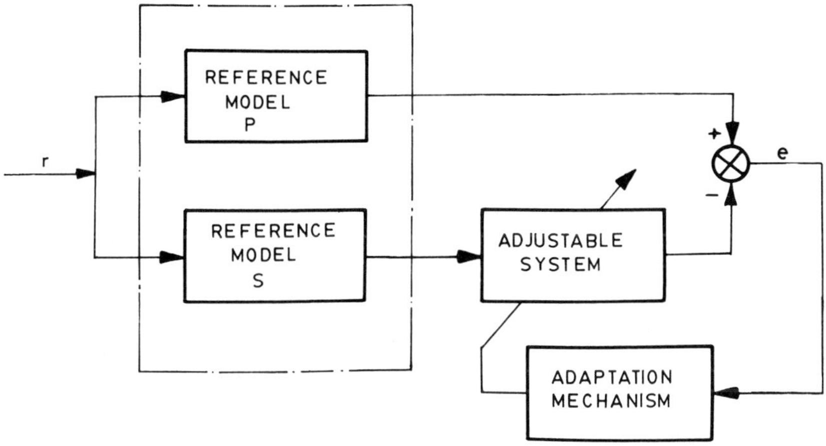

FIG. 2.14 Series-parallel MRAS. (From Ref. 2.)

When used for identification purposes, the parallel structure is called the output error method, the series-parallel structure with an adjustable model is called the equation error method, and the series structure is called the input error method.

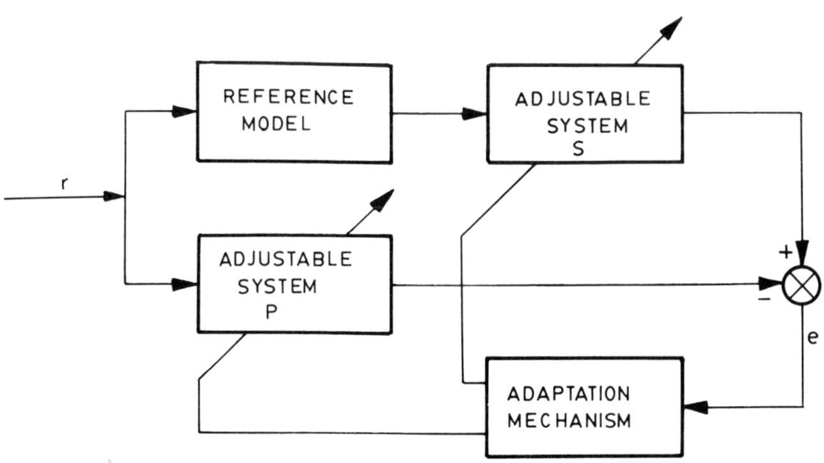

FIG. 2.15 Series-parallel MRAS. (From Ref. 2.)

2.3.2 Classification with Respect to Other Criteria

The MRAS may also be classified according to the following criteria:

1. The adaptation principle: that is, parameter adaptation, signal synthesis adaptation or a combination of both.
2. The conditions of operation: that is, (a) with test signals applied to the input of the system or to the parameters of the adjustable system (Fig. 2.16a, b); or (b) without test signals, in which case adaptation is done with the signals already present in the system.
3. The index of performance: for example, minimization of a norm of the error e and its derivatives. There can be other indexes. (We should not confuse this index of performance with the "desired" performance index determined by the model. The index cited here only serves the purpose of designing the adaptation mechanism and this is a measure of the difference between the "desired" performance index of the model and the "actual" performance index of the adjustable system.)
4. The type of application: that is, AMFC, identification, state observation (estimation), self-adaptive regulation, or extremal control. (The choice of the index of performance used for the design of the adaptation mechanism mentioned earlier in item 3 will depend on the field of application).
5. The type of parameter disturbances: that is, the parameters may be unknown but constant or the parameters may have frequent nonmeasurable changes (in both cases, either for a model or part of the plant).
6. The nature of the equations: that is, continuous-time, discrete-time, or hybrid MRAS.

2.4 MATHEMATICAL DESCRIPTION

We will obtain the equations for different MRAC configurations, following Landau [2].

2.4.1 Parallel MRAC Systems (see Figs. 2.3, 2.4, and 2.12)

State Variable Description

The reference model is considered stable and completely controllable. The state equation for the reference model is

$$\dot{x} = A_M x + B_M r \qquad x(0) = x_0 \tag{2.1}$$

The adjustable system can have either adjustable parameters (Fig. 2.3) or an auxiliary input signal (Fig. 2.4).

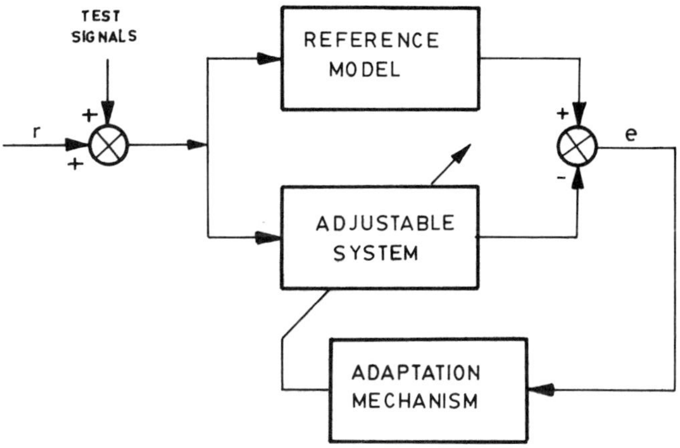

(a)

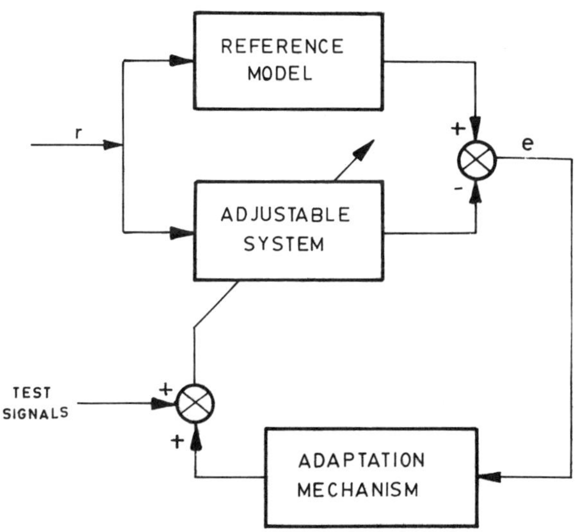

(b)

FIG. 2.16 Test signals in MRAS. (From Ref. 2.)

1. For parameter adaptation

$$
\begin{aligned}
&\dot{y} = A_P(\epsilon, t)y + B_P(\epsilon, t)r \\
&y(0) = y_0 \\
&A_P(0) = A_{P_0} \\
&B_P(0) = B_{P_0}
\end{aligned}
\tag{2.2}
$$

where

$$\epsilon \triangleq (x - y) \tag{2.3}$$

and ϵ is the generalized state error. A_P and B_P are time-varying matrices depending on the generalized state error through the adaptation law. They may depend on other variables also.

The design objective is to determine an adaptive law that modifies $A_P(\epsilon, t)$ and $B_P(\epsilon, t)$ so that ϵ tends to zero as t tends to infinity, for any input r.

It is also possible to include memory in the adaptation mechanism so that it memorizes the good values of the parameters once determined. Then the values of the parameters at any time t will depend not only on $\epsilon(t)$ but also on its past values, $\epsilon(\tau)$, $\tau \leq t$.

The adaptation law will then be

$$
\begin{aligned}
&A_P(\epsilon, t) = A(\epsilon, \tau, t) + A_P(0), \qquad 0 \leq \tau \leq t \\
&B_P(\epsilon, t) = B(\epsilon, \tau, t) + B_P(0), \qquad 0 \leq \tau \leq t
\end{aligned}
\tag{2.4}
$$

where A and B denote a functional relationship between $A_P(\epsilon, t)$, $B_P(\epsilon, t)$ and the values of ϵ during $0 \leq \tau \leq t$.

2. For signal synthesis adaptation

$$
\begin{aligned}
&\dot{y} = A_P y + B_P r + u_a(\epsilon, t) \\
&y(0) = y_0 \\
&u_a(0) = u_{a_0}
\end{aligned}
\tag{2.5}
$$

u_a is the auxiliary signal which depends on ϵ through the adaptation law, which is given by

$$u_a(\epsilon, t) = \mathcal{U}(\epsilon, \tau, t) + u_a(0) \tag{2.6}$$

where $\mathcal{U}$ denotes the functional relationship between $u_a(\epsilon, t)$ and the values of ϵ during $0 \leqslant \tau \leqslant t$.

Input/Output Description

An input/output description is necessary whenever all the state vectors are inaccessible.

The reference model is described by

$$N_M(p)y_M = M_M(p)r \tag{2.7}$$

where $p = d/dt$ and

$$N_M(p) = \sum_{i=0}^{n} a_i p^i \tag{2.8}$$

$$M_M(p) = \sum_{i=0}^{m} b_i p^i \tag{2.9}$$

where r is the input to the model; y_M is the output of the model; and a_i, b_i are constant coefficients.

1. For parameter adaptation the adjustable system is described by

$$N_P(t, p)y_P = M_P(t, p)r \tag{2.10}$$

where

$$N_P(t, p) = \sum_{i=0}^{n} f_i(e, t)p^i \tag{2.11}$$

$$M_P(t, p) = \sum_{i=0}^{m} g_i(e, t)p^i \tag{2.12}$$

$$e \triangleq y_M - y_P \tag{2.13}$$

f_i and g_i are time-varying coefficients depending on the error e through the adaptation law, which takes the form:

$$f_i(e, t) = F_i(e, \tau, t) + f_i(0), \qquad \tau \leqslant t \tag{2.14}$$

$$g_i(e, t) = G_i(e, \tau, t) + g_i(0), \qquad \tau \leqslant t \tag{2.15}$$

2. For signal synthesis adaptation the adjustable system is described by

$$N_P(p)y_P = M_P(p)[r + u_a(e,t)] \tag{2.16}$$

where

$$N_P(p) = \sum_{i=0}^{n} f_i p^i \tag{2.17}$$

$$M_P(p) = \sum_{i=0}^{m} g_i p_i \tag{2.18}$$

$$u_a(e,t) = \mathcal{U}(e,\tau,t) + u_a(0), \qquad \tau \leqslant t \tag{2.19}$$

2.4.2 Series-Parallel MRAC Systems [2] (see Figs. 2.7, 2.14, and 2.15)

State Variable Description

Case (i) (Fig. 2.7). The adjustable system is in two parts—one in series with the reference model and the other in parallel.

In state variable description the reference model is described by Eq. (2.1) and the adjustable system by

$$\dot{y} = A_P(\epsilon,t)x + B_P(\epsilon,t)r \tag{2.20}$$

where

$$\epsilon \triangleq x - y \quad \text{as in (2.3)}$$

$$y(0) = x_0$$

$$A_P(0) = A_{P_0}$$

$$B_P(0) = B_{P_0}$$

The parallel part of the adjustable system is given by $B_P(\epsilon,t)$ and the series part by $A_P(\epsilon,t)$.

Case (ii) (Fig. 2.14). The reference model is in two parts—one in series with the adjustable system and the other in parallel.

In this case, the adjustable system is given by Eq. (2.2) and the reference model by

$$\dot{x} = A_M y + B_M r \tag{2.21}$$

$$x(0) = y_0$$

Input/Output Description

For case (i):

$$e = y_{PS} - y_{PP} \tag{2.22}$$

where

$$y_{PS} = \sum_{i=0}^{n} f_i(e, t)p^i \; y_M \tag{2.23}$$

$$y_{PP} = \sum_{i=0}^{m} g_i(e, t)p^i \; r \tag{2.24}$$

For case (ii):

$$e = y_M - y_P \tag{2.25}$$

$$y_P = -\sum_{i=0}^{n} f_i(e, t)p^i y_M + \sum_{i=0}^{m} g_i(e, t)p^i r \tag{2.26}$$

As can be seen from the preceding equations, the series-parallel configurations require pure derivative operators acting on y_M and r in both cases. To avoid these in view of the problems of noise, state variable filters are introduced in both the paths. The state variable filter is an asymptotically stable low-pass filter having n integrators, allowing one to obtain filtered derivatives up to the order n.

2.4.3 Series MRAC Systems [2] (Fig. 2.13)

The realization of a series MRAC system depends on the invertibility properties of the reference model. This is relatively simple for single-input, single-output (SISO) systems, but rather complicated for multivariable systems. The existence conditions for inverse systems and the relevant algorithms are given in Silverman [13] and Singh and Liu [14].

Considering only SISO systems, the reference model is given by Eq. (2.7) and the series adjustable system by

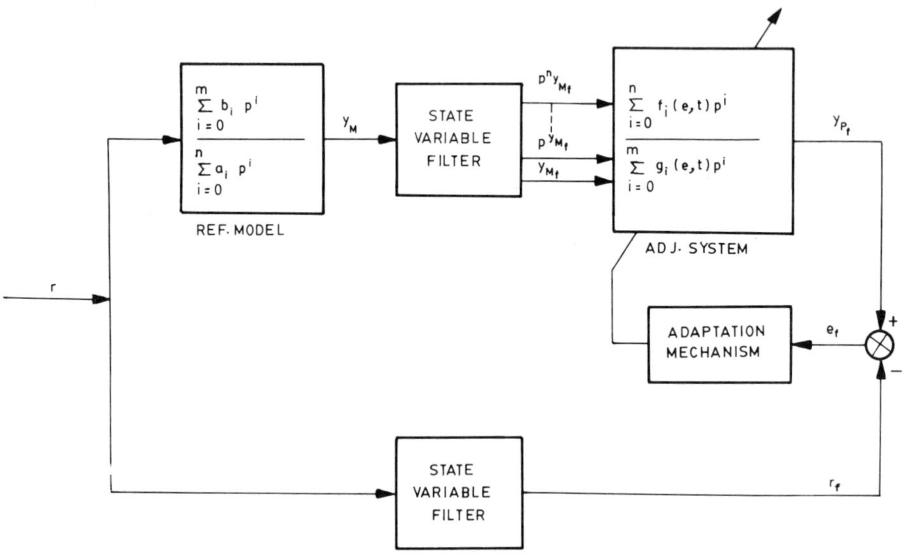

FIG. 2.17 Series MRAS. (From Ref. 2.)

$$\sum_{i=0}^{m} g_i(e,t)p^i y_P = \sum_{i=0}^{n} f_i(e,t)p^i y_M \tag{2.27}$$

where

$$e = y_P - r \tag{2.28}$$

To avoid the pure derivative operators, state variable filters have to be introduced as in the series-parallel case in both the paths, as shown in Fig. 2.17. Equations (2.27) and (2.28) will be of the same form but y_M and r will be replaced by their filtered values y_{M_f} and r_f respectively, and the output of the series adjustable system by y_{P_f}.

2.5 EQUIVALENT REPRESENTATION AS A NONLINEAR TIME-VARYING SYSTEM

A model reference adaptive system can be represented by an equivalent nonlinear time-varying system such that it is reduced to a standard feedback system having a linear part in the feedforward path and a nonlinear time-varying part in the feedback path. Such a representation is necessary for the design of MRA systems using the hyperstability concepts that will be discussed in Section 3.4. The following formulation of equations and representations is obtained following Landau [2].

Let us consider a parallel MRAC system that has accessible state vectors. Such a system is described by Eqs. (2.1)-(2.4). We will obtain an equation involving ϵ, as this is an important factor in design [2].

Subtracting (2.2) from (2.1), we get

$$\begin{aligned} \dot{\epsilon} &= \dot{x} - \dot{y} \\ &= A_M x + [B_M - B_P(\epsilon, t)]\, r - A_P(\epsilon, t) y \end{aligned} \tag{2.29}$$

By adding and subtracting $A_M y$ on the right-hand side of (2.29) and using (2.3) and (2.4), we get

$$\begin{aligned} \dot{\epsilon} = {} & A_M \epsilon + [A_M - A_P(0) - A(\epsilon, \tau, t)] y \\ & + [B_M - B_P(0) - B(\epsilon, \tau, t)] r \end{aligned} \tag{2.30}$$

We have therefore obtained one equation describing the MRAC system. This enables the equivalent representation as a nonlinear time-varying system (Fig. 2.18).

The design requirement for perfect asymptotic adaptation is that $\epsilon(t)$ tend to zero as t tends to infinity. We can therefore state the design problem as follows:

> Find $A(\epsilon, \tau, t)$ and $B(\epsilon, \tau, t)$ such that the equivalent feedback system represented by Eq. (2.30) will be asymptotically stable for any $\epsilon(0)$, $[A_M - A_P(0)]$, and $[B_M - B_P(0)]$.

Let us next consider representing the preceding nonlinear time-varying system as decomposed into a linear part and a nonlinear, time-varying part, so that the stability of the feedback system is determined only by the characteristics of the linear part, provided the nonlinear part satisfies certain conditions given by Popov.

Instead of using the generalized state error ϵ directly, let another vector v be introduced by processing ϵ through a linear compensator D, which is so chosen that the linear part satisfies the requirements for stability:

$$v = D\epsilon \tag{2.31}$$

If, for all t, $r \not\equiv 0$ and $y \not\equiv 0$, then the condition $\lim_{t \to \infty} \epsilon(t) = 0$ implies that the adaptation mechanism has memory. This means that once $A(\epsilon, \tau, t)$ and $B(\epsilon, \tau, t)$ are determined, these values must be conserved when ϵ becomes null, to assure that $A_P(\epsilon, t) = A_M$ and $B_P(\epsilon, t) = B_M$. This memory

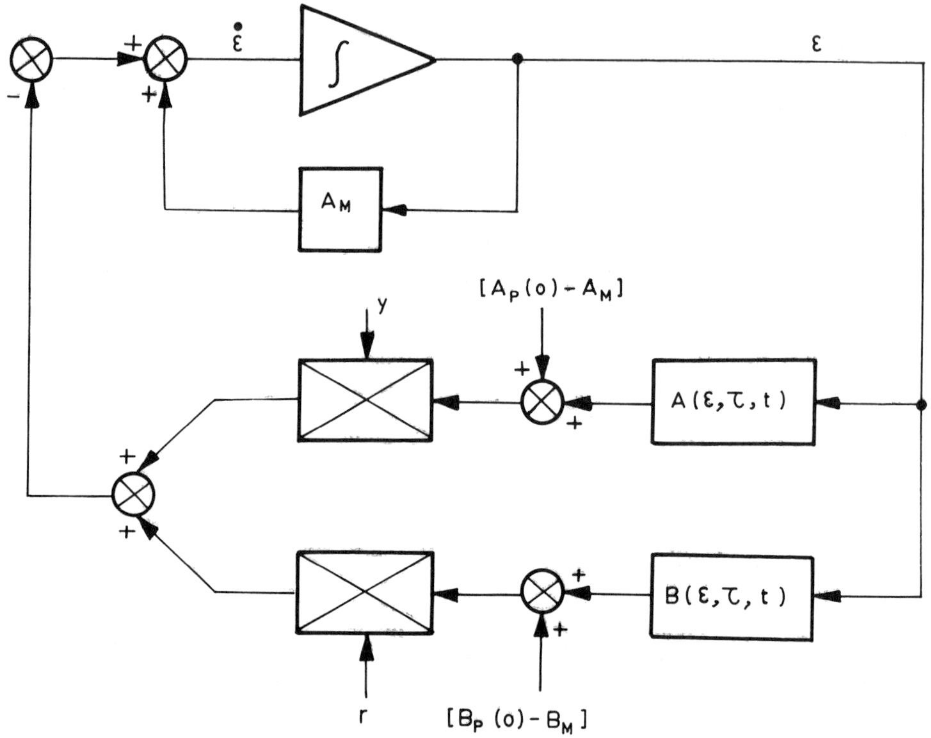

FIG. 2.18 Equivalent representation of generalized state error ϵ. (From Ref. 2.)

can be obtained by using an integrator in the adaptation mechanism. We can rewrite the adaptation laws for A_P and B_P as follows:

$$
\begin{aligned}
A_P(\epsilon, t) &= A_P(v, t) = \int_0^t w_1(v, \tau, t)\, d\tau + w_2(v, t) + A_P(0) \\
B_P(\epsilon, t) &= B_P(v, t) = \int_0^t w_3(v, \tau, t)\, d\tau + w_4(v, t) + B_P(0)
\end{aligned}
\tag{2.32}
$$

where the first terms refer to the memory in the adaptation mechanism and the second terms, which are functions of v(t), become null when $v = 0$, $\epsilon = 0$.

w_1 and w_3 are $n \times n$ and $m \times n$ matrices respectively and denote the relationship between $A_P(\epsilon, t)$, $B(\epsilon, t)$ and the values of v for $0 \leqslant \tau \leqslant t$.

w_2 and w_4 are also matrices of the same dimension and denote the relationship between $A_P(\epsilon, t)$, $B_P(\epsilon, t)$ and the values of v(t) and are null for v = 0.

We can obtain the set of adaptation laws by using Eqs. (2.32) in (2.30)

$$\dot{\epsilon} = A_M \epsilon + I h_1 \tag{2.33}$$

$$v = D\epsilon \tag{2.34}$$

$$h = -h_1 = \left\{ \left[\int_0^t w_1(v, \tau, t)\, d\tau + w_2(v, t) + A_P(0) - A_M \right] y \right\} + \left\{ \left[\int_0^t w_3(v, \tau, t)\, d\tau + w_4(v, t) + B_P(0) - B_M \right] r \right\} \tag{2.35}$$

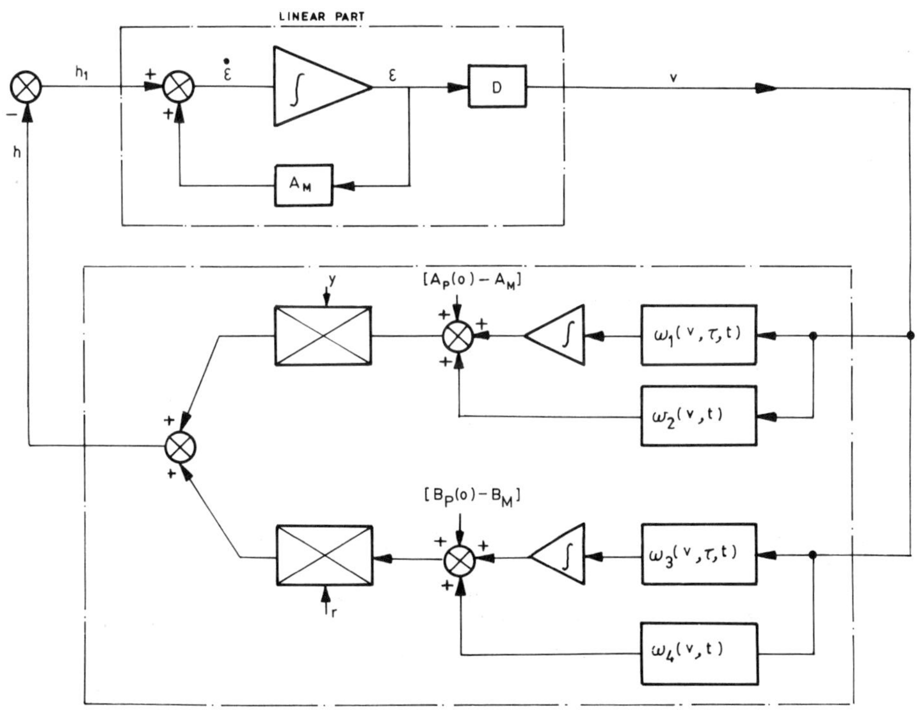

FIG. 2.19 Equivalent representation as a nonlinear time-varying system. (From Ref. 2.)

Equations (2.33)-(2.35) give the set of adaptation laws for the adaptation mechanism and these are shown by an equivalent representation in Fig. 2.19.

2.6 DIRECT AND INDIRECT MRAC

2.6.1 Introduction

Two principal approaches to the model reference adaptive control problem are introduced by Narendra et al. [15,16] considering an unknown plant, namely, indirect and direct control, as pointed out in Section 1.5.3.

1. Indirect control: The basic philosophy of this approach is to estimate the parameters of the unknown plant from input/output data, and, in turn, to use these estimates to generate a feedback control function to adjust the parameters of the controller.
2. Direct control: Unlike the previous case, no explicit plant identification is involved in direct control prior to generating the feedback control signal to adjust the parameters of the controller.

The adaptive control problem using indirect or direct control may be stated as follows:

Given a plant with an input-output pair $\{u, y_P(t)\}$, find a suitable control input u(t) such that

$$\lim_{t \to \infty} |e(t)| = \lim_{t \to \infty} |y_M(t) - y_P(t)| = 0$$

where $y_M(t)$ is the desired output.

The main difference between indirect and direct adaptive control lies in the following two facts: First, a model of the desired behavior is explicitly used in direct control whereas a model of the plant identified on-line is used in indirect control. It is imbedded in the observer structure and is used implicitly to generate the adaptive laws. Second, identification error in indirect control and the control error in direct control are used to update the controller parameters. In both cases it is required that the controller be free of differentiators.

We next develop equations pertaining to these two classes of MRAC system, following Narendra and Valavani [15]. The ideas presented here will be of importance in the subsequent chapters.

2.6.2 Direct Control

Figure 2.20 shows a direct control scheme. The plant is described by the equation

$$\dot{y}_P = -A_P y_P + Bu(t) \tag{2.36}$$

where A_P is unknown and assumed to be positive and B is a known constant parameter. State feedback is used so that the output behavior of the system asymptotically approaches that of a specified stable reference model described by the equation

$$\dot{y}_M = -A_M y_M + Br(t) \tag{2.37}$$

where A_M is positive.

Using state feedback, we have

$$u(t) = r(t) - k(t)y_P(t) \tag{2.38}$$

Then Eq. (2.36) becomes

$$\dot{y}_P = -[A_P + Bk(t)]y_P(t) + Br(t) \tag{2.39}$$

Define $e(t) \triangleq y_M(t) - y_P(t)$. Then we have

$$\dot{e}(t) = -A_M e + \phi y_P(t) \tag{2.40}$$

where the parameter error

$$\phi(t) = A_P + Bk(t) - A_M \tag{2.41}$$

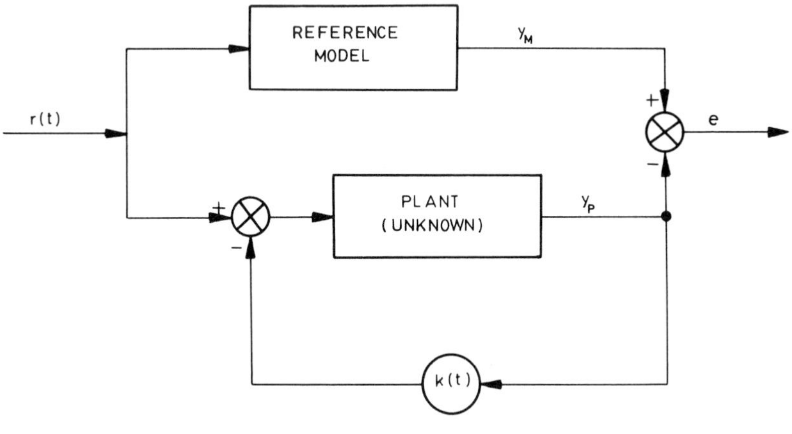

FIG. 2.20 Direct control. (From Ref. 15. Copyright © 1976 IEEE.)

$\phi(t)$ is unknown; but $\dot{\phi}(t)$ can be adjusted. An adaptive law is to be determined so that $\lim_{t \to \infty} e(t) = 0$ and $\lim_{t \to \infty} \phi(t) = 0$.

It may be mentioned at this stage that an adaptive law which accomplishes this requires that the input be "sufficiently rich" or "persistently exciting." For the time being, we may define this condition as follows:

DEFINITION. An input r(t) is considered to be "sufficiently rich" or "persistently exciting" if it contains a sufficient number of distinct frequencies, that is,

$$r(t) = \sum C_i \sin(\omega_i t + b_i)$$

The concept of persistent excitation is of great significance in adaptive systems and is discussed in Section 8.11.

2.6.3 Indirect Control

Figure 2.21 shows an indirect control scheme. The plant is described by a scalar differential equation, as before:

$$\dot{y}_P = -A_P y_P + Bu(t) \tag{2.42}$$

The parameter A_P of the plant is identified by a simple observer of the type given by Luders and Narendra [17,18].

k(t) is adjusted using the law:

$$\dot{k}(t) = \gamma_1 e(t) y_P(t), \qquad \gamma_1 > 0 \tag{2.43}$$

where $e_1(t)$ is the identification error between the plant and the model. If the input to the plant is "sufficiently rich," k(t) tends to $(A_P - \lambda_0)$, so that $k(t) + \lambda_0$ may be used as an estimate of A_P at any time t.

As in the direct control case, let us assume that a model having the desired behavior is given by

$$\dot{y}_M = -A_M y_M + Br(t), \qquad A_M > 0 \tag{2.44}$$

If state feedback is used in the plant with a feedback parameter $\rho(t)$, then

$$\dot{y}_P = -[A_P + B\rho(t)] y_P + Br(t) \tag{2.45}$$

Define the identification parameter error

$$\phi(t) \triangleq A_P - k(t) - \lambda_0 \tag{2.46}$$

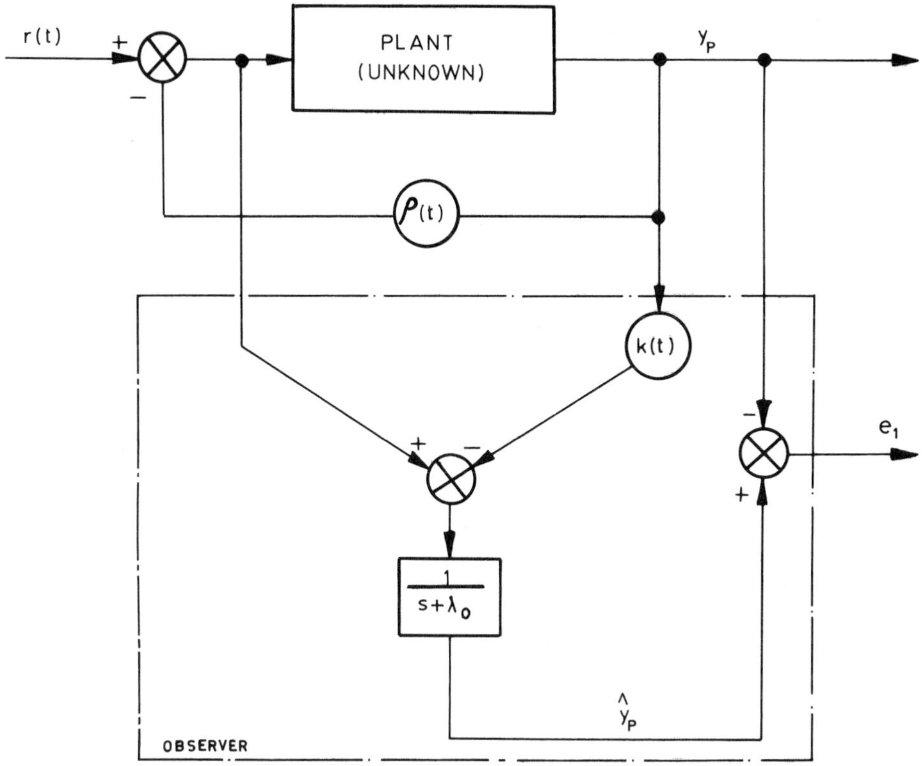

FIG. 2.21 Indirect control. (From Ref. 15. Copyright © 1976 IEEE.)

Then we have

$$\dot{e}_1 = -\lambda_0 e_1 + \phi y_P \tag{2.47}$$

$$\dot{\phi} = -\gamma_1 e_1 y_P \tag{2.48}$$

Now we have to decide the manner in which the feedback gain $\rho(t)$ is to be updated. There are a number of choices. One choice is

$$\rho(t) = \frac{A_M - k(t) - \lambda_0}{B} \tag{2.49}$$

where $\rho(t)$ is adjusted assuming that the estimate $[k(t) + \lambda_0]$ is the true value of A_P. In such a case we have the following equations:

Plant with feedback:

$$\dot{y}_P = -A_M y_P - \phi y_P + Br(t) \tag{2.50}$$

Identification error:

$$\dot{e}_1 = -\lambda_0 e_1 + \phi y_P \tag{2.51}$$

Parameter error in identification:

$$\dot{\phi} = -\gamma_1 e_1 y_P \tag{2.52}$$

Several approaches exist for stability analysis of the preceding equations and for proving the convergence of the error, that is, $e_1(t)$ and $\phi(t)$ tend toward zero, resulting in $|y_P(t) - y_M(t)| \longrightarrow 0$ as $t \longrightarrow \infty$. However, in most adaptive control situations the parameter error does not tend to zero. This may be accounted for qualitatively by the fact that, as the adaptation goes on, the input u(t) to the plant will not continue to be "sufficiently rich." These issues are discussed in Sections 3.9 and 8.5.

It may be mentioned at this stage that Narendra and Valavani [16] have shown that the error equations for both direct and indirect control are the same. Consequently, the adaptive laws for updating the parameters are also identical. This leads to a close connection between the two philosophically different approaches. Such a unified view is dealt with in subsequent chapters.

2.7 CONCLUDING REMARKS

Model reference adaptive systems have experienced a series of developments since they were first introduced for adaptive autopilots as early as 1958. Their popularity has been caused by their simplicity and quickness of response. However, only in 1980 was there a significant breakthrough in relation to the problem of global asymptotic stability. There are still problems that have to be resolved, such as the design of systems in the presence of stochastic disturbances.

MRAC systems form the subject matter of this and the succeeding two chapters. This chapter treated the different configurations and classifications of MRAC systems and gives a mathematical description as well as an equivalent representation of an MRAC system as a nonlinear time-varying system that are of significant use in the succeeding chapters. Direct and indirect control schemes are next dealt with and the concept of "sufficiently rich" or "persistently exciting" signals, which have an important bearing in many adaptive control problems, is also developed.

REFERENCES

1. Landau, I. D., A survey of model reference adaptive techniques (theory and applications), Automatica, 10: 353-379 (1974).
2. Landau, I. D., Adaptive Control: The Model Reference Approach, Marcel Dekker, New York, 1979.
3. Gilbart, J. W., and Winston, G. C., Adaptive compensation for an optical tracking telescope, Automatica, 10: 125-131 (1974).
4. van Amerongen, J., and Udink Ten Cate, A. J., Model reference adaptive autopilots for ships, Automatica, 11: 441-449 (1975).
4a. Raju, G. V. S., and Riad, K. A., Signal synthesis MRAC systems, Proceedings IEEE Int. Conf. on Systems, Man and Cybernetics, New Delhi, India, pp. 934-939 (1983).
5. Landau, I. D., Muller, F., Dolle, G., and Bianchi, G., A new method for carbon control in basic oxygen furnace, Proceedings of IFAC Symp. on Automation in Mining, Mineral and Metal Processing, Johannesburg (1976).
6. Åström, K. J., and Eykoff, P., System identification—A survey, Automatica, 7: 123-162 (1971).
7. Richalet, J., Rault, A., and Poulinquen, R., Identification des Processus par la Methode du Modele, Gordon & Breach, London, 1971.
8. Mendel, J. M., Discrete Techniques of Parameter Estimation: The Equation Error Formulation, Marcel Dekker, New York, 1973.
9. Keviczky, L., and Haber, R., Adaptive dual extremem control by Hammerstein models, Proc. of IFAC Symp. Stochastic Control, Budapest, pp. 333-341 (1974).
10. Bamberger, W., and Isermann, R., Adaptive on-line steady state optimization of slow dynamic processes, Automatica, 14: 223-230 (1978).
11. Sternby, J., Extremum control systems—An area for adaptive control, Proc. JACC (1980).
12. Sternby, J., Adaptive control of extremum systems, in Methods and Applications in Adaptive Control (H. Unbehauen, ed.), Springer-Verlag, Berlin, pp. 151-160, 1980.
13. Silvermann, L. M., Inversion of multivariable systems, IEEE Trans. Auto. Contr., AC-14: 270-276 (1968).
14. Singh, S. P., and Liu, R., A survey of inverse systems, Proc. Nat. Electro. Conf., U.S.A., pp. 380-384 (1970).
15. Narendra, K. S., and Valavani, L. S., Stable adaptive observers and controllers, IEEE Proc., 64: 1198-1208 (1976).
16. Narendra, K. S., and Valavani, L. S., Direct and indirect adaptive control, Automatica, 15: 653-664 (1979).
17. Luders, G., and Narendra, K. S., A new canonical form for an adaptive observer, IEEE Trans. Auto. Contr., AC-19: 117-119 (1974).
18. Luders, G., and Narendra, K. S., Stable adaptive schemes for state estimation and identification of linear systems, IEEE Trans. Auto. Contr., AC-19: 841-847 (1974).

3

Continuous-Time MRAC Systems

3.1 INTRODUCTION

In the preceding chapter, various general aspects of MRAC systems, including classification and mathematical description, were discussed. This chapter deals with the design aspects of continuous-time MRAC systems. Most of the techniques proposed for design reduce the adaptive problem to that of determining the controller structure and finding how the controller parameters have to be adjusted, that is, obtaining a parameter adaptation law to guarantee global asymptotic stability.

This chapter is organized as follows: Section 3.2 introduces the MIT rule, which is an important early landmark in the design of MRAC systems. The Lyapunov approach, which ensures both adaptation and stability, is discussed in Section 3.3. The hyperstability approach is dealt with in Section 3.4, which includes a comparison between Lyapunov and hyperstability methods. Section 3.5 outlines Monopoli's augmented error approach, which is conceptually a novel idea and which led to other important techniques such as Narendra's error model approach, described in Section 3.6. The recent trend has been to evolve a unified approach and the one by Egardt is the subject matter of Section 3.7. Section 3.8 deals with some aspects of multivariable MRAC systems. Section 3.9 addresses itself to the important problems of stability and convergence considerations. Concluding remarks are included in Section 3.10.

3.2 MIT Rule

3.2.1 Introduction

The model reference adaptive control scheme was first suggested by Whitaker and his co-workers at MIT [1] as far back as 1958 in order to develop an adaptive autopilot. Further work was reported by them in Refs. 2-5. This work was based on the minimization of a performance index and led to the design rule referred to as the MIT rule. This work was more experimental than theoretical. Since then it has been further developed on a rather more theoretical basis by Donalson [6], Dressler [7], Price [8], Kokotovic et al. [9, 10] and Winsor et al. [11, 12]. Other designs have been developed by Pearson [13, 14], and Serdyukov [15].

It may be noted that the MIT design rule does not necessarily lead to a globally stable system, and a good compromise between the stability and the speed of response is to be struck after a good deal of simulation studies.

3.2.2 Derivation of the MIT Rule

The MIT rule is based on minimizing the integral of the square of the error between the desired output generated by passing the input r(t) of the system through a carefully designed reference model and the actual output of the plant.

Consider the generalized model reference adaptive control system shown in Fig. 3.1. The performance index is given by

$$J = \int_0^T \|y_M(t) - y_P(t)\|^2 \, dt \tag{3.1}$$

The following two assumptions are made:

1. The speed of adaptation is low.
2. We start in the neighborhood of the model parameter value.

Considering a linear system, the relationships may be written in terms of transfer functions. Following Saridis [16],

$$Y_M(s) = G_M(s)R(s)$$

$$Y_P(s) = \frac{G_P(s,a)G_R(s,b)R(s)}{1 + G_P(s,a)G_R(s,b)H(s,b)}$$

$$= G(s,a,b)R(s) \tag{3.2}$$

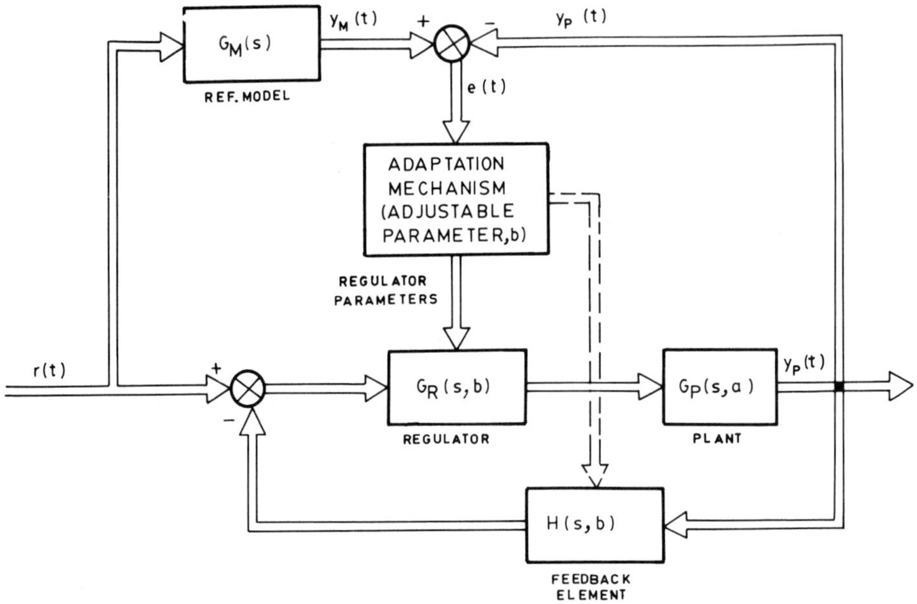

FIG. 3.1 Generalized MRAC system. (From Ref. 16.)

where

$G_M(s)$ = transfer matrix of the reference model
$G_P(s, a)$ = transfer matrix of the plant with unknown parameters vector a
$G_R(s, b)$ = transfer matrix of the regulator with the vector of adjustable parameters b

For an exact adjustment, the order of the model should be the same as that of the system. However, it has been conjectured that if the order of the model is smaller than that of the system, then the reference model acts in the best way to match the response of the system. This may lead to pole-zero cancellation between the plant and the regulator. Special consideration is necessary when the plant is inherently unstable.

The minimization of the performance index is to be performed on-line. It is, therefore, not necessary to have the performance evaluation over the entire interval of time. A "per-interval" criterion may be specified for sufficiently small Δt as:

$$J_p(b) = \frac{1}{2} \| y_M(t) - y_P(t) \|^2 \tag{3.3}$$

where b is the adjustable parameter. $J_P(b)$ is dependent on t and is to be minimized with respect to b by using a search technique, after it has attained a steady-state value, or, for practical purposes, when it changes much more slowly than the rate of search technique used. The latter case is termed a tracking system.

The following are some of the search techniques to generate adaptation algorithms:

1. The gradient method
2. The steepest descent method
3. The conjugate gradient method

The implementation of these methods requires the use of sensitivity functions, which must be generated on-line. We shall discuss the first method here.

The standard gradient search algorithm may be used to generate estimates of the parameters:

$$\dot{b}(t) = -B\frac{\partial J}{\partial b} = +B\frac{\partial y_P}{\partial b}e(t) \tag{3.4}$$

where B is an arbitrary positive constant called the adaptive gain and

$$e(t) \triangleq [y_M(t) - y_P(t)] \tag{3.5}$$

Equation (3.4) gives the MIT rule.

The MIT rule can also be written in an alternative form, by using the relationships:

$$\frac{\partial e}{\partial b} = \frac{\partial y_M}{\partial b} - \frac{\partial y_P}{\partial b} = -\frac{\partial y_P}{\partial b} \tag{3.6}$$

since y_M is not a function of the adjustable parameter b.

Substituting (3.6) in (3.4), we have

$$\dot{b}(t) = -B\frac{\partial e}{\partial b}e(t) \tag{3.7}$$

This form of the MIT rule is also frequently used.

In Eq. (3.4) the sensitivity matrix $(\partial y_P/\partial b)$ of the output y_P with respect to the variation of b may be got by differentiating G(s, a, b) with respect to b. From (3.2), we get

$$\frac{\partial Y_P(s)}{\partial b} = \frac{\partial G(s, a, b)}{\partial b} R(s)$$

In the preceding equation, since the parameter vector a is not known, the sensitivity matrix cannot be evaluated exactly. Some approximations must, therefore, be made.

Whitaker et al. [5] assumed that the performance of the system is sufficiently close to that of the reference model. It was further assumed that the order of the model is the same as that of the plant. Then the sensitivity matrix may be obtained from the model. See Refs. 4, 5, 6, and 17 for further details.

3.2.3 Application of the MIT Rule [18]

Consider a simple MRAC system, shown in Fig. 3.2. The objective is to adjust K_c such that $K_c K_V$ will eventually be equal to the model gain K. Applying the MIT rule given in (3.7), we have

$$\dot{K}_c = -B\left(\frac{\partial e}{\partial K_c}\right)e \tag{3.8}$$

where B is a positive constant.

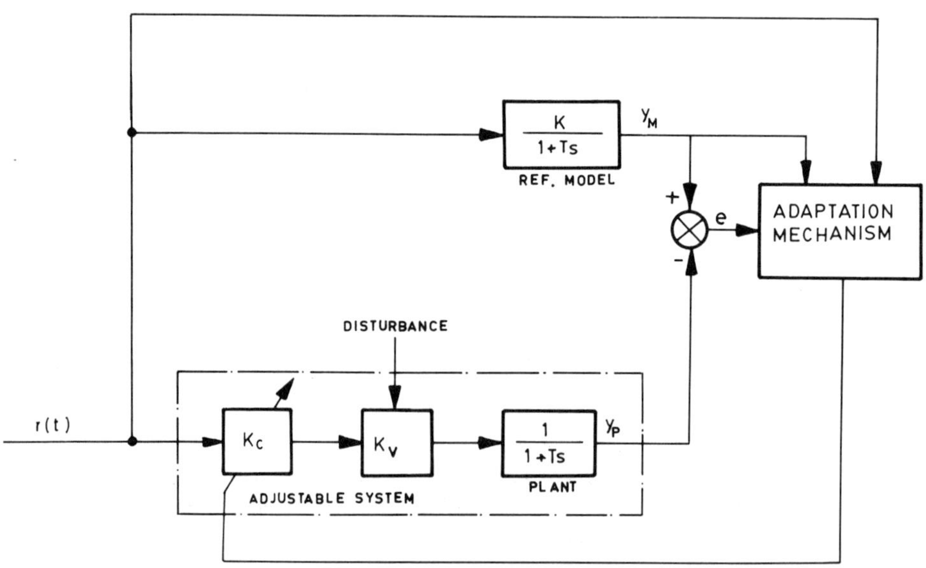

FIG. 3.2 Gain adaptation MRAC. (From Ref. 18. Copyright © 1966 IEEE.)

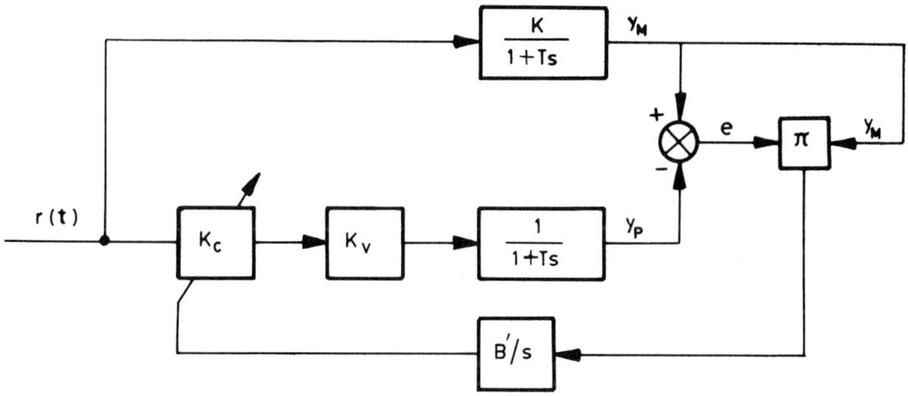

FIG. 3.3 First-order MIT gain-adaptation scheme. (From Ref. 18. Copyright © 1966 IEEE.)

$$\left(\frac{\partial e}{\partial K_c}\right) = \frac{-K_v \cdot r}{(1 + Ts)} \tag{3.9}$$

This is obtained by partially differentiating

$$e = \frac{(K - K_c K_v) r}{(1 + Ts)} \tag{3.10}$$

with respect to K_c. From (3.9) we see that $(\partial e/\partial K_c)$ is proportional to y_M.

The signal $(\partial e/\partial K_c)$ is usually obtained by additional circuitry. The system is shown in Fig. 3.3, in which $K_c = B'ey_M$, where the constant B' is the adaptive gain.

The equations for the system in Fig. 3.3 are

$$\begin{aligned} T\dot{e} + e &= (K - K_v K_c) r(t) \\ T\dot{y}_M + y_M &= Kr(t) \\ \dot{K}_c &= B'ey_M \end{aligned} \tag{3.11}$$

Parks [18] shows that the analysis of these equations is not easy even for simple sinusoidal inputs. However, let us take, for illustration, the case when r(t) = +V. At t = 0, $K_c K_v \neq K$. K_v remains constant subsequently and K_c is to be adjusted as per (3.11). Then we get the following equations:

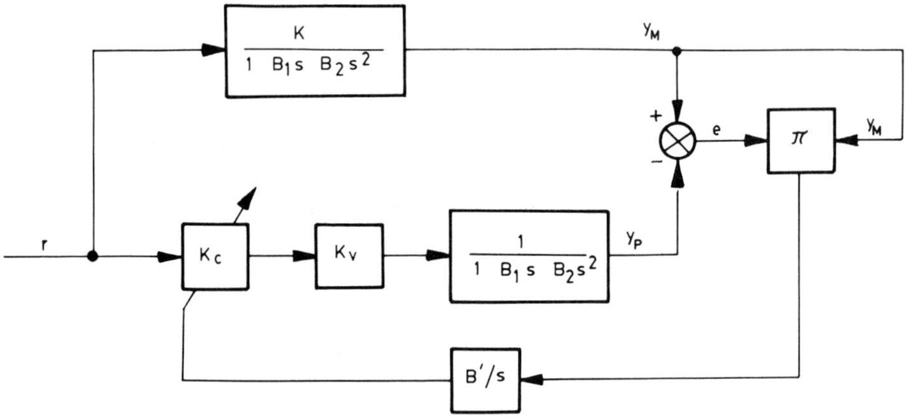

FIG. 3.4 Second-order MIT gain-adaptation scheme. (From Ref. 18. Copyright © 1966 IEEE.)

$$y_M = KV[1 - \exp(-\frac{t}{T})] \tag{3.12}$$

$$T\ddot{e} + \dot{e} + KK_vV^2B'e[1 - \exp(-\frac{t}{T})] = 0 \tag{3.13}$$

In (3.13), the third coefficient approaches KK_vV^2B' as $t \longrightarrow \infty$. The equation

$$T\ddot{e} + \dot{e} + KK_vV^2B'e = 0$$

is asymptotically stable. By an extension of the Dini-Hukuhara theorem, Eq. (3.13) is also asymptotically stable. Thus, as $t \longrightarrow \infty$,

$$e \longrightarrow 0$$

$$K_c \longrightarrow \frac{K}{K_v}$$

This is what one expects.

However, Parks [18] points out that the analysis for higher-order systems is difficult and a quick answer about stability is not possible.

Consider a second-order system, shown in Fig. 3.4. The equations of this system are

$$B_2\ddot{e} + B_1\dot{e} + e = (K - K_vK_c)r(t)$$

$$\dot{K}_c = B'ey_M \tag{3.14}$$

$$B_2\ddot{y}_M + B_1\dot{y}_M + y_M = Kr(t)$$

Let us assume that r(t) = V, which is a constant, and that the adaptive loop is closed when y_M and y_P have attained steady-state values. Parks [18] has shown that instability is possible if the input V or the gain B' are large enough.

The MIT rule was very popular because of its simplicity in practical implementation, although it may require a large number of sensitivity filters for multiparameter adjustments.

Further consideration is not given here to this rule since Parks [18] and Hang and Parks [19] have conclusively shown that the analysis is not simple for higher-order systems even for simple sinusoidal inputs and also, more importantly, global stability is not guaranteed.

3.2.4 Other Related Design Methods

Donalson [6,20] has used a more general performance index than that of Whitaker and obtained an improved design rule with respect to the speed of response. However, additional filters and measurement of the state vectors are required.

Dressler [7] has developed a gradient method by which the need for sensitivity filters can be avoided. The resulting controller is very simple. The disadvantages are that the damping of the response suffers at large loop gains and that global stability is not guaranteed.

Price [8] has applied an accelerated gradient method which is easier to implement and is also capable of achieving faster adaptation compared to the other gradient techniques. The controller is similar to that of Dressler except for the addition of a feedforward term. This term has the effect of improving the damping and the stability of the response. This stabilizing effect will be affected if the order of the system increases. Global stability cannot be guaranteed.

Kokotovic and his colleagues [9,10,21] have applied sensitivity analysis resulting in the simplification of the design. The rule obtained is somewhat similar to the MIT rule. In this case, with further approximation, only one sensitivity filter is required for multiparameter adjustments.

Winsor et al. [11,12] have modified the MIT rule to reduce the sensitivity of the response to the loop gain, though additional instrumentation is necessary.

All the design rules formulated by the preceding researchers are not globally stable and hence the adaptive gain which governs the speed of response is restricted. A good compromise between the stability and the speed of adaptation is to be struck after time-consuming simulation studies.

Green [22] has extended the work of Dressler [7] to obtain a stable maximum descent method. However, though the problem of global stability is solved, the rule is not entirely suitable from the practical point of view since one has to have the first derivatives of the state vectors for its implementation.

As Hang and Parks [19] point out, the MIT rule and other gradient methods that have been preferred by many designers should be used with great caution, in view of the possible instability. Furthermore, some other drawbacks [17] are as follows:

1. The initial difference between the parameters of the reference model and those of the adjustable system is assumed to be small.
2. The adaptation process is also assumed to be slow.
3. There is no indication given regarding the selection of the adaptation gains in order to assume the convergence of the adaptation process and the stability of the MRAC system.

The next section discusses the application of Lyapunov's second method for obtaining the design equations such that the system stability is guaranteed for all inputs.

3.3 LYAPUNOV APPROACH

3.3.1 Introduction

Stability is an extremely important factor which must be taken into consideration in the design of MRAC, especially since these systems behave like nonlinear, time-varying systems. We have seen how in earlier design methods, instability can be caused by large inputs or large gains. To achieve a good design, a stability aspect should invariably be included and this can be done by using the Lyapunov approach or the hyperstability approach.

In the Lyapunov approach first we obtain the differential equation that describes the error between the output of the reference model and that of the plant. The objective is to obtain parameter-adjusting equations which assure that the differential equation describing the error is asymptotically stable. To do this, a positive-definite Lyapunov function is formulated for the error equation. The adaptation mechanism equations are then selected so as to cause the time derivative of the Lyapunov function to be negative-definite.

Lyapunov's stability theorems make possible a method of synthesizing the control laws to result in a design that guarantees stability. The type of solution depends on the form of the Lyapunov function selected. The system will be stable in the sense that the tracking error goes to zero or remains bounded, as when disturbances or time-varying parameters are involved.

Butchart and Shackcloth [23] were the first to propose the use of a quadratic Lyapunov function. This was later taken up by Parks [18] to redesign systems which were formerly designed by the MIT rule. The use of a different Lyapunov function by Philipson [24] and Gilbart et al. [25] resulted in the addition of feedforward loops which improved the damping of the response. The main difficulty of the Lyapunov approach is that the entire state vector must be available for measurement and this is not always possible. If all the state variables of the plant under control are not available, the adaptive law is inadequate since ignorance of parameters and inherent system noise may not allow their generation by an observer or by differentiation. To overcome this problem, investigators have attempted to find adaptive laws requiring the minimum of state variable measurements. The first such attempt was made by Parks [18] using Kalman's lemma [26] to obtain an adaptive law that adjusts the gain so that the error vanishes. It turns out that the gain adjustment using e is possible if, and only if, the plant and model transfer functions are positive real functions.

Monopoli et al. [27] have also applied this idea of a positive real transfer function, which resulted in the elimination or reduction of the number of differentiators required for implementing the design rule. Reference may be made to Section 3.7.3, which discusses the way in which the positive real condition can be eliminated by following a procedure suggested by Egardt. To avoid the use of derivative networks, Currie and Stear [28] suggest the use of a Kalman filter, which would also handle the measurement noise problem. However, the use of state observers to eliminate the states of an unknown time-varying plant remains a problem. Another disadvantage is that the Lyapunov design rule may not be applicable to cases where the plant parameters cannot be directly adjusted. Such a case was indicated by Winsor and Roy [29]. Gilbart et al. [25] give a solution, though rather complex, to this case.

Shahein et al. [30] combine the Lyapunov approach with the steepest-descent design technique resulting in the acceleration of convergence and improvement of transient response due to disturbance. Lindorff [31] and Narendra et al. [32, 33] have considered the effect of disturbance on stability. Lindorff has shown that even though the error is bounded, disturbance can cause the adaptive gains to be unbounded. Narendra et al. give a modified scheme guaranteeing boundedness of errors when disturbance is present [34]. Udink Ten Cate and Verstoep [35] have presented an improvement of performance of MRAC in the presence of noisy environment. New adaptive algorithms for Lyapunov synthesis have been given by Carroll [36]. Gupta and Chalam [37] have applied an improved Lyapunov function to obtain increased damping of the system response.

Other improvements in the application of the Lyapunov approach were reported by Monopoli et al. [38-41], Gilbart et al. [42, 43], Kudva and Narendra [34, 44], Luders and Narendra [45, 46], Oliver et al. [47], and Lindorff [48].

Hang and Parks [19] have done exhaustive comparative studies and demonstrated the merits of the Lyapunov approach compared with earlier methods. Lindorff and Carroll [49] give a critical survey of various Lyapunov design methods, especially with respect to relative stability, reduction of order techniques, design with disturbance, design with time-variable parameters, multivariable systems, identification and an adaptive observer. Narendra and Kudva [50] and Landau [17, 51] give a survey of work done in Lyapunov design of MRAC.

The MKY Lemma

The Kalman lemma referred to earlier is also called the Meyer-Kalman-Yacubovitch (MKY) lemma, which is of great importance in a number of MRAC problems. It may be stated as follows:

> Given a real number Y, two real n-vectors g, k and a real $n \times n$ matrix A. Let Y be larger than zero and let A be stable and completely controllable. Then a real vector q satisfying
>
> (a) $A^T P + PA = -qq^T$
>
> (b) $Pg - k = \sqrt{Y}\ q$
>
> exists if and only if,
>
> $$\frac{1}{2}Y + \text{Re}\{k^T(i\omega I - A)^{-1}g\} > 0$$
>
> for all real ω.

It may be noted that the preceding condition becomes the condition for the transfer function to be a positive real transfer function.

3.3.2 A Lyapunov Scheme

Consider the following configuration of a MRAC system in Fig. 3.5 for which a Lyapunov scheme will be worked out, following Lindorff and Carroll [49]. The state equation of the stable reference model is given by

$$\begin{aligned} \dot{y}_M &= A_M y_M + B_M r \\ y_M(0) &= y_M^0 \end{aligned} \qquad (3.15)$$

and that of the time-invariant plant by

$$\dot{y}_P = A_P y_P + B_P u$$
$$y_P(0) = y_P^0 \tag{3.16}$$

In the preceding equations $y_M = \{y_{M_i}\}$ and $y_P = \{y_{P_i}\}$ are n-dimensional state vectors, and r, u are m-dimensional control inputs. A_P, B_P contain unknown coefficients.

$$\text{Error} = e = y_M - y_P$$

The differential equation for the error is given by

$$\dot{e} = A_M e + f \tag{3.17}$$

where

$$f = (A_M - A_P)y_P + B_M r - B_P u$$

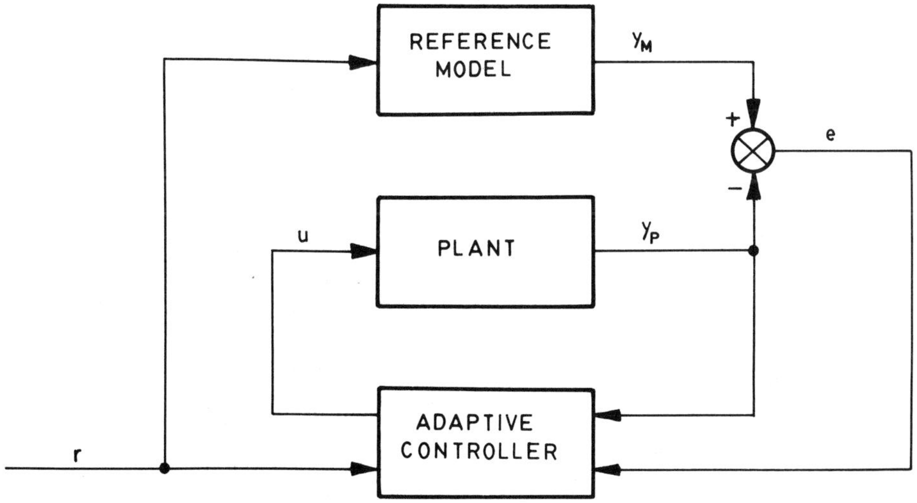

FIG. 3.5 Basic configuration of MRAC. (From Ref. 49.)

The objective is to manipulate f in such a way that

$$\lim_{t \to \infty} e(t) = 0$$

To achieve this let us introduce a positive-definite function

$$V = e^T Pe = h(\phi, \Psi) \tag{3.18}$$

where ϕ, Ψ are matrices of parameter vectors ϕ_i, Ψ_i to be defined. Then we obtain the time derivative of V

$$\dot{V} = -e^T Qe + 2e^T Pf + \dot{h} \tag{3.19}$$

where

$$-Q = A_M^T P + PA_M \tag{3.20}$$

By a well-known theorem of Lyapunov [52], with any $Q^T = Q > 0$, it follows that $P = P^T > 0$ is a unique solution to (3.20) if and only if A_M is a stable matrix, as assumed.

According to Philipson [24], there can be two methods for causing $e \longrightarrow 0$. These will be considered next.

Input Modification

The input modification method is the same as signal synthesis adaptation, in contrast to parameter adaptation. This method uses the solution obtained with

$$h \equiv 0$$

Although it is not in itself useful in synthesizing an adaptive law, that is, one in which a set of parameters is adjusted automatically to eliminate or minimize error, it has been used with advantage in solving some types of design problems.

With $h \equiv 0$, asymptotic stability in e will be satisfied if [53]

$$\begin{aligned} e^T Pf &= 0 \quad \text{for } A_M - A_P = 0 \\ &\leqslant 0 \quad \text{otherwise} \end{aligned} \tag{3.21}$$

This inequality cannot be satisfied except in special cases depending on the system structure. In some restricted cases, such as if (3.15) and (3.16) are in phase variable form, a solution exists. In this case u becomes a scalar, $B_M = b$, and b_n, f_n are the only nonzero elements of b and f respectively. Then we have the simplification

$$e^T Pf = e^T p_n f_n$$

where p_n is the nth column of P.

Writing

$$f_n = f'_n + b_n u$$

and assuming that $b_n > 0$, we see that the conditions (3.21) are fulfilled if

$$|u| \geqslant f'_n |b_n|$$

$$\operatorname{sgn} u = \operatorname{sgn} e^T p_n, \qquad e \neq 0$$

This is realized by a switching function.

Feedback Synthesis

This method differs from input modification as the parameters of the system are adjusted continuously so that in the simplified case treated here $e \longrightarrow 0$. In this case we write for (3.18)

$$V = e^T Pe + \sum_{i=1}^{n} \phi_i^T \phi_i + \sum_{i=1}^{m} \Psi_i^T \Psi_i \tag{3.22}$$

where ϕ_i, Ψ_i are parameter misalignment vectors to be defined in terms of the elements of matrices $A_M - A_P$ and $B_M - B_P$ which express the misalignment between the reference model and the plant. In the present case, (3.19) becomes

$$\dot{V} = -e^T Qe + 2\left(e^T Pf + \sum_{i=1}^{n} \phi_i^T \phi_i + \sum_{i=1}^{m} \Psi_i^T \Psi_i\right) \tag{3.23}$$

Let E define the whole state space with

$$\xi \in E$$

where ξ is defined by

$$\xi^T = [e^T, \phi_1^T, \ldots, \phi_n^T, \Psi_1^T, \ldots, \Psi_m^T]$$

Let $E_1 \subset E$, where E_1 is the n-dimensional subspace with $e \in E_1$. Then V in (3.22) is positive definite in E.

The basic principle in feedback synthesis is to specify (ϕ_i, Ψ_i) in (3.23) such that

$$\left(e^T Pf + \sum_{i=1}^{n} \phi_i^T \phi_i + \sum_{i=1}^{m} \Psi_i^T \Psi_i \right) = 0 \tag{3.24}$$

Then

$$\dot{V} = -e^T Qe \tag{3.25}$$

Since $\dot{V}$ is only negative semidefinite in E, but negative definite in E_1, it can be said from the theorems of Lyapunov [52] that

1. The equilibrium at $e = 0$ is asymptotically stable.
2. The equilibrium at $\xi = 0$ is stable.

It follows from this formulation that $e \longrightarrow 0$ and that the misalignment vectors are bounded.

We can consider two different schemes, namely, direct and indirect adaptation, to implement the controls to satisfy (3.24). Direct adaptation implies that plant parameters are directly adjustable. Indirect adaptation requires that the adjustment takes place external to the plant.

Direct Adaptation

In this scheme

$$u = r \quad \text{in (3.16)}$$

ϕ and Ψ are in turn defined by

$$\phi = A_M - A_P$$

$$\Psi = B_M - B_P$$

Then (3.17) becomes

$$\dot{e} = A_M e + f \tag{3.26}$$

where

$$f = \phi y_P + \Psi r$$

It is seen that (3.24) can be satisfied if

$$\begin{aligned} \dot{\phi}_i^T &= -e^T P y_{P_i}, \qquad i = 1, \ldots, n \\ \dot{\Psi}_i^T &= -e^T P r_i, \qquad i = 1, \ldots, m \end{aligned} \tag{3.27}$$

If adaptive control is the objective, the elements of A_P, B_P should be adjusted. Then (3.27) becomes

$$\begin{aligned} \dot{a}'^T_i &= e^T P y_{P_i}, \qquad i = 1, \ldots, n \\ \dot{b}'^T_i &= e^T P r_i, \qquad i = 1, \ldots, n \end{aligned} \tag{3.28}$$

It is assumed that A_M and B_M are constant. In the problem of identification we note that the matrices A_M, B_M should be modified in place of A_P, B_P.

Indirect Adaptation

In some systems the elements of A_P and B_P are not directly adjustable. Then it becomes necessary to introduce some modifications.

Consider a single-input/single-output (SISO) plant defined by

$$\frac{y_{P_1}}{u} = \frac{K'}{s^n + a'_n s^{n-1} + \cdots + a'_1} \tag{3.29}$$

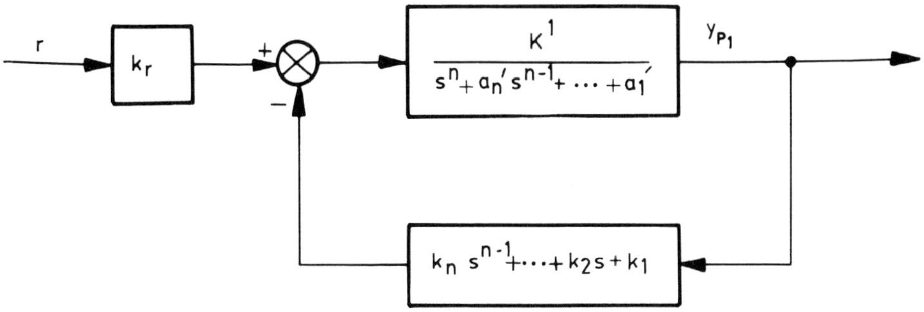

FIG. 3.6 Configuration for indirect adaptation. (From Ref. 49.)

Let us introduce k_1, k_2, ..., k_n and k_r as adjustable parameters as in Fig. 3.6. Then the compensated system is described by the equation

$$\dot{y}_P = A_P y_P + b'r \tag{3.30}$$

where

$$A_P = \left[\begin{array}{ccc} 0 & \vdots & I \\ \hline a'_{n1} & \cdots & a'_{nn} \end{array}\right] \tag{3.31}$$

$$b'^T = [0, \ldots, 0, b'_n] \tag{3.32}$$

with

$$b'_n = k_r k' \tag{3.33}$$

$$a'_{ni} = -a'_i + k_i, \quad i = 1, \ldots, n \tag{3.34}$$

Thus the problem is amenable to direct adaptation by the application of (3.28). For the SISO case only n coefficients in A_P can be adjusted. In multivariable systems, certain conditions are imposed on the form of (3.15) and (3.16) in order that Lyapunov design may yield a unique set of controls [48].

Some Synthesis Procedures

Let us consider the following error equation for further discussion:

$$e = \begin{bmatrix} 0 & 1 \\ -a_0 & -a_1 \end{bmatrix} e + \begin{bmatrix} 0 \\ (k - k') \end{bmatrix} r \tag{3.35}$$

where k' is the only adjustable parameter.

Gilbart and Monopoli [42] give a synthesis procedure by considering a V function different from (3.22):

$$V = e^T Pe + \frac{\Psi^2}{h_1} \tag{3.36}$$

where h_1 is a constant. Then (3.19) becomes

$$\dot{V} = -e^T Qe + 2\left(e^T P_2(k - k')r + \frac{\dot{\Psi}\Psi}{h_1}\right) \tag{3.37}$$

where

$$P_2^T = [p_{12} \ \ p_{22}]$$

and P is defined according to (3.20).

In this case, with $\Psi = k - k'$, it is seen that (3.24) is satisfied if

$$k' = (p_{12}e_1 + p_{22}e_2)h_1 r \tag{3.38}$$

For improving the relative stability with increasing inputs, Philipson [24] modified the adaptive rule (3.38) by setting

$$u = k'r + u_1 \tag{3.39}$$

and using u_1 to insert input modification. With $u_1 = \gamma \dot{k}'r$, we have

$$\dot{V} = -e^T Qe - 2(e^T P_2)^2 h_1 r^2 \tag{3.40}$$

The effect has been to make $\dot{V}$ more negative with r^2.

Gilbart and Monopoli [42] introduced a modified V function to get the same result as (3.40)

$$V = e^T Pe + \frac{(\Psi + h_1 z)^2}{h_2} \tag{3.41}$$

Then

$$\dot{V} = [-e^T Qe + 2e^T P_2(k - k')r + \frac{2}{h_2}(\Psi\dot{\Psi} + h_1 \Psi z + h_1 z\dot{\Psi} + h_1^2 z\dot{z})] \tag{3.42}$$

By taking

$$z = e^T P_2 r$$
$$\Psi = k - k' \tag{3.43}$$
$$\dot{\Psi} = -h_2 z - h_1 \dot{z}$$

we get

$$\dot{V} = -e^T Qe - 2h_1 (e^T P_2 r)^2 \tag{3.44}$$

This is the same as (3.40).

From (3.43), we can obtain

$$k' = h_1 e^T P_2 r + h_2 \int_{t_0}^{t} e^T P_2 r \, dt + k(t_0) \tag{3.45}$$

The reader is also referred to Hang and Parks [19] and Lindorff and Carroll [49], who have extensively dealt with Lyapunov design of MRAC systems; the latter consider also the problems of identification and adaptive observers.

3.3.3 Redesign Using an Improved Lyapunov Function

A method for the generation of improved Lyapunov functions developed by Nagaraja and Chalam [54] has been applied to Lyapunov redesign of MRAC systems [37]. Some examples considered earlier by Parks [18] and Philipson [24] have been studied and improved results are obtained.

Consider the second-order system discussed earlier (Fig. 3.4) whose equations are given by (3.14). For this system the redesign by Gupta and Chalam [37], following the method of Nagaraja and Chalam [54], yields

$$V = \left[\frac{e^2}{2B_2} + \frac{\dot{e}^2}{2} + \frac{x^2 r^2}{2B_2} + \frac{x^2}{2} \frac{B_1}{B_2} \right] \tag{3.46}$$

and

$$\dot{V} = -\frac{B_1}{B_2} \dot{e}^2 - \ddot{e}^2 - \dot{x}^2 - \frac{B_1}{B_2} x^2 r^2 - \frac{B_1}{B_2^2} e^2 r^2 \tag{3.47}$$

where

$$x = K - K_v K_c \tag{3.48}$$

$$\dot{K}_c = B'\dot{e}r$$

where

$$B' = \frac{B_1}{B_2 k_v} \tag{3.49}$$

The V and $\dot{V}$ functions obtained for this system by Parks [18] are

$$V = \frac{B_1}{B_2^2}e^2 + \frac{B_1}{B_2}\dot{e}^2 + \lambda x^2 \tag{3.50}$$

$$\dot{V} = -2\left(\frac{B_1}{B_2}\right)^2 \dot{e}^2 \tag{3.51}$$

$$\dot{K}_c = B'\dot{e}r$$

where

$$B' = \frac{B_1}{\lambda B_2 K_v} \tag{3.52}$$

Comparing the preceding results with (3.46) and (3.47) we note that the V function in (3.46) is more general as it includes r along with other parameters of the plant, whereas (3.50) does not include r. Also $\dot{V}$ in (3.47) includes more negative terms, which is indicative of increased damping of the system. However, although V and $\dot{V}$ functions are modified, the structure of the redesigned systems remains similar to the one given in Parks [18].

Similar results are obtained for the system studied by Philipson [34]. It must, however, be pointed out as Parks [18] states that these schemes employ the derivative of the error. The error rate might be available directly from the comparison of the system rate gyro output with the corresponding model rate and it is not necessary to obtain the same by analog differentiation of the error signal. However, the use of derivatives of error is avoidable and Parks [18] uses Kalman's lemma [26] and positive real transfer functions to obtain a Lyapunov design of MRAC without the need for derivatives of error. Further work has been done by Monopoli et al. [27] in this direction.

3.3.4 Discussion

The results obtained by the Lyapunov approach are very favorable as there are no restrictions on the values of gains in the adaptive loop as with the MIT rule or on the form and magnitude of the input signals. Hence this approach has gained a wide following in recent years. Some of the difficulties encountered in this method have already been pointed out. Further, Lyapunov's method can be applied effectively to autonomous systems. But as pointed out by Narendra and Valavani [55], this method is less decisive when the equations are nonautonomous. The difficulty arises because the V-function candidate yields a negative semidefinite $\dot{V}$ function. This is not adequate to assure asymptotic stability, as La Salle's theorem [52] for autonomous systems cannot be extended to this case.

Further, for a good design we should be able to choose the most suitable adaptation law from a wide class of adaptation laws that lead to global asymptotic stability. For this, we should have a wide class of alternative Lyapunov functions and this is not always possible.

An answer to this search for a large class of stable adaptation algorithms lies in the hyperstability approach [17], which will also avoid some other shortcomings of the Lyapunov approach. The hyperstability approach can be used for cases in which one cannot find a Lyapunov function. This approach is being increasingly used as an alternative to the Lyapunov approach. The next section deals with hyperstability concepts and their application in MRAC systems.

3.4 HYPERSTABILITY APPROACH

3.4.1 Introduction

Hyperstability concepts are applied to the MRAC system by recasting it as a "standard" system with a linear time-invariant operator in the forward path and a passive operator in the feedback path. This structure results in greater flexibility in the choice of the adaptive laws. In this method it is only required to make the feedback block satisfy some passivity conditions in order that the system be hyperstable. It may, however, be stated, at this stage, that the Lyapunov and hyperstability approaches, under certain conditions, are equivalent [55]. This will be discussed in Section 3.4.5.

Popov [56] introduced the concept of hyperstability as a natural extension of absolute stability. His book [57] gives an extensive treatment of this subject. Anderson [58] gives a lucid presentation of the important results of Popov's theory.

Landau [59,60] proposed the application of the hyperstability concept for the design of adaptive control systems. The resulting design is asymptotically hyperstable and the structure of the adaptive controller is simple. However, a series of differentiating networks is often necessary in this scheme [19]. This approach has been extensively studied by Landau et al. [17, 51], Calise [61], Giles and Sinner [62], and Hang [63]. An exhaustive treatment of the subject including same case studies and examples is given by Landau [17].

3.4.2 Hyperstability Concepts [17, 58, 60]

Standard Feedback System

The hyperstability concept deals mainly with the stability properties of those feedback control systems that can be represented by the standard form, consisting of two blocks as shown in Fig. 3.7. In this figure the feedback block may be linear or nonlinear and/or time varying which satisfies an input/output relationship,

$$\int_0^t u^T v \, d\tau \geq -\gamma_0^2 \quad \text{for all } t \geq 0 \tag{3.53}$$

where v is the input vector of the feedback clock, u is the output vector of the feedback block, and γ_0^2 is a finite positive constant, which does not depend on t. It depends on the initial state of the system. The preceding integral inequality (3.53) is called the Popov Integral Inequality.

The state equations for the multivariable standard system represented in Fig. 3.7 with n inputs and m outputs are given by

$$\dot{x} = Ax + Bu \tag{3.54}$$

$$v = Cx + Du \tag{3.55}$$

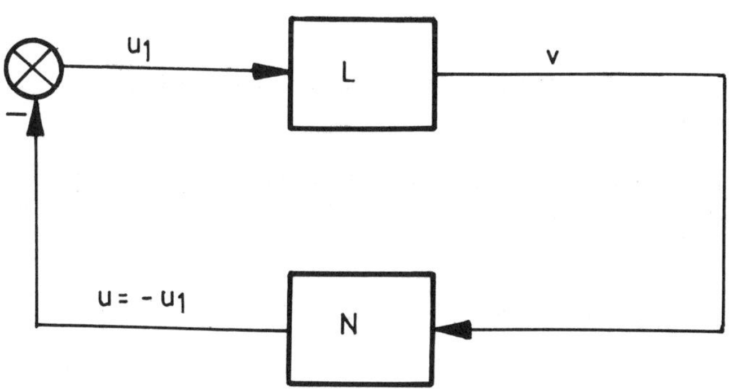

FIG. 3.7 Standard feedback system.

$$u_1 = -u \tag{3.56}$$

$$u(t) = F(v(\tau), t), \qquad \tau \leq t \tag{3.57}$$

where A, B, C, and D are constant matrices of appropriate dimensions, x is an $(n \times 1)$ state vector, u and v are $(m \times 1)$ vectors. It is assumed that the system is completely observable and completely controllable. F denotes a nonlinear dependence (functional dependence) between u and the values of v in the interval $\tau \leq t$. Further we consider the subset of u, which satisfies the Popov Integral Inequality given in (3.53).

Definitions

DEFINITION 1. The system described by (3.54)-(3.57) is termed hyperstable if for any $u(\cdot)$ defined by (3.53) the following inequality holds good for some positive constants k and δ and for all t

$$||x(t)|| \leq k(||x(0)|| + \delta) \tag{3.58}$$

where the symbol $||\cdot||$ denotes the Euclidean norm.

DEFINITION 2. The preceding system is termed asymptotically hyperstable if for any $u(\cdot)$ in the subset defined by (3.53) which is also bounded, the inequality (3.58) holds together with

$$\lim_{t \to \infty} x(t) = 0 \tag{3.59}$$

The conditions for hyperstability of a system depend on "positivity" properties of the transfer matrix of the feedforward block, namely,

$$Z(s) = \{D + C(sI - A)^{-1}B\} \tag{3.60}$$

according to Popov theorems which will be considered later. Hence we should define the following:

DEFINITION 3. For a transfer function Z(s) to be strictly positive real it must have the following properties:

a. Z(s) is real for real s.
b. The poles of Z(s) should lie in $\mathrm{Re}[s] < 0$ (i.e., no poles should lie in the RHP).
c. For all real ω,

$$\mathrm{Re}[Z(j\omega)] > 0, \qquad -\infty < \omega < \infty$$

It may be noted that if the transfer function Z(s) is reduced to a positive gain, then all the preceding properties are satisfied.

DEFINITION 4. For a transfer function Z(s) to be positive real it must have the following properties:

a. Z(s) is real for real s.
b. Z(s) has no poles in the open RHP.
c. The eventual poles of Z(s) on the axis $\text{Re}[s] = 0$ (i.e., $s = j\omega$) are distinct and the associated residues are real and positive (or null).
d. For all real ω, for which $s = j\omega$ is not a pole of Z(s),

$$\text{Re}[Z(j\omega)] \geqslant 0$$

It may be noted that the difference between "strictly positive real" and "positive real" transfer functions lies in the fact that in the former case the poles on $\text{Re}[s] = 0$ are not tolerated and that $\text{Re}[Z(j\omega)]$ is greater than zero for all real ω (and not greater than or equal to zero, as in the latter case).

Popov's Theorems

Popov's two results, generalized to the multiple input systems are stated in the following two theorems:

THEOREM 1 (Hyperstability). A necessary and sufficient condition for the transfer function matrix Z(s) given by (3.60) to define a hyperstable system is that Z(s) is positive real.

THEOREM 2 (Asymptotic Hyperstability). A necessary and sufficient condition for the transfer function matrix Z(s) given by (3.60) to define an asymptotically hyperstable system is that Z(s) is strictly positive real.

These two theorems have interesting interpretations in terms of energy constraints in passive systems and these interpretations enable us to provide proofs of the two theorems. The reader is referred to Anderson [58] for the same.

The stability of the systems with nonlinear memoryless feedback can be studied using the above theorems. The problem must be cast as a stability problem related to a feedback system. Then it should be possible to isolate one part that satisfies the Popov integral inequality, while the remainder should satisfy a corresponding positivity condition that will assure the hyperstability of the whole system.

The hyperstability approach enables us to look at the stability of the whole system by considering the properties of its components. We can state the following [17]: "If the feedforward block is such that the feedback system is globally (asymptotically) stable for all feedback blocks satisfying (3.53), then the feedback system is said to be (asymptotically) hyperstable and that the feedforward block is a hyperstable block."

3.4.3 Application to MRAC Systems [60]

Consider a multivariable MRAC system with m inputs and r outputs represented schematically in Fig. 3.8. This system is described by the following equations.

$$\dot{x} = A_M x + B_M u \tag{3.61}$$

$$y_M = Cx \tag{3.62}$$

$$\dot{y} = A_P(t)y + B_P(t)u \tag{3.63}$$

$$y_P = Cy \tag{3.64}$$

$$e = y_M - y_P \tag{3.65}$$

$$v = Ge \tag{3.66}$$

$$\dot{A}_P(t) = \phi(v(\tau), t), \qquad \tau \leq t \tag{3.67}$$

$$\dot{B}_P(t) = \Psi(v(\tau), t), \qquad \tau \leq t \tag{3.68}$$

where A_M and B_M are $n \times n$ and $n \times m$ dimensional constant matrices that define the model; $A_P(t)$ and $B_P(t)$ are respectively $n \times n$ and $n \times m$ dimensional time-variable matrices which define the plant, C is a $r \times n$ dimensional constant output matrix, G is a $r \times r$ dimensional constant matrix which defines the linear processing block; y_M, y_P, e are vectors that represent the model output, plant output, and generalized error respectively; $\phi(t)$ and $\Psi(t)$ represent a nonlinear dependence between the elements of the matrices $\dot{A}_P(t)$ and $\dot{B}_P(t)$ and the values of v in the interval $\tau \leq t$.

Let us consider the adaptive response of the system described by (3.61)-(3.68) for a given input u(t), starting from the initial situation,

$$[A_M - A_P(t_0)] \neq 0$$

$$[B_M - B_P(t_0)] \neq 0$$

at $t = t_0$, with the assumption that $A_P(t_0)$ and $B_P(t_0)$ depend only on the adaptation mechanism for $t > t_0$.

Hyperstability conditions for this system can be obtained by using Popov's theorems, given earlier.

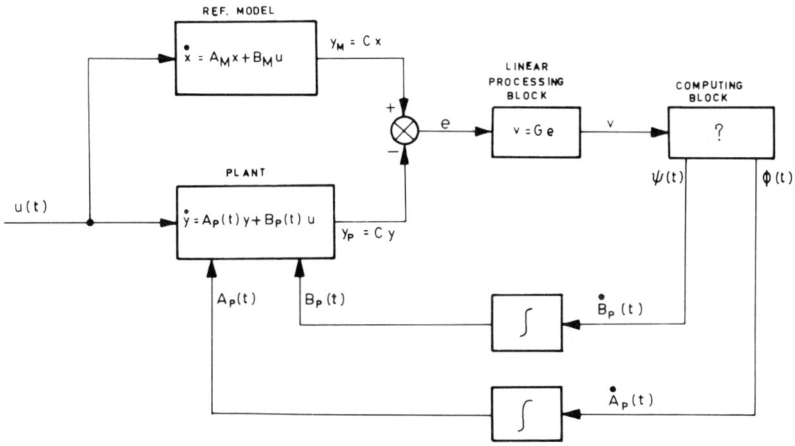

FIG. 3.8 Multivariable MRAC system. (From Ref. 60. Copyright ©1969 IEEE.)

First, we should represent the system described by (3.61)-(3.68) as a standard feedback system described by (3.54)-(3.57). The transfer matrix of the linear block in the standard system is

$$Z(s) = GC(sI - A_M)^{-1} \tag{3.69}$$

From (3.62), (3.64), and (3.65), we get

$$e = C(x - y) = C \cdot z \tag{3.70}$$

where we introduce a new vector z defined by

$$z = x - y \tag{3.71}$$

From (3.61) and (3.63), we get

$$\dot{z} = A_M z + [A_M - A_P(t)]y + [B_M - B_P(t)]u \tag{3.72}$$

Let

$$\omega_1 = [A_M - A_P(t)]y + [B_M - B_P(t)]u \tag{3.73}$$

Then the set of equivalent equations is obtained:

$$\dot{z} = A_M z + I\omega_1 \tag{3.74}$$

$$v = GCz \tag{3.75}$$

From (3.67), (3.68), and (3.73), we get

$$\omega(t) = -\omega_1(t) = F(v(\tau), t), \qquad \tau \leqslant t \tag{3.76}$$

Equations (3.74)-(3.76) obtained from (3.61)-(3.68) are similar to Eqs. (3.54)-(3.57) of the standard feedback system with the linear block having Z(s) given by (3.69). For asymptotic hyperstability, Z(s) must be strictly positive real.

The inequality (3.53) corresponding to the system described by (3.74)-(3.76) can be written as

$$\int_0^t \omega^T(\tau)v(\tau)\, d\tau \geqslant -\gamma_0^2 \tag{3.77}$$

The first condition to satisfy this inequality is that the vectors ω and v must have the same dimensions; that is to say, the vectors $[A_M - A_P(t)]y$, $[B_M - B_P(t)]u$, and v must be of the same dimension. This condition may also be interpreted as follows: The dimension of v (which is identical to the dimension of e) must be equal to the number of lines of $A_P(t)$ and $B_P(t)$ respectively that contain elements not identical to the corresponding elements of A_M and B_M.

Another condition to satisfy the inequality (3.53) is that the computing block of the matrices $\dot{A}_P(t)$, $\dot{B}_P(t)$ must introduce functions with the following form:

$$\Phi(t) = [\Phi_{ij}(t)] = [\alpha_{ij} v_i y_j] \tag{3.78}$$

$$\Psi(t) = [\Psi_{ij}(t)] = [\beta_{ij} v_i u_j] \tag{3.79}$$

where α_{ij} and β_{ij} are positive constants,

$$1 \leqslant i \leqslant r \quad \text{and} \quad 1 \leqslant j \leqslant n$$

That this is also a necessary and sufficient condition is proved in [60], which also considers two particular cases of MRAC systems and obtains the necessary and sufficient conditions for their asymptotic hyperstability.

The preceding results may be summarized as follows:

THEOREM. Necessary and sufficient conditions for those MRAC systems described by (3.61)-(3.68) to be an asymptotically hyperstable system are that

a. The transfer matrix Z(s) be strictly positive real.
b. The vectors $[A_M - A_P(t)]y$, $[B_M - B_P(t)]u$, and v are of the same dimension.
c. The computing block of the matrices $\dot{A}_P(t)$, $\dot{B}_P(t)$ must introduce functions of the form:

$$\phi(t) = [\phi_{ij}(t)] = [\alpha_{ij} v_i y_j]$$

$$\Psi(t) = [\Psi_{ij}(t)] = [\beta_{ij} v_i u_j]$$

where α_{ij} and β_{ij} are positive constants.

This theorem, given in Landau [60], includes all the results obtained earlier [18,23,29], using the Lyapunov approach, and in Dressler [7], using the sensitivity approach. The procedure given above may be used for solving problems in which one cannot find a Lyapunov function, but in which it is possible to evaluate an integral of the form (3.53).

It is also possible to get other results for the problem of hyperstability of MRAC systems by finding new solutions of integral equations (3.77).

3.4.4 Landau's Procedure

The procedure given above can be applied directly to the synthesis of MRAC systems. A step-by-step procedure for the design of MRAC systems is given by Landau [17].

Step 1. Transform the MRAC system into an equivalent standard feedback system composed of two blocks as in Fig. 3.7. This is always possible.

Step 2. Find the solutions for the part of the adaptation laws which appear in the feedback path of the equivalent system such that the Popov integral inequality is satisfied.

Step 3. Find the solutions for the remaining part of the adaptation law which appears in the feedforward path such that the feedforward path is a hyperstable block. This ensures global stability of the entire system.

Step 4. Return to the original MRAC system in order to specify the adaptation law explicitly, that is, the structure of the adaptation mechanism.

(Note: A similar four-step design is also applicable to discrete-time systems and has been discussed in some detail with reference to an example in Section 4.3.)

Landau treats the subject extensively and gives a number of designs to which the reader is referred. He also gives many important results on positivity, hyperstability, and solutions for integral inequality in the Appendixes. These are necessary in solving the design problems.

Implementation by this approach requires the use of the derivatives of the error. In other words, we should have access to the full state vectors of the reference model and of the adjustable system. This is a significant difficulty in implementing asymptotically stable MRAC systems, when only some of the states of the model and of the adjustable system are directly available. We have earlier seen that the same difficulty arises in the Lyapunov approach. This problem and various solutions are discussed by Landau [17].

Section 3.5 considers Monopoli's method for designing asymptotically stable adaptive schemes which do not require the use of derivatives of the plant output signal, which is a significant advance.

3.4.5 Discussion

The potentials of the Lyapunov approach or the hyperstability approach are theoretically the same. However, more general adaptation laws (and algorithms, in the discrete-time case) can be obtained using the hyperstability method. This is probably because the search for a suitable Lyapunov function for the equivalent system is replaced by positivity problems to be solved independently. We can always find a Lyapunov function to a MRAC system designed by the hyperstability approach, because for a globally asymptotically stable system, a Lyapunov function always exists. The Lyapunov function will consist of two parts:

1. A quadratic form in terms of the state variables of the linear feed-forward block
2. Terms depending on the form of the feedback block

A comparison of both these approaches to the design of continuous MRAC systems is given by Narendra and Valavani [55]. They examine the conditions which have to be satisfied for the two approaches to be successfully applied to adaptive controllers and observers and demonstrate that they are entirely equivalent. Using a typical error model (this topic is to be considered in Section 3.6), they show that hyperstability and asymptotic stability are achieved under exactly the same conditions as stability and asymptotic stability in the Lyapunov sense when all the signals in the plant are uniformly bounded. They state that in cases where the adaptive control problem remains unresolved, the difficulties encountered in the Lyapunov approach are shown to have their counterparts in the hyperstability approach as well. When the plant signals cannot be assumed to be uniformly bounded (as in the general control problem), neither approach works directly and a special analysis of the feedback system is needed to prove the boundedness of the plant output. However, as mentioned earlier, in many adaptive situations, hyperstability approach may be easier to apply than Lyapunov approach.

3.5 MONOPOLI'S AUGMENTED ERROR APPROACH

3.5.1 Introduction

Monopoli [64] suggested an ingenious scheme which involves the use of an augmented error signal so that the design does not require derivatives of the plant output signal. This is an extremely attractive feature from an engineering point of view and was attainable previously only for a limited class of plants [18, 40, 41]. In this novel scheme, instead of nulling the true error between the plant output and model output directly, an augmented error is introduced and the design assures that the augmented error approaches zero. The true error signal as well is driven to zero. Monopoli proposed an adaptive algorithm for a continuous-time system of known relative degree, the relative degree of a system being the difference between the number of poles and zeros of the transfer function of the system. A somewhat similar algorithm was suggested by Åström and Wittenmark [65] for application to discrete-time linear systems with known delays; the delay of a discrete-time linear system as defined in Ref. 65 is the same as the relative degree of the system.

Monopoli's algorithm incorporates certain new concepts, most important being the augmented error and a special parametric structure. These are the essential components of any adaptive control system for which global stability can be assured [66-69]. Monopoli gives a systematic procedure for the design of adaptive controllers. He combines the augmented error concept (which was indicated [70] earlier) and the state variable filter concept so that a large class of systems may be put into a form where the Meyer-Kalman-Yacubovich (MKY) lemma [71, 72], may be applied. For cases where the plant has n poles and (n - 1) or (n - 2) zeros, the controller is shown to be uniformly asymptotically stable. However, in cases where the plant has less than or equal to (n - 3) zeros, the necessary modification raises questions regarding the boundedness of the outputs of the model and the plant even if the output error is bounded [68]. In fact, even when the relative stability is greater than 1, doubts have been expressed [73] whether

Monopoli's algorithm provides stability, except for the very special case when only one process model parameter is to be estimated [74, 75].

Ever since the publication of Monopoli [64], a good deal of importance has been given to the problem of designing stable adaptive controllers, with more and more refinements, as will be seen in Sections 3.6 and 3.7.

Monopoli's idea has also been extended to discrete-time systems by Ionescu and Monopoli [76]. Goodwin et al. [77] consider discrete-time multivariable systems.

3.5.2 Design Objectives

The general system layout is given in Fig. 3.9. The plant is described by the following nonlinear nonautonomous differential equation [64]:

$$D_p(p)y_p(t) = D_u(p)u(t) + c\cdot f(y_p, t) \tag{3.80}$$

where

$$\begin{aligned} u(t) &= \text{plant input} \\ y_p(t) &= \text{plant output} \\ f(y_p, t) &= \text{a nonlinear, time-varying function of known form} \\ p &= d/dt \\ D_p(p) &= p^n + a_1 p^{n-1} + \cdots + a_n \\ D_u(p) &= b_0 p^m + b_1 p^{m-1} + \cdots + b_m \end{aligned}$$

It is assumed that the coefficients a_i, b_i, and c are unknown and constant or slowly varying. The term $c\cdot f(y_p, t)$ may be replaced by a sum of terms of the same type. There is no loss of generality in considering only one term, since each is treated in the same way. The following assumptions are made:

1. The function f satisfies the continuity conditions necessary for solutions of (3.80) to exist and be unique.
2. All roots of $D_u(s)$ are in the open LHP only.
3. $b_0 \neq 0$.
4. $m \leq n - 1$.

Assumption 2 is required to ensure a bounded control input. For a similar reason, u is not included in the argument of f. However, $f(y_p, u, t)$ may be allowed in certain cases, though it is not known what restrictions to impose on these functions to ensure that u is bounded.

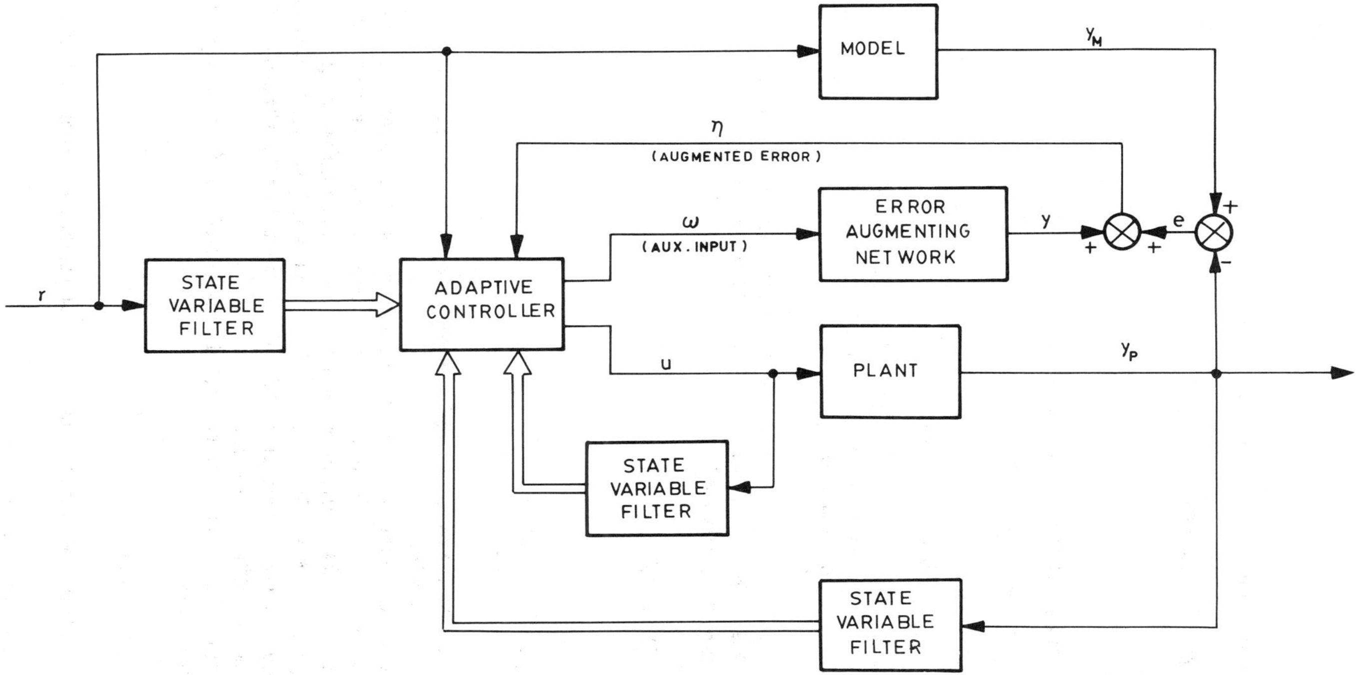

FIG. 3-9 Augmented error signal scheme. (From Ref. 64. Copyright © 1974 IEEE.)

The reference model described by the equation

$$D_M(p)y_M(t) = K_0 r(t) + g(x_M, r, t) \tag{3.81}$$

where

$r(t)$ = reference input

$y_M(t)$ = model output

$D_M(p) = p^n + a_{d_1} p^{n-1} + \cdots + a_{d_n}$

x_M = model state vector

$g(x_M, r, t)$ = nonlinear, time-varying function with the smoothness properties required to ensure the existence and uniqueness of solutions to (3.81)

It is assumed that $D_M(p) = 0$ has roots in open LHP only. In some cases it may be desirable to have a model that differentiates the reference input. Then $r(t)$ may be replaced by

$$r(t) = D_r(p)r'(t)$$

where

$$D_r(p) = p^\alpha + R_1 p^{\alpha-1} + \cdots + R_\alpha$$

If $\alpha \leqslant m$, no derivatives of r' will be required in the design, but if $\alpha > m$, these will be required.

The design objective is to synthesize a parameter-adaptive control system that will cause the error e(t) to approach zero. As stated earlier, in the procedure that will follow no derivatives of the plant output will be used, as, in practice, they may be noisy. This is possible because of the use of an augmented error signal along with state variable filters and the application of the MKY lemma.

The augmented error signal $\eta(t)$ is obtained by adding an augmenting signal $y(t)$ to $e(t)$:

$$\eta(t) = e(t) + y(t) \tag{3.82}$$

The signal y(t) is the output of the error augmenting network defined by

$$D_M(p)y = D_\omega(p) \cdot \omega(t) \tag{3.83}$$

where

$\omega(t)$ = auxiliary input (to be determined along with u(t) as part of the design)

$$D_\omega(p) = p^{n-1} + c_1 p^{n-2} + \cdots + c_n$$

The coefficients c_i in $D_\omega(p)$ are so chosen that the transfer function, $D_\omega(s)/D_M(s)$ is a positive real function of s in order to use the MKY lemma.

Subtracting (3.80) from (3.81) gives the following differential equation for e:

$$D_M(p)e = K_0 r + g - D_u(p)u - cf + D_\Delta(p)y_P \tag{3.84}$$

where

$$D_\Delta(p) = D_P(p) - D_M(p)$$

$$= \Delta a_1 p^{n-1} + \cdots + \Delta a_n$$

$$\Delta a_i = a_i - a_{di} \quad \text{for } i = 1 \text{ to } n$$

Next (3.83) is added to (3.84) to yield the equation for the augmented error signal η:

$$D_M(p)\eta = K_0 r + g - D_u(p)u - cf + D_\Delta(p)y_P + D_\omega(p)\cdot\omega \tag{3.85}$$

Equation (3.85) is the starting point for the synthesis procedure.

3.5.3 Design Solutions

The design problem is to find u and ω independent of derivatives of y_P and in such a way that $e \longrightarrow 0$ as $t \longrightarrow \infty$. The augmenting signal y acts as a sort of a catalyst only during the adaptive transient period and goes to zero once adaptation is completed.

This is achieved by use of the MKY lemma, which is demonstrated by Monopoli [64] in great detail to obtain design rules to achieve the objectives for the following cases:

1. A special class of plants that can be represented by the product of an unknown gain and a positive real transfer function. This approach is extended to the general case.

2. The class of plants that can be represented by the product of an unknown gain and a known, nonpositive real transfer function. Two approaches are given for this adaptation problem:
   a. when no bounds are known for the unknown plant parameters.
   b. when bounds are known for these parameters. (This enables the designer to keep all the adjustable gain within known bounds. Constraining the magnitude of the adjustable gains in this way is helpful in determining the stability of the differential operator which arises in the course of solution for the general case.)
3. The general class of plants described by (3.80) and the following equations up to (3.85). Again, both the approaches a and b in item 2 have been treated separately.

In the following paragraphs the general solutions are sketched out to illustrate this method, which, as pointed out earlier, though novel and powerful has certain limitations that have been to some extent avoided by later work by Narendra et al. [68, 69, 78], Morse et al. [66, 79], and Monopoli et al. [67]. However, the reader is referred to Monopoli [64], as this paper is the starting point of all this work, for a complete analysis of all the above cases which require the use of some identities, some alternate expressions, and the MKY lemma.

Solution by Approach a

The equation for augmented error $\eta(t)$ given by (3.85) may be written as

$$D_M(p)\eta = D_{\omega}(p)[-D_u(p)z_0 + K_0z_1 + D_{\Delta}(p)z_2 - cz_3 + z_4 + \omega] \tag{3.86}$$

where

$$z_0 = D_{\omega}^{-1}(p)u$$

$$z_1 = D_{\omega}^{-1}(p)r$$

$$z_2 = D_{\omega}^{-1}(p)y_P$$

$$z_3 = D_{\omega}^{-1}(p)f$$

$$z_4 = D_{\omega}^{-1}(p)g$$

The additional filtering required is associated with the operator $D_f(p)$ where

$$D_f(p)y_{P_f} = (p^{n-m-1} + F_1p^{n-m-2} + \cdots + F_{n-m-1})y_{P_f} = y_P \tag{3.87}$$

Similarly,

$$D_f(p)u_f = u$$

It is assumed that $D_f(p)$ is a Hurwitz polynomial. It may be noted that if $m = n - 1$, this filter is not necessary.

The terms, $D_u(p)z_0$ and $D_\Delta(p)z_2$ in (3.86) may be replaced by alternate expressions (given in Appendix II of Monopoli [64]):

$$\begin{aligned} D_u(p)z_0 &= D_u(p)D_\omega^{-1}(p)u \\ &= b_0 D_f^{-1}(p)u + [A_0 p^{n-2} + \cdots + A_{n-2}]D_f^{-1}(p)z_0 \end{aligned} \tag{3.88}$$

where constants A_0 to A_{n-2} are unknown.

$$\begin{aligned} D_\Delta(p)z_2 &= \sum_{i=0}^{n-1} B_i p^{(n-2-i)} D_f^{-1}(p)z_2 + \sum_{i=0}^{n-2} C_i p^{(n-2-i)} D_f^{-1}(p)z_0 \\ &\quad + \sum_{i=0}^{n-m-2} D_i p^{(n-m-2-i)} D_f^{-1}(p)cz_3 \end{aligned} \tag{3.89}$$

where B_0 to B_{n-1}, C_0 to C_{n-2}, and D_0 to D_{n-m-2} are unknown.

It is important to note that the constants A_i, B_i, C_i, and D_i need not actually be found. All that is needed is the fact that they exist.

Substituting (3.88) and (3.89) in (3.86), we get

$$D_M(p)\eta = D_\omega(p)\left[-b_0 u_f + \omega + \sum_{i=0}^{N-1} \beta_i \phi_i\right] \tag{3.90}$$

where

$$N = 3n - m + 1$$

$$\beta_0 = K_0, \quad \beta_1 = -c, \quad \beta_2 = 1$$

$$\phi_0 = z_1, \quad \phi_1 = z_3, \quad \phi_2 = z_4$$

$$\left.\begin{aligned} \beta_{3+i} &= A_i + C_i \\ \phi_{3+i} &= p^{(n-2-i)} D_f^{-1}(p)z_0 \end{aligned}\right\} \quad \text{for } i = 0 \text{ to } (n-2)$$

$$\left.\begin{aligned} \beta_{n+2+i} &= B_i \\ \phi_{n+2+i} &= p^{(n-1-i)} D_f^{-1}(p) z_2 \end{aligned}\right\} \text{ for } i = 0 \text{ to } (n-1)$$

and

$$\left.\begin{aligned} \beta_{2n+2+i} &= cD_i \\ \phi_{2n+2+i} &= p^{(n-m-2-i)} D_f^{-1}(p) z_3 \end{aligned}\right\} \text{ for } i = 0 \text{ to } (n-m-2)$$

Proceeding according to Monopoli [64], choose

$$\begin{aligned} \omega &= (1 + k_n(t))\omega_1 \\ u_f + \phi_N &= \sum_{i=0}^{N-1} k_i(t)\phi_i \end{aligned} \tag{3.91}$$

where

$$\phi_N = \omega_1$$

These lead to an adaptive algorithm which assures $n \longrightarrow 0$.

Following the procedure adopted by Monopoli [64], the final design equations are obtained as

$$u = \sum_{i=0}^{N-1} k_i D_f(p)\phi_i \tag{3.92}$$

$$\omega_1 = D_f^{-1}(p)\left[\sum_{i=0}^{N-1} \sum_{j=0}^{n-m-2} D_{f_j}(p)(pk_i)(p^{(n-m-2-i)}\phi_i\right] \tag{3.93}$$

These inputs can be generated without derivatives of y_p. These will cause $e \longrightarrow 0$ as required.

Solution by Approach b

In certain cases it is possible to use an alternate method for avoiding the derivatives of y_P. In this method the derivatives of Y_M are substituted for those of y_P, resulting in a much simpler controller, depending on the knowledge of parameter bounds.

Substituting only (3.88) and not (3.89) in (3.86), we have

$$D_M(p)\eta = D_\omega(p)\left[-b_0 u_f + \omega + K_0 z_1 - c z_3 + z_4 + \sum_{i=0}^{n-2} A_0 p^{(n-2-i)} D_f^{-1}(p) z_0 + D_\Delta(p) z_2\right] \tag{3.94}$$

Following the procedure adopted for obtaining (3.92) and (3.93) for u and ω_1 results in the need for derivatives of y_P. To avoid this, a modification is made by letting $\omega_1 = \omega_2 + \omega_3$. Then Eq. (3.91) may be rewritten as

$$u_f + \omega_2 + \omega_3 = \sum_{i=0}^{n+1} k_i \phi_i + \sum_{i=n+2}^{2n+1} k_i p^{(2n+1-i)} z_5 - \sum_{i=n+2}^{2n+1} k_i p^{(2n+1-i)} (z_5 - z_2)$$

where

$$z_5 = D_\omega^{-1}(p) y_M \tag{3.95}$$

Then, let us define ω_3 by

$$\omega_3 = -\sum_{i=n+2}^{2n+1} k_i p^{(2n+1-i)} D_\omega^{-1}(p) e \tag{3.96}$$

This leads to the following equation:

$$u_f + \omega_2 = \sum_{i=0}^{n+1} k_i \phi_i + \sum_{i=n+2}^{2n+1} k_i p^{(2n+1-i)} z_5 \tag{3.97}$$

Application of $D_f(p)$ operator to (3.97) results in equations like (3.92) and (3.93) and for u and ω_2. Here we see that instead of derivatives of y_P, we have the derivatives of y_M, which are available in the model. It is shown [64] that $y \longrightarrow 0$ and $e \longrightarrow 0$ as $\eta \longrightarrow 0$.

This approach may result in the reduction of the required number of adjustable parameters. This is a significant advantage of this method.

3.5.4 Discussion

Monopoli gives the results of digital computer simulation of a third-order plant for which parameter bounds are known. Approach b requires only two adjustable gains, whereas six would have been needed if Approach a had been followed. The simulation results show how the error converges quite rapidly, after which excellent model following is achieved.

The concept of an augmented error signal enables us to design stable adaptive control systems without the use of plant output derivatives, which is an important improvement over earlier methods. This concept has led to a considerable amount of research to design globally asymptotically stable adaptive systems. One of the refinements is Narendra's error models approach to be discussed in the next section.

In a later paper by Monopoli et al. [67] the idea is further extended by including one additional auxiliary input signal which is much simpler to implement than the more complicated ones suggested by Narendra et al. [69] and Morse [66].

In the latter methods, generation of such a function would require as many as 2n signal multiplications where n is the order of the plant. This is undesirable from the point of view of either the number of analog multipliers required for the continuous-time case or the computation time required in the discrete case. In contrast to this, in the approach given by Monopoli et al. [67] only one multiplication is required instead of 2n multiplications. This simpler auxiliary input will guarantee global asymptotic stability as the more complex ones. Further, the simpler algorithm may be applied in other problems also.

3.6 NARENDRA'S ERROR MODEL APPROACH

3.6.1 Introduction

Monopoli's innovative approach discussed in Section 3.5 led many researchers to consider the design of globally asymptotically stable MRAC systems. As already mentioned in Section 3.5.4, Narendra's error model approach is one of the efforts in this direction. The error models relate the output error to the parameter error vector.

Such error models were first proposed by Narendra [80] for continuous-time systems. These were dealt with by Narendra et al. [68, 69, 81] from the point of view of stable adaptive controller design. Lin and Narendra [82] consider one of the error models which has a wide application and present a unified treatment for both discrete-time and continuous-time cases. In many cases, in recent years, a computer is a part of the controller, especially in view of the availability of a variety of microprocessors, and so a discrete version of the error model is necessary. In the following paragraphs a unified approach is therefore given, though discrete-time MRAC systems will be taken up in the next chapter.

3.6.2 Assumptions

In all approaches concerning stable controller design, the following assumptions regarding the plant should be made [78]:

Consider a plant with a transfer further $G_P(s)$ and a model with $G_M(s)$. Let $G_P(s)$ and $G_M(s)$ have n_1 and n_2 poles and m_1 and m_2 zeros respectively. The relative degrees of the plant and the model are $n_1^* = (n_1 - m_1)$ and $n_2^* = (n_2 - m_2)$ respectively. For discrete cases, n_1^* and n_2^* represent the effective delays of the plant and the model respectively.

The following assumptions are made:

1. n_1^* should be known exactly.
2. $n_2^* > n_1^*$
3. An upper bound on n_1 should be known.
4. If

$$G_P(s) = k_P \frac{p(s)}{q(s)}$$

   where p(s) and q(s) are monic polynomials, the sign of k_P should be known.
5. The zeros of the plant must be in the open LHP of the complex plane. For the discrete-time case, the zeros of $G_P(z)$ must lie in the interior of the unit circle.
6. An upper bound on the gain k_P must be known for the discrete-time case.

It must be noted that these restrictions cannot be satisfied in many cases. Further, this approach cannot be applied to distributed parameter systems. There is, therefore, scope for evolving a new theory to overcome these limitations.

3.6.3 Error Model

The error vector e(t) between the plant and the model may in general be described by a nonlinear, time-varying differential equation of the form

$$\dot{e}(t) = f_1[e(t), \phi(t), t] \tag{3.98}$$

or by the difference equation,

$$e(k+1) = f_1[e(k), \phi(k), k] \tag{3.99}$$

where $\phi(\cdot)$ is the parameter error vector. The error depends explicitly on t (or k). If θ^* is the desired (but unknown) controller parameter vector and $\theta(t)$ is its true value, then

$$\phi(t) \triangleq \theta(t) - \theta^* \tag{3.100}$$

The objective of adaptive control is to obtain an adaptive law for updating $\phi(\cdot)$, in the form:

$$\dot{\phi}(t) = f_2[e(t), t] \tag{3.101}$$

or

$$\Delta\phi(k) = f_2[e(k), k] \tag{3.102}$$

Although $\phi(\cdot)$ is unknown, $\phi(t)$ or $\Delta\phi(k)$ can be adjusted so that all signals are bounded and e(t) [or e(k)] tends to zero as $t \longrightarrow \infty$ [or $k \longrightarrow \infty$].

Figure 3.10 represents the error model that can be used for both continuous- and discrete-time systems [82]. Let it be designated as error model C. (This error model is labeled as one of the third type by Lin and Narendra [82].)

In the continuous case the equations describing the error model are

$$\begin{aligned} \dot{e}(t) &= Ae(t) + bv(t) \\ e_1(t) &= c^T e(t) \\ v(t) &= \phi(t)u(t) - \beta u^T(t)\Gamma u(t)e_1(t) \end{aligned} \tag{3.103}$$

where

$$\beta > 0$$

$$\Gamma = \Gamma^T > 0$$

(Note: The symbol β has been used here in place of α in Lin and Narendra [82].)

In the discrete-time case the corresponding equations are

$$\begin{aligned} e(k+1) &= Ae(k) + bv(k) \\ e_1(k) &= c^T e(k) + dv(k) \\ v(k) &= \phi^T(k)u(k) - \beta u^T(k)\Gamma u(k)e_1(k) \end{aligned} \tag{3.104}$$

where

$$\beta > \frac{1}{2}$$

$$\Gamma = \Gamma^T > 0$$

In (3.103) and (3.104), $e(\cdot)$ and $u(\cdot)$ are n- and m-dimensional vectors and $v(\cdot)$ and $e_1(\cdot)$ are scalars. A is a matrix, b and c are vectors of appropriate dimensions, and d is a positive scalar. The following transfer functions are strictly positive real (S.P.R.):

$[c(sI - A)^{-1}b]$ in the continuous case

$[d + c^T(zI - A)^{-1}b]$ in the discrete case

We note from Fig. 3.10 that the error model contains

1. A known linear time-invariant part with a S.P.R. transfer function.
2. A known time-varying gain $[\beta u^T(\cdot)\Gamma u(\cdot)]$ which depends on the input signal $u(\cdot)$ in the feedback path.
3. A time-varying vector gain $\phi(\cdot)$ in the feedforward path. This vector is unknown; however, using the measured signals $u(\cdot)$ and $e_1(\cdot)$, changes in $\phi(\cdot)$ can be made ($\dot{\phi}$ in the continuous-time case and $\Delta\phi$ in the discrete-time case).

As indicated earlier, the aim of adaptive control is to find adaptation laws for updating $\phi(\cdot)$ such that in the limit $e_1(t)$ [or $e_1(k)$] $\longrightarrow 0$ as $t \longrightarrow \infty$ [or $k \longrightarrow \infty$].

The error model developed in Lin and Narendra [82] to achieve this is a simpler scheme than the one proposed by Landau et al. [83] for discrete-time systems. The main results of [82] are that the adaptive laws:

$$\dot{\phi}(t) = -\Gamma e_1(t)u(t) \tag{3.105}$$

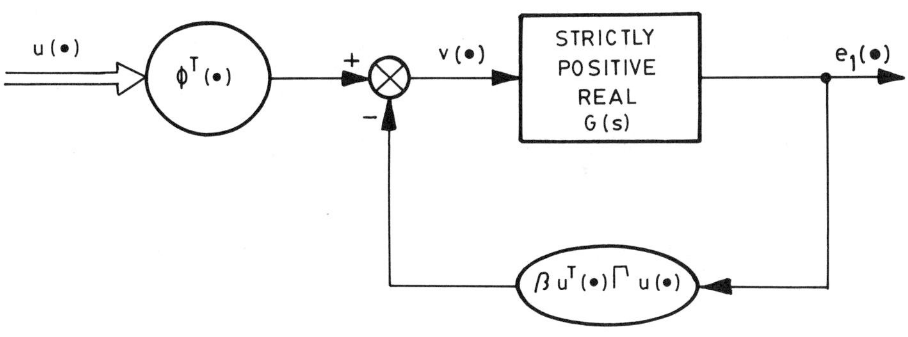

FIG. 3.10 Error model C. (From Ref. 82. Copyright ©1980 IEEE.)

and

$$\Delta\phi(k) \triangleq \phi(k+1) - \phi(k)$$
$$= -\Gamma e_1(k)u(k) \qquad (3.106)$$

will yield a stable error model in which $e(\cdot)$ and $\phi(\cdot)$ remain uniformly bounded and $\lim_{t\to\infty} e_1(t) = 0$ [or $\lim_{k\to\infty} e_1(k) \to 0$] if the input $u(\cdot)$ is uniformly bounded. It has also been shown that in the discrete case both $e_1(k)$ and $\Delta\phi(k)$ tend to zero as $k \to \infty$ whether or not $u(k)$ is uniformly bounded. In both the cases we cannot conclude directly about the convergence of $\phi(\cdot)$ to a constant vector.

Let us analyze the stability of the continuous-time case given by Eqs. (3.103) and (3.105). These may be combined and the $(n + m)$ equations may be written [55]:

$$\begin{bmatrix} \dot{e}(t) \\ \dot{\phi}(t) \end{bmatrix} = \begin{bmatrix} A - \beta b c^T u^T(t) \Gamma u(t) & b u^T(t) \\ -\Gamma u(t) c^T & 0 \end{bmatrix} \begin{bmatrix} e(t) \\ \phi(t) \end{bmatrix} \qquad (3.107)$$

Choose a Lyapunov function

$$V(e, \phi) = \frac{1}{2}[e^T P e + \phi^T \Gamma^{-1} \phi] \qquad (3.108)$$

Then we can write

$$\dot{V}(e, \phi) = \{\frac{1}{2} e^T(t)[A^T P + PA]e(t) + e^T(t)Pb[\phi^T(t)u(t)$$
$$- \beta u^T(t) \Gamma u(t) e_1(t)] + \phi(t) \Gamma^{-1} \dot{\phi}(t)\} \qquad (3.109)$$

Using the Kalman-Yacubovich lemma [84, pp. 49-50], referred to also in Section 3.3.1, if $c^T(sI - A)^{-1} b$ is strictly positive real, we can show that

$$\dot{V}(e, \phi) = \{-\frac{1}{2} e^T(t)[qq^T + kL]e(t) - \beta u^T(t) \Gamma u(t) e_1^2(t)\} \qquad (3.110)$$

where q is a vector, $L = L^T > 0$, and k is a positive constant. Hence, V is a nonincreasing function of time that is bounded below and hence converges to a finite value V_∞.

Case A: Bounded Input Vector u(t)

Let us first consider the case when the input u(t) is uniformly bounded. In such a case we have

$$\lim_{t \to \infty} \int_0^t \dot{V}(t)\, dt = V_\infty - V(0) \tag{3.111}$$

This is a finite number and $\dot{V}(t)$ is uniformly continuous since $\dot{e}(t)$ and hence $\ddot{V}(t)$ is bounded. Hence, by the Lemma of Lin and Narendra [82], we have

$$\lim_{t \to \infty} \dot{V}(t) = \lim_{t \to \infty} \left\{ \frac{1}{2} e^T(t)\left[qq^T + kL\right]e(t) - \beta u^T(t)\Gamma u(t) e_1^2(t) \right\} = 0 \tag{3.112}$$

Hence

$$\lim_{t \to \infty} e(t) = 0; \quad \lim_{t \to \infty} e_1(t) = 0 \tag{3.113}$$

By (3.105), since u(t) is bounded, it follows that

$$\lim_{t \to \infty} \dot{\phi}(t) = 0 \tag{3.114}$$

Hence $\phi(t)$ is a bounded vector whose time derivative approaches zero as $t \to \infty$.

Morgan and Narendra [85] have shown that if u(t) in (3.107) is "sufficiently rich"

$$\lim_{t \to \infty} \phi(t) = 0$$

See Section 8.11 for a discussion of "sufficiently rich" signals.

The results obtained above may be summarized as follows:

1. If $V(e, \phi)$ is positive definite and $\dot{V}(e, \phi)$ is negative semi-definite, e(t), $\phi(t)$ are bounded, if e(0), $\phi(0)$ are bounded.
2. If the input u(t) is bounded, $V(e, \phi)$ is bounded and hence

$$\lim_{t \to \infty} e(t) = 0; \quad \lim_{t \to \infty} \dot{\phi}(t) = 0$$

3. If u(t) is bounded and "sufficiently rich,"

$$\lim_{t \to \infty} \phi(t) = 0$$

Case B: Unbounded Input Vector u(t)

The problem is much more complicated if any finite time truncation of u(t) is bounded but u(t) is not uniformly bounded. We cannot conclude from the preceding analysis that e(t) and $\dot{\phi}(t)$ behave as shown in conclusion 2 above. However, Morgan and Narendra [86] have shown that even when the inputs are unbounded, the system described by (3.107) can be uniformly asymptotically stable, provided the inputs are "uniformly exciting."

Narendra and Valavani [55] have also applied hyperstability theory to the preceding problem and have shown that this approach also yields identical results. This was discussed earlier in Section 3.4.5.

3.6.4 Other Error Models

In the preceding section an error model was considered. In the following paragraphs four other error models [78] will be taken up for continuous-time systems. However, equivalent error models for the discrete case may be obtained by replacing $\dot{\phi}$ by $\Delta\phi$ and t by k.

Error Model A (Fig. 3.11a)

If $x_P(t)$ and $x_M(t)$ are the state vectors of the plant and the model respectively,

$$e(t) \triangleq x_P(t) - x_M(t)$$

If $e_1(t) = c^T e(t)$ is a scalar output error, then this error model is represented by the following equations:

Output error equation:

$$\phi^T(t)u(t) = e_1(t) \tag{3.115}$$

Adaptive law:

$$\dot{\phi}(t) = \frac{-\alpha \Gamma e_1(t) y(t)}{1 + u^T(t) \Gamma u(t)} \tag{3.116}$$

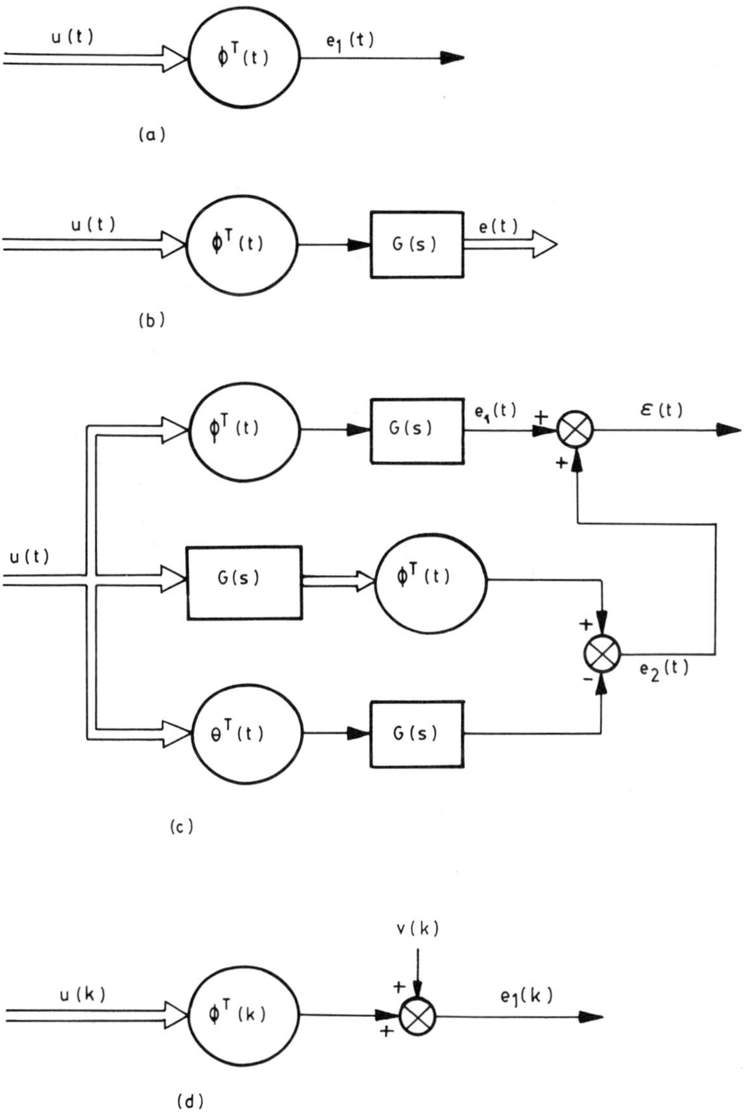

FIG. 3.11 Error models of different types. (a) Error model A. (b) Error model B, G(s) stable. (c) Error model D. (d) Error model E. (From Ref. 78.)

where

$$\alpha > 0$$

$$\Gamma = \Gamma^T > 0$$

$u(t)$ and $\phi(t)$ are $m \times 1$ vectors.

This error model has been analyzed for various classes of inputs $u(\cdot)$. It has been shown that for any input $u(\cdot)$, $v(t) = \phi^T(t)\Gamma^{-1}\phi(t)$ is a Lyapunov function so that $\|\phi\|$ is bounded, and $\lim_{t \to \infty} \phi^T(t)\Gamma^{-1}\phi(t) = V_\infty < \infty$. However, very little can be said about the convergence of $\phi(t)$ as $t \to \infty$.

When $u(t)$ is uniformly bounded, $e_1(t)$ is also bounded and tends to zero as $t \to \infty$. From Eq. (3.116) it follows that $\dot{\phi}(t) \to 0$ as $t \to \infty$. But $\phi(t)$ may not converge to a constant.

When $u(t)$ is "sufficiently rich" (persistently exciting or satisfies a mixing condition) it has been shown [87-89] that $\phi(t) \to 0$ as $t \to \infty$ and the system is asymptotically stable. In Bitmead and Anderson [90] it is shown that for stochastic inputs instead of deterministic inputs, the parameter error vector converges to zero. It may be pointed out that when $G(s) \equiv 1$, error model C is identical to error model A.

Error Model B (Fig. 3.11b)

The state error equation is given by

$$e(t) = G(s)\phi^T(t)u(t) \tag{3.117}$$

where $G(s)$ is a known stable linear operator. (It may be kept in mind that "s" is used both as an operator d/dt and as the Laplace transform variable).

If $G(s) = (sI - A)^{-1}b$, it can be shown that the adaptive law is

$$\dot{\phi}(t) = -e^T(t)Pbu(t) \tag{3.118}$$

where

$$P = P^T > 0$$

$$A^TP + PA = -Q, \qquad Q > 0$$

This results in uniformly bounded parameter and state error vectors. The remarks made in the case of error model A when $u(\cdot)$ is uniformly bounded or sufficiently rich are also valid for error model B.

Error Model D (Fig. 3.11c)

This model becomes necessary when G(s) is not strictly positive real. Such a situation may sometimes arise. In this case it is necessary to have error augmentation to generate stable adaptive control laws.

The following equations describe this model:

Error equation:

$$e_1(t) = G(s)\phi^T(t)u(t) \tag{3.119}$$

Auxiliary signal:

$$\begin{aligned} e_2(t) &= \left[G(s)\theta^T(t) - \theta^T(t)G(s)\right]u(t) \\ &= \left[G(s)\phi^T(t) - \phi^T(t)G(s)\right]u(t) \end{aligned} \tag{3.120}$$

Augmented error (as in [64])

$$\begin{aligned} \epsilon(t) &= e_1(t) + e_2(t) \\ &= \phi^T(t)G(s)u(t) \\ &= \phi^T(t)y(t) \end{aligned} \tag{3.121}$$

Adaptive law can be determined from error model A by inspection:

$$\dot{\phi}(t) = \frac{-\Gamma\epsilon(t)y(t)}{1 + y^T(t)\Gamma y(t)} \tag{3.122}$$

This law assures that $\epsilon(t) \longrightarrow 0$ as $t \longrightarrow \infty$. Also, if u(t) is uniformly bounded, $\dot{\phi}(t) \in \mathcal{L}^2$. This will assure that the true error, $e_1(t) \longrightarrow 0$ as $t \longrightarrow \infty$, as proved by Narendra et al. [69].

Error Model E (Fig. 3.11d)

Error model E is useful when external disturbances v(k) are present. Then the equations are nonhomogeneous. In the case of error models A to D, the equations are homogeneous.

The equations for error model E may be obtained for the discrete case from error model A with an extra input v(k), which is an external scalar disturbance.

Discrete error equation:

$$e_1(k) = \phi^T(k)u(k) + v(k) \tag{3.123}$$

Adaptive law can be written from the equation for the noise-free case (3.116):

$$\phi(k+1) = \left\{ \left[I - \frac{\alpha \Gamma u(k)u(k)^T}{1 + u(k)^T \Gamma u(k)} \right] \phi(k) + \frac{\alpha \Gamma u(k)v(k)}{1 + u^T(k)\Gamma u(k)} \right\} \tag{3.124}$$

From studying (3.124), the following conclusions can be drawn:

1. $\phi(\cdot)$ will be uniformly bounded, provided $u(\cdot)$ and $v(\cdot)$ are uniformly bounded and the homogeneous equation is uniformly asymptotically stable [i.e., $u(\cdot)$ is "sufficiently rich"].
2. $\phi(\cdot)$ can grow unbounded, if $u(\cdot)$ is not "sufficiently rich" and if $u(\cdot)$ and $v(\cdot)$ are correlated.
3. If in item 1, $u(\cdot)$ and $v(\cdot)$ are uncorrelated, the expected value of $\phi(\cdot)$ will go asymptotically to zero with a finite variance. However, $\phi(\cdot)$ will tend to zero in the mean square sense, if α is made time varying such that

$$\sum_{k=0}^{\infty} \alpha(k) = \infty \quad \text{and} \quad \sum_{k=0}^{\infty} \alpha^2(k) < \infty$$

In view of the preceding conclusions, this error model E is useful for application in stochastic adaptive control systems.

3.6.5 Application of Error Models [78]

Consider the MRAC system given in Fig. 3.12. In the figure, F(s) is a known filter, y(t) an (m × 1) vector output of F, and $\theta^T(t)y(t)$ a feedback signal.

It is known that if $\theta(t) \equiv \theta^*$, a constant vector, then the transfer function of the plant together with the controller matches exactly the transfer function of the model, $G_M(s)$.

Let the parameter error vector, $\phi(t) \triangleq \theta(t) - \theta^*$ and the output error, $e_1 \triangleq y_P(t) - y_M(t)$. The two are related by the equation:

$$e_1(t) = G_M(s)\phi^T(t)y(t) \tag{3.125}$$

Consider an additional feedback signal $[y^T(t)\Gamma y(t)e_1(t)]$ shown in the figure. Then the resulting error equation is

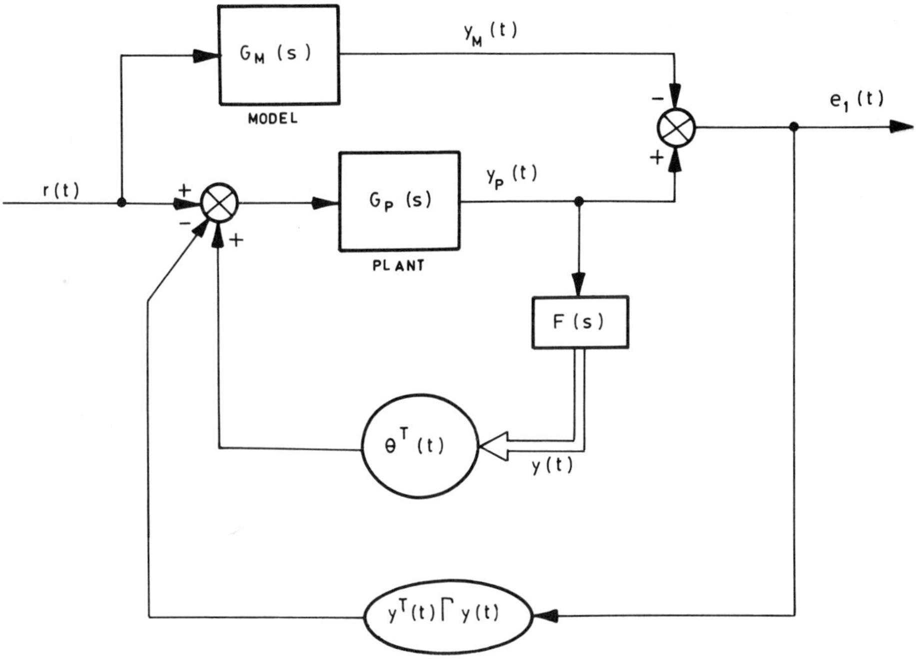

FIG. 3.12 MRAC system with an error model C. (From Ref. 78.)

$$e_1(t) = G_M(s)v(t) \tag{3.126}$$

where

$$v(t) = \phi^T(t)y(t) - y^T(t)\Gamma y(t)e_1(t)$$

as in (3.103).

If $G_M(s)$ is strictly positive real and strictly proper, then (3.126) corresponds to error model C. In such a case the adaptive law is given by

$$\dot{\phi}(t) = -\alpha \Gamma e_1(t)y(t) \tag{3.127}$$

where $\alpha > 0$ and $\Gamma = \Gamma^T > 0$.

As discussed earlier, this results in $e_1(t)$ and $\phi(t)$ being bounded. Since $y_M(t)$ is uniformly bounded and $e_1(t) \triangleq (y_P(t) - y_M(t))$, both $y_P(t)$ and y(t) are also uniformly bounded. Hence $e_1(t) \longrightarrow 0$ as $t \longrightarrow \infty$, and all signals in the system are uniformly bounded.

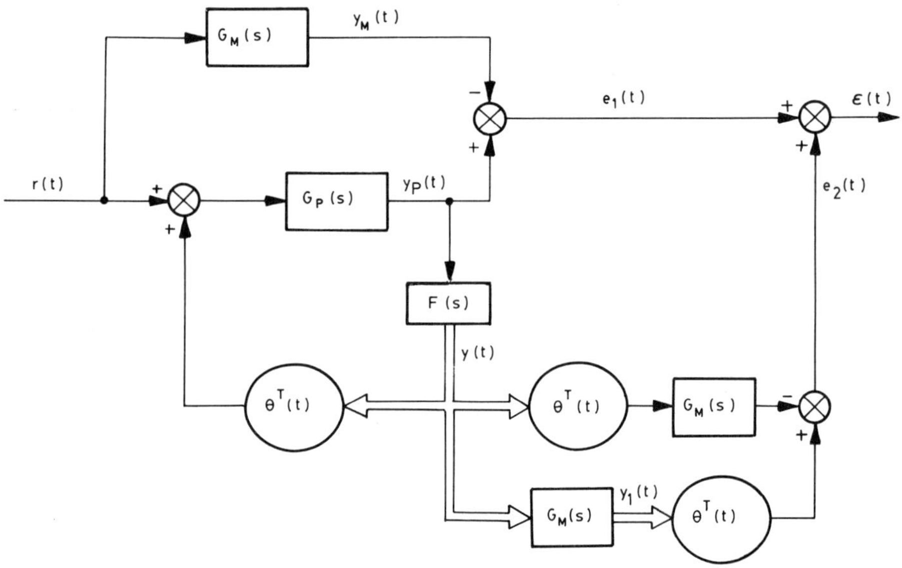

FIG. 3.13 MRAC system with an error model D. (From Ref. 78.)

We should note that error model C cannot be used if $G_M(s)$ is not strictly positive real. In that case, error model D is required. This involves the use of an augmented error signal. As in Section 3.6.4, the auxiliary signal $e_2(t)$ is

$$e_2(t) = [\phi^T(t)G_M(s) - G_M(s)\phi^T(t)]y(t)$$

This is added to $e_1(t)$ to obtain the augmented error $\epsilon(t)$ (Fig. 3.13). The adaptive control law is, as in (3.122),

$$\dot{\phi}(t) = \frac{-\alpha\Gamma(t)y_1(t)}{1 + y_1^T(t)\Gamma y_1(t)} \tag{3.128}$$

where

$$y_1(t) = G_M(s)y(t)$$

This law assumes that $\epsilon(t)$ is bounded and $\dot{\phi}(\cdot) \in \mathcal{L}^2$. This will assure that $e_1(t) \longrightarrow 0$ as $t \longrightarrow \infty$, as stated earlier.

3.6.6 Discussion

The error model approach is a powerful method applicable to not only transfer functions that are strictly positive real, but also those that are not S.P.R. This approach is valid for continuous-time and also discrete-time adaptive control systems, thus giving rise to a unified treatment. In fact, the proofs of stability for continuous-time and discrete-time cases given by Narendra et al. [69, 91] involve the use of these error models. Furthermore, error model E is useful even in cases where external disturbances are present.

3.7 EGARDT'S UNIFIED APPROACH

3.7.1 Introduction

Egardt [92] introduces a general adaptive algorithm and on this basis shows that many proposed schemes [64, 68, 79, 93] are special cases of this general algorithm. This unified approach is also extended by reinterpreting the concept of augmented error to show that there are no essential differences between the MRAC algorithms and the self-tuning regulators. This unified approach is an extremely important feature of Egardt's work; similar study is also made for discrete-time schemes by him [94].

3.7.2 Design Method for Known Plants

A design scheme given by Åström [95] for the known parameter case is first considered. The plant is described by the equation

$$y_P(t) = \frac{b_0 B_P(s)}{A_P(s)} u(t) = \frac{b_0(s^m + b_1 s^{m-1} + \cdots + b_m)}{(s^n + a_1 s^{n-1} + \cdots + a_n)} u(t) \tag{3.129}$$

The desired closed-loop response $y_M(t)$ is represented by

$$y_M(t) = \frac{B_M(s)}{A_M(s)} r(t) \tag{3.130}$$

where r(t) is the command input.

To avoid differentiators in the control law, we assume that the relative degree of the model is greater than or equal to the relative degree of the plant.

Consider a controller configuration given in Fig. 3.14, where R_1, R_2, R_3 are polynomials in the differential operator. It must be borne in mind that the actual realization of the controller is not exactly the same as in Fig. 3.14 as this would require differentiators in the control law.

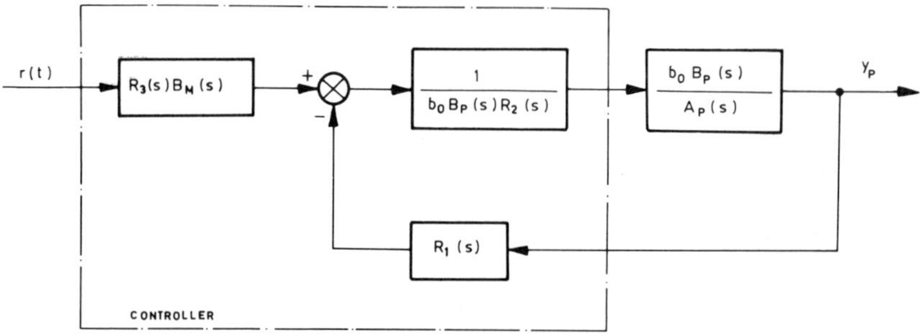

FIG. 3.14 Controller configuration. (From Ref. 92. Copyright © 1979 IEEE.)

The desired closed-loop transfer function is obtained if the polynomials R_1, R_2, and R_3 are chosen to satisfy the equation

$$R_3(s)A_M(s) = A_P(s)R_2(s) + R_1(s) \tag{3.131}$$

By doing so, the polynomials $b_0B_P(s)$ and $R_3(s)$ are cancelled in the closed-loop transfer function. These polynomials are neither observable nor controllable. The cancellation of the zeros of the plant would, therefore, require that only minimum phase systems be dealt with.

The polynomial $R_3(s)$ may be freely chosen, without changing the closed-loop transfer function. Once $R_3(s)$ has been chosen, Eq. (3.131) has many solutions for $R_1(s)$ and $R_2(s)$. However, we assume that the degree of $R_1(s)$ is less than or equal to $(n - 1)$, so that the equation has a unique solution [95]. Since the polynomials $A_P(s)$ and $A_M(s)$ are of the same degree n, $R_2(s)$ and $R_3(s)$ will have the same degree, n_1. To avoid any derivatives in the control law, we choose n_1 greater than or equal to $(n - m - 1)$. Furthermore, $R_2(s)$ and $R_3(s)$ are made monic.

Steps in the Design Scheme

Step 1. Choose a monic polynomial $R_3(s)$:

$$R_3(s) = s^{n_1} + k_1 s^{n_1-1} + \cdots + k_{n_1} \tag{3.132}$$

where

$$n_1 \geq (n - m - 1)$$

Step 2. Solve the equation

$$R_3(s)A_M(s) = A_P(s)R_2(s) + R_1(s) \tag{3.133}$$

for the unique solutions of $R_1(s)$ and $R_2(s)$, defined by

$$R_1(s) = \ell_0 s^{n-1} + \cdots + \ell_{n-1} \tag{3.134}$$

$$R_2(s) = s^{n_1} + p_1 s^{n_1-1} + \cdots + p_{n_1} \tag{3.135}$$

If n_1 in (3.132) and (3.135) is chosen to be equal to n, then it can be shown [95] that the controller will be a standard solution consisting of a Kalman filter and state estimate feedback, augmented with a zero cancellation and adding of zeros. We can also show that the choice $n_1 = (n - 1)$ makes the controller an input/output counterpart to a state space solution with a Luenberger observer and state estimate feedback.

It must be noted that the first step, the choice of $R_3(s)$ (including its degree), does not affect the closed-loop transfer function. But it is of importance for the transient behavior and the effect of disturbances. Since only the deterministic design case is considered, the importance of noise color for the choice of the observer will not influence the discussion in this section.

Step 3. The following assumptions are made:

1. The degrees n and m are known and $m \leq n - 1$. (It is sufficient to know the relative degree and an upper bound on the number of poles in order to write Eq. (3.129) with known n and m).
2. The parameter b_0 is nonzero and its sign is known. Without loss of generality, b_0 is assumed to be positive.
3. The plant is minimum phase, as mentioned already.

Step 4. From the results in the earlier steps, obtain a set of equations in terms of the unknown controller parameters. Using (3.131), (3.129), and (3.130) and the expression for the error

$$e(t) = y_P(t) - y_M(t)$$

we get

$$\begin{aligned} R_3A_Me(t) &= R_3A_My_P(t) - R_3A_My_M(t) \\ &= (A_PR_2 + R_1)y_P(t) - R_3A_My_M(t) \\ &= b_0B_PR_2u(t) + R_1y_P(t) - R_3B_Mr(t) \end{aligned} \tag{3.136}$$

Let $P_1(s)$, $P_2(s)$, and $Q(s)$ be monic polynomials of degree $(n - m - 1)$, $(m + n_1)$ and $(n + n_1 - 1)$ respectively and let

$$P(s) = P_1(s)P_2(s) \tag{3.137}$$

Next define the filtered error,

$$e_f(t) \triangleq \frac{Q(s)}{P(s)}\, e(t) \tag{3.138}$$

By introducing the polynomials P and Q and the filtered error, we obtain some flexibility. Different choices of the polynomials will generate different MRAC schemes.

Using (3.136), we can write

$$\begin{aligned} e_f(t) &= \frac{Q}{R_3A_M}\left[\frac{b_0B_PR_2}{P}\,u(t) + \frac{R_1}{P}\,y_P(t) - \frac{R_3B_M}{P}\,r(t)\right] \\ &= \frac{Q}{R_3A_M}\left[\frac{b_0(P_2 + B_PR_2 - P_2)}{P}\,u(t) + \frac{R_1}{P}\,y_P(t) - \frac{R_3B_M}{P}\,r(t)\right] \\ &= \frac{Q}{R_3A_M}\left[b_0\,\frac{u(t)}{P_1} + b_0(B_PR_2 - P_2)\,\frac{u(t)}{P}\right. \\ &\qquad \left. + R_1\,\frac{y_P(t)}{P} - \frac{R_3B_M}{P}\,r(t)\right] \end{aligned} \tag{3.139}$$

Define

$$u_f(t) \triangleq \frac{u(t)}{P_1} \tag{3.140}$$

Introduce θ as a vector containing the unknown parameters of the polynomials $(B_PR_2 - P_2)$ of degree $(m + n_1 - 1)$ and (R_1/b_0) of degree $(n - 1)$ and the constant $1/b_0$ as the last element. It is seen that θ contains essentially all the parameters of the controller. In addition, define the vector

$$\phi^T(t) \triangleq \left[\frac{s^{m+n_1-1}}{P} u(t), \ldots, \frac{1}{P} u(t), \right.$$
$$\left. \frac{s^{n-1}}{P} y_P(t), \ldots, \frac{1}{P} y_P(t), - \frac{R_3 B_M}{P} r(t) \right] \tag{3.141}$$

We can now rewrite (3.139) in terms of the newly defined quantities:

$$e_f(t) = \frac{Q}{R_3 A_M} [b_0 u_f(t) + b_0 \theta^T \phi(t)] \tag{3.142}$$

Step 5. Estimate the unknown parameters b_0 and θ, so that from these estimates we can determine the control law. Using Eq. (3.142), the estimate of $e_f(t)$, that is, $\hat{e}_f(t)$, is defined as

$$\hat{e}_f(t) \triangleq \frac{Q}{R_3 A_M} [\hat{b}_0(t) u_f(t) + \hat{b}_0(t) \hat{\theta}^T(t) \phi(t)] \tag{3.143}$$

where $\hat{b}_0(t)$ and $\hat{\theta}(t)$ are estimates of b_0 and θ.

Define the difference between the filtered error and its estimate as

$$\epsilon(t) \triangleq e_f(t) - \hat{e}_f(t) \tag{3.144}$$

Then we get the following expression

$$\epsilon(t) = \frac{Q}{R_3 A_M} \{[b_0 - \hat{b}_0(t)][u_f(t) + \hat{\theta}(t)\phi(t)] + b_0[\theta - \hat{\theta}(t)]^T \phi(t)\} \tag{3.145}$$

The following parameter updating is used in the constant gain case:

$$\begin{bmatrix} \dot{\hat{b}}_0(t) \\ \dot{\hat{\theta}}(t) \end{bmatrix} = F \begin{bmatrix} u_f(t) + \hat{\theta}(t)\phi(t) \\ \phi(t) \end{bmatrix} \epsilon(t) \tag{3.146}$$

where F is a constant, positive-definite matrix.

Note: It is possible to have other forms of algorithm (3.146) with F replaced by time-varying gain. This becomes necessary in the presence of noise.

If the transfer function $(Q/R_3 A_M)$ is strictly positive real, and if we use the parameter updating algorithm (3.146), it is possible to apply Lyapunov theory and the Kalman-Yakubovich lemma to prove that $\epsilon(t)$ tends to zero, provided all the closed loop signals are bounded, as in Monopoli [64]. We note that (Q/R_3A_M) can always be made S.P.R. by choosing Q appropriately, because R_3 and A_M are known and the degree of Q is 1 less than the degree of R_3A_M.

The estimation part of the controller with the MRAS structure is defined by (3.142), (3.144), and (3.146).

Step 6. The choice of control law has one difficulty. It is natural to determine the control signal such that the estimate of the error, namely, $\hat{e}_f(t)$, is equal to zero. According to (3.143), we see that $\{u_f(t) + \hat{\theta}^T(t)\phi(t)\}$ must be zero or from (3.140) we can write

$$u(t) = -P_1(s)[\hat{\theta}^T(t)\phi(t)] \tag{3.147}$$

The control law is identical to the one described earlier if $\hat{\theta}(t)$ is equal to the true parameter vector θ.

This control law uses derivatives of the parameter estimates, except for the trivial case when $m = n - 1$, that is, when the relative degree is 1. In this case P_1 is a constant. However, it is also possible to use (3.147) without differentiators in the case when the relative degree is 2, since $\hat{\theta}(t)$ is in general obtained by integration of known signals as in (3.146).

In the general case we should modify the control law to avoid differentiators. Different solutions are suggested by different researchers. For example, Monopoli [64] chooses a control signal that corresponds to the choice of the control law:

$$u(t) = -\hat{\theta}^T[P_1(s)\phi(t)] \tag{3.148}$$

This control law is asymptotically equivalent to the control law (3.147). From the definition of $\phi(t)$, we see that $P_1(s)\phi(t)$ contains filtered input, output, and reference signals, without any differentiators. The choice (3.148) does not guarantee that $\hat{e}_f(t) = 0$. But we can conclude that $e_f(t)$ tends to zero because of the fact that $\epsilon(t) = [e_f(t) - \hat{e}_f(t)]$ tends to zero. This problem is closely related to the problem of boundedness of the closed-loop signals. This is also indicated in Narendra et al. [68, 79].

3.7.3 Elimination of the Positive-Real Condition

The positive-real condition is considered to be an essential condition in order to guarantee the convergence of the estimation error. This condition has been imposed on MRAS since Parks introduced the idea [18]. However, we can eliminate this condition if the polynomials P and Q are specially chosen as in the following case.

When using (3.139), the positive-real condition on (Q/R_3A_M) comes in. But let us modify the polynomial degrees as follows:

Degree of $P_1 = (n - m)$

Degree of Q = degree of $P = (n + n_1)$

then it is natural to choose

$$Q = P = R_3A_M$$

Then the transfer function $(Q/R_3A_M) = 1$ and this is strictly positive real. However, since the degree of P_1 is $(n - m)$, it is possible to set $\hat{e}_f(t)$ to zero by the control law only in the case $(n - m) = 1$. This was also discussed in the preceding subsection.

3.7.4 Comparison with Other Algorithms

The important aspect of Egardt's work in Ref. 92 lies in the fact that he has shown that from the general algorithms developed by him, it is possible—by an appropriate choice of the polynomials—to obtain all the MRAS algorithms proposed by Monopoli [64]. Bénéjean [93], Feur and Morse [79], and Narendra and Valavani [68]. For further details the reader is referred to Egardt [92] and for stability analysis to Egardt [96, 97].

Note: In this section the symbols y_p, B_p, A_p, r, R_1, R_2, R_3, k_{n1}, n_1, r_{n-1}, and p_{n_1} have been used in place of y, B, A, u_m, R, S, T, t_{n_T}, n_T, 1_{n-1}, and s_{n-1}, respectively used in Egardt [92].

3.8 MULTIVARIABLE MRAC SYSTEMS

3.8.1 Introduction

There has been significant progress in the design of globally stable model reference adaptive control schemes for single-input, single-output systems, as can be judged from the earlier sections. However, not much work has been done in extending these ideas to the design of multiinput, multioutput (i.e., multivariable) MRAC systems except by Monopoli et al. [98] and Elliott et al. [99] for continuous-time systems and Goodwin et al. [77, 100] for discrete-time systems. One reason for this is that the type of information regarding the multivariable plant structure required for design is not as readily apparent in the multivariable case in the SISO case. In [98] and [77], this problem was somewhat simplified by limiting the class of multivariable plants to which these designs were applicable. Elliott [99] also deals with a class of linear multivariable continuous-time plants which can be represented by (m × m) strictly proper, full-rank, transfer matrices whose zeros lie strictly in the left half of the complex plane.

3.8.2 Monopoli's Augmented Error Approach

This design technique [98] can be applied to a broad class of nonlinear, time-varying, multivariable plants. The plants considered are those that can be split into a set of multiple-input, single-output subplants and that have at least as many inputs as outputs. Each subplant is used to derive one input signal that will make its output approach asymptotically the corresponding output of a reference model. The Lyapunov approach yields a stable adaptive algorithm and derivatives of the plant output are avoided by the use of state variable filters. The augmented error approach, introduced earlier in Section 3.5, enables a large class of multivariable systems to be designed with the use of the MKY lemma discussed in Section 3.3.1. Even though the augmented error signals are introduced, the asymptotic convergence to zero of the true system response errors is still guaranteed.

The multivariable plant is described by a set of nonlinear, nonautonomous differential equations

$$D_i(p)y_{P_i}(t) = \sum_{j=1}^{\ell} N_{ij}(p)u_j(t) + C_i f_i(y_{P_1}, \ldots, y_{P_q}, t), \quad i = 1, 2, \ldots, q \tag{3.149}$$

where

$u_j(t)$, $j = 1$ to ℓ = plant inputs

$y_{P_i}(t)$, $i = 1$ to q = plant outputs

$f_i(y_{P_1}, y_{P_2}, \ldots, y_{P_q}, t)$, $i = 1$ to q = nonlinear time-varying functions of known form

$$D_i(p) = p^{n_i} + a_i p^{n_i-1} + \cdots + a_{in_i}$$

$$N_{ij}(p) = b_{ij0}p^{m_{ij}} + b_{ij1}p^{m_{ij}-1} + \cdots + b_{ijm_{ij}}, \quad p = d/dt,$$

$$i = 1 \text{ to } q, \quad \text{and} \quad j = 1 \text{ to } \ell$$

Assumptions

1. The number of inputs is not less than the number of outputs, that is, $\ell \geq q$.
2. a_{ij}, b_{ijk}, and C_i are unknown constant coefficients (but may vary with time slowly compared with the response time of the system).

3. There exist q inputs, u_j, j = 1 to q, such that the integers m_{ij} can be made to satisfy the inequality $m_{ij} \geqslant \max_k \{m_{jk}\} - 1$, for k = 1, 2, ..., ℓ with different values of i (i = 1, 2, ..., q) for each value of j.

The reference model is described by the differential equations

$$D_{Mi}(p)y_{Mi}(t) = b_{di}r_i(t), \qquad i = 1 \text{ to } q \tag{3.150}$$

where

(a) $D_{Mi}(p) = p^{n_i} + a_{di1}p^{n_i-1} + \cdots + a_{din}$ has the same order as D (p).
(b) $r_i(t)$, i = 1 to q, are the reference model inputs.
(c) $y_{Mi}(t)$, i = 1 to q, are the reference model outputs.

It is assumed that the operator $D_{Mi}(p)$ for i = 1 to q is stable; that is, $D_{Mi}(p) = 0$ has roots in the open left-half s-plane only.

The design objective is to synthesize a parameter adaptive control system for the multivariable plant given in (3.149) that will drive the response error vector between the plant and model outputs to approach zero, keeping in view that no derivatives of the plant outputs are used.

The MKY lemma is the link between the existence of the solutions to a set of matrix and vector equations and the stability conditions in the frequency domain. We apply this lemma to a set of subsystems characterized by the differential equations

$$D_{Mi}(p)x_i(t) = D_{\omega_i}(p) \sum_{j=1}^{N_i} \delta_{ij}(t)y_{ij}(t), \qquad i = 1, 2, \ldots, q \tag{3.151}$$

where

(a) Every transfer function $[D_{\omega_i}(p)/D_{Mi}(p)]$ is positive real.
(b) $\delta_{ij}(t)$ is linearly related to only one adaptive parameter.
(c) The signal $y_{ij}(t)$ is available without differentiating.

Equation (3.151) can be rewritten in the vector matrix form:

$$\begin{aligned} \bar{x}_i(t) &= F_i\bar{x}_i(t) + \bar{b}_i \sum_{j=1}^{N_i} \delta_{ij}(t)y_{ij}(t), \qquad i = 1, 2, \ldots, q \\ x_i(t) &= \bar{h}_i^T\bar{x}_i(t) \end{aligned} \tag{3.152}$$

The stability criteria for each of these subsystems can be obtained by applying Lyapunov's direct method and the MKY lemma.

Choosing, for the ith subsystem,

$$V_i = \bar{x}_i^T P_i \bar{x}_i + \sum_{j=1}^{N_i} \lambda_{ij} \delta_{ij}^2, \qquad i = 1, 2, \ldots, q \tag{3.153}$$

(where λ_{ij} are arbitrary constants), it has been shown [98] that the adaptive laws for the ith subsystem in (3.152) are given by

$$\dot{\delta}_{ij} = - \frac{1}{\lambda_{ij}} x_i(t) y_{ij}(t), \qquad \text{with } i = 1 \text{ to } q, \quad \text{and } j = 1 \text{ to } N_i \tag{3.154}$$

These laws will ensure $\bar{x}_i(t) \longrightarrow 0$ as $t \longrightarrow \infty$.

The function $V = \Sigma_{i=1}^{q} V_i$ will be the Lyapunov function for the complete system, if $P_i \neq P_j$ for $i \neq j$, so that a linear combination of the $\bar{x}_i$ vectors does not appear in V. This requirement can be met by having different realizations for each subsystem.

Let us next consider the derivation of the control input u_1, for the first subplant. If $n_1 - 1 \geqslant m_{11} \geqslant m_{12} \cdots \geqslant m_{1\ell}$ in (3.149), then the differential equation for the first subplant is

$$D_1(p) y_{P_1}(t) = \sum_{j=1}^{\ell} N_{ij}(p) u_j(t) + C_1 f_1(y_{P_1}, \ldots, y_{P_q}, t) \tag{3.155}$$

The design objective is to find $u_1(t)$ which will casue $y_{P_1}(t)$ to follow the model output $y_{M_1}(t)$, satisfying the first model equation which is written from (3.150):

$$D_{M_1}(p) y_{M_1}(t) = b_{d_1} r_1(t) \tag{3.156}$$

Next define the augmented error signal $\eta_1(t)$ as

$$\eta_1(t) = e_1(t) + y_1(t) \tag{3.157}$$

where $e_1(t)$ is the system tracking error $e_1 = (y_{M_1} - y_{P_1})$ and $y_1(t)$ is the augmenting signal which is the output of the error augmenting filter defined by the differential equation

$$D_{M_1}(p)y_1(t) = D_{\omega_1}(p)\omega_1(t) \tag{3.158}$$

where

(a) $\omega_1(t)$ is an auxiliary input to be determined along with $u_1(t)$ as part of the design.

(b) $D_{\omega_1}(p) = p^{n_1-1} + C_{11}p^{n_1-1} + \cdots + C_{1,n_1-1}$, with n_1 being the order of the operators $D_{M_1}(p)$ and $D_1(p)$.

The coefficients C_{1i} in $D_{\omega_1}(p)$ are so chosen that $D_{\omega_1}(s)$ and $D_{M_1}(s)$ have no common roots and the transfer function $D_{\omega_1}(s)/D_{M_1}(s)$ is a positive-real function of s.

Subtracting (3.155) from (3.156), we get the differential equation for $e_1(t)$:

$$D_{M_1}(p)e_1(t) = \left[bd_1r_1(t) - N_{11}(p)u_1(t) - \cdots - N_{11}(p)u_\ell(t) - C_1f_1 - D_{\Delta_1}(p)y_{P_1}(t)\right] \tag{3.159}$$

where

$$D_{\Delta_1}(p) = \sum_{i=1}^{n_1} (a_{d1i} - a_{1i})p^{n_1-1}$$

Next (3.158) is added to (3.159) to get the equation for the augmented error signal $\eta_1(t)$:

$$D_{M_1}(p)\eta_1(t) = \left[b_{d_1}r_1(t) - N_{11}(p)u_1(t) - \cdots - N_{11}(p)u_\ell(t) - C_1f_1 - D_{\Delta 1}(p)y_{P_1}(t)\right] + D_{\omega_1}(p)\omega_1(t) \tag{3.160}$$

The auxiliary input ω_1 is expressed as

$$\omega_1(t) = \left[1 + k_{\omega_1}(t)\right]\omega_{10}(t) \tag{3.161}$$

where k_{ω_1} is one of the gains to be adjusted, and the signal $\omega_{10}(t)$ is to be found.

The design objective is to generate $u_1(t)$ and $\omega_{10}(t)$ independent of derivatives of $x_1(t)$ and in such a way that $e_1(t) \longrightarrow 0$ as $t \longrightarrow \infty$.

This is achieved by using state variable filters and by reducing the equations to a form (3.151) so that the MKY lemma can be applied. Finally, Monopoli and Hsing [98] give the adaptive laws, as in (3.154):

$$\begin{aligned} \dot{\delta}_{11} &= -\dot{k}_{11} = -\frac{1}{\lambda_{11}}\eta_1 y_{11} \\ \dot{\delta}_{1i} &= -b_1\dot{k}_{1i} = -\frac{1}{\lambda_{1i}}\eta_1 y_{1i}, \qquad i = 2, 3, \ldots, N_1 \end{aligned} \tag{3.162}$$

where $k_{1i}(t)$ are the adjustable parameters. The preceding laws will guarantee $\eta_1 \to 0$ as $t \to \infty$. It is seen that these adaptive laws are very simple and involve only η_1 and none of its derivatives. The signals η_1 and y_{1i} are available without differentiating.

The expressions for the inputs $u_1(t)$ and $\omega_{10}(t)$ are shown [98] to be of the form:

$$u_1(t) = \sum_{i=2}^{N_1} k_{1i}(t) D_{f1}(p) y_{1i}(t) \tag{3.163}$$

$$\omega_{10}(t) = D_{f1}^{-1}(p)\left[\sum_{i=2}^{N_1}\sum_{j=0}^{n_1-m_{11}-2} D_j(p)[\dot{k}_{1i}(t)]p^{n_1-m_{11}-2-j} y_{1i}(t)\right] \tag{3.164}$$

Both these signals can be generated without using derivatives of the plant's output signal.

As $\eta_1(t) \to 0$, $\omega_1(t) \to 0$ since each term in the expression for $\omega_{10}(t)$ in (3.164) has $\eta_1(t)$ as a factor (via $\dot{k}_{1i}(t)$). $\omega_1(t) \to 0$ implies that $y_1(t) \to 0$. Hence the control laws that ensure $\eta_1 \to 0$ asymptotically will also guarantee $e_1(t) \to 0$ asymptotically, as is the requirement of the design.

All other control and auxiliary inputs can be derived in the same manner as u_1 and ω_{10} by using a different subplant in each case. Since it is necessary to derive only q control inputs, the remaining $(\ell - q)$ control inputs may be any arbitrary known signals.

In the special cases where $m_{11} = (n_1 - 1)$ or $m_{11} = (n_1 - 2)$, we do not have to use augmented error signals and we can still derive u_1 without the need for derivative signals.

3.8.3 Discussion

Monopoli et al. [98] give the results of digital computer simulations for two different multivariable plants, including the implementation of the adaptive control laws and the evolution of the system response errors. Two different adaptive gains were used to show their influence on the speed of adaptation. The following observations have been made:

1. Quite rapid convergence can be achieved as well as excellent model following.
2. The frequencies contained in the response errors depend on the values of the adaptive loop gains. Higher gains may excite the higher frequency modes.
3. Varying the fixed adaptive loop gains, a trade-off between small error overshoot and minimum settling time is effected.

Also see Section 8.8, which deals with some recent trends in multivariable systems.

Note: The symbols x_i, x_m, $\eta_i(t)$, and $\xi_{ij}(t)$ used by Monopoli and Hsing [98] are replaced by y_P, y_M, $x_i(t)$, and y_{ij} respectively in this section.

3.9 STABILITY AND CONVERGENCE STUDIES

An important problem in adaptive control theory is the question of existence of simple, globally convergent adaptive control algorithms. Such algorithms will cause the outputs of a given system to track asymptotically a desired output, for all initial system states, and achieve this with bounded input signals.

3.9.1 Stability Considerations

Stability conditions are crucial in the convergence of the output error. The global stability of MRAC systems had remained a difficult question since adaptive control systems belong basically a special class of nonlinear time-varying systems. The Lyapunov and Popov methods cannot be directly applied to prove stability, because of the constraints on the structure of the controllers and the adaptive algorithms. However, this seemingly formidable problem has been recently resolved by researchers, notably by Narendra et al. [69,101,102], Morse [66], and Egardt [96] for continuous-time systems, and by Goodwin et al. [77], Narendra et al. [91], and Egardt [96] for discrete-time systems. These investigations are for deterministic schemes. The following assumptions are made in the proofs of global stability:

1. (a) The upper bound on the order n, (b) the relative degree $n^* = (n - m)$, and (c) the sign of the high frequency gain k_P of the transfer function of the plant are known.
2. The zeros of the plant transfer function are in the open left-half plane.
3. The plant parameters are constant (though unknown).
4. The system is disturbance-free.

These conditions are rather restrictive and so questions have been raised as to whether the preceding sufficient conditions are also necessary. There is a need to find methods to relax these conditions since they are not always fulfilled in actual practical systems. There have also been efforts to determine similar sufficient conditions for multivariable systems. Stability studies for systems under bounded disturbances have also been made by Narendra et al. [103, 104], Goodwin et al. [105], and Samson [106].

Studies of Narendra et al.

A conjecture was made by Narendra et al. [68] that the plant along with the controller would be stable if the augmented error is bounded. In Narendra et al. [69] a complete proof is given for the global stability of continuous-time systems. This results in a structure similar to that considered by Monopoli [64] and Narendra et al. [68] but contains an additional feedback term. In this approach use is made of error models, discussed in Section 3.6. The introduction of error models is particularly useful in proving stability. The proof of stability in Narendra et al. [69] is by contradiction and is dependent on characterizing the behavior of unbounded signals in the adaptive loop. It has been shown that two unbounded signals in the system cannot grow at different rates. It is necessary to bring together the results of two independent analyses, one regarding signals in the adaptive loop and the other regarding the adaptive scheme. In both the cases the unbounded signals are characterized and the effect of specific transformations on them is determined. This approach proposed in Narendra et al. [69] for the ideal case is applied to the cases where bounded disturbances are present [103, 104]. A brief discussion is given in Section 8.7. The same approach may also be applicable to other adaptive contexts. A correction is reported by Narendra [101]. For details see references 69 and 101 and also 46, 50, 68, 107, and 108, which give a clear exposition of the evolution of the stability studies.

A general approach to the stability analysis of adaptive systems is given by Narendra et al. [102]. This general approach is based on order relationships between inputs and outputs of linear systems with particular reference to a basic structure used in adaptive control. This approach has a certain flexibility since it is not problem specific and can be suitably applied to prove many of the recent results reported in the literature. Stability is proved by contradiction as in Narendra et al. [69]. It is first shown, using the feedback structure of the adaptive system, that when the control parameters are bounded and the signals in the adaptive loop are assumed to grow without bound, specific signals in the system grow at the same rate. From the adaptive schemes it is next proved that the same signals grow at different rates, thus leading to a contradiction.

The same methodology can also be directly applied to the case of bounded disturbances to establish boundedness of all the signals in the system. Narendra et al. [102] only deal with continuous-time systems, but all the results can also be applied to discrete-time systems. The results can also be used to derive relations between signals in feedback systems with arbitrary structure.

Proof of Stability

We shall briefly outline the proof of stability for the case of the ideal system [109]. The plant to be controlled is described by the state equations:

$$\dot{x}_P = A_P x_P + B_P u \tag{3.165}$$

$$y_P = C_P^T x_P \tag{3.166}$$

The reference model is described by

$$\dot{x}_M = A_M x_M + B_M r \tag{3.167}$$

$$y_M = C_M^T x_M \tag{3.168}$$

where x_P, u, and y_P are the state, input, and output of the plant respectively and x_M, r, and y_M are the state, input, and output of the model respectively.

The transfer functions of the plant and model are given by

$$G_P(s) = C_P^T (sI - A_P)^{-1} B_P = \frac{k_P Z_P(s)}{R_P(s)} \tag{3.169}$$

$$G_M(s) = C_M^T (sI - A_M)^{-1} B_M = \frac{k_M}{R_M(s)} \tag{3.170}$$

We have to make the following assumptions about $G_P(s)$ and $G_M(s)$:

1. $Z_P(s)$, $R_P(s)$, and $R_M(s)$ are monic polynomials of degree, m, n, and $n^* = (n - m)$.
2. $Z_P(s)$ and $R_M(s)$ are unknown.
3. $Z_P(s)$ and $R_M(s)$ are strictly stable polynomials.
4. r is a piecewise continuous uniformly bounded reference input.

The control objective is to make the output error between the plant and model, that is, $e_1 = (y_P - y_M)$, tend to zero asymptotically, with all the signals and parameters in the system remaining uniformly bounded.

As mentioned already, we should have prior knowledge of

(a) n^*, the relative degree of the plant, which enables us to construct the model.
(b) An upper bound n on the plant order, which determines the order of the controller to be used.
(c) The sign of the gain k_P.
(d) The condition that $Z_P(s)$ be Hurwitz, as stated in item 1 above.

We need (c) and (d) in order to prove the stability of the overall system.

To simplify the analysis, we assume that k_P is known and that $k_P = k_M = 1$.

The control objective can be met by using an adaptive controller described by the following equations:

$$\begin{aligned} \dot{\omega}^{(1)} &= F\omega^{(1)} + gu \\ \dot{\omega}^{(2)} &= F\omega^{(2)} + gy_P \\ u &= \theta^T\omega + r \end{aligned} \tag{3.171}$$

where F is an asymptotically stable $n \times n$ matrix, (F, g) is controllable, $\omega^T = [\omega^{(1)T}, \omega^{(2)T}]$, and $\theta(t)$ is a 2n-dimensional parameter vector that is to be adjusted adaptively.

The aim of the adaptive law is to adjust $\theta(t)$ such that

$$\lim_{t \to \infty} e_1(t) = 0$$

It is shown in Narendra and Valavani [68] that a unique constant θ^* exists such that the transfer function of the plant together with the controller matches that of the model exactly, when $\theta(t) = \theta^*$.

We shall discuss the general case when $G_P(s)$ has a relative degree $n^* \geq 2$. However, the problem is much simplified when the model transfer function is strictly positive real and $n^* = 1$.

Let us define the parameter error vector $\phi(t)$ as

$$\phi(t) \triangleq \theta(t) - \theta^* \tag{3.172}$$

Then the problem reduces to finding the adaptive law for adjusting $\dot{\phi}(t)$ and whence $\dot{\theta}(t)$.

The output of the plant may be expressed as

$$y_P(t) = G_M(s)[r(t) + \phi^T(t)\omega(t)] \tag{3.173}$$

We generate an augmented error $\epsilon_1(t)$ by adding an auxiliary error signal $e_a(t)$ to $e_1(t)$. Let

$$e_a(t) \triangleq [\theta^T(t)G_M(s)I - G_M(s)\theta^T(t)]\omega(t) \tag{3.174}$$

Then

$$\epsilon_1(t) = e_1(t) + e_a(t) = \phi^T(t)\xi(t) \tag{3.175}$$

where

$$\xi(t) = G_M(s)I\omega(t) \tag{3.176}$$

The adaptive law for updating $\theta(t)$ depends on the augmented error $\epsilon_1(t)$ and the signal $\xi(t)$. The adaptive law is chosen as

$$\dot{\phi}(t) = \dot{\theta}(t) = \frac{-\epsilon_1(t)\xi(t)}{1 + \xi^T(t)\xi(t)} \tag{3.177}$$

This has been shown to result in global stability of the adaptive loop [107]. Let

$$V(\phi) = \frac{1}{2}\phi^T\phi \tag{3.178}$$

Them, from the adaptive law (3.177), we have

$$\dot{V}(\phi) = \frac{-\epsilon_1^2(\phi)}{1 + \xi^T(t)\xi(t)} \tag{3.179}$$

From this it follows that

$$\phi \text{ and } \dot{\phi} \text{ are uniformly bounded} \tag{3.180a}$$

$$\dot{\phi} \in \mathcal{L}^2 \tag{3.180b}$$

$$\epsilon_1(t) = \nu(t)[1 + \xi^T(t)\xi(t)]^{1/2}, \qquad \nu \in \mathcal{L}^2 \tag{3.180c}$$

We give only the outline of the main steps involved in the proof, since the complete proof is very lengthy and involved. Reference may be made to Narendra et al. [102].

1. Since the parameter vector is bounded according to (3.180a), it has been shown [102] that

$$\sup_{\tau\leq t}|y_p(\tau)| \sim \sup_{\tau\leq t}\|\omega^{(2)}(\tau)\| \sim \sup_{\tau\leq t}\|\omega(\tau)\| \sim \sup_{\tau\leq t}\|\xi(\tau)\| \tag{3.181}$$

In (3.181) the symbol $\sim$ represents an equivalent relation and implies that the corresponding signals grow at the same rate.

2. Since $\dot{\phi} \in \mathcal{L}^2$, it can be shown [102] that $e_a(t)$ grows at a slower rate than $\sup_{\tau\leq t}\|\omega(\tau)\|$. This is denoted by

$$e_a(t) = o[\sup_{\tau\leq t}\|\omega(\tau)\|] \tag{3.182}$$

3. From (3.173), (3.180c), and (3.182), it follows that

$$e_1(t) = G_M\phi^T\omega = \nu[1 + \xi^T\xi]^{1/2} + o[\sup_{\tau\leq t}\|\omega(\tau)\|] \tag{3.183}$$

4. Since $\nu \in \mathcal{L}^2$, using (3.171), we conclude that

$$\sup_{\tau\leq t}\|\omega^{(2)}(\tau)\| = o[\sup_{\tau\leq t}\|\omega(\tau)\|] \tag{3.184}$$

This contradicts (3.181).

Because of this contradiction, we conclude that all the signals in the system are uniformly bounded and

$$\lim_{t\to\infty} e_1(t) = 0$$

Once the boundedness of all the signals in the adaptive system is established, it can be shown that the parameter vector $\theta(t)$ converges to its desired value θ^* or equivalently that $\phi(t)$ tends to the null vector. The conditions that have to be imposed on $\xi(t)$ is of importance. From the results of [88], it has been shown that if

$$\frac{\xi(t)}{\sqrt{1 + \xi^T(t)\xi(t)}} \quad \text{is persistently exciting,}$$

then $\lim_{t \to 0} \phi(t) = 0$ and the convergence is exponential. Since by (3.176) $\xi(t) = G_M(s)I\omega(t)$, a sufficient condition for $\xi(\cdot)$ to be persistently exciting is that $\omega(\cdot)$ be persistently exciting. A brief discussion about this important problem of persistent excitation as investigated by several researchers recently is given in Section 8.11.

We have noted above that the convergence of the parameter error vector to the null vector is dependent on the persistent excitation of $\omega(t)$. But we can say very little about its persistent excitation directly, since $\omega(t)$ is a dependent variable within the adaptive loop. Hence it is desirable to express this condition in terms of the persistent excitation of signals in the model which can be chosen by the designer. We have seen that when $\theta(t) = \theta^*$, the model and plant transfer functions are identical. Hence we can parameterize the model such that a signal $\omega^*(t)$ in it would correspond to the signal $\omega(t)$ in the adaptive loop. We can then find the condition on r(t) that would assure the persistent excitation of $\omega^*(t)$. Hence in the ideal case this would result in the persistent excitation of $\omega(t)$ and this in turn would lead to the convergence of the parameter vector $\theta(t)$ to its true value. From several investigations it is now concluded that an almost periodic reference input with n-distinct frequencies will give rise to the persistent excitation of $\omega(t)$.

The stability analysis considered above is applicable only to an ideal case in which the plant is linear and time invariant and there are no external disturbances. Further, we can set up a reference model and obtain the stable adaptive laws only if we have considerable prior knowledge about the plant transfer characteristics, as discussed already. However, we must note that these ideal conditions are never met in practice, since mostly plants are not strictly linear or finite dimensional and parameters vary. In fact, it is due to the extreme changes in the values of the parameters that one takes recourse to adaptive control. Furthermore, in actual systems, external inputs and output disturbances are commonly present. Hence there is a need to develop a theoretical framework to extend the theory developed for the ideal case to real situations. Robustness studies are therefore necessary to consider modeling errors and external disturbances. See Sections 8.3 and 8.7, where these issues are considered.

Studies of Morse et al.

Feuer and Morse [79] have developed a generalized algorithm for a globally stable adaptive system controller for any relative stability, $n^* \geq 1$. Unfortunately, the complexity of the controller increases rapidly with increasing n^* and precludes its use in practical situations. Later Morse [66], motivated by the work of Goodwin et al. [77] in connection with a relatively uncomplicated algorithm applicable to globally stable discrete-time systems with $n^* \geq 1$, examined the possibility of simpler algorithms for a continuous-time system. He developed two relatively uncomplicated algorithms, each capable of providing global closed-loop stability while causing the closed-loop process output to approach and track a suitably defined reference signal. Morse's approach, with suitable modifications, can also be extended to discrete systems. As in the algorithms previously developed, it is assumed that only the process input and output can be measured and the controller does not require any derivatives of any signals. It may be mentioned that the proof of stability given by Morse [66] also involves the use of an additional quadratic feedback signal, as in the case of Narendra et al. [69]. This feedback signal is required to generate the "augmented error." This additional feedback signal was not used in the algorithm of Feuer and Morse [79]. Further, in this algorithm a complicated auxiliary feedback law is used to ensure stability and this is not used in Morse [66]. Apart from these two differences, the algorithm of [79] is the same as the first of the two algorithms of [66]. Further, this algorithm is also essentially similar to the one proposed by Monopoli [64], again if the quadratic term is deleted. The inclusion of such a term was suggested by Egardt [110]. Similar parameter adjustment laws have been proposed by Ionescu and Monopoli [76] and Lin and Narendra [82] for discrete-time systems. Researchers may refer to Morse [66] to fully appreciate the subtle and complex problems involved in this analysis.

Studies of Egardt

Egardt [110] proposed somewhat more complicated algorithms both for continuous-time and discrete-time systems. His stability analysis is more involved. This work also includes results when disturbances are present. Also see Egardt's stability studies for continuous systems [97] and for discrete systems [111]. Interesting work has been done regarding unification of different adaptive control schemes for continuous systems [92] and for discrete systems [94]. Furthermore, these two papers and also [96, 112] bring out the essential equivalence of model reference adaptive control systems and the self-tuning regulators, which will be discussed in Chapter 5.

In Egardt [96], stability analysis of both MRAC systems and STRs is done in a general framework, and a fairly general algorithm is formulated. The stability properties of this algorithm are analyzed. Sufficient conditions for boundedness of closed-loop signals are obtained. Furthermore, these studies also include the effects of disturbances. The stability properties of the algorithms are investigated by applying the L^{∞}-stability concept. Emphasis is given to algorithms with a stochastic approximation type of estimation scheme. The principal conclusion is that the closed-loop signals remain bounded under some assumptions. We can omit the main assumption, namely, boundedness of parameter estimates, if the algorithms are slightly modified. If there are no disturbances present in the plant, we can use the stability results to prove the convergence of the output error to zero. This result holds good without the need a priori for the closed loop to be stable.

The positive-real condition has been seen to be an essential condition to guarantee convergence of the estimation error. However, it has been shown that this condition can be eliminated by a slightly modified estimation algorithm in the deterministic case by including filtering. In the stochastic case we cannot dispense with the positive-real condition in the same way, since the filter is unknown. Both continuous-time and discrete-time cases are studied, although more emphasis is given to the discrete-time systems, since adaptive regulators are now mostly implemented using digital computers.

3.9.2 Convergence Analysis

By convergence we mean that the control objective is asymptotically achieved and all system variables remain bounded for the given class of initial conditions. The convergence analysis of an adaptive algorithm is necessary for the following reasons [113]:

1. To lend credibility to the practical application of the algorithm
2. To distinguish between good and bad algorithms
3. To formulate improved algorithms

There have been several studies of this important question, in spite of the complexities of the convergence problem, in view of the fact that the adaptive controller is a special nonlinear device and this essential nonlinearity of the algorithms has proved to be a difficulty in proving the necessary convergence properties. Even for simple algorithms, the convergence analysis is very difficult. Considerably more formidable will be the cases of nonlinear, time-varying algorithms that operate in a stochastic environment.

Goodwin et al. [113] present an overview of the current status of convergence analysis with particular reference to discrete-time cases. The conceptual common ground among the different approaches is emphasized and also some areas for further work is identified.

It is desirable to establish not only convergence but also exponential convergence of adaptive algorithms. This is because systems that are exponentially convergent are generally tolerant of modeling errors, noise, and various inadequacies of realization. Furthermore, in a slowly time-varying plant, if the rate of convergence is slower than exponential, there will not be satisfactory behavior.

The earliest study of exponential convergence was done by Lion [114], who showed that signals should be periodic so that Lyapunov theorems on periodic linear systems could be applied. This result was extended to almost periodic cases, that is, when the signals comprised linear combination of sinusoids, by Anderson [115, 116] who applied a Lyapunov time-varying lemma which can be used to find exponential stability precisely when a uniform observability property holds. Morgan and Narendra [88] derived the same result and give many interesting examples, proving the necessity and sufficiency of the exponential stability condition rather than just sufficiency. Anderson [117] later extended his earlier results [116] to include the main results of Morgan and Narendra [85, 88]. Anderson [118] applied the results of [117] to problems of adaptive identification. Sondhi and Mitra [89] and Weiss and Mitra [119] establish results for adaptive filters regarding conditions for convergence and rates of convergence and explain the importance of exponential convergence. Kreisselmeier [120] obtains in an alternative approach many results of Anderson [115-118] and Morgan and Narendra [85, 88]. Yuan and Wonham [87] deal with the design of system input signals to assure exponential convergence. Kreisselmeier [121] discusses the rates of convergence in continuous-time MRAC systems. He proposes a new adaptive scheme that allows us to increase the exponential rate of parameter convergence to an arbitrary, desired level subject to a sufficiently rich reference input and a sufficiently large adaptive gain, while retaining the global stability of the overall system. However, noise considerations impose an upper bound on the feasible gain, in practice. A proof of stability, which is different from that in references 66 and 69 is also given. Dugard et al. [122-124] discuss convergence analysis of model reference adaptive systems used for adaptive state estimation. Ljung [125] studies convergence in parametric identification methods. Undink ten Cate [126], Kim [127], and Gilbart et al. [25] have worked on methods of design for improved convergence in adaptive systems. All the above references deal mostly with adaptive situations where the plant is excited by deterministic signals.

Bitmead and Anderson [90, 128, 129] consider the convergence properties of certain algorithms arising in stochastic adaptive estimation problems and operating in random environments of engineering significance. Farden et al. [130] and Jones [131] have developed new techniques to handle the presence of random signals.

Anderson and Johnson [132] have shown that certain adaptive identifiers and adaptive inverse control algorithms are exponentially convergent when excited by deterministic or stochastic signals. The novelty of this paper arises from two distinct grounds. First, the paper reports results for the exponential stability of output and equation error identification algorithms and of adaptive control algorithms. Second, the main requirement for exponential convergence is shown to be a persistently exciting or spanning condition only on the system input (or reference signal) for adaptive identification (or control), together with some standard conditions.

Exponential convergence implies robustness of adaptive algorithms, in the presence of noise or slowly varying parameters. Proving exponential convergence opens up the possibility of results on approximate identification and adaptive control, including results applicable when the unknown plant order is underestimated. At the worst, bounded errors will result. Similar work can be extended to multivariable systems. This paper [132] simply proves the existence of an exponential convergence rate but does not provide a formula for that rate, and so further work concerning the actual convergence rate can open up new directions for the designer.

The important problem of parameter convergence is dealt with briefly in Section 8.5.

3.10 CONCLUDING REMARKS

In this chapter we have considered design aspects of continuous-time MRAC systems, starting from the so-called MIT rule. Lyapunov and hyperstability approaches, the augmented error method, and the error model technique have been studied. A unified approach has also been given to indicate the connections between the various approaches. Some aspects of multivariable systems have been considered. Problems of stability and convergence have been dealt with. The next chapter extends some of these methods to discrete-time MRAC systems.

REFERENCES

1. Whitaker, H. P., Yamron, J., and Kezer, A., Design of model reference adaptive control systems for aircraft, Report No. R-164, Instrumentation Lab., MIT (1958).
2. Whitaker, H. P., Massachusetts Institute of Technology Presentation, Proc. Self-adaptive Flight Control Systems Symp., WADC Tech. Report No. 59-49, pp. 16-78 (1959).
3. Whitaker, H. P., Design capabilities of model reference adaptive systems, Proc. Nat. Electro. Conf., 18: 241-249 (1962).
4. Whitaker, H. P., and Rediess, H. A., A new model performance index for engineering design of flight control systems, AIAA J. Aircraft, 7 (1970).
5. Osburn, P. V., Whitaker, H. P., and Kezer, A., New developments in the design of model reference adaptive control system, ISA Paper No. 61-39, 1961.
6. Donalson, D. D., and Leondes, C. T., A model referenced parameter tracking technique for adaptive control system, IEEE Trans. Appln. Ind., pp. 241-262 (1963).
7. Dressler, R. M., An approach to model reference adaptive control systems, IEEE Trans. Auto. Contr., AC-12: 75-80 (1967).
8. Price, C. F., An accelerated gradient method for adaptive control, Proc. IEEE Conf. Decision & Control, pp. IV.4.1-4.9 (1970).
9. Kokotovic, P. V., Medanic, J. V., Vuskovic, M. J., and Bingulac, S. P., Sensitivity method in the experimental design of adaptive control systems, Proc. IFAC Congress, London, pp. 45B.1-12 (1966).

10. Medanic, J. V., and Kokotovic, P. V., Some problems in the development of adaptive systems using the sensitivity operator, Proc. IFAC Symp. on Adaptive Control, Teddington (1966).
11. Winsor, C. A., and Roy, R. J., The application of specific optimal control to the control of desensitized model following control systems, IEEE Trans. Auto. Contr., AC-15: 326-333 (1970).
12. Winsor, C. A., Model reference adaptive design, Report No. NASA-CR-98453 (1968).
13. Pearson, A. E., and Vanguir, V. R., A synthesis procedure for parameter adaptive control systems, IEEE Trans. Auto. Contr., AC-16: 440-449 (1971).
14. Pearson, A. E., An adaptive control algorithm for linear systems, IEEE Trans. Auto. Contr., AC-14: 497-503 (1969).
15. Serdyukov, V. A., Synthesis of a generalized adaptation algorithm for a non-search self-adjusting system with reference model, Auto. Rem. Contr., 31: 1078-1084 (1970).
16. Saridis, G. N., Self-Organizing Control of Stochastic Systems, Marcel Dekker, New York, 1977.
17. Landau, I. D., Adaptive Control: The Model Reference Approach, Marcel Dekker, New York, 1979.
18. Parks, P. C., Lyapunov redesign of model reference adaptive control, IEEE Trans. Auto. Contr., AC-11: 362-367 (1966).
19. Hang, C. C., and Parks, P. C., The comparative studies of model reference adaptive control systems, IEEE Trans. Auto. Contr., AC-18: 419-428 (1973).
20. Donalson, D. D., and Kishi, F. H., Review of adaptive control system theories and techniques, in Modern Control Systems Theory (C. T. Leondes, ed.), McGraw-Hill, New York, pp. 228-284, 1965.
21. Kokotovic, P. V., and Rutman, R. S., Sensitivity of automatic control systems, Auto. Rem. Contr., 26: 730-750 (1965).
22. Green, J. W., Adaptive control of multiloop speed control systems with particular reference to the Ward-Leonard system, Doctoral Thesis, EEE Dept., Leeds University, England (1969).
23. Butchart, R. L., and Shackcloth, B., Synthesis of model reference adaptive control systems by Lyapunov's second method, Proc. IFAC Symp. on Adaptive Control, Teddington, England, pp. 144-152 (1965).
24. Philipson, P. H., Design methods for model reference adaptive systems, Proc. Inst. Mech. Eng., 183, Pt. I: 695-706 (1968).
25. Gilbart, J. W., Monopoli, R. V., and Price, C. F., Improved convergence and increased flexibility in the design of model reference adaptive control systems, Proc. IEEE Symp. on Adaptive Processes, Austin, pp. IV.3.1-3.10 (1970).
26. Kalman, R. E., Liapunov functions for the problem of Luré in automatic control, Proc. Nat. Acad. Sci., 49: 201-205 (1963).

27. Monopoli, R. V., Gilbart, J. W., and Thayer, W. D., Model reference adaptive control based on Lyapunov-like techniques, Proc. IFAC Symp. Syst. Sensitivity and Adaptivity, Dubrovnik, Yugoslavia, pp. F.24-F.26 (1968).
28. Currie, M. G., and Stear, E. D., State space structure of model reference adaptive control for a noisy system, Proc. ASILOMAR Conf. Circuits & Systems, pp. 401-405 (1968).
29. Winsor, C. A., and Roy, R. J., Design of model reference adaptive control systems by Lyapunov's second method, IEEE Trans. Auto. Contr., AC-13: 204 (1968).
30. Shahein, H. I. H., Ghonaimy, M. A. R., and Shen, D. W. C., Accelerated model reference adaptation via Lyapunov and steepest descent design techniques, IEEE Trans. Auto. Contr., AC-17: 125-128 (1972).
31. Lindorff, D. P., Regarding the effect of disturbances on stability of model reference adaptive control systems, Preprints 13th JACC (1972).
32. Narendra, K. S., and Tripathi, S. S., The choice of adaptive parameters in model reference control problems, Proc. 5th ASILOMAR Conf., pp. 596-600 (1971).
33. Narendra, K. S., Tripathi, S. S., Lüders, G., and Kudva, P., Adaptive control using Lyapunov's direct method, Tech. Report No. CT-43, Becton Center, Yale University (1971).
34. Kudva, P., and Narendra, K. S., An identification procedure for linear multivariable systems, Tech. Report No. CT-48, Becton Center, Yale University (1972).
35. Udink ten Cate, A. J., and Verstoep, N. D. L., Improvement of Liapunov model reference adaptive control systems in a noisy environment, Int. J. Contr., 20: 977-996 (1974).
36. Carroll, R. L., New adaptive algorithms in Lyapunov synthesis, IEEE Trans. Auto. Contr., AC-21: 246-249 (1976).
37. Gupta, R. R., and Chalam, V. V., Lyapunov redesign of model reference adaptive control systems, Pts. I and II, J. Inst. Electr. Telecom. Eng., 25: 456-461 (1979).
38. Monopoli, R. V., Lyapunov's method for adaptive control system design, IEEE Trans. Auto. Contr., AC-12: 334-335 (1967).
39. Monopoli, R. V., and Gilbart, J. W., Model reference adaptive control systems with feedback and pre-filter adjustable gains, Proc. 4th Princeton Conf. Inform. Sci. & Syst., pp. 452-456 (1970).
40. Monopoli, R. V., Adaptive control and identification via Lyapunov's direct method, Proc. Allerton Conference (1972).
41. Monopoli, R. V., The Kalman-Yacubovich lemma in adaptive control system design, IEEE Trans. Auto. Contr., AC-18: 527-529 (1973).
42. Gilbart, J. W., and Monopoli, R. V., A modified Lyapunov design for model reference adaptive control systems, Proc. Allerton Conference (1969).

43. Gilbart, J. W., Winston, G. C., and Monopoli, R. V., Combination open-loop, closed loop, and adaptive compensation to an optical telescope, Proc. Int. Conf. Cybernetics & Society, Washington, D.C. (1972).
44. Kudva, P., and Narendra, K. S., Synthesis of an adaptive observer using Lyapunov's direct method, Int. J. Contr., 18: 1201-1210 (1973).
45. Lüders, G., and Narendra, K. S., Lyapunov functions for quadratic differential equations with applications to adaptive control, IEEE Trans. Auto. Contr., AC-17: 798-801 (1972).
46. Lüders, G., and Narendra, K. S., Stable adaptive schemes for state estimation and identification of linear systems, IEEE Trans. Auto. Contr., AC-19: 841-847 (1974).
47. Oliver, W. K., Seborg, D. E., and Fisher, D. G., Model reference adaptive control based on Lyapunov's direct method, Pts. I and II, Chem. Eng. Commun., 125-140 (1974).
48. Lindorff, D. P., Adaptive control of model reference adaptive systems, Proc. 3rd ASILOMAR Conf. on Circuits & Syst., pp. 121-124 (1969).
49. Lindorff, D. P., and Carroll, R. L., Survey of adaptive control using Lyapunov design, Int. J. Contr., 18: 897-914 (1973).
50. Narendra, K. S., and Kudva, P., Stable adaptive schemes for system identification and control, Pts. I and II, IEEE Trans. Syst., Man and Cybernetics, SMC-4: 542-560 (1974).
51. Landau, I. D., A survey of model reference adaptive techniques (theory and applications), Automatica, 10: 353-379 (1974).
52. La Salle, J. P., and Lefschetz, S., Stability of Lyapunov's Direct Method, Academic Press, New York, 1961.
53. Grayson, L. P., The status of synthesis by Lyapunov's method, Automatica, 3: 91-121 (1967).
54. Nagaraja, T., and Chalam, V. V., Generation of Lyapunov functions— A new approach, Int. J. Contr., 19: 781-787 (1974).
55. Narendra, K. S., and Valavani, L. S., A comparison of Lyapunov and hyperstability approaches to adaptive control of continuous systems, IEEE Trans. Auto. Contr., AC-25: 243-247 (1980).
56. Popov, V. M., The solution of a new stability problem for controlled systems, Auto. Rem. Contr., 24: 1-23 (1963).
57. Popov, V. M., Hyperstability of Control Systems, Springer-Verlag, Berlin, 1973.
58. Anderson, B. D. O., A simplified viewpoint of hyperstability, IEEE Trans. Auto. Contr., AC-13: 292-294 (1968).
59. Landau, I. D., Analyse et synthèse des commandes adaptives à l'aide d'un modèle par des méthodes d'hyperstabilité, Automatisme, 14: 301-309 (1969).
60. Landau, I. D., A hyperstability criterion for model reference adaptive control systems, IEEE Trans. Auto. Contr., AC-14: 552-555 (1969).

61. Calise, A. J., Hyperstability in variable structure systems, Proc. IEEE Conf. Decision and Control, Albuquerque (1980).
62. Giles, G., and Sinner, E., Synthèse d'un système d'identification en temps réel fondé sur la théorie de l'hyperstabilité, Proc. Colloque Informatique et Petroleochimie (1971).
63. Hang, C. C., The design of model reference parameter estimation systems, using hyperstability theories, Proc. 3rd IFAC Symp. System Parameter Estimation, North Holland-Elsevier, The Hague, pp. 741-744 (1973).
64. Monopoli, R. V., Model reference adaptive control with an augmented error signal, IEEE Trans. Auto. Contr., AC-19: 474-484 (1974).
65. Åström, K. J., and Wittenmark, B., On self-tuning regulators, Automatica, 9: 185-199 (1973).
66. Morse, A. S., Global stability of parameter-adaptive control systems, IEEE Trans. Auto. Contr., AC-25: 433-439 (1980).
67. Monopoli, R. V., and Subbarao, V. N., A simplified algorithm for model reference adaptive control, IEEE Trans. Auto. Contr., AC-25: 579-581 (1980).
68. Narendra, K. S., and Valavani, L. S., Stable adaptive controller design—Direct control, IEEE Trans. Auto. Contr., AC-23: 570-583 (1978).
69. Narendra, K. S., Lin, Y.-H., and Valavani, L. S., Stable adaptive controller design—Part II: Proof of stability, IEEE Trans. Auto. Contr., AC-25: 440-448 (1980).
70. Carroll, R. L., and Lindorff, D. P., An adaptive observer for SISO linear systems, IEEE Trans. Auto. Contr., AC-18: 428-435 (1973).
71. Lefschetz, S., Stability of Nonlinear Control Systems, Academic Press, New York, p. 115, 1965.
72. Meyer, K. R., On the existence of Lyapunov functions for the problem of Luré, SIAM J. Contr., Ser. A, 3: 373-383 (1966).
73. Feuer, A., Barmish, B. R., and Morse, A. S., An unstable dynamical system associated with model reference adaptive control, IEEE Trans. Auto. Contr., AC-23: 499-500 (1978).
74. Feuer, A., Adaptive control of SISO linear systems, Doctoral Thesis, Yale University (1978).
75. Feuer, A., and Morse, A. S., Local stability of parameter-adaptive control systems, Proc. Johns Hopkins Conf. Inf. Sci. Syst. (1978).
76. Ionescu, T., and Monopoli, R. V., Discrete-model reference adaptive control with an augmented error signal, Automatica, 13: 507-518 (1977).
77. Goodwin, G. C., Ramadge, P. J., and Caines, P. E., Discrete-time multivariable adaptive control, IEEE Trans. Auto. Contr., AC-25: 449-456 (1980).
78. Narendra, K. S., and Peterson, B. B., Recent developments in adaptive control, in Methods and Applications in Adaptive Control (H. Unbehauen, ed.), Springer-Verlag, Berlin, pp. 84-101 (1980).

79. Feuer, A., and Morse, A. S., Adaptive control of a SISO linear system, IEEE Trans. Auto. Contr., AC-23: 557-569 (1978).
80. Narendra, K. S., Stable identification schemes, in System Identification: Advances and Case Studies (R. K. Mehra and D. G. Lainiotis, eds.), Academic Press, New York, 1976.
81. Narendra, K. S., and Valavani, L. S., Direct and indirect adaptive control, Automatica, 15: 653-664 (1979).
82. Lin, Y.-H., and Narendra, K. S., A new error model for adaptive systems, IEEE Trans. Auto. Contr., AC-25: 585-587 (1980).
83. Landau, I. D., and Silveria, H. M., A stability theorem with applications to adaptive control, IEEE Trans. Auto. Contr., AC-24: 305-311 (1979).
84. Narendra, K. S., and Taylor, J. H., Frequency Domain Criteria for Absolute Stability, Academic Press, New York, 1973.
85. Morgan, A. P., and Narendra, K. S., On the stability of nonautonomous differential equations $\dot{x} = [A + B(t)]x$, with skew-symmetric matrix B(t), SIAM J. Contr. Optim., 15: 163-176 (1977).
86. Morgan, A. P., and Narendra, K. S., On the uniform asymptotic stability of certain linear time-varying differential equations with unbounded coefficients, S & IS Report No. 7807, Yale University (1978).
87. Yuan, J. S.-C., and Wonham, W. M., Probing signals for model reference identification, IEEE Trans. Auto. Contr., AC-22: 530-538 (1977).
88. Morgan, A. P., and Narendra, K. S., On the uniform asymptotic stability of certain linear non-autonomous differential equations, SIAM J. Contr. Optim., 15: 5-24 (1977).
89. Sondhi, M. M., and Mitra, D., New results on the performance of a well-known class of adaptive filters, Proc. IEEE, 64: 1583-1597 (1976).
90. Bitmead, R. R., and Anderson, B. D. O., Exponentially convergent behavior of simple stochastic adaptive estimation algorithms, Proc. IEEE Conf. Decision and Control, San Diego (1978).
91. Narendra, K. S., and Lin, Y. H., Stable discrete adaptive control, IEEE Trans. Auto. Contr., AC-25: 456-461 (1980).
92. Egardt, B., Unification of some continuous-time adaptive control schemes, IEEE Trans. Auto. Contr., AC-24: 588-592 (1979).
93. Bénéjean, R., La commande adaptive à modèle de référence evolutif (application à la commande sans accés aux variables d'état), Thése 3me cycle, Institut National Polytechnique de Grenoble (1977).
94. Egardt, B., Unification of some discrete-time adaptive control schemes, IEEE Trans. Auto. Contr., AC-25: 693-697 (1980).
95. Åström, K. J., Reglerteori, second ed. (in Swedish), Almquist & Wiksell, Stockholm, 1976.
96. Egardt, B., Stability of Adaptive Controllers, Springer-Verlag, Berlin, 1979.

97. Egardt, B., Stability analysis of continuous-time adaptive control systems, SIAM J. Contr. Optim., 18: 540-558 (1980).
98. Monopoli, R. V., and Hsing, C. C., Parameter adaptive control of multivariable systems, Int. J. Contr., 22: 313-327 (1975).
99. Elliott, H., and Wolovich, W. A., A parameter adaptive control structure for linear multivariable systems, IEEE Trans. Auto. Contr., AC-27: 340-352 (1982).
100. Goodwin, G. C., and Long, R. S., Generalization of results on multivariable adaptive control, IEEE Trans. Auto. Contr., AC-25: 1241-1245 (1980).
101. Narendra, K. S., Correction to "Stable adaptive controller design—Part II: Proof of stability," IEEE Trans. Auto. Contr., AC-29: 640-641 (1984).
102. Narendra, K. S., Annaswami, A. M., and Singh, R. P., A general approach to the stability analysis of adaptive systems, Int. J. Contr., 41: 193-216 (1985).
103. Kreisselmeier, G., and Narendra, K. S., Stable model reference adaptive control in the presence of bounded disturbances, IEEE Trans. Auto. Contr., AC-27: 1169-1176 (1982).
104. Peterson, B. B., and Narendra, K. S., Bounded error adaptive control, IEEE Trans. Auto. Contr., AC-27: 1161-1169 (1982).
105. Goodwin, G. C., and Chan, S. W., Model reference adaptive control of systems having purely deterministic disturbances, IEEE Trans. Auto. Contr., AC-28: 855-858 (1983).
106. Samson, C., Stability analysis of adaptively controlled systems subject to bounded disturbances, Automatica, 19: 81-86 (1983).
107. Narendra, K. S., and Lin, Y.-H., Design of stable model reference adaptive controllers, Proc. Workshop on Applications of Adaptive Control, Yale University (1979).
108. Narendra, K. S., and Valavani, L. S., Stable adaptive observers and controllers, Proc. IEEE, 64: 1198-1208 (1976).
109. Narendra, K. S., and Annaswamy, A. M., Robust adaptive control, Tech. Report, Center for Systems Science, Yale University, New Haven, Conn. (1985).
110. Egardt, B., A unified approach to model reference adaptive systems and self-tuning regulators, Tech. Report No. TFRT-1017, Dept. of Auto. Contr., Lund Inst. of Tech., Lund, Sweden (1978).
111. Egardt, B., Stability analysis of discrete-data adaptive control schemes, IEEE Trans. Auto. Contr., AC-25: 710-716 (1980).
112. Egardt, B., Stability of model reference adaptive control systems and self-tuning regulators, Doctoral Thesis, Lund Inst. of Tech., Lund, Sweden (1978).
113. Goodwin, G. C., Hill, D. L., and Palaniswami, M., A perspective on convergence of adaptive control algorithms, Automatica, 20: 519-531 (1984).

114. Lion, P. M., Rapid identification of linear and nonlinear systems, AIAA J., 5: 1835-1842 (1967).
115. Anderson, B. D. O., Adaptive identification of MIMO plants, Proc. IEEE Conference Decision and Control, Phoenix, pp. 273-281 (1974).
116. Anderson, B. D. O., Multivariable adaptive identification, Tech. Report No. EE-7402, University of Newcastle, Australia (1974).
117. Anderson, B. D. O., Exponential stability of linear equations arising in adaptive identification, IEEE Trans. Auto. Contr., AC-22: 83-88 (1977).
118. Anderson, B. D. O., An approach to multivariable system identification, Automatica, 13: 401-408 (1977).
119. Weiss, A., and Mitra, D., Digital adaptive filters: Conditions for convergence, rates of convergence, effects of noise and errors arising from the implementation, IEEE Trans. Inf. Th., IT-25: 637-652 (1979).
120. Kreisselmeier, G., Adaptive observers with exponential rate of convergence, IEEE Trans. Auto. Contr., AC-22: 2-8 (1977).
121. Kreisselmeier, G., and Joos, D., Rate of convergence in model reference adaptive control, IEEE Trans. Auto. Contr., AC-27: 710-713 (1982).
122. Dugard, L., and Landau, I. D., Convergence analysis of model reference adaptive schemes used for recursive identification, Proc. 5th IFAC Symp. on Identification and System Parameter Estimation, Darmstadt (1979).
123. Dugard, L., and Landau, I. D., Convergence analysis of model reference adaptive schemes used for adaptive state estimation, in Methods and Applications in Adaptive Control (H. Unbehauen, ed.), Springer-Verlag, Berlin, pp. 112-121, 1980.
124. Dugard, L., Landau, I. D., and Silveria, H. M., Adaptive state estimation using MRAS techniques—convergence analysis and evaluation, Report No. 79-06, Institut National Polytechnique de Grenoble (1979).
125. Ljung, L., Convergence analysis of parametric identification methods, IEEE Trans. Auto. Contr., AC-23: 770-783 (1978).
126. Udink ten Cate, A. J., Improved convergence of Lyapunov MRAS by a parameter misalignment function, IEEE Trans. Auto. Contr., AC-20: 132-134 (1975).
127. Kim, C., Convergence studies for an improved adaptive observer, Doctoral Thesis, University of Connecticut, Storrs (1975).
128. Bitmead, R. R., and Anderson, B. D. O., Lyapunov techniques for the exponential stability of linear difference equations with random coefficients, IEEE Trans. Auto. Contr., AC-25: 782-787 (1980).
129. Bitmead, R. R., and Anderson, B. D. O., Performance of adaptive estimation algorithms in dependent random environments, IEEE Trans. Auto. Contr., AC-25: 788-794 (1980).

130. Farden, D. C., Goding, Jr., J. C., and Sayood, K., On the "desired behaviour" of adaptive signal processing algorithms, Proc. IEEE Conf. on Acoustics, Speech and Signal Processing (1979).
131. Jones, S. K., Adaptive filtering with correlated training samples: A mean square error analysis, Doctoral Thesis, Southern Methodist University (1973).
132. Anderson, B. D. O., and Johnson, Jr., C. R., Exponential convergence of adaptive identification and control algorithms, Automatica, 18: 1-13 (1982).

BIBLIOGRAPHY

Asseo, S. J., Application of optimal control to perfect model following, Proc. JACC, Ann Arbor, Mich., pp. 1056-1070 (1968).

Chen, Y. T., Perfect model following with real model, Proc. JACC, pp. 287-293 (1973).

Courtiol, B., Etude et application de la Commande Adaptive avec modele, Doctoral Thesis, Institut National Polytechnique de Grenoble (1976).

Curran, R. T., Equi-controllability and the model following problem, Tech. Report No. 6303-2, Center for Systems Research, Stanford University (1971).

de Larminat, Ph., On overall stability of certain adaptive control systems, Proc. 5th IFAC Symp. on Identification and System Parameter Estimation, Darmstadt (1979).

Devaud, F. M., and Caron, J. Y., Asymptotic stability of model reference systems with bang-bang control, IEEE Trans. Auto. Contr., AC-20: 694-696 (1975).

Elliott, H., A model reference adaptive controller with arbitrarily small error, Proc. JACC (1978).

Erzberger, H., Analysis and design of model following systems by state space techniques, Proc. JACC, Ann Arbor, Mich., pp. 572-581 (1968).

Foulard, C., and Zoydo-Crespo, P., Commande modèle et commande découplée multidimensionnelle par des techniques de poursuite d'un modèle de reference, Proc. 4th IFAC/IFIP Int. Conf. on Digital Computer Applications to Process Control, Zurich, Vol. 1, pp. 88-100 (1974).

Goodwin, G. C., and Sin, K. S., Effect of model, system, and controller order on adaptive control, Proc. IEEE Conf. Decision and Control, Florida (1979).

Guile, R. W., A variational technique and its application to model reference control, Doctoral Thesis, University of Connecticut, Storrs (1970).

Hahn, V., Modelladaptive Regelung von Mehrgrößensystemen mit beliebigen totzeiten, Report, Ruhr University, Bochum, F.R.G. (1983).

Hang, C. C., Design studies of model reference adaptive control and identification systems, Doctoral Thesis, University of Warwick (1973).

James, D. J. G., Stability analysis of model reference adaptive control with sinusoidal inputs, Int. J. Contr., 9: 311-321 (1969).

James, D. J. G., Stability of a model reference control system, AIAA J., 9: 950-952 (1971).

James, D. J. G., Stability of model reference adaptive systems with random inputs, Proc. Int. Symp. on Stability of Stoch. Dyn. Systems, pp. 147-159 (1972).

James, D. J. G., Application of stochastic stability theory to model reference systems, Int. J. Contr., 16: 1169-1192 (1972).

Johnstone, R. M., Shah, S. L., and Fisher, D. G., An extension of hyperstable adaptive control to non-minimum phase systems, Int. J. Contr., 31: 539-545 (1980).

Kreisselmeier, G., and Joos, D., Stable model reference adaptive control with rapid adaptation, Report No. DFVLR, Institut für Dynamic der Flugsysteme, F.R.G. (1981).

Kudva, P., and Gourishankar, V., On the stability problem of multivariable model-following systems, Int. J. Contr., 24: 801-805 (1976).

Landau, I. D., Hyperstability and identification, Proc. IEEE Conf. Decision and Control, Austin, Texas, pp. x.4.1-x.4.8 (1970).

Landau, I. D., Les systèmes adaptifs avec modèle (théorie, mise en oeuvre, applications), Automatisme, 15: 272-292 (1971).

Landau, I. D., A generalization of the hyperstability conditions for MRAC systems, IEEE Trans. Auto. Contr., AC-17: 246-247 (1972).

Landau, I. D., Algorithmes d'adaptation hyperstables pour une classe de systèmes adaptifs echantillonnés, C. R. Acad. Sci. (Paris), 275 (Series A): 1391-1394 (1972).

Landau, I. D., Sur une méthode de synthèse des systèmes adaptifs avec modèles utilisés pour la commande et l'identification d'une classe de procédés physiques, Doctoral Thesis, University of Grenoble, France (1973).

Landau, I. D., Elimination of the real positivity condition in the design of parallel model reference adaptive systems, IEEE Trans. Auto. Contr., AC-23: 1015-1020 (1978).

Landau, I. D., and Courtiol, B., Design of multivariable adaptive model-following control systems, Automatica, 10: 483-494 (1974).

Lawson, C. L., and Hanson, R. J., Unbiased recursive identification using model reference adaptive techniques, IEEE Trans. Auto. Contr., AC-21: 194-202 (1976).

Matko, D., and Bremsak, F., On the equivalence of parameter adaptive and model reference systems, Int. J. Contr., 30: 203-211 (1979).

Morse, A. S., Structure and design of linear model following control systems, IEEE Trans. Auto. Contr., AC-18: 346-354 (1973).

Narendra, K. S., and Lin, Y.-H., Design of stable model reference adaptive controllers in Applications of Adaptive Control (K. S. Narendra and R. V. Monopoli, eds.), Academic Press, New York, pp. 69-130, 1980.

Newell, R. B., and Fisher, D. G., Experimental evaluation of optimal, multivariable regulatory controllers with model following capabilities, Automatica, 8: 247-262 (1972).

Nishikama, Y., Sannomiya, N., Itakura, H., and Yang, H. W., Model reference adaptive control for multivariable continuous systems, Proc. IFAC Symp. on Identification and System Parameter Estimation, Washington, D.C., pp. 463-468 (1982).

Parks, P. C., Stability and convergence of adaptive controllers—continuous time systems, IEE Proc., 128 Pt. D: 195-200 (1982).

Parks, P. C., and James, D. J. G., Stability of model reference adaptive systems with random impulsive inputs, Electron. Lett., 6: 131-132 (1970).

Porter, B., and Tatnall, M. L., Stability analysis of a class of multivariable MRAS having time-varying process parameters, Int. J. Contr., 11: 325-332 (1970).

Serdyukov, V. A., Design of adaptive systems using the Lyapunov's direct method, Autom. Rem. Contr., 33: 1138-1144 (1972).

Silveira, H. M., Contributions a la synthèse de systèmes adaptifs avec modèle sans acces aux variables d'état, Doctoral Thesis, Institut Polytechnique de Grenoble (1978).

Sobel, K., Model reference adaptive control of MIMO plants without the need for explicit parameter identification, Doctoral Thesis, Rensselaer Polytechnic Institute, Troy, N.Y. (1980).

Sobel, K., Kaufman, H., and Mabias, L., Model reference output adaptive control systems without parameter identification, Proc. IEEE Conf. Decision and Control, Ft. Lauderdale, Fla. (1979).

Ten Hacken, G. V., Adaptive control with model reference principle, Master's Thesis, Lab. for Control Eng., Delft University of Technology, Netherlands (1976).

Tomizuka, M., Parallel MRAS without compensation block, IEEE Trans. Auto. Contr., AC-27: 505-506 (1982).

Tyler, J. S., The characteristics of model following systems synthesized by optimal control, IEEE Trans. Auto. Contr., AC-9: 485-498 (1984).

Wang, S. H., and Desoer, C. A., The exact model matching of linear multivariable systems, IEEE Trans. Auto. Contr., AC-17: 347-349 (1972).

Wenk, C. J., and Bar Shalom, Y., A multiple model adaptive dual control algorithm for stochastic systems with unknown parameters, IEEE Trans. Auto. Contr., AC-25: 703-710 (1980).

Winsor, C. A., Model reference control system design, Doctoral Thesis, Rensselaer Polytechnic Institute, Troy, N.Y. (1969).

Wolovich, W. A., The use of state feedback for exact model matching, SIAM J. Contr., 10: 512-523 (1972).

4

Discrete-Time MRAC Systems

4.1 INTRODUCTION

Model reference adaptive control systems are making ever-increasing use of digital computers or microprocessor-based hardware. This requires the derivation of discrete-time adaptation laws. Although it may be possible to discretize a linear, time-invariant continuous-time system, we cannot easily do so for a MRAC system because a MRAC is usually a nonlinear, time-varying system and also there is an inherent delay of the sample in the adaptation loop. It is therefore desirable to develop adaptation laws directly for discrete-time MRAC systems.

We can use Lyapunov approach and hyperstability concepts as in the design of continuous-time MRAC system, keeping in mind, however, the presence of the delay in a discrete-time MRAC system, which gives rise to an additional condition the adaptation algorithm should satisfy. Moreover, having a digital computer as part of the system results in a certain flexibility which is absent in a continuous-time system in that time-varying adaptive gains can also be considered in the development of adaptation algorithms. The digital computer makes it possible for adaptation gains to be computed recursively at each step. Thus, many problems which could not be solved using constant adaptation gains can now be tackled [1].

This chapter is organized as follows: Section 4.2 deals with Lyapunov design of discrete-time MRAC. The hyperstability approach is discussed in Section 4.3. Section 4.4 considers briefly the augmented error approach and Section 4.5 the error model approach, both of which were also dealt with in some detail in Chapter 3. Section 4.6 discusses unified approaches, also showing the relation to other adaptation algorithms including self-tuning algorithms. Multivariable systems are considered in Section 4.7. Section 4.8 describes a recent development of hybrid adaptive control in which discrete-time adaptive control is proposed for continuous-time systems, to provide the advantages of both methods. Section 4.9 gives the concluding remarks.

4.2 LYAPUNOV DESIGN

Although there are a number of papers on the Lyapunov approach to design of continuous-time MRAC systems, as already discussed in Section 3.3, there are not many regarding the application of Lyapunov technique to discrete-time MRAC systems. Sebakhy [2] considers such a design in 1976, followed by Ionescu and Monopoli [3], who apply the augmented error approach, which is an extension of the method developed for continuous-time systems already dealt with in Section 3.5. Sebakhy [2] discusses an adaptive scheme for discrete-time model reference systems in which the Lyapunov method is used to prove the scheme's asymptotic stability. It has also been shown that this method can be used to identify the parameters of the plant, when the input is sufficiently general. It is necessary to assume that the state of the plant is accessible and the measurements are free of noise. An illustrative example shows the feasibility of this scheme. Although there exists a unique performance index which is optimized in such designs, this performance index cannot be chosen a priori; it cannot even be defined explicitly in these design procedures. In some cases, Lyapunov design becomes identical to the MIT rule design, indirectly establishing the fact that a performance index of ISE type is implicitly considered in such Lyapunov designs. This, however, cannot be extended to a general Nth order plant with a performance index other than ISE. Because in some situations ISE is not a suitable performance index, a method is needed which can include different performance indexes.

Banerjee and Chalam [5] have used a method of generating Lyapunov functions for discrete-data systems due to Nagaraja and Chalam [4], in conjunction with the Lagrange multiplier technique to obtain a design rule for a SISO system such that stable adaptation is guaranteed and a performance index is minimized during adaptation procedure. The performance index chosen is "square of gain and/or parameter misalignment." The adaptation equation is obtained by choosing a positive-definite V function and constraining the corresponding ΔV function to be negative definite.

The model is completely known and the plant gain G_V is unknown. $G_C(k)$ is the controllable gain in the forward path. The design objective is to have the overall plant gain $G_V G_C(k)$ follows the model gain G. Problems of gain adaptation for first- and second-order plants have been considered and the results verified by computer simulation [6]. This approach can

can also be extended to the problem of two-parameter adaptation involving gain and pole misalignment.

Further discussion of the Lyapunov method is not given here because this method, and its disadvantages, has been discussed in Section 3.3 and because recently developed and more generalized methods are to be considered later.

4.3 HYPERSTABILITY APPROACH

4.3.1 Popov's Hyperstability Theorem

Hyperstability concepts and Popov's theorems applicable to continuous-time systems have been discussed in Section 3.4. A similar approach to the design of discrete-time MRAC systems involves applying of Popov's theorem, following Landau [1, 7].

THEOREM. Consider the (feedforward) discrete time, linear time-invariant block described by

$$x(k+1) = Ax(k) + Bu(k) \tag{4.1}$$

$$\nu(k+1) = Cx(k) + Ju(k) \tag{4.2}$$

where the pair (A, B) is completely controllable and the pair (C, A) is completely observable.

Consider a feedback block described by

$$h(k) = f(\nu, k, 1), \qquad 1 \leqslant k$$

satisfying the inequality

$$\eta(k_0, k_1) \triangleq \sum_{k=k_0}^{k_1} h^T(k)\nu_k \geqslant -\gamma_0^2 \qquad \text{for all } k_1 \geqslant k_0 \tag{4.3}$$

where γ_0^2 is a finite positive constant.

The necessary and sufficient condition for the preceding feedback system to be (asymptotically) hyperstable is that the discrete transfer matrix

$$G(z) = J + C(zI - A)^{-1}B \tag{4.4}$$

must be (strictly) positive real.

The preceding Popov's hyperstability theorem can be applied after the discrete-time MRAC system has been transformed into an equivalent standard feedback system composed of two blocks, as shown in Section 3.4.

4.3.2 Design Procedure

Let us consider the design of a SISO, parallel, discrete-time MRAC system, following Landau [1]. The mathematical description of the reference model and the adjustable system is as follows:

Reference Model

$$y_M(k) = a_{M_1} y_M(k-1) + A_{M_2} y_M(k-2) + b_{M_1} u(k-1) \tag{4.5}$$

where

$u(\cdot)$ = input to the reference model

$y_M(\cdot)$ = output of the reference model

$a_{M_1}, a_{M_2}, b_{M_1}$ = parameters of the model

Parallel Adjustable System

$$y_P^o(k) = a_{P_1}(k-1)y_P(k-1) + a_{P_2}(k-1)y_P(k-2) + b_{P_1}(k-1)u(k-1) \tag{4.6}$$

where $y_P^o(k)$ is the a priori output of the adjustable system, using the values of the adjustable parameters at the instant $(k-1)$.

$$y_P(k) = a_{P_1}(k)y_P(k-1) + a_{P_2}(k)y_P(k-2) + b_{P_1}(k)u(k-1) \tag{4.7}$$

where $y_P(k)$ is the a posteriori output of the adjustable system computed using the values of the adjustable parameters at the instant k, that is, after adaptation has taken place.

The expressions $y_P^o(k)$ and $y_P(k)$ indicate the delay of one sample which occurs when discrete adaptation is used. There is a necessity to introduce the a priori variables, as there is a one-step delay inherent in the adaptation loop.

Output Error

$$e^o(k) = y_M(k) - y_P^o(k) \tag{4.8}$$

$$e(k) = y_M(k) - y_P(k) \tag{4.9}$$

Adaptation Mechanism

The adaptation mechanism will contain a linear compensator generating a signal ν_k:

$$\nu_k^o = e^o(k) + \sum_{i=1}^{r} d_i e(k - i) \tag{4.10}$$

$$\nu_k = e(k) + \sum_{i=1}^{r} d_i e(k - i) \tag{4.11}$$

It is required, as part of the design, to find r and d_i.

Adaptation Algorithms

The signal ν_k^o will be used in the adaptation algorithms of the form:

$$a_{P_i}(k) = a_{P_i}(k - 1) + \phi_i(\nu_k^o), \qquad i = 1, 2 \tag{4.12}$$

$$b_{P_1}(k) = b_{P_1}(k - 1) + \psi_1(\nu_k^o) \tag{4.13}$$

We will use, alternatively, adaptation algorithms in terms of ν_k:

$$a_{P_i}(k) = a_{P_i}(k - 1) + \phi_i'(\nu_k), \qquad i = 1, 2, \tag{4.14}$$

$$b_{P_1}(k) = b_{P_1}(k - 1) + \psi_1'(\nu_k) \tag{4.15}$$

It is, of course, necessary to find the relationship between ν_k^o and ν_k, which will be done in the final stage of design.

4.3.3 Design Steps

The design is done in four steps as indicated in the continuous-time case in Section 3.4.4.

Step 1. Equivalent representation is achieved as follows: Subtracting (4.7) from (4.5) and using (4.9), we obtain

$$e(k) = \{a_{M_1} e(k - 1) + a_{M_2} e(k - 2) + [a_{M_1} - a_{P_1}(k)]y_P(k - 1) + [a_{M_2} - a_{P_2}(k)]y_P(k - 2) + [b_{M_1} - b_{P_1}(k)]u(k - 1)\} \tag{4.16}$$

Using (4.11), (4.14), and (4.15), we get the following equivalent feedback system,

$$a(k) = a_{M_1} e(k - 1) + a_{M_2} e(k - 2) + h_1(k) \tag{4.17}$$

$$\nu_k = e(k) + \sum_{i=1}^{r} d_i e(k - i) \tag{4.18}$$

$$\begin{aligned} h(k) = -h_1(k) = &\sum_{i=1}^{2}\left[\sum_{l=0}^{k} \phi_i'(\nu_l) + a_{P_i}(-1) - a_{M_i}\right] y_P(k - i) \\ &+ \left[\sum_{l=0}^{k} \psi_1'(\nu_l) + b_{P_1}(-1) - b_{M_1}\right] u(k - 1) \end{aligned} \tag{4.19}$$

Equations (4.17) and (4.18) describe the linear time-invariant feedforward block and Eq. (4.19) describes the nonlinear time-varying block.

Step 2. Popov's hyperstability theorem for discrete-time systems, mentioned earlier, is applied to the equivalent feedback block described by (4.17)-(4.19). Before doing this, we should first determine $\phi_i'(\nu_l)$ and $\psi_1'(\nu_l)$ such that the feedback block described by (4.19) satisfies the inequality mentioned in the theorem, which can be written as

$$\eta(0, k_1) = \sum_{k=0}^{k_1} h(k)\nu_k \geq -\gamma_0^2 \quad \text{for all } k_1 \geq 0 \tag{4.20}$$

This inequality becomes, on substituting for h(k) from (4.19),

$$\begin{aligned} \eta(0, k_1) = &\left\{\sum_{i=1}^{2} \sum_{k=0}^{k_1} \nu_k y_P(k - i)\left[\sum_{l=0}^{k} \phi_i'(\nu_l) + a_{P_i}(-1) - a_{M_i}\right]\right. \\ &+ \sum_{k=0}^{k_1} \nu_k u(k - 1)\left[\sum_{l=0}^{k} \psi_1'(\nu_l) + b_{P_1}(-1) - b_{M_i}\right] \geq -\gamma_0^2 \end{aligned} \tag{4.21}$$

where k = 0 corresponds to the beginning of adaptation. However, it may be noted that input r can be there before the adaptation starts.

Also, the inequality (4.21) will be satisfied if each of the terms on the left-hand side satisfy an inequality of the same type.

Using the following relation [8], we can find particular solutions for ϕ_i' and ψ_1' which satisfy such inequalities.

$$\sum_{k=0}^{k_1} x_k\left(\sum_{l=0}^{k} x_l + c\right) = \frac{1}{2}\left(\sum_{k=0}^{k_1} x_k + c\right)^2 + \frac{1}{2}\sum_{k=0}^{k_1} x_k^2 - \frac{c^2}{2} \geqslant -\frac{c^2}{2} \tag{4.22}$$

Using (4.22), we get the particular solutions:

$$\phi_i'(\nu_k) = \alpha_i \nu_k y_P(k - i), \qquad \alpha_i > 0,\ i = 1,\ 2 \tag{4.23}$$

$$\psi_1'(\nu_k) = \beta_1 \nu_k u(k - 1), \qquad \beta_1 > 0 \tag{4.24}$$

We see that ν_k, which appears in the expression for $\eta(0, k_1)$, should appear in ϕ_i' and ψ_1' in order to satisfy (4.21). Thus the feedback block described by (4.19) satisfies the inequality (4.20).

Step 3. Next we consider the remaining part in the feedforward path and apply the Popov theorem. The discrete transfer function of the equivalent forward path described by (4.17) and (4.18) is given by

$$G(z) = \frac{1 + \sum_{i=1}^{r} d_i z^{-1}}{1 - a_{M_1} z^{-1} - a_{M_2} z^{-2}} \tag{4.25}$$

This should be strictly positive real to assure asymptotic hyperstability of the system. This means that the poles of G(z) should lie in $|z| < 1$. This condition is satisfied if a_{M_1} and a_{M_2} lie within the triangle in Fig. 4.1. The stability domain in Fig. 4.1 is obtained from the inequalities

$$\begin{aligned} 1 + a_{M_1} - a_{M_2} &> 0 \\ 1 - a_{M_1} - a_{M_2} &> 0 \\ 1 + a_{M_2} &> 0 \end{aligned} \tag{4.26}$$

[This can be observed from the Eq. (4.27).]

We can compute the coefficients d_i by applying the transformation $z = (1 + s)/(1 - s)$ so that the problem becomes a continuous-time one.

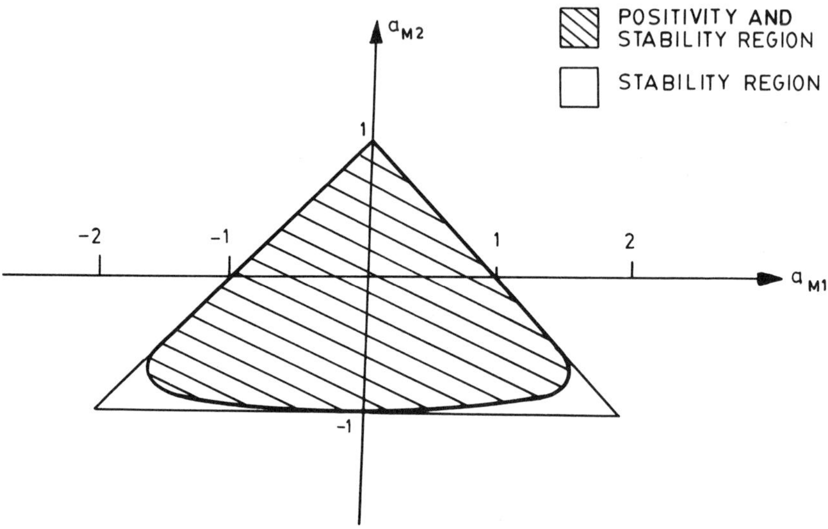

FIG. 4.1 Positivity and stability regions. (From Ref. 1.)

Applying this transformation to (4.25) with r = 0 (i.e., $d_i \equiv 0$), we get

$$G(s) = \frac{1 + 2s + s^2}{(1 - a_{M_1} - a_{M_2}) + 2s(1 + a_{M_2}) + s^2(1 + a_{M_1} - a_{M_2})} \tag{4.27}$$

$$\mathrm{Re}[G(j\omega)] = \frac{(1 - a_{M_1} - a_{M_2}) + 2(1 + 3a_{M_2})\omega^2 + (1 + a_{M_1} - a_{M_2})\omega^4}{[1 - a_{M_1} - a_{M_2} - \omega^2(1 + a_{M_1} - a_{M_2})]^2 + 4\omega^2(1 + a_{M_2})} \tag{4.28}$$

$\mathrm{Re}[G(j\omega)]$ will be strictly positive real for any real ω and therefore G(s) will be strictly positive real if, in addition to the condition of (4.26), a_{M_1} and a_{M_2} satisfy the following two conditions:

$$\begin{gathered} 1 + 3a_{M_2} \geqslant 0 \\ (1 + 3a_{M_2})^2 - (1 + a_{M_1} - a_{M_2})(1 - a_{M_1} - a_{M_2}) < 0 \end{gathered} \tag{4.29}$$

The hatched area in Fig. 4.1 shows the strictly real positivity region [9]. This is smaller than the stability region.

If a_{M_1} and a_{M_2} are outside this positivity region, but still within the stability region, we should find a transfer function which is strictly positive real. For this, we consider either $r = 1$ or $r = 2$.

Let us take $r = 1$ and apply the same procedure as above, with $-1 < d_1 < 1$ (for the numerator to be asymptotically stable). In this case we find that d_1, a_{M_1}, and a_{M_2} should satisfy either the condition

$$1 - d_1 a_{M_1} + 3a_{M_2} \geqslant 0 \tag{4.30}$$

or the condition

$$(1 - d_1 a_{M_1} + 3a_{M_2})^2 - (1 - d_1)^2(1 + a_{M_1} - a_{M_2})(1 - a_{M_1} - a_{M_2}) < 0 \tag{4.31}$$

By an appropriate choice of d_1, the condition (4.31) can be satisfied. It is seen that if we choose

$$d_1 = -0.5a_{M_1} \tag{4.32}$$

this condition is satisfied for any value of a_{M_1} and a_{M_2}.

We can work this out in a similar way for the case when $r = 2$.

Step 4. Finally we return to the original system. The adaptation laws in (4.23) and (4.24) are given in terms of ν_k. We should get these in terms of ν_k^o. For this it is necessary to find the relationship between ν_k^o and ν_k. From (4.18), we get, using (4.16)

$$\begin{aligned} \nu_k &= e(k) + d_1 e(k - 1) + d_2 e(k - 2) \\ &= \{a_{M_1} e(k - 1) + a_{M_2} e(k - 2) + [a_{M_1} - a_{P_1}(k)]y_P(k - 1) \\ &\quad + [a_{M_2} - a_{P_2}(k)]y_P(k - 2) + [b_{M_1} - b_{P_1}(k)]u(k - 1) \\ &\quad + d_1 e(k - 1) + d_2 e(k - 2)\} \end{aligned} \tag{4.33}$$

Substituting for $a_{P_1}(k)$, $a_{P_2}(k)$, $b_{P_1}(k)$ from (4.14), (4.15), (4.23), and (4.24), we get

$$\nu_k = \{a_{M_1} e(k-1) + a_{M_2} e(k-2) + [a_{M_1} - a_{P_1}(k-1) - \alpha_1 \nu_k y_P(k-1)] y_P(k-1) + [a_{M_2} - a_{P_2}(k-1) - \alpha_2 \nu_k y_P(k-2)] y_P(k-2) + [b_{M_1} - b_{P_1}(k-1) - \beta_1 \nu_k u(k-1)] u(k-1) + d_1 e(k-1) + d_2 e(k-2)\} \tag{4.34}$$

The expression for ν_k^o is obtained from (4.5), (4.6), (4.8), and (4.9):

$$\nu_k^o = \{a_{M_1} e(k-1) + a_{M_2} e(k-2) + [a_{M_1} - a_{P_1}(k-1)] y_P(k-1) + [a_{M_2} - a_{P_2}(k-1)] y_P(k-2) + [b_{M_1} - b_{P_1}(k-1)] u(k-1) + d_1 e(k-1) + d_2 e(k-2)\} \tag{4.35}$$

Subtracting (4.34) from (4.35), we have

$$\nu_k^o - \nu_k = \sum_{i=1}^{2} \alpha_i \nu_k y_P^2(k-i) + \beta_1 \nu_k u^2(k-1)$$

or

$$\nu_k = \frac{\nu_k^o}{1 + \sum_{i=1}^{2} \alpha_i y_P^2(k-i) + \beta_1 u^2(k-1)} \tag{4.36}$$

We can get adaptation laws using (4.36) from (4.14), (4.15), (4.23), and (4.24),

$$a_{P_i}(k) = a_{P_i}(k-1) + \frac{\alpha_i y_P(k-i)}{1 + \sum_{i=1}^{2} \alpha_i y_P^2(k-i) + \beta_1 u^2(k-1)} \nu_k^o \tag{4.37a}$$

$$b_{P_1}(k) = b_{P_1}(k-1) + \frac{\beta_1 u(k-1)}{1 + \sum_{i=1}^{2} \alpha_i y_P^2(k-i) + \beta_1 u^2(k-1)} \nu_k^o \quad (4.37b)$$

where ν_k^o can be found from (4.5), (4.6), and (4.10) as

$$\nu_k^o = \left[y_M(k) - \sum_{i=1}^{2} a_{P_i}(k-i) y_P(k-i) - b_{P_1}(k-1)u(k-1) + \sum_{i=1}^{2} d_i e(k-i) \right] \quad (4.38)$$

The preceding example illustrates the design of discrete-time MRAC systems and can be generalized.

4.3.4 Discussion

Landau [1] discusses, in detail, design for two main classes of problems:

Class A: When there are frequent changes in values of the parameters either in the model or in the plant, as for example, in real-time identification or adaptive model-following control (AMFC).

Class B: When there is only an unknown initial difference between the parameters of the model and those of the plant, as, for example, in identification of linear time-invariant systems or AMFC of linear time-invariant systems with unknown parameters.

In most cases, class A algorithms use constant adaptive gains (or, in general, linear time-invariant positive filters). Class B algorithms use adaptive gains which decrease with time, as in this case adaptation is unnecessary once equilibrium is reached.

We can also have problems of slowly time-varying linear systems which are a class between A and B. Adaptive algorithms can be developed with time-varying gains.

Landau [1] considers the design of PI-type and integral-type adaptation for both class A and class B and also deals with experimental results with special reference to improving convergence of the parameters and of the error. Case studies are also given on (1) an adaptive speed controller for dc electric drives, (2) an aircraft longitudinal control problem, and (3) real-time identification and adaptive control of a static dc/ac converter.

Hyperstable-design schemes are also given by Suzuki et al. [10] and Unbehauen [11] for discrete MRAC systems. Unbehauen's systematic procedure [11] can be applied to stable plants of arbitrarily high order and dead time and is based on the work of Schmid [12]. This design method allows the characteristic dynamics of the error signal driving the adaptation of the controller parameters to be chosen arbitrarily fast to ensure that the adaptive control scheme allows fast computation of the plant input. This general approach results in a fairly complicated adaptive control law. However, this can be simplified for plants without dead time; simplification is also possible by splitting the reference model into parallel and series models. It is also possible to show that Monopoli's augmented error approach is a special case of this general approach. A simulation example shows the effectiveness of the general adaptive control law which can be readily implemented on a process computer.

It must be noted that the unified approach by Landau discussed in Section 4.6 leads to a much more generalized algorithm and a simplified design.

4.4 AUGMENTED ERROR APPROACH

The concept of augmented error introduced by Monopoli [13] for continuous-time MRAC systems (and discussed in some detail in Section 3.5) has been extended in a straightforward manner to discrete-time MRAC systems by Ionescu and Monopoli [3]. This approach arose from the usefulness of augmented error in the stability analysis of the nonlinear, time-varying difference equations describing the entire MRAC system. The augmented error concept circumvents the problems related to system delay and to the inherent one-step delay existing in discrete-time MRAC systems. This approach to design is based on the stability point of view via the use of Lyapunov functions and results in an adaptation algorithm with constant adaptation gains. Adaptation algorithms use filtered variables. The design is based on the assumption that the sign of the leading coefficient of the plant numerator polynomial is known. This assumption is rather restrictive and is avoided in a unified approach by Landau and Lozano [14] based on a theorem given by Landau and Silveria [15] and an extension to the same by Landau [16]. Further, the parallel configuration considered by Ionescu and Monopoli [3] is applicable for tracking purposes only and the design for regulation purposes has not been investigated, whereas other configurations (say, the series-parallel one) can be applied for regulation [1,17].

Ionescu's adaptation algorithm with constant adaptation gains, though simple, has been found by extensive simulations to have two main disadvantages:

1. It is not clear how the best values of the gains can be chosen.
2. The system performance is sensitive to the values of the gains.

It is therefore considered necessary to consider adaptation algorithms with time-varying gains so that we can eliminate the need to choose the values of the adaptation gains, thus providing better performance. Such algorithms will be discussed in Section 4.6. However, such algorithms make the stability analysis rather complicated. The stability theorems [15,16] can be applied in the analysis.

In view of these disadvantages, and also because the concept of augmented error has been dealt with in Section 3.5, we will not discuss this further. However, the reader will do well to refer to Ionescu and Monopoli [3]. In fact, the general design procedure developed by Landau [14] on the basis of the stability theorem [16] and discussed in Section 4.6 is dependent on the concept of augmented error.

4.5 NARENDRA'S ERROR MODEL APPROACH

4.5.1 Introduction

Narendra's error model approach was discussed in Section 3.6 where a unified treatment is given for both continuous-time and discrete-time systems. Narendra and Lin [18] present a proof of stability of discrete-time MRAC systems on the lines of the proof given for continuous-time systems by Narendra et al. [19]. In this section we consider some aspects of the specific technique used for the design of a discrete-time adaptive controller, as given in Narendra and Lin [18].

4.5.2 Problem Statement

A SISO discrete linear time-invariant plant is described by the transfer function

$$G_P(z) = k_P \frac{Z_P(z)}{R_P(z)} \tag{4.39}$$

where $G_P(z)$ is proper, with $R_P(z)$ a monic polynomial of degree n, $Z_P(z)$ is a monic polynomial (with all zeros inside the unit circle) of degree $m \leq n$, and k_P is a constant-gain parameter. $(n - m)$ is the relative degree of the plant. We assume that m, n and the sign of k_P as well as the upper bound on $|k_P|$ are known, while the coefficients of Z_P and R_P are unknown. $\{u(k), y_P(k)\}$ is the input/output pair.

Consider a reference model, whose output $y_M(k)$ is the desired response of the plant, is represented by the transfer function

$$G_M(z) = k_M \frac{Z_M(z)}{R_M(z)} \tag{4.40}$$

where $R_M(z)$ and $Z_M(z)$ are monic polynomials (with all zeros inside the unit circle) of degrees n and $r \leq m$, respectively, and k_M is a constant. The relative degree of the model is therefore greater than or equal to that of the plant. The reference input r(k) to the model is specified and is uniformly bounded.

The objective of adaptive control is to find a suitable control u(k) such that

$$y_P(k) - y_M(k) \longrightarrow 0 \quad \text{as} \quad k \longrightarrow \infty \tag{4.41}$$

For the sake of simplicity we assume that r = m.

4.5.3 Adaptive Controller

We can consider two cases, namely, m = n and $m \leq n - 1$ and so two different structures are needed.

<u>Case of m = n</u>

The model transfer function is assumed to be positive real. (It may be noted that by prefiltering, the model can be made positive real). The structure is shown in Fig. 4.2. The physical realization of an algebraic loop involved in Fig. 4.2 is given in Fig. 4.3.

The controller consists of (2n + 1) adjustable parameters which are the elements of a parameter vector $\bar{\theta}(k)$:

$$\bar{\theta}(k) \triangleq [k_0(k), c_1(k), \ldots, c_n(k), d_1(k), d_2(k), \ldots, d_n(k)] \tag{4.42}$$

Two identical auxiliary signal generators of dimension n having state variables $v^{(1)}(k)$ and $v^{(2)}(k)$ and inputs u(k) and $y_P(k)$, respectively, as shown in Fig. 4.2 form part of the controller.

Let a vector $\bar{\omega}(k)$ be defined as

$$\bar{\omega}(k)^T = [r(k), v^{(1)}(k)^T, v^{(2)}(k)^T] \tag{4.43}$$

Then the signal feedback into the plant is given by

$$\{\bar{\theta}^T(k)\bar{\omega}(k) - \alpha\bar{\omega}(k)^T\Gamma\bar{\omega}(k)e_1(k)\} \tag{4.44}$$

The first term in (4.44) is a linear combination of the elements of $\bar{\omega}(k)$ and the second term depends on $e_1(k)$ and is essential to determine the stability of the error equations. If $\bar{\omega}(k)$ is bounded, as the adaptation proceeds, this term is seen to tend to zero with $e_1(k)$.

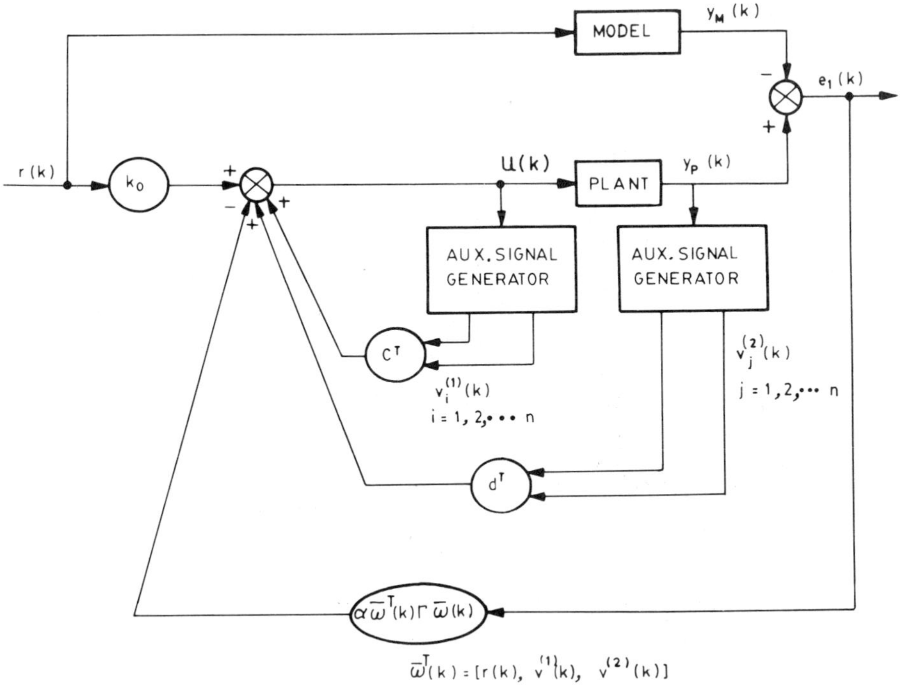

FIG. 4.2 Adaptive controller for m = n. (From Ref. 18. Copyright © 1980 IEEE.)

It can be shown, following the results of Narendra and Valavani [20], that a constant vector $\bar{\theta}^*$ exists such that when $\bar{\theta}(k) \equiv \bar{\theta}^*$, the transfer function of the plant together with the controller matches that of the reference model.

The adaptive control objective is to find the law for updating $\bar{\theta}(k)$ such that $\bar{\theta}(k) \longrightarrow \bar{\theta}^*$ as $k \longrightarrow \infty$, while maintaining the overall system stability.

Let the parameter vector $\bar{\theta}(k)$ be represented by

$$\bar{\theta}(k) = \bar{\theta}^* + \bar{\Phi}(k) \tag{4.45}$$

where $\bar{\Phi}(k)$ is the parameter error vector at time k.

The output error $e_1(k)$ in Fig. 4.2 is given by

$$e_1(k) = \frac{k_P}{k_M} [G_M(z)\{\bar{\Phi}^T(k)\bar{\omega}(k) - \alpha \bar{\omega}^T(k) \Gamma \bar{\omega}(k) e_1(k)\}] \tag{4.46}$$

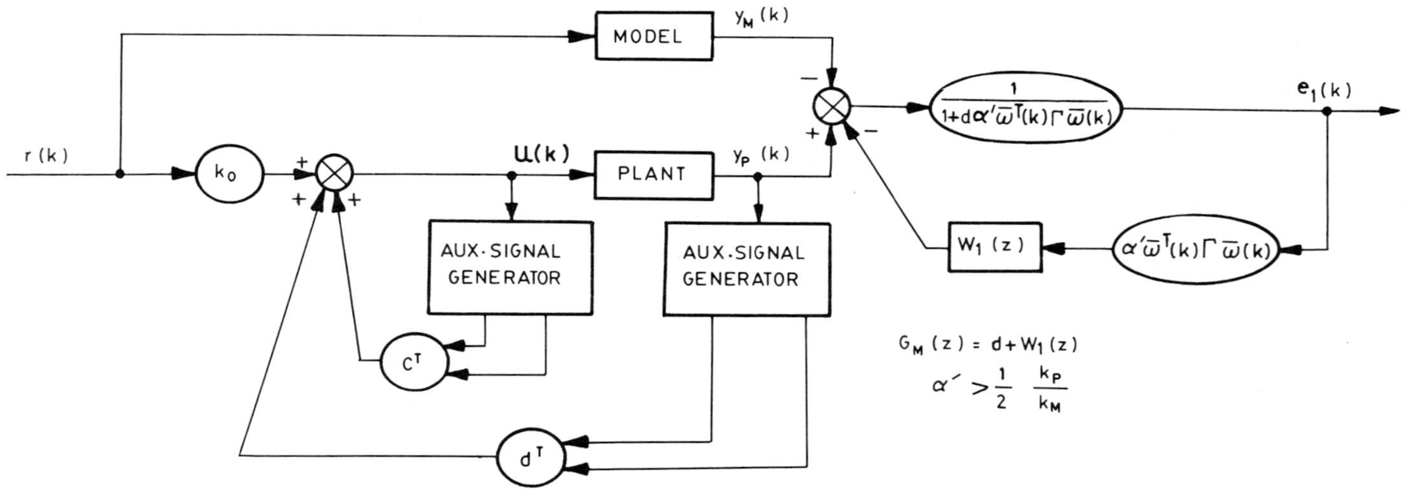

FIG. 4.3 Physical realization of the adaptive control for m = n. (From Ref. 18. Copyright © 1980 IEEE.)

where $G_M(z)$ is a strictly positive real operator. From the results of Lin and Narendra [21], it is seen that if $\bar{\phi}(k)$ is updated according to the law:

$$\bar{\phi}(k+1) = \bar{\phi}(k) - \Gamma e_1(k)\bar{\omega}(k) \tag{4.47}$$

then $e_1(k) \longrightarrow 0$ whether or not $\bar{\omega}(k)$ is bounded. Since in this case the desired output $y_M(k)$ of the model is uniformly bounded, the plant output will also be uniformly bounded and approach the desired output asymptotically.

Case of $m \leqslant n - 1$

In this case an auxiliary signal has to be fed into the model, as described in Narendra and Valavani [20]. The corresponding structure is shown in Fig. 4.4. An equivalent representation of the same is given in Fig. 4.5.

We can assume, with no loss of generality, that L(z), a rational function in z [with $L^{-1}(z)$ a strictly proper minimum phase function], exists such that $G_M(z)L(z)$ is strictly positive real. However, since L(z) is not physically realizable, some modification (as in the case of continuous-time systems) is necessary. This involves feeding back signals into both plant and model such that the error equations have the same from as in the case of $m = n$.

Figure 4.5 shows the structure of the controller when the plant gain k_P is known. For simplicity we assume $k_M = k_P = 1$. [If $k_M \neq k_P$, then an additional adjustable parameter has to be used to generate the auxiliary signal $u_M(k)$].

There are 2n parameters to be adjusted. We therefore define

$$\bar{\theta}^T(k) = [c_1(k), c_2(k), \ldots, c_n(k), d_1(k), \ldots, d_n(k)] \tag{4.48}$$

$$\bar{\omega}^T(k) = [v^{(1)}(k)^T, v^{(2)}(k)^T] \tag{4.49}$$

The signal fed into the plant is

$$u_P(k) = \bar{\theta}^T(k)\bar{\omega}(k) \tag{4.50}$$

The signal fed back to the model is

$$u_M(k) = L(z)[\{L^{-1}(z)\bar{\theta}^T(k) - \bar{\theta}^T(k)L^{-1}(z)\}\bar{\omega}(k) + \alpha\bar{\xi}^T(k)\Gamma\bar{\xi}(k)e_1(k)] \tag{4.51}$$

where

$$L^{-1}(z)\bar{\omega}(k) = \bar{\xi}(k)$$

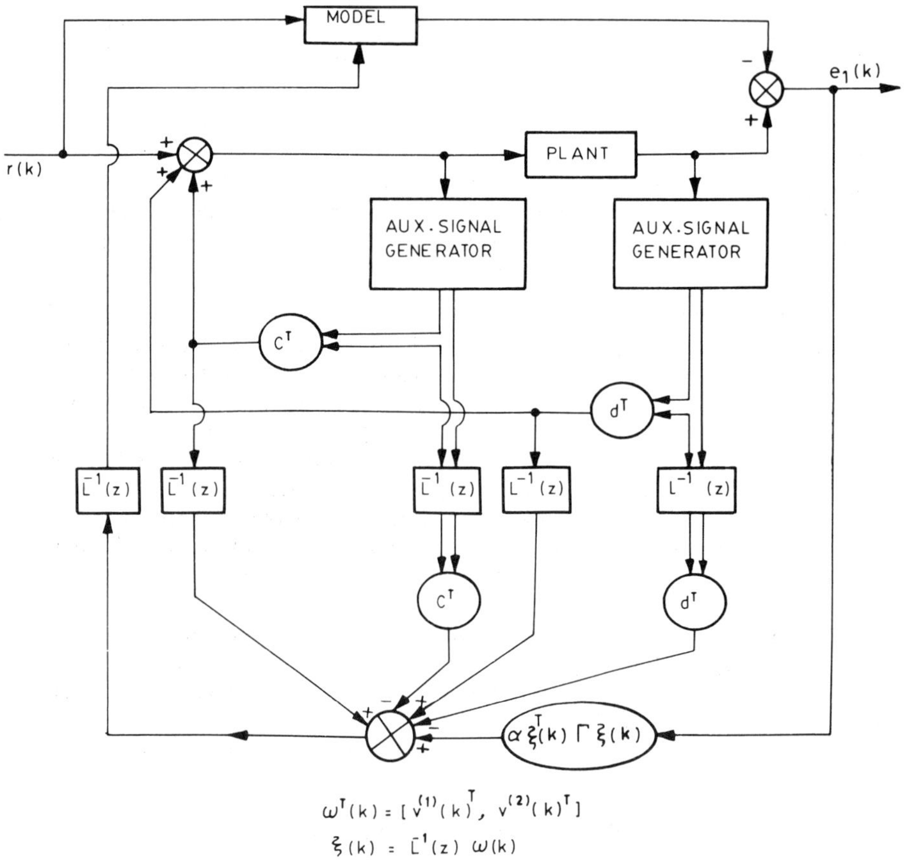

FIG. 4.4 Adaptive controller for $m \leqslant n - 1$. (From Ref. 18. Copyright © 1980 IEEE.)

Then the resulting error equations have the form required to generate stable adaptive laws, to be discussed later.

Although L(z) is not physically realizable, $G_M(z)L(z)$ can be realized. Hence the overall system is as shown in Fig. 4.5.

The augmented error $e_1(k)$ in Figs. 4.4 and 4.5 satisfies the error equation

$$e_1(k) = [G_M(z)L(z)]\{\phi^T(k)\bar{\xi}(k) - \alpha\bar{\xi}^T(k)\Gamma\bar{\xi}(k)e_1(k)\} \tag{4.52}$$

where $G_M(z)L(z)$ is a strictly positive real transfer function.

As defined earlier,

$$\xi(k) = L^{-1}(z)\bar{\omega}(k) \tag{4.53}$$

As in the case of m = n, the following adaptive law

$$\bar{\phi}(k+1) = \bar{\phi}(k) - \Gamma e_1(k)\bar{\xi}(k) \tag{4.54}$$

results in

$$e_1(k) \longrightarrow 0 \quad \text{as} \quad k \longrightarrow \infty$$

4.5.4 Discussion

In the present case we cannot simply say that the plant output will be bounded, since the model output (which is due to both the reference input

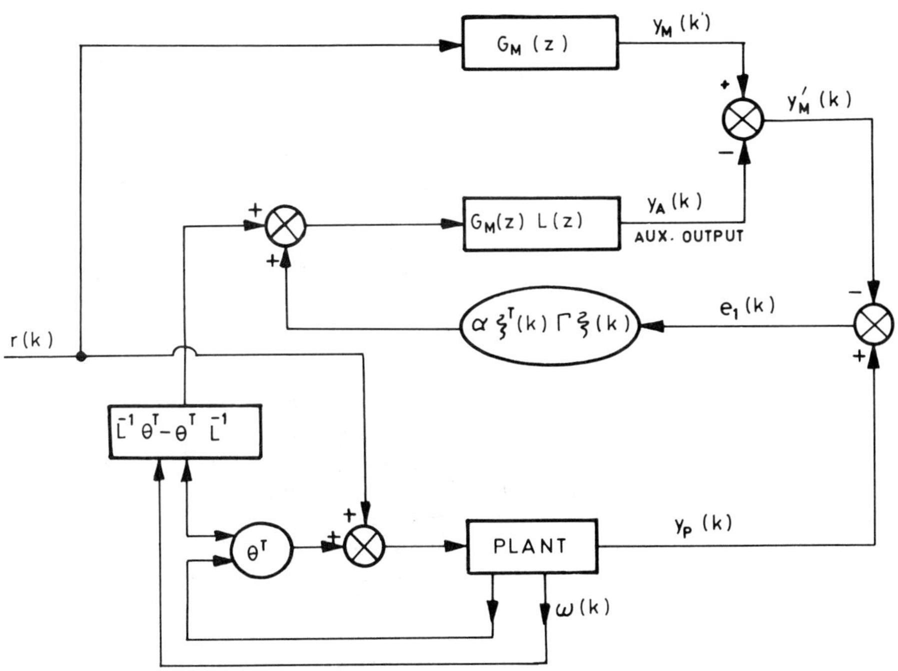

FIG. 4.5 Equivalent representation of Fig. 4.4 (From Ref. 18. Copyright © 1980 IEEE.)

and the auxiliary input) may be unbounded. The proof for global asymptotic stability of the system, showing that neither the plant output nor the model output can be unbounded, is given in Narendra and Lin [18], to which reference may be made. The same proof may be used to show that the discrete-time controller proposed by Ionescu and Monopoli [3] is also stable.

4.6 UNIFIED APPROACH

4.6.1 Introduction

Section 4.4 mentioned the need to develop adaptation algorithms which do not have some restrictions the augmented error approach has. Landau and Lozano [14] present a unified approach based on the stability theorems [15,16].

Egardt [22] has also given a unified approach (as for the continuous-time systems discussed in Section 3.7). From this general algorithm it is possible to obtain, as special cases, the schemes proposed by Ionescu and Monopoli [3] and Bénéjean [23], as well as the self-tuning regulator algorithm proposed by Åström and Wittenmark [24,25] and Clarke and Gawthrop [26] and the algorithm based on input matching by Johnson and Tse [27]. Egardt's papers [23,28] dealing with discrete-time systems closely follow his work on continuous-time systems discussed earlier in Section 3.7.

In this section we deal with the unified approach of Landau and Lozano [14], which includes the adaptive algorithms of Egardt as particular cases. This will be dealt with in Section 4.6.6. In this approach the tracking and regulation objectives can be specified independently, which is not the case for most of the previous schemes given in the literature. Earlier MRAC algorithms developed by others can be obtained as special cases of the general adaptive algorithm. Similarities between MRAC and certain types of stochastic self-tuning regulators have also been established.

4.6.2 Stability Theorem

The stability theorem [14-16] can be considered to be a basic tool for designing MRAC systems from the stability point of view, as the earlier approaches based on Lyapunov functions or on hyperstability concepts. One of the advantages of this approach is that it simplifies the design considerably. Furthermore, a unified view is possible.

The following is the statement of the theorem concerning the asymptotic stability of the equivalent feedback system representation of the MRAC system.

THEOREM. Assume that the following adaptation algorithms for updating the parameter vector $\hat{p}(k)$ is used:

$$\hat{p}(k) = \hat{p}(k-1) + F_k \phi(k-d) \nu_k \tag{4.55}$$

where

$$F_{k+1}^{-1} = \lambda_1(k)F_k^{-1} + \lambda_2(k)\phi(k - d)\phi^T(k - d) \tag{4.56}$$

with

$$\begin{aligned} F_0 > 0, \quad & 0 < \lambda_1(k) \leqslant 1 \\ & 0 < \lambda_2(k) < 2, \ \forall\, k \end{aligned} \tag{4.57}$$

Assume that the relation between $\phi(k - d)$ and ν_k is given by

$$\nu_k = H(q^{-1})[p - \hat{p}(k)]^T\phi(k - d) \tag{4.58}$$

where $\{q^{-1}\}$ is the backward shift operator, $\{d\}$ is the plant time delay, p is a constant parameter vector, $\phi(k - d)$ is a bounded or unbounded vector sequence, $H(z^{-1})$ is a rational discrete transfer function normalized under the form:

$$H(z^{-1}) = \frac{1 + h_1' z^{-1} + \cdots + h_\alpha' z^{-\alpha}}{1 + h_1 z^{-1} + \cdots + h_\beta z^{-\beta}} \tag{4.59}$$

Then, if the transfer function

$$H'(z^{-1}) = H(z^{-1}) - \frac{1}{2}\lambda \tag{4.60}$$

is strictly positive real, where

$$2 > \lambda \geqslant \max_{0 \leqslant k < \infty} [\lambda_2(k)] \tag{4.61}$$

the following properties are applicable, for any ν_0 and $\hat{p}(0)$ bounded:

$$\lim_{k \to \infty} \nu_k = 0 \tag{4.62}$$

$$\lim_{k \to \infty} x_k = 0 \tag{4.63}$$

where x_k is the state vector of any of the state realizations of $H(z^{-1})$.

$$\lim_{k \to \infty} \phi^T(k - d)\{[\hat{p}(k - 1) - p] + F_k \phi(k - d)\nu_k\} = 0 \tag{4.64}$$

$$\lim_{k \to \infty} [1 - \lambda_1(k)] \, || [\hat{p}(k - 1) - p] + F_k \phi(k - d)\nu_k ||^2_{F_k^{-1}} = 0 \tag{4.65}$$

$$\lim_{k \to \infty} || \phi(k - d)\nu_k ||^2_{F_k} = \lim_{k \to \infty} \Delta\hat{p}(k)^T F_k^{-1} \Delta\hat{p}(k) = 0 \tag{4.66}$$

where $\Delta\hat{p}(k) = \hat{p}(k) - \hat{p}(k - 1)$.

$$[\hat{p}(k - 1) - p]^T F_k^{-1} [\hat{p}(k - 1) - p] < M_1 < \infty, \qquad k > 0 \tag{4.67}$$

$$\lim_{k \to \infty} [\hat{p}(k - 1) - p]^T F_k^{-1} [\hat{p}(k - 1) - p] = \text{constant} \tag{4.68}$$

COROLLARY. If in addition to the hypothesis of the preceding theorem, the following conditions are also considered,

$$F_k^{-1} > \epsilon F_0^{-1}, \quad F_0 > 0, \quad \epsilon > 0, \quad k > 0, \qquad F^{-1} \text{ nondecreasing for } k \geq k_0 \tag{4.69}$$

then, in addition to the preceding results, we have

$$\lim_{k \to \infty} \Delta\hat{p}(k) = \lim_{k \to \infty} F_k \phi(k - d)\nu_k = 0 \tag{4.70}$$

$$||\hat{p}(k)|| \leq M_2 < \infty, \qquad k \geq 0 \tag{4.71}$$

The proofs of the theorem and the corollary are found in Landau [16].

Remark 1. The adaptation algorithm considered corresponds to an integral adaptation with the general form of the time-varying gain matrix F_k [15]. The results can be extended to the case of proportional and integral adaptation.

Remark 2. Equations (4.55) and (4.58) describe a feedback system which has a linear block [defined by $H(z^{-1})$] and a time-varying nonlinear block [1,15].

4.6.3 Relation to Other Algorithms

Depending on the choices for $\lambda_1(k)$ and $\lambda_2(k)$, different adaptation algorithms can be obtained, thus showing that the approach based on the preceding theorem gives a unified point of view:

Case a. The choice of $\lambda_1(k) = 1$ and $\lambda_2(k) = 0$ corresponds to a constant adaptation gain used in Ionescu and Monopoli [3] and Bénéjean [23].

Case b. The choice of $\lambda_1(k) = 1$ and $\lambda_2(k) = \lambda$, $0 < \lambda < 2$, corresponds to a time-decreasing adaptation gain [1,29].

Case c. The choice of $\lambda_1(k) = \lambda_1$ and $\lambda_2(k) = \lambda_2$ corresponds to a time-varying adaptation gain, useful for slowly time-varying plants (usually with $0.95 < \lambda_1 < 0.99$) [30].

Case d. The choice of $\lambda_1(k)$ and $\lambda_2(k)$ such that trace F_k = constant, corresponds to a real-time adaptation algorithm for tracking time-varying plants [17].

Remarks. The conditions (4.69) are automatically satisfied for all the particular algorithms mentioned above, except for Case c. For this algorithm, when $\lambda_2(k) = 0$ and $\lambda_1(k) < 1$, in order to satisfy (4.69), $\lim_{k\to\infty} \lambda_1(k) = 1$ or more specifically $\lambda_1(k) = 1$ for $k > k_1$ where k_1 is determined by

$$\prod_{i=0}^{k_1} \lambda_1(i) = \epsilon > 0$$

4.6.4 General Design Procedure

Before considering the general design procedure for a MRAC system, let us consider a linear model-following control system, from the results of which an extension to MRAC design can be obtained.

Let a SISO, discrete linear time-invariant plant be described by

$$A(q^{-1})y_P(k) = q^{-d}B(q^{-1})u(k), \quad d > 0, \quad y(0) \neq 0 \tag{4.72}$$

$$A(q^{-1}) = 1 + a_1 q^{-1} + \cdots + a_{n_A} q^{-n_a} \tag{4.73}$$

$$B(q^{-1}) = b_0 + b_1 q^{-1} + \cdots + b_{n_B} q^{-n_b}, \quad b_0 \neq 0 \tag{4.74}$$

$\{u(k)\}$ and $\{y_P(k)\}$ are the plant input and output, respectively.

We assume that the zeros of $B(z^{-1})$ are all in $|z| < 1$. They can therefore be cancelled without leading to an unbounded control input.

The objectives of control in this particular case are the following:

Objective 1. For tracking, the control should be such that the output of the process satisfies the equation

$$C_1(q^{-1})y_P(k) = q^{-d}D(q^{-1})r(k)$$

where $C_1(q^{-1})$, $D(q^{-1})$ are polynomials in $\{q^{-1}\}$, $C_1(q^{-1})$ is asymptotically stable, and r(k) is a bounded reference sequence.

Objective 2. For regulation $[r(k) \equiv 0]$, the control should be such that an initial disturbance $[y_P(0) \neq 0]$ is eliminated with the dynamics defined by

$$C_2(q^{-1})y_P(k + d) = 0, \quad k \geq 0 \tag{4.76}$$

where

$$C_2(q^{-1}) = 1 + C_1^2 q^{-1} + \cdots + C_{n_{c_2}} q^{-n_{c_2}} \tag{4.77}$$

is an asymptotically stable polynomial.

A solution is obtained by using an explicit reference model described by

$$C_1(q^{-1})y_M(k) = q^{-d}D(q^{-1})r(k) \tag{4.78}$$

where $\{y_M(k)\}$ and $\{r(k)\}$ are the model output and input, respectively. They are both bounded.

The output error is given by

$$e(k) = y_P(k) - y_M(k) \tag{4.79}$$

It can be seen that the objectives of the control can be fulfilled if the following equation is satisfied:

$$C_2(q^{-1})e(k + d) = 0, \quad k > 0 \tag{4.80}$$

Using the identity

$$C_2(q^{-1}) = A(q^{-1})S(q^{-1}) + q^{-d}T(q^{-1}) \tag{4.81}$$

where

$$S(q^{-1}) = 1 + s_1 q^{-1} + s_2 q^{-2} + \cdots + s_{n_S} q^{-n_S} \tag{4.82}$$

$$T(q^{-1}) = t_0 + t_1 q^{-1} + t_1 q^{-2} + \dots + t_{n_T} q^{-n_T} \tag{4.83}$$

It has been shown [14] that the preceding identity has a unique solution and

$$n_S = d - 1 \quad \text{and} \quad n_T = \max(n_A - 1, n_{c_2} - d)$$

Equation (4.80) can be written as

$$\begin{aligned} C_2(q^{-1})e(k + d) &= [B(q^{-1})S(q^{-1})u(k) + T(q^{-1})y_P(k) - C_2(q^{-1})y_M(k + d)] \\ &= b_0 u(k) + p_o^T \phi_0(k) - C_2(q^{-1})y_M(k + d) \\ &= p^T \phi(k) - C_2(q^{-1})y_M(k + d) \end{aligned} \tag{4.84}$$

where

$$\phi_0^T(k) = [u(k - 1), \dots, u(k - d - n_B + 1);\ y_P(k), \dots, y_P(k - n_T)] \tag{4.85}$$

$$p_0^T = [b_0 s_1 + b_1, b_0 s_2 + b_1 s_1 + b_2, \dots, b_{n_B} s_{d-1};\ t_0, \dots, t_{n_T}] \tag{4.86}$$

$$\phi^T(k) = [u(k);\ \phi_0^T(k)] \tag{4.87}$$

$$p^T = [b_0;\ p_0^T] \tag{4.88}$$

Equating the right-hand side of (4.84) with zero, the control objective of (4.80) is achieved with the control input

$$u(k) = \frac{1}{b_0}[C_2(q^{-1})y_M(k + d) - T(q^{-1})y_P(k) - B_S(q^{-1})u(k)] \tag{4.89}$$

where

$$B_S(q^{-1}) = B(q^{-1})S(q^{-1}) - b_0 \tag{4.90}$$

whence we get

$$u(k) = \frac{1}{b_0}[C_2(q^{-1})y_M(k + d) - p_0^T \phi_0(k)] \tag{4.91}$$

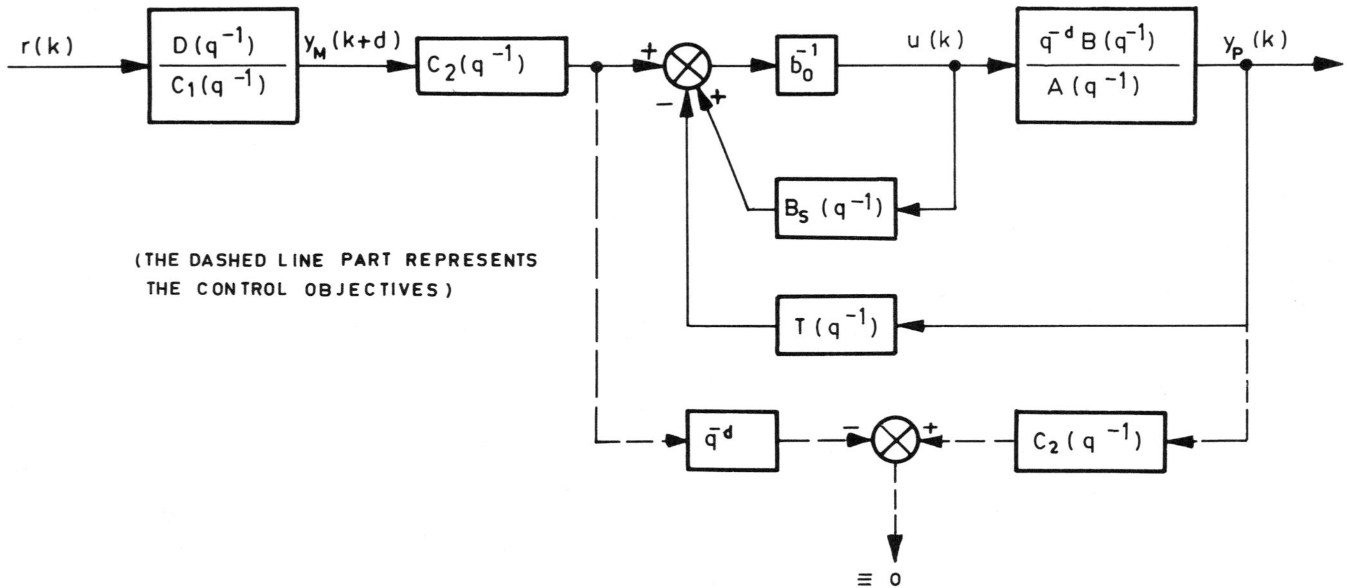

FIG. 4.6 Tracking and regulation control scheme for known plants. (From Ref. 14. Reprinted with permission from Pergamon Press, copyright 1981.)

The linear model-following control system depicting the preceding scheme is shown in Fig. 4.6.

4.6.5 General MRAC Design

When the plant parameters a_i and b_i are unknown, we have to replace the vector p_0 and the coefficient b_0 in (4.91) by the adjustable parameters $\hat{p}_0(k)$ and $\hat{b}_0(k)$, which will be updated by the adaptation mechanism. In this design we use a filter which acts on the plant input and output, and also a linear compensator which acts on the output error. This is done to obtain design flexibility and also to be able to connect this design with previously developed designs. In addition, it enables the positive-real condition to be satisfied in some cases, apart from improving the performance of the system in the presence of disturbances [31].

The filtered variables are defined as follows:

$$
\begin{aligned}
&L(q^{-1})y_{P_f}(k) = y_P(k)\\
&L(q^{-1})u_f(k) = u(k)\\
&L(q^{-1})y_{M_f}(k) = y_M(k)\\
&L(q^{-1})\phi_{0f}(k) = \phi_0(k)
\end{aligned}
\tag{4.92}
$$

where

$$L(q^{-1}) = 1 + l_1 q^{-1} + \dots + l_{n_L} q^{-n_L} \tag{4.93}$$

is an asymptotically stable polynomial. [The choice of n_L will be considered after Eq. (4.113)].

Using the preceding filtered variables, Eq. (4.84) may be rewritten as

$$C_2(q^{-1})e(k + d) = L(q^{-1})[b_0 u_f(k) + p_0^T \phi_{0f}(k) - C_2(q^{-1})y_{M_f}(k + d) \tag{4.94}$$

From (4.94) we see that for known parameters, the control input is

$$u(k) = L(q^{-1})u_f(k) \tag{4.95}$$

where

$$u_f(k) = \frac{1}{b_0}\left[C_2(q^{-1})y_{M_f}(k+d) - p_0^T\phi_{0f}(k)\right] \tag{4.96}$$

Then the objective of (4.80) is achieved.

If the plant parameters are unknown, then p_0 and b_0 are replaced by $\hat{p}_0(k)$ and $\hat{b}_0(k)$, which are adjustable.

$$u_f(k) = \frac{1}{\hat{b}_0(k)}\left[C_2(q^{-1})y_{M_f}(k+d) - \hat{p}_0^T(k)\phi_{0f}(k)\right] \tag{4.97}$$

Equivalently, we can write

$$\hat{p}^T(k)\phi_f(k) = C_2(q^{-1})y_{M_f}(k+d) \tag{4.98}$$

with

$$\phi_f^T(k) = \left[u_f(k), \phi_{0f}^T(k)\right] \tag{4.99}$$

and

$$\hat{p}^T(k) = \left[\hat{b}_0(k), \hat{p}_0^T(k)\right] \tag{4.100}$$

Let us define the filtered error as

$$e_f(k) = C_2(q^{-1})e(k) \tag{4.101}$$

If we substitute (4.98) in (4.94) and use (4.87) and (4.88), we get

$$\begin{aligned} e_f(k) &= C_2(q^{-1})e(k) \\ &= L(q^{-1})\left\{\left[p - \hat{p}(k-d)\right]^T\phi_f(k-d)\right\} \end{aligned} \tag{4.102}$$

The design objective is to find an adaptation mechanism which will ensure that the output error converges to zero and the plant input and output are bounded; that is,

$$\lim_{k \to \infty} e(k) = 0, \quad \forall e(0) \neq 0, \quad \hat{p}(0) \in \mathcal{R}, \quad \|\phi(k)\| \leq M < \infty, \quad \forall k \tag{4.103}$$

We cannot solve this problem directly, on account of the form of (4.102). However, we can apply the stability theorem discussed earlier. For this, we have to introduce an augmented filtered error, $\eta(k)$, which is obtained by adding an auxiliary error $y(k)$ to $e_f(k)$:

$$\eta(k) = e_f(k) + y(k) \tag{4.104}$$

We define the auxiliary error $y(k)$ as follows:

$$y(k) = L(q^{-1})\{[\hat{p}(k - d) - \hat{p}(k)]^T \phi_f(k - d)\} \tag{4.105}$$

Then we can find an adaptation mechanism which will ensure that

$$\lim_{k \to 0} \eta(k) = 0 \tag{4.106}$$

Then, through additional analysis, it can be shown that the objectives of (4.103) are achieved.

Substituting (4.102) and (4.105) in (4.104), we get

$$\eta(k) = L(q^{-1})\{[p - \hat{p}(k)]^T \phi_f(k - d)\} \tag{4.107}$$

Next we define the processed augmented error as

$$\nu(k) = \frac{H_1(q^{-1})}{H_2(q^{-1})} \eta(k) \tag{4.108}$$

where $H_1(q^{-1})$ and $H_2(q^{-1})$ are monic polynomials in $\{q^{-1}\}$. Then we have

$$\nu(k) = \frac{H_1(q^{-1}) L(q^{-1})}{H_2(q^{-1})} \{[p - \hat{p}(k)]^T \phi_f(k - d)\} \tag{4.109}$$

We can now directly apply the Stability Theorem.

Consider the adaptation algorithm:

$$\hat{p}(k) = \hat{p}(k - 1) + F_k \phi_f(k - d)\nu(k) \tag{4.110}$$

with

$$F_{k+1} = \frac{1}{\lambda_1(k)}\left[F_k - \frac{F_k\phi_f(k-d)\phi_f(k-d)^T F_k}{\lambda_1(k)/\lambda_2(k) + \phi_f(k-d)^T F_k\phi_f(k-d)}\right] \tag{4.111}$$

Then

$$\lim_{k\to\infty} \nu(k) = 0 \tag{4.112}$$

$H_1(q^{-1})$, $H_2(q^{-1})$, and $L(q^{-1})$ should be asymptotically stable polynomials satisfying:

$$\frac{H_1(q^{-1})L(q^{-1})}{H_2(q^{-1})} - \frac{\lambda}{2} = \text{S.P.R.} \tag{4.113}$$

with λ as in Eq. (4.61). Thus $H_1(q^{-1})$, $H_2(q^{-1})$, and $L(q^{-1})$ are any finite-dimensional asymptotically stable polynomials satisfying the positivity condition (4.113). These polynomials are important in stochastic MRAC [31].

It can be shown [14] that if $C_2(q^{-1})$ is an asympototically stable polymial, then $\lim_{k\to\infty} e(k) = 0$ and $\phi_f(k)$ is a bounded vector. Thus the objectives of (4.103) are satisfied.

To make the algorithm implementable, we should express $\nu(k)$ in terms of parameters estimated up to (k - 1). By using Eqs. (4.87), (4.98), (4.102), (4.104), (4.105), (4.108), and (4.110) we obtain

$$\nu(k) = \frac{C_2(q^{-1})y_p(k) - p(k-1)\phi_f(k-d) + \alpha(k)}{1 + \phi_f^T(k-d)F_k\phi_f(k-d)} \tag{4.114}$$

where

$$\alpha(k) = [1 - H_2(q^{-1})]\nu(k) + [H_1(q^{-1}) - 1]\eta(k) + [L(q^{-1}) - 1] \times \{[\hat{p}(k-d) - \hat{p}(k-1)]^T\phi_f(k-d) - \phi_f^T(k-d)F_k\phi_f(k-d)\nu(k)\} \tag{4.115}$$

If $L(q^{-1}) = H_1(q^{-1}) = H_2(q^{-1}) = 1$, then (4.114) corresponds to a simplified MRAC system, which is shown in Fig. 4.7.

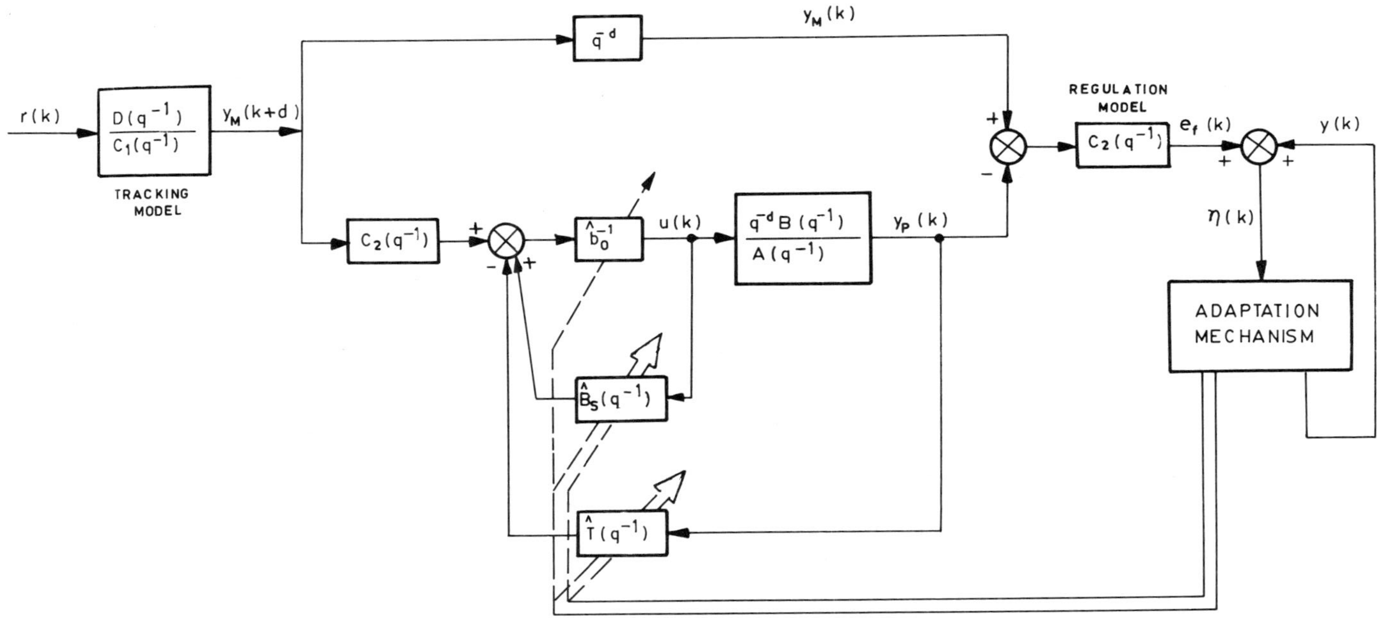

FIG. 4.7 Simplified MRAC system. (From Ref. 14. Reprinted with permission from Pergamon Press, copyright 1981.)

Remark 1. When using the filter $L(q^{-1})$, u(k) is effectively computed using (4.95) where $u_f(k)$ is given by (4.97). It is noted that u(k) computed when $L(q^{-1}) \neq 1$ is different transiently from u(k) when $L(q^{-1}) = 1$. However, stability analysis shows that the control objectives are asymptotically achieved.

Remark 2. The same control algorithm can be obtained by using the concept of the a priori and a posteriori reference model [1]. The a priori output is designated by $\{y_M^o(k)\}$ and the a posteriori output by $\{y_{M1}(k)\}$. $\{u^o(k)\}$ is a transient signal equal to the auxiliary error y(k) given by (4.105) and tends asymptotically to zero (as the auxiliary error). Then

$$y_M^o(k) = -C_1^*(q^{-1})y_{M1}(k) + q^{-d}D(q^{-1})r(k)$$

$$y_{M1}(k) = y_{M0}(k) + u^o(k)$$

where

$$C_1^*(q^{-1}) = C_1(q^{-1}) - 1$$

The a priori and a posteriori reference model is an extension of the a priori and a posteriori adjustable system considered by Landau [1,29,32] for discrete-time MRAS identifiers.

4.6.6 Relation to Other Algorithms

It is possible to establish the correspondence of this algorithm with various adaptation algorithms given in the literature:

Algorithms Given by Ionescu and Monopoli [3] and Bénéjean [23]

Consider the algorithm with $H_1(q^{-1}) = 1$, $C_1(q^{-1}) = C_2(q^{-1}) = H_2(q^{-1})$, and $L(q^{-1})$ such that the positivity condition (4.113) is satisfied. Then (4.114) can be written as

$$\nu(k) = \frac{H_2(q^{-1})[y_P(k) - y_{Mf}(k)] + [\hat{p}(k-d) - \hat{p}(k-1)]^T \phi_f(k-d) + \alpha'(k)}{1 + \phi_f^T(k-d)F_k\phi_f(k-d)} \tag{4.116}$$

where

$$\alpha'(k) = L^*(q^{-1})\{[\hat{p}(k-d) - \hat{p}(k-1)]^T\phi_f(k-d)\} - H_2^*(q^{-1})\nu(k) \tag{4.117}$$

with $y_{Mf}(k)$ given by (4.92) and (4.78) and $L^*(q^{-1}) = L(q^{-1}) - 1$.

This is a generalization of the algorithm given by Ionescu and Monopoli [3] and Bénéjean [23].

Algorithms Given by Egardt [22, 28]

The algorithm (4.114) corresponds to that considered by Egardt, with the choice $C_2 = TC_1$, and $H_2 = 1$ (where T is the observer polynomial introduced by Egardt) and with the notation

$$C_2 = TA^M, \quad C_1 = A^M, \quad L = P, \quad H_1 = Q, \quad H_2 = 1$$

Egardt has used a stochastic approximation algorithm.

Generalization of the Scheme of Goodwin et al. [33]

If we choose $H_1(q^{-1}) = H_2(q^{-1}) = L(q^{-1}) = C_2(q^{-1}) = 1$, then we get a generalization of the algorithm using recursive least squares given by Goodwin et al. where $y^*(t+d) = y_M(t+d)$.

Minimum-Variance Self-Tuning Regulator of Åström and Wittenmark [24]

The algorithm (4.114) is identical to the minimum-variance STR with the choice of $C_2(q^{-1}) = H_2(q^{-1}) = H_1(q^{-1}) = L(q^{-1}) = 1$ (for regulation $y_M(k) \equiv 0$), and for $\lambda_1(k) = 1$ and $\lambda_2(k) = \lambda =$ constant. A reference can be made also to the work of Dugard and Landau [31].

It is also possible to obtain from the algorithms (4.114), different algorithms, such as (1) a series-parallel MRAC algorithm, (2) a parallel MRAC algorithm with a linear compensator, and (3) an independent tracking and regulation algorithm. The regulation behavior of the various MRAC algorithms may also be studied. Extensive simulations of various algorithms have been made by Landau and Lozano [14] for both tracking and regulation and a comparison is made of the performance of the various types of adaptation algorithms.

4.7 MULTIVARIABLE SYSTEMS

4.7.1 Introduction

Goodwin et al. [33] introduce a new approach which does not use augmented errors or auxiliary inputs. The algorithms are relatively simple and are applicable to multi-input/multi-output systems with rather general assumptions. Global convergence has been established for several adaptive control algorithms. One of the important assumptions is that a minimum-delay output predictor possessed a stable inverse. This assumption is rather restrictive. It is simpler if we assume that the system transfer function possesses a stable inverse. In a subsequent paper [34], Goodwin and Long generalize the results of Ref. 33 to include the stably invertible case. This extension is based on knowledge of the "system interactor matrix" [35] (to be defined later). This interactor matrix must be Hurwitz. The results of Ref. 33 are shown to be a special case of this general approach, when the interactor matrix is diagonal. In fact, in practice the interactor matrix is often diagonal. In the following section we consider the approach given in Goodwin and Long [34].

4.7.2 Key Technical Lemmas

Consider a linear time-invariant finite-dimensional system, having a strictly proper transfer function T(z).

The objective is to find an adaptive control algorithm which is globally convergent in the sense that for any initial system state and for any initial parameter values, the system input and output remain bounded for all time, and the output vector $y_P(t)$ asymptotically tracks a desired output vector $y_M(t)$.

Although we do not assume that the system is stable, we require that the system be stably invertible.

We require the following three lemmas for the analysis:

LEMMA 1. Output function controllability is possible if and only if the rank of T(z) is equal to the number of outputs.

In view of this, we assume, for simplicity, that the number of inputs is equal to the number of outputs. However, if this is not true, we select only those inputs for which the corresponding columns of T(z) are linearly independent. This leads to the following lemma:

LEMMA 2. For any full rank $m \times m$ transfer matrix T(z), there exists a unique nonsingular $m \times m$ lower left triangular matrix $\xi_T(z)$, known as the <u>interactor matrix</u>, which is of the form:

$$\xi_T(z) = H_T(z)\,\mathrm{diag}[z^{f_1}, \ldots, z^{f_m}] \tag{4.118}$$

where

$$H_T(z) = \begin{bmatrix} 1 & \cdots & 0 \\ h_{21}(z) & \cdots & \\ \vdots & & \\ h_{m1}(z) & \cdots & 1 \end{bmatrix} \tag{4.119}$$

and $h_{ij}(z)$ is divisible by z (or is zero) such that

$$\lim_{z \to \infty} \xi_T(z)T(z) = K_T \tag{4.120}$$

where K_T is nonsingular.

LEMMA 3. (1) In the discrete-time case, $\xi_T(z)^{-1}$ is a stable operator, and (2) for a strictly proper transfer matrix T(z), f_1 to f_m in (4.118) are nonzero.

The following assumptions are made:

Assumption a. $\xi_T(z)$ is known for the system.

Assumption b. T(z) has a stable inverse.

In Goodwin et al. [33] it was further required that $\xi_T(z)$ be diagonal. However, the approach here is more general, since $\xi_T(z)$ need not be diagonal.

4.7.3 Adaptive Control Algorithms

Since $\xi_T(z)$ is known for the system, we define the discrete-time model:

$$y_P(t) = \xi_T(q)y_P(t) \tag{4.121}$$

where q denotes a forward-shift operator such that $\{qy_P(t)\} = \{y_P(t+1)\}$.

We note that $\{y_P(t)\}$ is related to $\{u(t)\}$ by the transfer function, $\xi_T(z)T(z)$. This leads to the following lemma:

LEMMA 4. There exists a pair of left prime polynomial matrices $\tilde{A}(z)$ and $\tilde{B}(z)$ such that

$$\xi_T(z)T(z) = \tilde{A}(z)^{-1}\tilde{B}(z) \tag{4.122}$$

$$\tilde{A}(z) \text{ is row proper} \tag{4.123}$$

$$\deg_{r_i}[\tilde{B}(z)] = \deg_{r_i}[\tilde{A}(z)] \tag{4.124}$$

where $\deg_{r_i}[\tilde{B}(z)]$ denotes the degree of the ith row of $\tilde{B}(z)$ and

$$\tilde{B}(z) \text{ is row proper} \tag{4.125}$$

The transfer function $\xi_T(z)T(z)$ relating $\{y_P(t)\}$ to $\{u(t)\}$ is stably invertible (4.126)

Next we require the use of the following theorem:

THEOREM 1. There exists a positive integer d and a pair of polynomial matrices $\alpha(q^{-1})$ and $\beta(q^{-1})$ in the unit delay operator q^{-1} such that

$$\hat{y}_P(t + d) = \alpha(q^{-1})\hat{y}_P(t) + \beta(q^{-1})u(t) \tag{4.127}$$

where

$$\hat{y}_P(t + d) = y_P(t) \tag{4.128}$$

$$\hat{y}_P(t) = \text{known function of } y_P(t),\ y_P(t - 1),\ \ldots \tag{4.129}$$

$$\beta(q^{-1}) = \beta_0 + \beta_1 q^{-1} + \cdots + \beta_{n_2} q^{-n_2}, \quad \text{with } \beta_0 \text{ nonsingular} \tag{4.130}$$

We shall use the model (4.127) to develop the adaptive control algorithms. We require another assumption:

Assumption c. The upper bounds n', n_2' are known for the orders of the polynomials $\alpha(q^{-1})$ and $\beta(q^{-1})$ in (4.127).

As stated earlier, our objective is to determine an adaptive control law such that $\{y_P(t)\}$ and $\{u(t)\}$ remain bounded for all time and $\{y_P(t)\}$ tracks a desired output sequence $\{y_M(t)\}$ asymptotically.

In view of Assumption a, we define

$$\hat{y}_M(t + d) = \xi_T(q)y_M(t) \tag{4.131}$$

We assume that $\hat{y}_M(t + d)$ is known or can be calculated at time t. We also note that $\hat{y}_P(t)$ is a known function of $y_P(t)$, $y_P(t - 1)$, ... as per (4.129).

There are many possible ways to develop adaptive control algorithms. However, we consider two projection algorithms, I and II, which are generalizations of those given in Goodwin et al. [33].

Projection Algorithm I

Let us rewrite (4.127) in the form

$$\hat{y}(t + d) = \theta_0 \psi(t) \tag{4.132}$$

where $[\theta_0]$ is an $m \times (n_1' + n_2')$ matrix consisting of the coefficients of $[\alpha(q^{-1}), \beta(q^{-1})]$ and $\psi(t)$ is an $(n_1' + n_2') \times 1$ column vector given by

$$\psi(t) = [\hat{y}_P(t), \hat{y}_P(t-1), \ldots, u(t), u(t-1), \ldots]^T \tag{4.133}$$

these components being measurable.

The above suggests the following adaptive control algorithm:

$$\hat{\theta}(t) = \hat{\theta}(t-1) + \frac{\alpha(t)[y_P(t) - \hat{\theta}(t-1)\psi(t-d)]\psi(t-d)^T}{1 + \psi(t-d)^T\psi(t-d)} \tag{4.134}$$

$$\hat{y}_M(t+d) = \hat{\theta}(t)\psi(t) \tag{4.135}$$

It may be noted that (4.135) represents a feedback control law, as can be seen by rearranging the equation such that u(t) is the subject of the formula. We then have the following lemma:

LEMMA 5. Provided $\hat{\theta}(0)$ is chosen so that the coefficient matrix of u(0) in (4.135) is nonsingular, then $\alpha(t)$ can be chosen in the range $\epsilon < \alpha(t) < 2 - \epsilon$, $0 < \epsilon < 1$, so that the coefficient of u(t) is nonsingular for all t.

Next, we have the following theorem:

THEOREM 2. Subject to Assumptions a, b, and c, if algorithm (4.134), (4.135) is used, then $\{y(t)\}$, $\{u(t)\}$ are bounded for all time and

$$\lim_{t \to \infty} |y_{P_i}(t) - y_{M_i}(t)| = 0, \qquad \text{for } i = 1, \ldots, m$$

Proofs of the lemmas and theorems stated in this section are given in Goodwin and Long [34].

Projection Algorithm II

Next we consider an algorithm where the control law is estimated directly, unlike the case of the preceding algorithm. This approach is also used in Feuer and Morse [36] and involves the factorization of β_0 in (4.127). A related procedure is adopted in the self-tuning regulator [24], where it is assumed that the value of β_0 is known.

In view of (4.130), we can factor β_0 from the right-hand side of (4.127):

$$\hat{y}_P(t+d) = \beta_0\{u(t) + D(q^{-1})u(t-1) + C(q^{-1})\hat{y}_P(t)\} \tag{4.136}$$

where

$$D(q^{-1}) = \beta_0^{-1}\beta(q^{-1}) - I \tag{4.137}$$

$$C(q^{-1}) = \beta_0^{-1}\alpha(q^{-1}) \tag{4.138}$$

Subtracting $\hat{y}_M(t + d)$ from both sides of (4.136), we have

$$e(t+d) = \beta_0\{u(t) + D(q^{-1})u(t-1) + C(q^{-1})\hat{y}_P(t) - \beta_0^{-1}\hat{y}_M(t+d)\} \tag{4.139}$$

where

$$e(t + d) = \hat{y}_P(t + d) - \hat{y}_M(t + d) \tag{4.140}$$

We can rewrite (4.139) in the form

$$e(t + d) = \beta_0\{u(t) - \theta_0'\phi(t)\} \tag{4.141}$$

where θ_0' is an $m \times (n_1' + n_2')$ matrix consisting of the coefficients in $[C(q^{-1}), D(q^{-1})]$, and $\phi(t)$ is an $(n_1' + n_2') \times 1$ column vector given by

$$\phi(t) = [\hat{y}_P(t), \hat{y}_P(t - 1), \ldots, u(t - 1), u(t - 2), \ldots, \hat{y}_M(t + d)]^T \tag{4.142}$$

these components being measurable.

This suggests the following algorithm:

$$\hat{\theta}(t + d) = \hat{\theta}(t) - Pe(t + d)[1 + \phi(t)^T\phi(t)]^{-1}\phi(t)^T \tag{4.143}$$

$$u(t) = \hat{\theta}(t)\phi(t) \tag{4.144}$$

where P is a real square matrix specified a priori. P represents an a priori estimate of β_0^{-1}.

The preceding algorithm is actually d interlaced algorithms, depending on d initial conditions, $\hat{\theta}(0), \ldots, \hat{\theta}(d - 1)$.

We have the following theorem:

THEOREM 3. Subject to Assumptions a, b, and c and provided that

$$(P\beta_0)^T + (P\beta_0) - (P\beta_0)^T(P\beta_0)$$

is positive definite, then the algorithm (4.143), (4.144) ensures that $\{y(t)\}$, $\{u(t)\}$ are bounded and

$$\lim_{t \to \infty} |y_{P_i}(t) - y_{M_i}(t)| = 0, \qquad \text{for } i = 1, \ldots, m$$

4.7.4 Discussion

The preceding approach is the generalization for multivariable discrete-time adaptive control systems, based on a knowledge of the system interactor matrix. Although the analysis has been carried out for deterministic linear systems, it can be extended to certain classes of nonlinear systems of known form, which satisfy certain conditions, as cone-bounded nonlinearities do. Reference may be made to Section 8.8, where some recent trends in multivariable systems are discussed.

4.8 HYBRID ADAPTIVE CONTROL

4.8.1 Elliott's Approach

A recent development [37, 38] is the discrete-time adaptive control of continuous-time systems. Instead of making the entire compensator discrete, the compensator is separated into two parts: a continuous-time adjustable control structure and a discrete-time adaptation mechanism for updating the controller parameters. Thus the overall adaptive system is hybrid. In the conventional sampled-data approach where the continuous-time plant is modeled as a discrete-time system, stability is dependent on the sampling rate and so an appropriate rate should be chosen to assure stability. This will require additional a priori information on the system bandwidth. An important advantage of the hybrid approach is that stability is not dependent on the sampling rate. The structure of such a hybrid adaptive structure is shown in Fig. 4.8.

In Fig. 4.8 a continuous-time adjustable controller is used to control the plant. The discrete adaptation block samples the control input and plant output and uses this information to update discretely the adjustable controller gains. Because of the availability of digital-to-analog multiplying integrating circuits, it is now possible for one computer to monitor and update many control loops simultaneously. Elliott has demonstrated the stability of such a structure for the SISO case [37], and the stability results are extendable to the multivariable case [38].

Using a random sampling scheme, Elliott has shown that the combination of discrete adaptation algorithm and continuous-time adjustable controller gives a stable system, with all internal signals bounded and the parameter identification error converging to zero. Since stability is unaffected by the average sampling rate, we are assured of asymptotic stability

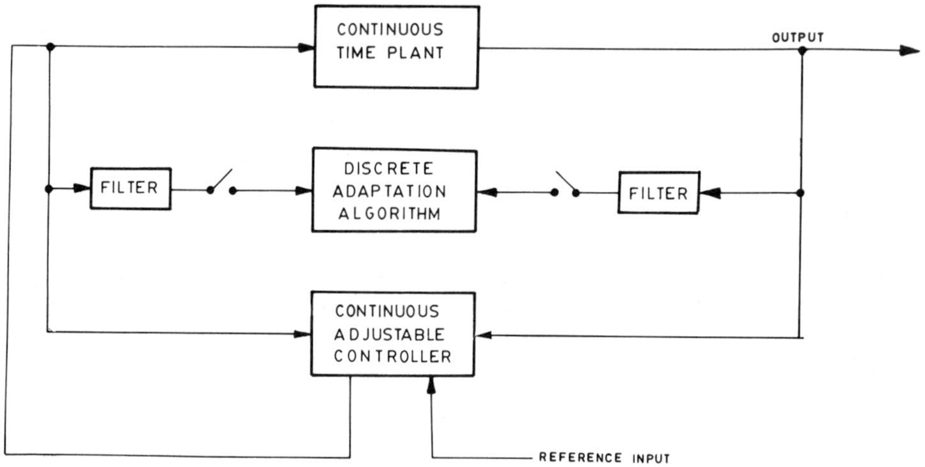

FIG. 4.8 Hybrid adaptive control. (From Ref. 37. Copyright © 1982 IEEE.)

even if the plant input and output are sampled infrequently. Stability analysis has been done utilizing some important results of Goodwin et al. [33] for discrete-time adaptive systems and of Narendra et al. [19] and Morse [39] for continuous-time systems.

Following Elliott [37], we can describe the dynamical behavior of the unknown plant (to be controlled) by a differential operator representation of the form:

$$\begin{aligned} D_P(p)x(t) &= u(t) \\ y_P(t) &= g_P N_P(p)x(t) \end{aligned} \tag{4.145}$$

where $u(t)$ and $y_P(t)$ are the plant input and output signals, respectively, $x(t)$ and its derivatives represent the state of the system, and $D_P(p)$ and $N_P(p)$ are monic, relatively prime polynomials. The degree of $D_P(p)$ is m_P and that of $N_P(p)$ is l_P, so that the relative degree is $(m_P - l_P)$. It is assumed that the relative degree $(m_P - l_P) \geq 1$, and $m_P \leq n$, the upper bound on the order of the plant. We also assume an upper bound g^u of the magnitude of g_P so that $g^u \geq |g_P|$. It is also required that $N_P(p)$ be Hurwitz. Without loss of generality, we assume $g_P > 0$.

The output y_P of the plant is required to track the output y_M of the model described by

$$D_M(p)y_M(t) = g_M N_M(p)r(t) \tag{4.146}$$

where $r(t)$ and $y_M(t)$ are the reference input to the model and the output of the model, respectively. The polynomials $D_M(p)$ and $N_M(p)$ are monic and of degrees m_M and l_M, respectively. The relative degree $(m_M - l_M) \geqslant (m_P - l_P)$. It is also assumed that $D_M(p)$ is Hurwitz and is factorable as $D_M(p) = D_{M_1}(p) \cdot D_{M_2}(p)$, where the degree of $D_{M_1}(p)$ equals $(m_P - l_P)$.

Consider the control law comprising the following three equations:

$$D_{M_2}(p)\bar{r}(t) = g_M N_M(p) r(t) \tag{4.147}$$

$$q(p)s(t) = k^*(p)u(t) + h^*(p)y_P(t) \tag{4.148}$$

$$u(t) = s(t) + g^* r(t) \tag{4.149}$$

where $q(p)$ is an arbitrary Hurwitz polynomial of degree $(n - 1)$,

$$h^*(p) = \sum_{i=0}^{n-1} h_i^* p^i \tag{4.150}$$

$$k^*(p) = \sum_{i=0}^{n-2} k_i^* p^i \tag{4.151}$$

These three satisfy the polynomial equation:

$$[h^*(p) g_P N_P(p) + k^*(p) D_P(p)] = q(p)[D_P(p) - D_{M_1}(p) N_P(p)] \tag{4.152}$$

If $m_P = n$, then solutions $h^*(p)$ and $k^*(p)$ for (4.152) exist and are unique. However, if $m_P < n$, solutions exist but are not unique [40].

When (4.152) holds good, the control law represents the combination of input dynamics (4.147) and a Luenberger observer (4.148), which produces closed loop poles corresponding to the stable polynomial $q(p)D_{M_1}(p)N_P(p)$ (where the zeros of $q(p)$ represent the uncontrollable observer poles) [40]. If g^* of (4.149) equals $1/g_P$, then $y_P(t)$ will converge exponentially to $y_M(t)$ as desired.

We can therefore implement adaptive model matching by producing a control law which converges to (4.147)-(4.149).

Consider a parameter adaptive law of the form given by the following equations:

$$q(p)\bar{u}(t) = u(t) \tag{4.153}$$

$$q(p)\bar{y}_P(t) = y_P(t) \tag{4.154}$$

$$D_{M_2}(p)\bar{r}(t) = g_M N_M(p)r(t) \tag{4.155}$$

$$u(t) = \sum_{i=0}^{n-1} h_i(t)p^i \bar{y}_P(t) + \sum_{i=0}^{n-2} k_i(t)p^i \bar{u}(t) + g(t)\bar{r}(t) \tag{4.156}$$

If

$$\lim_{t\to\infty} h_i(t) = h_i^*, \qquad i = 0, 1, \ldots, n-1 \tag{4.157}$$

$$\lim_{t\to\infty} k_i(t) = k_i^*, \qquad i = 0, 1, \ldots, n-2 \tag{4.158}$$

$$\lim_{t\to\infty} g(t) = g^* \equiv \frac{1}{g_P} \tag{4.159}$$

then the control law given by (4.153)-(4.156) will converge to that given by (4.147)-(4.149).

Elliott [37] next gives a method of obtaining an adaptation algorithm for adjusting the parameters $h_i(t)$, $k_i(t)$, and $g(t)$ contained in the control law given in (4.153)-(4.156). Specifically, this method involves constructing a measure $\epsilon(t)$ of the controller parameter errors directly from filtered input/output data without explicit implementation of a reference model. Since this error signal is linear in the parameter estimates, the problem of adaptation becomes a problem of linear parameter estimation.

Elliott proposes the use of two alternative adjustment schemes of a projection type and a least-squares type. Simulation studies show that considerably improved convergence is possible with the least-squares algorithm. A stability analysis has also been done to prove the boundedness of all internal controller signals and the convergence to zero of the continuous-time identification error $\epsilon(t)$. Although it has been shown that the adaptive algorithm results in a stable system, it has not been possible to show that Eqs. (4.157)-(4.159) hold good, that is, that the parameter estimates converge to the optimal values. In general, this will be the case if the external input is persistently exciting. Furthermore, this analysis does not guarantee that the plant output converges to that of the model, although simulation studies are indicative of this. There is therefore scope for further work.

This approach is very useful in extending to multivariable case where there are many more parameters to be adjusted. An important advantage of this approach is that it results in rapid convergence, whereas in traditional continuous adaptive control of continuous plants, we require considerably more complex control structures for implementing rapidly converging schemes, and the hardware complexity grows rapidly with an increasing number of parameters [38]. For this reason the hybrid scheme involving discrete estimation algorithms seems appropriate, since digital hardware is becoming faster and cheaper. The adjustable controller remains analog in nature and the digital computer is used only to sample filtered input and output data and discretely update the adjustable controller gains. This area appears very promising and more work needs to be done.

Note: The symbols $D_P(p)$, $N_P(p)$, $y_P(t)$, $D_M(p)$, $N_M(p)$, and r(t) are used in the present section in place of $p_p(D)$, $r_p(D)$, y(t), $p_M(D)$, $r_M(D)$, and v(t), respectively, used by Elliott [37].

4.8.2 Error Models for Hybrid Adaptive Control

Narendra et al. [41] have developed analytical models of hybrid systems in which control parameters are adjusted in discrete time even as the continuous plant signals are processed in real time. The same algorithms can be also extended to control discrete-time plants so that the overall discrete-time system operates on two time scales: a fast time scale in which the system operates and a slow time scale in which the control parameters are updated. Continuous algorithms of an integral type [42] are also shown to belong to this class. From the practical viewpoint, these algorithms are attractive for their robustness; from a theoretical point of view, they provide a unified framework for the design of continuous, discrete, and hybrid adaptive systems.

In Narendra et al. [41], using a Lyapunov approach, different adaptive laws are developed for various error models which assure the boundedness of the parameter error vector. The conditions under which the output and parameter error vectors tend to zero asymptotically are obtained. These results are useful in the identification problem. Further, the behavior of the outputs of the various error models is studied when their inputs grow in an unbounded manner. These results will find a direct application in the adaptive control problem.

The main contribution of this paper lies in the proof of the global stability of a hybrid adaptive control system and in the fact that the same algorithms when suitably modified are known to be applicable to both discrete- and continuous-time systems with two time scales. It is also shown that the use of time-varying adaptive gain matrices in the algorithms does not affect the stability.

Two conceptually different approaches are used to develop the adaptive algorithms in the three error models. The two approaches are based on the manner in which the parameter error vector is adjusted. Simulation results for three typical examples are given to compare the effectiveness of these two adaptive approaches. In all cases the parameters are adjusted periodically with a period T. Simulations also show the effect of T on the speed and accuracy of the responses. It is seen that the first approach is better when the sampling rate is high compared to the dominant frequencies of the input, while the second approach is better when the sampling rate is reduced. It is also found that for a large sampling period, there is only a single adjustment of the adaptive parameter vector, after which it remains essentially constant; this behavior is usually observed in continuous-time algorithms with large adaptive gain.

Further work is needed for a better understanding of the relationship between the sampling period and the effectiveness of the two approaches. More work is also needed to combine the two approaches in time-varying and noisy situations.

Narendra et al. [43] consider the effect of bounded external disturbances and show that the hybrid schemes should result in more robust systems, and simulation results are presented to show that this is indeed the case. It is also shown that the sampling interval can itself be time-varying provided it is within a bounded interval $[T_{min}, T_{max}]$, with $T_{min} > 0$. In practical systems it seems to be possible to adjust the sampling interval on line to improve the transient response of the system.

4.9 CONCLUDING REMARKS

Because of the advent of digital implementation of many schemes, discrete-time MRAC systems have become important. Different methods for designing such systems are considered in this chapter, including a unified approach which shows the relationships among the different adaptation algorithms, including self-tuning algorithms. Multivariable systems and the recent development of hybrid adaptive control are also discussed. Hybrid adaptive control has great potential as the advantages of both continuous-time and discrete-time methods are exploited.

REFERENCES

1. Landau, I. D., Adaptive Control: The Model Reference Approach, Marcel Dekker, New York, 1979.
2. Sebakhy, O. A., A discrete model reference adaptive system design, Int. J. Contr., 23: 799-804 (1976).
3. Ionescu, T., and Monopoli, R. V., Discrete model reference adaptive control with an augmented error signal, Automatica, 13: 507-518 (1977).
4. Nagaraja, T., and Chalam, V. V., On the generation of Lyapunov function for discrete data systems, Proc. IEEE, 65: 1417-1418 (1977).
5. Banerjee, D., and Chalam, V. V., Lyapunov design of discrete data model reference adaptive control systems, Proceedings of International Conf. on System Science, Wroclaw, Poland (1983).
6. Banerjee, D., Lyapunov design of discrete data model reference adaptive control systems, Master of Tech. Thesis, Banaras Hindu University, Varanasi, India (1977).

7. Landau, I. D., Hyperstability concepts and their application to discrete control systems, Proc. JACC, Stanford University, Stanford, Calif. (1972).
8. Landau, I. D., Synthesis of hyperstable discrete model reference adaptive systems, Proc. Conf. on Circuits and Systems, University of Santa Clara, Santa Clara, Calif. (1971).
9. Ljung, L., On positive real transfer functions and the convergence of some recursive schemes, IEEE Trans. Auto. Contr., AC-22: 539-550 (1977).
10. Suzuki, T., and Takashima, S., A hyperstable scheme for discrete MRAC system, Int. J. Contr., 28: 245-252 (1978).
11. Unbehauen, H., Systematic design of discrete model reference adaptive systems, in Self-Tuning and Adaptive Control (C. J. Harris and S. A. Billings, eds.), Peter Peregrinus, London, 1981.
12. Schmid, Ch., Ein Beitrag zur Realisierung adaptiver Regelungsysteme mit dem Prozeßrechner, Doctoral Thesis, Ruhr-Universität, Bochum, F.R.G. (1979).
13. Monopoli, R. V., Model reference adaptive control with an augmented error signal, IEEE Trans. Auto. Contr., AC-19: 474-484 (1974).
14. Landau, I. D., and Lozano, R., Unification of discrete-time explicit MRAC designs, Automatica, 17: 593-611 (1981).
15. Landau, I. D., and Silveria, H. M., A stability theorem with applications to adaptive control, IEEE Trans. Auto. Contr., AC-24: 305-311 (1979).
16. Landau, I. D., An extension of a stability theorem applicable to adaptive control, IEEE Trans. Auto. Contr., AC-25: 814-817 (1980).
17. Irving, E., Barret, J. P., Charcossey, C., and Monville, J. P., Improving power network stability and unit stress with adaptive generator control, Automatica, 15: 31-46 (1979).
18. Narendra, K. S., and Lin, Y.-H., Stable discrete adaptive control, IEEE Trans. Auto. Contr., AC-25: 456-461 (1980).
19. Narendra, K. S., Lin, Y.-H., and Valavani, L. S., Stable adaptive controller design—Pt. II: Proof of stability, IEEE Trans. Auto. Contr., AC-25: 440-448 (1980).
20. Narendra, K. S., and Valavani, L. S., Stable adaptive controller design—Direct control, IEEE Trans. Auto. Contr., AC-23: 570-582 (1978).
21. Lin, Y.-H., and Narendra, K. S., A new error model for adaptive systems, IEEE Trans. Auto. Contr., AC-25: 585-587 (1980).
22. Egardt, B., Unification of some discrete-time adaptive control schemes, IEEE Trans. Auto. Contr., AC-25: 693-697 (1980).
23. Bénéjean, R., La commande adaptive à modèle de référence evolutif (application à la commande sans accès aux variables d'état), Doctoral Thesis, 3me cycle, Institut National Polytechnique de Grenoble, France (1977).

24. Åström, K. J., and Wittenmark, B., On self-tuning regulators, Automatica, 9: 185-199 (1973).
25. Åström, K. J., and Wittenmark, B., Self-tuning controllers based on pole-zero placement, Proc. IEE, 127 Pt. D: 120-130 (1980).
26. Clarke, D. W., and Gawthrop, P. J., Self-tuning controller, Proc. IEE, 122: 929-934 (1975).
27. Johnson, C. R., Jr., and Tse, E., Adaptive implementation of one-step-ahead optimal control via input matching, IEEE Trans. Auto. Contr., AC-25: 865-872 (1978).
28. Egardt, B., Stability analysis of discrete data adaptive control schemes, IEEE Trans. Auto. Contr., AC-25: 710-716 (1980).
29. Landau, I. D., Unbiased recursive identification using model reference adaptive techniques, IEEE Trans. Auto. Contr., (a) AC-21: 194-202 (1976); (b) AC-23: 97-99 (1978).
30. Wittenmark, B., A self-tuning regulator, Doctoral Thesis, Department of Auto. Contr., Lund Inst. of Tech., Lund, Sweden (1973).
31. Dugard, L., and Landau, I. D., Stochastic model reference adaptive controllers, Proceedings of IEEE-CDC, Albuquerque, N.M. (1980).
32. Landau, I. D., Adaptive controllers with explicit and implicit reference models and stochastic self-tuning regulators—Equivalence and duality aspects, Proceedings of IEEE-CDC, San Diego, Calif. (1979).
33. Goodwin, G. C., Ramadge, P. J., and Caines, P. E., Discrete-time multivariable adaptive control, IEEE Trans. Auto. Contr., AC-25: 449-456 (1980).
34. Goodwin, G. C., and Long, R. S., Generalization of results on multivariable adaptive control, IEEE Trans. Auto. Contr., AC-25: 1241-1245 (1980).
35. Wolovich, W. A., and Falb, P. L., Invariants and canonical forms under dynamic compensation, SIAM J. Contr. Optim., 14: 996-1008 (1976).
36. Feuer, A., and Morse, A. S., Adaptive control of a SISO linear system, IEEE Trans. Auto. Contr., AC-23: 557-569 (1978).
37. Elliott, H., Hybrid adaptive control of continuous-time systems, IEEE Trans. Auto. Contr., AC-27: 419-426 (1982).
38. Elliott, H., and Wolovich, W. A., A parameter adaptive control structure for linear multivariable systems, IEEE Trans. Auto. Contr., AC-27: 340-352 (1982).
39. Morse, A. S., Global stability of parameter-adaptive control systems, IEEE Trans. Auto. Contr., AC-25: 433-439 (1980).
40. Wolovich, W. A., Linear Multivariable Systems, Springer, New York, 1974.
41. Narendra, K. S., Khalifa, I. H., and Annaswamy, A. M., Error models for stable hybrid adaptive systems, IEEE Trans. Auto. Contr., AC-30: 339-347 (1985).

42. Kreisselmeier, G., On adaptive state regulation, IEEE Trans. Auto. Contr., AC-27: 3-17 (1982).
43. Narendra, K. S., and Annaswamy, A. M., Robust adaptive control, Tech. Report, Center for Systems Science, New Haven, Conn. (1985).

BIBLIOGRAPHY

Bethoux, G., and Courtiol, B., A hyperstable discrete model reference control system, Proceedings 3rd IFAC Symp. on Sensitivity, Adaptivity and Optimality, Ischia, pp. 282-289 (1973).

Colburn, B. K., and Boland, J. S., III, Analysis of error convergence rate for a discrete model reference adaptive system, IEEE Trans. Auto. Contr., AC-21: 773-775 (1976).

Cristi, R., and Monopoli, R. V., Computational aspects of discrete-time adaptive control, IEEE Trans. Auto. Contr., AC-27: 722-725 (1982).

Elliott, H., and Wolovich, W. A., Adaptive control of stably invertible discrete multivariable systems: The pole-placement approach, Tech. Report No. FB 82-DELENG-1, Colorado State University (1982).

Fuchs, J. J.-J., Commande adaptive directe des systemes linearies discrets, Doctoral Thesis, University of Rennes, France (1979).

Fuchs, J. J.-J., Discrete adaptive control: A sufficient condition for stability and applications, IEEE Trans. Auto. Contr., AC-25: 940-946 (1980).

Johnson, C. R., Jr., Rephrasing the discrete model reference adaptive control problem, Proceedings 13th Conf. on Information Science and Systems, Baltimore, Md., pp. 351-355 (1979).

Kim, C. G., and Gourishankar, V., A model reference adaptive technique of identification for discrete systems, Automatica, 17: 637-640 (1981).

Landau, I. D., Design of discrete model reference adaptive systems using positivity concept, Proceedings 3rd IFAC Symp. on Sensitivity, Adaptivity, and Optimality, Ischia, pp. 307-314 (1973).

Landau, I. D., and Bethoux, G., Algorithms for discrete-time model reference adaptive systems, Proceedings 6th IFAC Congress, Boston, Mass., Part ID, 58.4.1-58.4.11 (1975).

Landau, I. D., and Lozano, R., Design and evaluation of discrete-time explicit model reference adaptive control for tracking and regulation, Report No. LA6, 79, 14, Institut National Polytechnique de Grenoble, France (1979).

Lozano, R., and Landau, I. D., Redesign of explicit and implicit discrete-time model reference adaptive control schemes, Int. J. Contr., 33: 247-268 (1980).

Matko, D., Some relations in discrete adaptive control, in Methods and Applications in Adaptive Control (H. Unbehauen, ed.), Springer-Verlag, Berlin, pp. 31-40, 1980.

Samson, C., Discrete model reference adaptive control of non-minimum phase systems, Opt. Contr. Appl. Meth., 3: No. 4 (1982).

Subbayyan, R., and Nagarajan, R., A discrete-time model reference adaptive control scheme with reduced on-line computations, Int. J. Contr., 30: 121-127 (1979).

Thathachar, M. A. L., and Gajendran, F., A global convergence criterion for discrete-time and multivariable adaptive systems, Int. J. Contr., 34: 127-142 (1981).

Wolovich, W. A., and Elliott, H., Discrete models for linear multivariable systems, Tech. Report No. ENG-MA 82, Brown University, Providence, R.I. (1982).

5

Self-Tuning Regulators

5.1 INTRODUCTION

Self-tuning controllers represent an important class of adaptive controllers; they are easy to implement and are applicable to complex processes with a wide variety of characteristics involving unknown parameters, the presence of time delay, time-varying process dynamics, and stochastic disturbances.

The original idea of self-tuning was first given by Kalman [1] in 1958, in his classic paper on the design of a self-optimizing control system. He tried to realize the algorithm in a special-purpose computer. However, because of the unavailability of sophisticated computers and inadequately developed theory, this technique was not taken up seriously until Peterka [2] extended the idea in his paper in 1970 on adaptive digital regulation of noisy systems. The major breakthrough came with the self-tuning regulator (STR) of Åström and Wittenmark [3] in 1973. This led to a series of significant developments in the industrial application of STRs, especially due to the advent of microprocessors, which make implementation of STR algorithms relatively easy. Theoretical developments have also occurred, giving different approaches to the formulation of STR algorithms.

Whereas in MRAC the basic idea is to asymptotically drive the output of an unknown plant to that of a reference model, in self-tuning the basic procedure is to select a design for known plant parameters and apply it to the unknown plant using recursively estimated values of these parameters. Section 5.2 deals with different approaches to self-tuning.

In developing STR algorithms, we require some knowledge of stochastic control-theoretic concepts, discussed in Section 4.3, and of recursive parameter estimation, discussed in Section 5.4. Section 5.5 deals with implicit self-tuning regulators and includes the minimum-variance STR and the generalized minimum-variance STR, as well as algorithms for the same. Explicit self-tuning regulators are described in Section 5.6 with special reference to the pole placement approach as well as the pole-zero placement method. Unification of different algorithms and equivalence between STR and MRAC are also discussed. LQG self-tuning is considered in Section 5.7, where explicit and implicit approaches are outlined. Section 5.8 gives two design procedures to handle plants displaying non-minimum-phase behavior which is commonly encountered in many industrial plants. The recent development of hybrid self-tuning is dealt with in Section 5.9. A brief account is given of convergence analysis in Section 5.10 and some aspects of self-tuning implementation are treated in Section 5.11. Concluding remarks are given in Section 5.12.

5.2 DIFFERENT APPROACHES TO SELF-TUNING

5.2.1 Minimum-Variance STR

The original STR developed by Åström and Wittenmark [3] is based on the stochastic minimum-variance principle. This is an approach using the optimal control objective. The major assumptions are that the system is minimum phase, that the time delay is known, and that a bound can be given to the order of the system. The first two assumptions lead to a simple algorithm, which can be easily implemented on a microcomputer. The feasibility was demonstrated by experiments on real processes in the paper and mining industries. The performance objective is the minimization of the variance of the process output measured at sampling instants. This objective is good enough for a restricted class of applications, as in paper making, where consistent thickness is to be maintained in spite of disturbances. The only modification the user can make is to vary the sample time. As no account is taken of the control effort required, there is a possibility of large control signals leading to instability. Instability may also result from fast sampling or large fractional delay. Hence such an STR should be used with caution. It must, however, be stated that this important paper gave rise to a new confidence in the applicability of adaptive control and was the starting point for a number of developments leading to more general STRs, and more and more industrial applications.

Clarke and Gawthrop [4,5] and Gawthrop [6] extended the preceding idea to cover a wide variety of performance objectives. This approach includes set-point variations so that both tracking and regulation are feasible. This involves minimization of a combination of control and output variances. Clarke's approach can also be used as a "cut and try" procedure to control non-minimum-phase plants, which is not possible with Åström's STRs. Further work on minimum-variance STRs was reported by Åström et al. [7]. Grimble [8] suggests weighted minimum-variance ST control. This allows a wider range of controls than is obtained by costing only output variance. The controller is also stable for ranges of control weighting where other control laws are unstable. When the system is minimum phase, the algorithm developed by Grimble is equivalent to the generalized minimum-variance ST controller developed by Clarke and Gawthrop. However, such algorithms lead to greater complexity to allow for the non-minimum-phase terms in the plant. A multivariable version of this type of ST controller was developed by Grimble and Fung [9]. A weighted minimum-variance controller suitable for use in nonlinear systems even when the exact form of the nonlinearities and dynamics are unknown was suggested by Grimble [10]. Anbumani et al. [11] investigate the self-tuning minimum-variance control of nonlinear systems of the Hammerstein model. Peterka [12] proposed predictor-based self-tuning control. Toivonen [13] discusses minimum-variance control with a constraint on the input amplitude.

5.2.2 Implicit and Explicit Self-Tuning

Minimum-variance self-tuners are called _implicit_ STRs because they are based on estimation of an implicit process model. In the implicit algorithms the parameters are updated directly. Such implicit STRs are developed on the basis of predictive control theory and require knowledge of the system time delay. To avoid this, _explicit_ STRs are developed in which the time delay can be estimated as part of the process dynamics. However, such explicit methods involve more computation. In explicit self-tuning, parameter identification is followed by separate control law calculation. Accuracy of the parameters is important for the control law calculation. Explicit STRs correspond to _indirect_ adaptive control [14]. An explicit self-tuning approach is discussed by Fuchs [15].

5.2.3 Pole-Placement STR

The explicit self-tuning strategy based on pole-placement design was studied by Edmunds [16] and Wouters [17]. This approach was further extended by Wellstead and his colleagues [18]. This method involves the placement of the closed-loop poles at prescribed locations which define a required transient response, while leaving the zeros in their open-loop positions. This approach is somewhat akin to classical control methods and has some immediate appeal. Robustness is the essential advantage of this method. Further, such STRs can also be applied to non-minimum-phase plants and this is a distinct advantage, as non-minimum-phase behavior is encountered in many practical applications. Another important feature of pole-assignment STRs is their ability to regulate systems with unknown and even varying transport delays which cannot be easily handled by other methods. Åström et al. [19] also consider pole-placement controllers, but this work differs from that of Wellstead et al. [18] in that it deals only with the servo problem and no stochastic disturbances are considered in deriving the controller. Further this method requires on-line polynomial factorization. Wellstead and his coworkers also report further work [20-23]. Wittenmark [24] discusses self-tuning PID controllers based on pole placement.

Allidina and Hughes [25] propose a generalized self-tuning controller, combining the pole-placement approach and Clarke's approach so as to yield robustness and ease of set-point tracking. Both unstable and non-minimum-phase characteristics can be handled. But varying time delays cannot be dealt with. It may be mentioned that none of the pole-placement techniques can deal efficiently with unknown disturbances, that is, by estimating the disturbance and compensating it directly.

A pole-zero-assignment procedure has been suggested by Åström and Wittenmark [26]. In this method, "safe" process zeros are also placed in the case of a deterministic servo. Vogel and Edgar [27,28] develop the adaptive pole-zero-placement, dead-time compensator which is effective with systems exhibiting varying time delay. It is also robust and is applicable to non-minimum-phase systems. It exhibits the ease of set-point tracking of the STC.

5.2.4 LQG Self-Tuning Regulators

Another approach to STR design is to transform the estimated model into its equivalent state space form so that the design procedure is that applied for linear-quadratic-Gaussian (LQG) control. LQG self-tuning is considered in Section 5.7. Such an LQG design was first presented in 1973 by Peterka and Åström [29]. Their process model excluded colored noise processes. A microcomputer implementation of a LQG self-tuner is discussed by Zhao-Ying and Åström [30], and further work was reported by them [31]. The LQG method demands the most computation and there are possibilities of minimizing the computing load, as proposed by Lam [32,33], avoiding the numerical problems involved in iterating the Riccati equation. An explicit LQG regulator has been described by Salcudean and Belanger [34], using both state space and input/output models. An interesting related linear quadratic adaptive control problem has been considered by Kumar [35] in which there is a finite set of parameterized systems, one of which (known) is the true system. Grimble [36] deals with implicit LQG self-tuners, which result in reduced computing load, as such schemes avoid the controller calculation stage involved in explicit LQG schemes. Grimble [37] considers a robust LQG design of discrete systems using a dual criterion. Reference to the LQG approach is also made by Grimble in other work [38,39].

5.2.5 STRs to Give PID Control

PID regulators are widely used in industry. It is common experience that many regulators are, in practice, poorly tuned. This is because simple robust methods are not available for tuning the regulators. Åström tackles this problem of finding automatic tuning methods for simple regulators with specifications on phase and amplitude margins [40,41]. The methods are based on the Ziegler-Nichols method [42], which is one of the classical approaches for tuning PID regulators. A simple identification method gives the critical points on the Nyquist plot. The key idea is a scheme which provides automatic excitation of the process which is nearly optimal for estimating the desired process characteristics. These algorithms are simple to implement using microprocessors and may be used in several ways. They may be included in single-loop controllers to provide an option for automatic tuning. They may also be used for safe initialization of self-tuning schemes. When combined with a bandwidth self-tuner, it is possible to obtain an adaptive regulator which sets a specified closed-loop bandwidth automatically.

Although conventional STRs may be configured to give PID control as proposed by Wittenmark and Åström [43] and Gawthrop [44], they have the disadvantage that information about the time scale of the process must be provided a priori to get a reasonable estimate of the sampling period in the regulator. The methods for tuning the sampling period automatically [45,46] work only for moderate changes in the process time constants. However, the method proposed by Åström and Hagglund [41] does not suffer from this disadvantage. It may be applied to processes with widely differing time scales. Results from laboratory and industrial experiments with the algorithm are also presented.

5.2.6 STRs for Non-Minimum-Phase Systems

As pointed out earlier, many industrial processes exhibit non-minimum-phase behavior and so STRs have to cope with this. Section 5.8 will deal with this problem. Studies of STRs for non-minimum-phase systems have been made by Åström and his colleagues [47-50], Samson et al. [51,52], Unbehauen et al. [53-55], Goodwin and Sin [56], Grimble [57], Elliott [58], Kumar and Moore [59], and others. Clarke [60] reviews various approaches for overcoming the problem of instability in non-minimum-phase systems and concludes that with appropriate design and engineering judgment, self-tuning control is also viable for non-minimum-phase processes, but points out the need for a study of robustness against unmodeled dynamics.

5.2.7 Hyrid Self-Tuning

Hybrid self-tuning combines the features of both the continuous-time and discrete-time systems to provide the advantages of both. Gawthrop [61,62] has developed the idea of hybrid self-tuning and there is considerable scope for further studies in this new area. Section 5.9 discusses this topic.

5.3 SOME STOCHASTIC CONTROL-THEORETIC CONCEPTS

The following concepts are relevant to the development of STR algorithms.

5.3.1 Dual Effect and Neutrality

The control is said to have a dual effect when, in addition to its effect on the state of the system, it affects the uncertainty of the state of the system [63]. That is, the present control might affect future state uncertainty.

If the control cannot affect this uncertainty, the system is called neutral [63]. In other words, the control system is neutral if the rate of reduction of uncertainty about states x is independent of the control u [64]. Hence in a neutral system a probing signal (discussed in Section 5.3.5) cannot affect the rate of reduction of uncertainty. A typical case of a neutral system is the linear-quadratic-Gaussian (LQG) system.

In a neutral system it is possible to "separate" the control part of the system from the estimation process. Further, the controller depends only on the past history of the system (i.e., past controls and states and not future ones) and is therefore implementable.

5.3.2 Separability

The property of separability implies that the system can be thought of as composed of two parts: one part estimates the state and the other part is the linear feedback from the estimated state to the control signal (Fig. 5.1). According to the separation principle, it is possible to make a separation between the identification of the parameters of the plant and the determination of the controller parameters. The parameters of the controller are also allowed to be functions of the uncertainties of the estimated parameters [65].

Some authors refer to the separable property in terms of a separation theorem. A version of the separation theorem was first published by Joseph and Tou [66], according to whom, a stochastic control problem is separable if the only information which needs to be transmitted is a point estimate $\hat{x}(i)$ of the current state, no information about accuracy being required. Therefore, separability implies neither probing to improve accuracy nor caution (discussed in Section 5.3.4) to allow for inaccuracy.

According to Saridis [67], the separation theorem is stated as follows: In linear systems with quadratic performance criteria and subjected to Gaussian inputs, the stochastic optimal feedback controller is synthesized by cascading an optimal estimator (say, Kalman-Bucy filter) with the deterministic optimal control.

The separation of estimation and control simplifies the implementation of the STR. This idea led a number of workers to approximate the solution of the general "nonneutral" problem by using the separation theorem. But such solutions do not guarantee a good approximation.

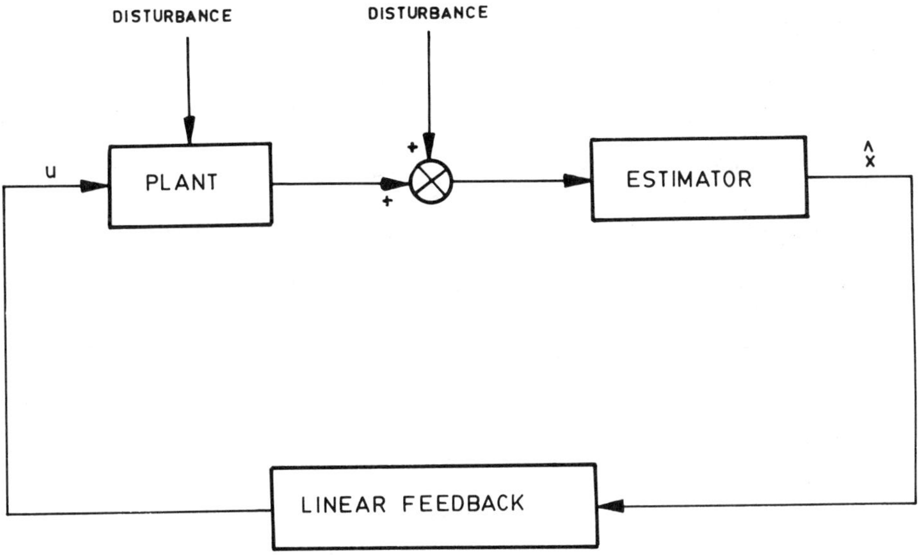

FIG. 5.1 A "separable" system.

5.3.3 Certainty Equivalence

The certainty equivalence principle makes it possible to first solve the deterministic problem with known parameters and then obtain the optimal controller for unknown parameters by substituting the true parameter values by the estimated values. Therefore, certainty equivalent control is that control which would be exercised if the estimated parameters were the actual ones [68, 69]. A separable problem is certainty equivalent [70, 71] when the form of the function is identical with that of the function specifying the optimal feedback control for an equivalent deterministic problem, having no uncertainties about the state. Certainty equivalence implies that the control law can be formulated without considering any stochastic effects [72] and does not require the knowledge of subsequent feedback or any reduction of the system uncertainties [67].

LQG problems are certainty equivalent. Some authors consider that certainty equivalence is a stronger property than either neutrality or separability [65, 73].

5.3.4 Caution

Because of the inherent uncertainties, the controller has to be "cautious" not to increase the effect of the existing uncertainties on the cost. A controller may exercise less "caution" if it "knows" that future observations will be available and corrective actions based upon them will be taken [63, 74].

According to Jacobs and Patchell [68], caution is that component of the control signal which recognizes that errors in the parameter estimates may cause excessive deviations in the control signal. This component takes into account the uncertainties in the estimated parameters and is a function of the estimated parameters and their accuracies. Therefore, cautious control exercises control less strongly than if there were no uncertainty [72].

Let u^* represent the optimal control and u_c the certainty equivalent control. Consider a hypothetical control u_n, called the neutral control [68]. This is the control which would be optimal if all interaction between current control and future uncertainty were ignored. Then the definition of "caution" [68] is given as the difference between the neutral control and certainty equivalent control, that is,

$$\text{Caution} \equiv u_n - u_c \tag{5.1}$$

5.3.5 Probing or Active Learning

A probing signal is that component of the control signal which is used to inject an optimal test signal into the process to improve the estimates of uncertain parameters [68]. When dual effect is present, the control can "help" in learning (estimation) by decreasing the uncertainty of the state. Therefore, that control which takes into account the future observation program has the capability of <u>probing</u> or <u>active learning</u> [63]. In terms of neutral control, the definition of "probing" [68] is given as the difference between optimal control and neutral control, that is,

$$\text{Probing} \equiv u^* - u_n \tag{5.2}$$

The significance of neutral control is that it provides a conceptual definition of probing and caution.

From (5.1) and (5.2), we get

$$u^* - u_c \equiv \text{caution} + \text{probing} \tag{5.3}$$

Equation (5.3) can be rewritten as

$$u^* = u_c + \text{caution} + \text{probing} \tag{5.4}$$

This equation shows [68] that the control signal can be decomposed into three components: certainty equivalent control, caution, and probing.

Caution and probing, as defined above, are theoretical concepts which cannot be generally evaluated for non-LQG problems, where they may be nonzero [73].

5.3.6 Sufficient Conditions

Consider a plant which can be represented by the following discrete-time equations relating the states x and outputs y_P to the inputs u and disturbances, g:

$$x(k+1) = F_1[x(k), u(k), g_1(k), k] \tag{5.5}$$

$$y_P(k) = F_2[x(k), g_2(k), k] \tag{5.6}$$

The argument of the vector functions F include k to show that the functions could be time varying. The noise variables g are "white" in the sense that their values at any instant are independent of their values at any other instant.

Consider that the following performance index is to be minimized:

$$J(k_1) = \sum_{k=k_1}^{k_2} H[x(k+1), u(k), k] \tag{5.7}$$

Necessary and sufficient conditions for the plant described by (5.5)-(5.7) to be neutral, or separable, or certainty equivalent are not known. However, as already stated, LQG problems have all these three properties. All the same, some sufficient conditions for each of these three properties are given in the literature [75]:

1. A system is neutral if the functions F of Eqs. (5.5) and (5.6) are linear [63].
2. A system is separable if, in addition to condition 1, all the random variables are Gaussian [76, 77].
3. A system is certainty equivalent if, in addition to condition 2, the function H of (5.7) is quadratic [78-80].
4. According to Wonham [77], the system is neutral and separable when Eq. (5.5) is written in a general linear form:

$$x(k+1) = Ax(k) + B[u(k)] + g_1(k) \tag{5.8}$$

In this case the control u is included as a nonlinear additive term; such a situation arises when $|u|$ is constrained.

5.3.7 Discussion

It is seen from the preceding considerations that linearity is an important part of the conditions above. Generally, adaptive control results in a nonlinear problem which does not satisfy the preceding sufficient conditions and so may not possess any of these properties. Hence, we should expect interactions between estimation and control, and these interactions must be taken into account when developing adaptive algorithms. That is why the properties of "caution" and "probing" are important in such designs. Jacobs and Patchell [68] have indicated that "caution" and "probing" may be needed for adaptive control. For instance, the self-tuning regulator shown in Fig. 1.5 is a certainty equivalent control because the fact that the current parameters estimates are not exact is disregarded and their uncertainties are ignored. It is therefore necessary to introduce modifications which also take into account the uncertainties of the parameter estimates, so that we have "cautious" control or introduce extra "probing" signals. However, the reason for not considering "caution" or "probing" is that although there may be poor control during the tuning phase, this can be minimized by taking adequate precautions, as the overall algorithm is considerably simplified. Many of the adaptive controls reported in the literature do not introduce either caution or probing and that may be why there are not many successful applications of adaptive control.

5.4 RECURSIVE PARAMETER ESTIMATION

5.4.1 Introduction

Self-tuning regulators are designed starting with a known system. The control algorithm is developed by introducing a recursive parameter estimator. Then the true parameter values are replaced by their estimated values to determine the control law. That is why this control strategy is a certainty equivalent control, considered in Section 5.3.

Many recursive estimation schemes can be used for parameter estimation, namely, stochastic approximation, least squares, extended and generalized least squares, maximum likelihood, instrumental variables, extended Kalman filtering, square-root algorithm, UD method. Some estimation methods will be discussed here. Several methods of parameter estimation are described by Eykoff [81], Sage and Melsa [82], Bierman [83], Åström and Eykoff [84] and Ljung and Söderström [85].

5.4. Plant Description

The input/output model of a single-input/single-output discrete-time system can be described by the difference equation:

$$
\begin{aligned}
y_P(t) &+ a_1 y_P(t-1) + \cdots + a_n y_P(t-n) \\
&= b_0 u(t-k) + b_1 u(t-k-1) + \cdots + b_n u(t-k-n) \\
&\quad + \xi(t) + c_1 \xi(t-1) + \cdots + c_n \xi(t-n)
\end{aligned}
\tag{5.9}
$$

where $y_P(t)$ is the output, u(t) is the input, and $\{\xi(t)\}$ is a sequence of independent equally distributed Gaussian variables with zero mean value and variance σ^2.

Let us introduce the following polynomials:

$$A(q^{-1}) = 1 + a_1 q^{-1} + \cdots + a_n q^{-n}$$

$$B(q^{-1}) = b_0 + b_1 q^{-1} + \cdots + b_n q^{-n} \tag{5.10}$$

$$C(q^{-1}) = 1 + C_1 q^{-1} + \cdots + C_n q^{-n}$$

where q is the forward shift operator,

$$q^k y(t) = y(t + k), \qquad \text{at sampling instant } t$$

Then we can write (5.9) in a compact form:

$$A(q^{-1})y_P(t) = B(q^{-1})u(t - k) + C(q^{-1})\xi(t) \tag{5.11}$$

We can also write in the following form:

$$A(q^{-1})y_P(t) = q^{-k}B(q^{-1})u(t) + C(q^{-1})\xi(t) \tag{5.12}$$

Equation (5.12) is a canonical form for describing SISO models, in the presence of noise. Some or all parameters of (5.12) may be unknown. For convenience, let us denote the vector of unknown parameters by θ.

The estimation of the plant parameters will be simplified if $C(q^{-1}) = 1$, so that $C_0 = 1$, $C_i = 0$, $i = 1, 2, \ldots, n$. Then the plant equation becomes

$$A(q^{-1})y_P(t) = B(q^{-1})u(t - k) + \xi(t) \tag{5.13}$$

If this plant has unit time delay, we get a simpler model:

$$A(q^{-1})y_P(t) = B(q^{-1})q^{-1}u(t) + \xi(t) \tag{5.14}$$

We can write in terms of the various parameters:

$$\begin{aligned} &y_P(t) + a_1 y_P(t - 1) + \cdots + a_n y_P(t - n) \\ &\quad = b_0 u(t - 1) + \cdots + b_{n-1} u(t - n) + \xi(t) \end{aligned} \tag{5.15}$$

The preceding equation is linear in parameters and the input is an uncorrelated zero-mean sequence. A least square estimator can be applied to get estimates of the parameters in the polynomials A and B.

Equation (5.15) may be rewritten as

$$y_P(t) = \phi(t-1)\theta(t-1) + \xi(t) \tag{5.16}$$

where

$$\phi(t-1) \triangleq [-y_P(t-1), -y_P(t-2), \ldots, u(t-1), u(t-2), \ldots] \tag{5.17}$$

$$\theta(t-1) \triangleq (a_1, a_2, \ldots, a_n, b_0, b_1 \cdots b_{n-1})^T \tag{5.18}$$

The parameter vector $\theta(\cdot)$ may be a function of t; if the plant parameters vary slowly, then

$$\theta(t) = \theta(t-1) + e(t-1) \tag{5.19}$$

where $e(\cdot)$ is a sequence of independently distributed normal random variables with zero mean and covariance matrix Q_P. Then Eq. (5.19) represents a dynamical system with unknown plant parameters $\theta(t)$ which may be chosen as the state variables. Thus the state and output equations are given by

$$\theta(t+1) = \theta(t) + e(t) \tag{5.20}$$

$$y_P(t) = \phi(t-1)\theta(t-1) + \xi(t) \tag{5.21}$$

The state transition matrix is unity and the noise is uncorrelated.

5.4.3 Recursive Least-Squares Method

Equations (5.20) and (5.21) for the plant estimators are in a form similar to a Kalman filtering problem and we can therefore use the following equation for state estimation [86]:

$$\hat{\theta}(t) = \hat{\theta}(t-1) + K_P(t)[y_P(t) - \phi(t-1)\hat{\theta}(t-1)] \tag{5.22}$$

where

$$K_P(t) = P(t-1)\phi(t-1)^T[\sigma^2 + \phi(t-1)P(t-1)\phi(t-1)^T] \tag{5.23}$$

$$P(t) = P(t-1) + Q_P - K_P(t)[\sigma^2 + \phi(t-1)P(t-1)\phi(t-1)^T]K_P^T(t) \tag{5.24}$$

where Q_P is the parameter noise covariance matrix. Generally, we do not have a priori knowledge of the parameter noise, and often we assume that the parameters are constant. In that case Q_P is zero. We can also assume that the variance σ^2 is unity. $K_P(t)$ is called the Kalman gain. Equation (5.22) can be interpreted thus:

(New estimate) = (old estimate) + (Kalman gain) × (prediction error)

The calculations are performed in the following order:

$$P(t - 1) \longrightarrow K_P(t) \longrightarrow \hat{\theta}(t) \longrightarrow P(t)$$

This method of parameter estimation is called the recursive least-squares (RLS) method, as the estimate of $\theta(t)$ is found by minimizing the following quadratic cost criterion with respect to the parameter vector θ:

$$V(t) = \sum_{s=1}^{t} ||y_P(s) - \hat{y}_P(s)||^2$$

$$= \sum_{s=1}^{t} ||y_P(s) - \phi(s - 1)\theta(s)||^2 \tag{5.25}$$

The parameter estimation equations (5.22)-(5.24) are in recursive form. The least-squares method is simple; but if $C(q^{-1}) \neq 1$ or if the actual system is described by (5.13) only approximately, then the parameter estimates will be biased. To overcome the problem of bias, the extended least-squares or recursive maximum-likelihood methods must be applied.

5.4.4 Extended Least-Squares Method

The extended least-squares (ELS) method was suggested first in Panuska [87] and Young [88] and has been applied widely; RLS can be shown to be a special case of ELS.

ELS is a straightforward extension of RLS, with an extended prediction model in place of (5.13) so that the estimated prediction errors are included as another input signal. In the ELS method the estimates are found as in the RLS method. However, it has been observed that ELS algorithms are not as robust as RLS algorithms. Further, Ljung [89] has pointed out one shortcoming in that the convergence of ELS depends on the positive-realness condition of the C polynomials, though this has not been found to be a serious handicap in practice.

5.4.5 Recursive Maximum-Likelihood Method

The recursive maximum-likelihood (RML) algorithm was originally proposed by Åström and is based on a recursive approximation of the maximum-likelihood estimate of the parameters of the plant given by Eq. (5.13). This algorithm has been shown to converge for all useful C polynomials, unlike the limitations of ELS method. RML algorithm has been discussed by Ljung et al. [90] and Söderström [91], and also in Eykoff [81] and Åström and Eykoff [84]. This method needs more computations than ELS because of the extra filtering and also requires extra data storage. The filtering of $\phi(t)$ is done to assure convergence to a local maximum of the likelihood function. The structure of the RML algorithm contains also RLS and ELS with a proper interpretation of the filtered new vector.

The RLS, ELS, and RML methods are called prediction error identification methods and their convergence properties have been discussed by Ljung [92].

Söderström et al. [93] have given a comparative study of recursive estimation methods. Isermann [94] gives a unified treatment for the RLS, ELS, and RML methods of parameter estimation; the relevant definitions for the various elements of the estimation algorithms are given in one consolidated table for RLS, ELS, and RML methods, illustrating certain similarities among these.

5.4.6 Other Estimation Methods

Other recursive parameter estimation schemes can be represented by equations such as (5.22). However, it has been found that RLS, ELS, and RML methods are best suited for adaptive control [95]. The recursive instrumental variable (RIV) method yields biased parameter estimates in closed loop. Stochastic approximation (STA) method converges slowly and unreliably [94]. A square-root algorithm developed by Peterka [96] may be used, if numerical ill-conditioning of the updating procedure for P(t) occurs. The UD algorithm due to Biermann [83] is computationally efficient to overcome this difficulty. Unton [97] has discussed a recursive estimator for the general case. Unbehauen [97a] has surveyed different system identification methods using parameter estimation. Theoretical and practical aspects including modern trends as well as still unsolved problems are considered. The parameter estimation algorithm is the principal computation burden in self-tuning and so we must choose the simplest and most robust algorithm.

5.4.7 Forgetting Factor

The algorithms discussed earlier would be acceptable if the "true" plant parameters were constant. But we find that in practice plant parameters vary slowly with time. Hence the algorithm should track such slowly varying parameters, so that the self-tuner stays in tune. This is possible with the use of a forgetting factor β, which is slightly less than unity. By using this, we obtain exponential forgetting of the past data. Instead of giving equal weight to the errors in the RLS, we give more weight to more recent data. This is achieved by artificially inflating the a priori covariance P(t) at each time instant.

Let us define, following [86]:

$$\bar{P}(t) \triangleq P(t) \cdot \beta/\sigma^2 \tag{5.26}$$

Then the recursive relations (5.23) and (5.24) become, for $Q_P = 0$:

$$K_P(t+1) = \bar{P}(t) \cdot \phi(t)^T/\bar{\mu}^2(t) = \bar{P}(t+1)\phi(t)^T \tag{5.27}$$

$$\bar{P}(t+1) = \beta^{-1}[\bar{P}(t) - \bar{P}(t)\phi(t)^T\phi(t)\bar{P}(t)/\bar{\mu}^2(t)] \tag{5.28}$$

where

$$\bar{\mu}^2(t) \triangleq \beta + \phi(t)\bar{P}(t)\phi^T(t) \tag{5.29}$$

with

$$0 < \beta \leqslant 1$$

The calculations are done in the following order:

$$\bar{P}(t) \longrightarrow K_P(t+1) \longrightarrow \hat{\theta}(t+1) \longrightarrow \bar{P}(t+1)$$

In this scheme we introduce another definition, the asymptotic sample length (ASL):

$$\text{ASL} \triangleq \frac{1}{1-\beta}$$

ASL indicates the number of previous samples contributing to $\hat{\theta}(t)$. Usually, the values of β range from 0.95 (for fast variations) to 0.999 (for slow variations), so that the corresponding ASLs are 20 to 1000 [98].

However, in some chemical processes we cannot use a constant forgetting factor. It has been proposed that β be chosen such that it is close to unity under steady-state conditions and has a lower value when the measurement covariance increases. Fortescue et al. [99] give the following expression:

$$\beta(t) = 1 - \{[1 - \phi^T(t-1)K_P(t)]e^2(t)\}/B_0 \tag{5.30}$$

where

$$e(t) = y_P(t) - \phi(t-1)\theta(t-1)$$

and B_0 is a constant. We may choose $B_0 = N_0\sigma_0^2$, where N_0 regulates the speed of adaptation under steady-state condition and σ_0^2 is the measurement covariance.

Fortescue's variable forgetting factor technique works quite well in slowly time-varying and nearly deterministic cases. In a noisy and quickly time-varying case, this technique is not quite satisfactory. Xianya and Evans [100] have proposed a method to track quickly time-varying parameters by using a small forgetting factor, which, in turn, causes the algorithm to be very sensitive to noise. This is an extension of their work [101] in which they used the least-squares algorithm with a resetting procedure to estimate the time-varying parameters in the deterministic case. Simulation results given by them show that these schemes are able to track time-varying parameters more effectively than some previous schemes.

Self-tuning begins when an appropriate parameter estimation procedure is combined with a suitable control strategy.

5.5 IMPLICIT STR

5.5.1 Minimum-Variance STR [3]

Consider the plant described by Eq. (5.12):

$$A(q^{-1})y_P(t) = q^{-k}B(q^{-1})u(t) + C(q^{-1})\xi(t) \tag{5.31}$$

The order of the plant is n, and k is the time delay of this controlled auto-regressive moving-average process. k is an integer number of sample intervals, and $k \geqslant 1$. The polynomials A, B, and C are given by:

$$\begin{aligned} A(q^{-1}) &= 1 + a_1q^{-1} + a_2q^{-2} + \cdots + a_{n_a}q^{-n_a} \\ B(q^{-1}) &= b_0 + b_1q^{-1} + b_2q^{-2} + \cdots b_{n_b}q^{-n_b} \\ C(q^{-1}) &= 1 + C_1q^{-1} + C_2q^{-2} + \cdots + C_nq^{-n_c} \end{aligned} \tag{5.32}$$

where n_a, n_b, and n_c are the upper bounds to the orders of the polynomials A, B, and C. Further, the definition of k implies that $b_0 \neq 0$. We assume that k, n_a, n_b, and n_c are known.

The description of the plant by (5.31) indicates that the discrete closed-loop system has at least one step delay and therefore $k \geq 1$. It can be shown that the delay k is equal to the magnitude of the pole excess, that is, $k = n_a - n_b$. It may be noted that if a backward shift operator is used, $(n_a - n_b)$ appears explicitly.

The discrete-time model of the plant [102] described by (5.31) is given in Fig. 5.2.

We note [3] that the following conditions are necessary:

1. The polynomial B has all zeros inside the unit circle. Then the plant (5.31) is minimum phase. Hence the minimum-variance controller does not lead to a stable system for non-minimum-phase plants.
2. The polynomial C has all zeros inside the unit circle.

The structure of the controller depends on the choice of the cost function for minimization. The minimum-variance cost criterion chosen by Åström and Wittenmark [3] is given by

$$J = E[y_P^2(t)] = \lim_{N \to \infty} E\left[\frac{1}{N}\sum_{i=1}^{N} y_P^2(i)\right] \tag{5.33}$$

where the expectation is unconditional and $y_P(i)$ is the plant output. The justification for the choice of this cost function is as follows. In many

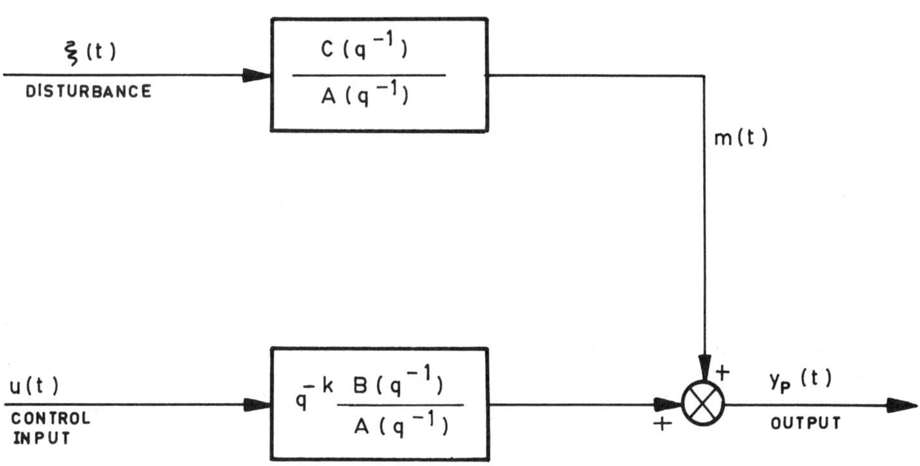

FIG. 5.2 Discrete model of the plant. (From Ref. 102.)

practical control problems it is generally necessary that only a certain specified small percentage of the product have a property (say, thickness of paper) outside the given range (say, to control the tearing strength); if there are large variations in this property, then we have to keep the set point giving the average value at a high value, which results in greater average product cost. But if these variations are minimized, then the set point can be kept at a lower value, resulting in the same rejection rate at a smaller average product cost. However, there are certain limitations, which have been pointed out earlier, in Section 5.2.1.

We next obtain the control law. The plant equation (5.31) may be rewritten as

$$y_p(t+k) = \frac{B(q^{-1})}{A(q^{-1})} u(t) + \frac{C(q^{-1})}{A(q^{-1})} \xi(t+k) \tag{5.34}$$

This equation shows that $y_p(t+k)$ is dependent on the control signal sequence up to and including t and on the disturbance sequence up to and including (t+k).

By time shifting and rearranging Eq. (5.34), we get

$$\xi(t) = \frac{A(q^{-1})y_p(t) - B(q^{-1})u(t-k)}{C(q^{-1})} \tag{5.35}$$

From Fig. 5.2, we note that the signal m(·) is given by

$$m(t+k) = \frac{C(q^{-1})}{A(q^{-1})} \xi(t+k) \tag{5.36}$$

We can consider m(t + k) to consist of two components:

$$m_1(t+k) = \text{function of } \{\xi(t+k), \ldots, \xi(t+1)\}$$

$$m_2(t+k) = \text{function of } \{\xi(t), \xi(t-1), \ldots\}$$

where $m_1(t+k)$ is unknown at time t and $m_2(t+k)$ can be calculated at time t. The random disturbance $\xi(t)$ has zero mean. Hence the best prediction m(t + k) can be shown to be the component $m_2(t+k)$. The minimum-variance predictor is therefore given by

$$\hat{m}(t+k|t) = m_2(t+k) \tag{5.37}$$

Let us introduce the following identity:

$$C(q^{-1}) = A(q^{-1})F(q^{-1}) + q^{-k}G(q^{-1}) \tag{5.38}$$

where F and G are polynomials of order (k - 1) and (n_a - 1), respectively.

Using (5.36)-(5.38), we have

$$\hat{m}(t + k|t) = \frac{G(q^{-1})}{A(q^{-1})}\xi(t) \tag{5.39}$$

Then the prediction error is given by

$$\begin{aligned}\tilde{m}(t + k|t) &= m(t + k) - \hat{m}(t + k|t) \\ &= F(q^{-1})\xi(t + k) \\ &= m_1(t + k)\end{aligned} \tag{5.40}$$

This signal is a moving average of order k and is uncorrelated with $\hat{m}$, u, and y_P.

The control law is derived now. The cost function is given by

$$J = E[y_P^2(t + k)|t] \tag{5.41}$$

Using (5.34), (5.36), and (5.38)-(5.40), we get

$$\begin{aligned}J &= E\left\{\left[\frac{B(q^{-1})}{A(q^{-1})}u(t) + \hat{m}(t + k|t) + \tilde{m}(t + k|t)\right]^2\right\} \\ &= E\left\{\left[\frac{B(q^{-1})}{A(q^{-1})}u(t) + \hat{m}(t + k|t)\right]^2 + [\tilde{m}(t + k|t)]^2\right\} \\ &= \left[\frac{B(q^{-1})}{A(q^{-1})}u(t) + \hat{m}(t + k|t)\right]^2 + E\{[\tilde{m}(t + k|t)]\}\end{aligned} \tag{5.42}$$

The first term of the last equation arises since u(t) and $\hat{m}(t + k|t)$ are known signals at time t. We note that the second term in the cost is not affected by the control signal. Hence, we have the minimum-variance control given by

$$u_0(t) = \frac{-A(q^{-1})}{B(q^{-1})} \hat{m}(t + k|t) \tag{5.43}$$

Using (5.39) and (5.35), we get

$$u_0(t) = \frac{-G(q^{-1})}{B(q^{-1})C(q^{-1})} [A(q^{-1})y_P(t) - B(q^{-1})u_0(t - k)]$$

whence we get

$$u_0(t) = \frac{-G(q^{-1})A(q^{-1})y_P(t)}{B(q^{-1})[C(q^{-1}) - q^{-k}G(q^{-1})]}$$

$$= \frac{-G(q^{-1})y_P(t)}{B(q^{-1})F(q^{-1})}$$

This is the optimal control strategy and this is to be combined with parameter estimation to produce a self-tuning regulator.

5.5.2 Generalized Minimum-Variance STR [5,102]

The cost criterion in this case is given by

$$J = E[\phi^2(t + k)|t] \tag{5.45}$$

where the expectation at (t + k) is conditioned on all input/output data acquired up to that time t. In (5.45) we note that the controller minimizes the variance of the auxiliary output $\phi(\cdot)$ (and not the actual plant output), which is given by:

$$\phi(t + k) = P(q^{-1})y_P(t + k) + Q(q^{-1})u(t) - R(q^{-1})w(t) \tag{5.46}$$

where $P(q^{-1}) = P_N(q^{-1})/P_D(q^{-1})$, w(t) is the known set-point signal at time t, and the transfer functions P, Q, and R can be freely chosen depending on the required response characteristic (e.g., integral action) or for minimizing the effects of disturbances. Further, by an appropriate choice of these polynomials, Gawthrop [6] has shown the relatedness of self-tuning control laws and model reference adaptive control laws. This is therefore a more flexible approach. Further, the generalized minimum-variance cost criterion (5.45) penalizes both the plant output and the plant input, whereas the cost function (5.33) chosen by Åström and Wittenmark includes only the plant output.

In Eq. (5.46), the signals Qu(t) and Rw(t) are known at (t + k). Hence the problem of predicting $\phi(t + k)$ becomes a problem of a least-squares prediction of the first term of (5.46), which is of the form:

$$\phi_y(t) = P(q^{-1})y_P(t) \tag{5.47}$$

This equation may be rewritten as

$$A(q^{-1})P_D(q^{-1})\phi_y(t) = P_N(q^{-1})B(q^{-1})q^{-k}u(t) + P_N(q^{-1})C(q^{-1})\xi(t) \tag{5.48}$$

Introduce an identity:

$$\frac{C(q^{-1})P_N(q^{-1})}{A(q^{-1})P_D(q^{-1})} = F'(q^{-1}) + \frac{q^{-k}G'(q^{-1})}{A(q^{-1})P_D(q^{-1})} \tag{5.49}$$

where F' is still of order (k - 1), although the coefficients are different from those of F in (5.38), and G' is of the order (n_a - 1 + degree of P_D).

From (5.48) and (5.49), we get

$$\phi_y(t + k) = F'(q^{-1})\xi(t + k) + \frac{P(q^{-1})B(q^{-1})}{A(q^{-1})}u(t) + \frac{G'(q^{-1})}{A(q^{-1})P_D(q^{-1})}\xi(t) \tag{5.50}$$

Substituting for $\xi(t)$ from (5.35), we have

$$\phi_y(t + k) = F'(q^{-1})\xi(t + k) + \frac{F'(q^{-1})B(q^{-1})}{C(q^{-1})}u(t) + \frac{G'(q^{-1})}{C(q^{-1})P_D(q^{-1})}y_P(t) \tag{5.51}$$

Then the k-step-ahead predictor becomes

$$\hat{\phi}_y(t + k|t) = \frac{F'(q^{-1})B(q^{-1})}{C(q^{-1})}u(t) + \frac{G'(q^{-1})}{C(q^{-1})P_D(q^{-1})}y_P(t) \tag{5.52}$$

Hence the prediction error is given by

$$\epsilon(t + k|t) = \phi_y(t + k) - \hat{\phi}_y(t + k|t) = F'(q^{-1})\ (t + k) \tag{5.53}$$

where if $k > 1$, the prediction error is not white.

To derive the feedback control law, the k-step-ahead prediction of ϕ is required:

$$\phi(t + k|t) = \phi_y(t + k|t) + Q(q^{-1})u(t) - R(q^{-1})w(t) \tag{5.54}$$

The control law is derived from a basic principle, namely, that the minimum-variance controller at t is chosen also as to set the k-step-ahead prediction of the auxiliary output to zero. The optimal control signal must therefore set $\hat{\phi}(t + k|t)$ to zero. Hence from (5.45) and (5.53), we get

$$C(q^{-1})\hat{\phi}(t + k|t) = \frac{G'(q^{-1})}{P_D(q^{-1})} y_P(t) + [F'(q^{-1})B(q^{-1}) + Q(q^{-1})C(q^{-1})]u(t)$$

$$- R(q^{-1})C(q^{-1})w(t) = 0 \tag{5.55}$$

and from (5.53),

$$\phi_y(t + k) = \hat{\phi}_y(t + k|t) + \epsilon(t + k|t) \tag{5.56}$$

Equations (5.55) and (5.56) determine the optimal control law.

We can obtain an expression for optimal control $u_0(t)$ from Eqs. (5.54) and (5.55):

$$u_0(t) = \frac{Rw(t) - \hat{\phi}_y(t + k|t)}{Q} \tag{5.57}$$

$$= \frac{CRw(t) - G'y_P(t)/P_D}{F'B + QC} \tag{5.58}$$

This is a feedback control law and includes the set-point signal also, unlike (5.44). This optimal control strategy must be combined with suitable parameter estimation to produce an STR.

We note from (5.57) and (5.58) the effect of nonzero Q in reducing the amplitude of the control signal. In practice, Q allows us to modify the control performance without altering the sample time.

Note: The symbols E, F, and G are used in this section in place of G, E, and F, respectively, used by Clarke and Gawthrop [5].

5.5.3 Implicit Self-Tuning Algorithms

The implicit schemes of Åström and Wittenmark [3] and Clarke and Gawthrop [4, 5] are based on the idea of avoiding separate steps of identification and control. The controller parameters are incorporated into the identification procedure (Fig. 5.3).

In contrast to the preceding schemes, explicit schemes involve system parameter identification followed by control as separate steps. This will be discussed in Section 5.6.2.

The actual derivation of the self-tuning algorithm is dependent on the type of control law involved. We have obtained two different laws and so two different derivations are required.

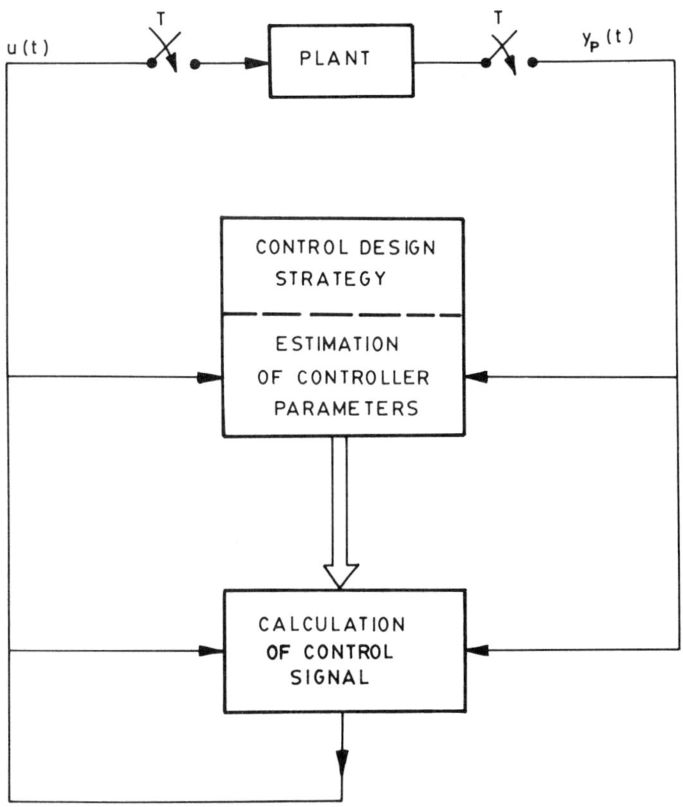

FIG. 5.3 Implicit self-tuning cycle. (From Ref. 86.)

Generalized Self-Tuning Algorithm

We first obtain the generalized Clarke-Gawthrop self-tuning algorithm, for which the following two equations determine the minimum-variance controller:

$$C(q^{-1})\hat{\phi}_y(t + k|t) = \frac{G'(q^{-1})}{P_D(q^{-1})} y_P(t) + F'(q^{-1})B(q^{-1})u(t) \tag{5.59}$$

[Note: This is the predictor part of Eq. (5.55) in which the signals Qu(t) and Rw(t) are known at (t + k), as already mentioned.]

$$\phi_y(t + k) = \hat{\phi}_y(t + k|t) + \epsilon(t + k|t) \tag{5.60}$$

This is the same as the Eq. (5.56).

Let us expand $C(q^{-1})$ to exhibit the parameters C_i, $i = 1, 2, \ldots, n$. Let us introduce

$$E'(q^{-1}) \triangleq F'(q^{-1})B(q^{-1}) \tag{5.61}$$

By this, we see that the plant parameters in $B(q^{-1})$ become part of the coefficients of the polynomial $E'(q^{-1})$. Then, from (5.59) and (5.60), we have

$$\begin{aligned}\hat{\phi}_y(t + k|t) &= \frac{G'(q^{-1})}{P_D(q^{-1})} y_P(t) + E'(q^{-1})u(t) - \sum_{i=1}^{n} C_i\hat{\phi}_y(t + k - i|t - i)\\ &= X(t)\theta\end{aligned} \tag{5.62}$$

where

$$X(t) \triangleq \left[\frac{y_P(t)}{P_D}, \frac{y_P(t - 1)}{P_D}, \ldots; u(t), u(t - 1), \ldots; \hat{\phi}_y(t + k - 1|t - 1), \hat{\phi}_y(t + k - 2|t - 2), \ldots\right] \tag{5.63}$$

$$\theta \triangleq [g_0, g_1, \ldots; e_0, e_1, \ldots; -C_1, -C_2, \ldots] \tag{5.64}$$

Then from (5.60) and (5.62), we have

$$\phi_y(t) = X(t - k)\theta + \epsilon(t|t - k) \tag{5.65}$$

This equation is similar to Eq. (5.21), which was obtained in connection with recursive parameter estimation. We can therefore use a least-squares estimation procedure to estimate θ. However, if $k > 1$, the prediction error is no longer white, and the matrix X(t) contains unknown quantities. To avoid this situation, we approximate $\hat{\phi}_y$ as in extended least-squares method and so we introduce $\hat{\phi}_{y*}$, which is a calculable approximation, and then we can use least-squares method as if $\hat{\phi}_{y*} = \hat{\phi}_y$ and the prediction error were white. Then

$$\hat{\phi}_{y*}(t + k|t) = \hat{X}(t)\hat{\theta}(t)$$
$$= [\hat{g}_0 \frac{y_P(t)}{P_D} + \hat{G}_1 \frac{y_P(t-1)}{P_D} + \cdots + e_0 u(t) + e_1 u(t-1)$$
$$+ \cdots - C_1 \hat{\phi}_{y*}(t + k - 1|t - 1) - \cdots] \tag{5.66}$$

where $\hat{X}(t)$ is an estimate of X(t) to be successively refined.

Using the recursive relations of Sections 5.4.3 and 5.4.7, we can obtain the following generalized self-tuning algorithm:

Step 1:

$$\hat{X}(t) = \left[u(t), u(t-1), \ldots; \frac{y_P(t)}{P_D}, \frac{y_P(t-1)}{P_D}, \ldots; \hat{\phi}_{y*}(t+k-1|t-1), \ldots\right] \tag{5.67}$$

Step 2:

$$\phi_y(t) = P(q^{-1})y_P(t) \tag{5.68}$$

Step 3:

$$\hat{\theta}(t) = \hat{\theta}(t-1) + K_P(t)[\phi_y(t) - \hat{X}(t-k)\hat{\theta}(t-1)] \tag{5.69}$$

where

$$K_P(t) = \frac{\bar{P}(t-1)X^T(t-k)}{1 + X(t-k)\bar{P}(t-1)X^T(t-k)} \tag{5.70}$$

$$\bar{P}(t) = \left[\bar{P}(t-1) - \frac{\bar{P}(t-1)X^T(t-k)X(t-k)\bar{P}(T-1)}{\beta + X(t-k)\bar{P}(t)X^T(t-k)}\right] \Big/ \beta \tag{5.71}$$

Step 4:

$$\hat{\phi}_{y*}(t + k|t) = \hat{X}(t)\hat{\theta}(t) \tag{5.72}$$

Step 5:

$$u_0(t) = [Rw(t) - \hat{\phi}_{y*}(t + k|t)]/Q \tag{5.73}$$

Repeat steps 1 to 5 at each sampling period.

This is the complete implicit generalized self-tuning algorithm corresponding to Fig. 5.3. We note that unknown parameters are not estimated; instead, the parameters of the prediction model are used directly to compute the control. This procedure avoids the necessity of the solution of the diophantine equation usually involved in control calculations. Thus the controller design calculations are eliminated.

Self-Tuning Algorithm

The implicit STR of Åström et al. [3, 7] consists of two operations:

1. The parameters in the polynomials of the following prediction model are estimated:

$$y_P(t + k) = G(q^{-1})y_P(t) + E(q^{-1})u(t) + \epsilon(t + k|t) \tag{5.74}$$

where

$$E(q^{-1}) \triangleq F(q^{-1})B(q^{-1}) \tag{5.75}$$

2. The control signal is obtained using these estimates in the certainty equivalent control law:

$$\hat{G}(q^{-1})y_P(t) + \hat{E}(q^{-1})u(t) = 0 \tag{5.76}$$

The algorithm for this form of STR may be obtained from the earlier equations (5.67)-(5.73) by setting $P = I$, $Q = R = 0$. With these values, the minimum-variance controller will result, with the cost function $J = E[y_P(t + k)^2]$ being minimized, with

$$\hat{\phi}_{y*}(t + k|t) = \hat{y}_{P*}(t + k|t) = 0 \tag{5.77}$$

The complete implicit STR algorithm is obtained, using Eqs. (5.67)-(5.73):

Step 1:

$$\hat{X}(t) = [u(t), u(t-1), \ldots; y_P(t), y_P(t-1), \ldots] \tag{5.78}$$

Step 2:

$$\hat{\theta}(t) = \hat{\theta}(t-1) + K_P(t)[y_P(t) - \hat{X}(t-k)\hat{\theta}(t-1)] \tag{6.79}$$

where

$$K_P(t) = \frac{\bar{P}(t-1)\hat{X}^T(t-k)}{1 + \hat{X}(t-k)\bar{P}(t-1)\hat{X}^T(t-k)} \tag{5.80}$$

$$\bar{P}(t) = \left[\bar{P}(t-1) - \frac{\bar{P}(t-1)X^T(t-k)X(t-k)\bar{P}(t-1)}{\beta + X(t-k)\bar{P}(t)X^T(t-k)}\right] \Big/ \beta \tag{5.81}$$

Step 3:

$$\hat{y}_{P^*}(t+k|t) = \hat{X}(t)\hat{\theta}(t) \tag{5.82}$$

where

$$\hat{\theta} \triangleq (\hat{g}_0, \hat{g}_1, \ldots; \hat{c}_0, \hat{c}_1, \ldots)^T \tag{5.83}$$

Step 4:

$$u_0(t) = \frac{-G(q^{-1})}{E(q^{-1})} y_P(t) \tag{5.84}$$

Repeat steps 1 to 4 at each sampling period.

This is the complete implicit self-tuner algorithm. As in the previous case, the unknown parameters are not estimated, but the parameters of the prediction model are used directly to obtain the control signal. As already mentioned, such algorithms are valid for minimum-phase plants only.

5.5.4 Discussion

Åström et al. [7] deal with the important problems of overall stability, convergence of the regulator, and properties of the possible limiting regulators by extensive simulation as well as by analysis. They also discuss some industrial applications, with special reference to the control of moisture content in a paper machine and to the control of input of the ore to an ore crusher to compensate for variations in crushability, lump size, and wear and tear. Ljung and Wittenmark [103] analyze the convergence properties of the minimum-variance STR algorithm.

Clarke and Gawthrop [5] also discuss aspects of stability, convergence, feedforward, and multivariable control. They also deal with some technical problems involved in the practical implementation of self-tuners using microprocessors. They [104] report a prototype microprocessor-based STR. In a later paper [105] they present the hardware and software details of portable microcomputer-based self-tuners which are applied by them for room temperature control, pH control of effluents, and chemical batch reactor control. They conclude that self-tuning controllers can cope well with nonlinearities encountered in practical processes, provided some precautions are taken. Reference may also be made to Clarke [106]. Some important interpretations of the self-tuning controller are given by Gawthrop [6], some of which are referred to in Clarke [98] and in Section 5.5.2.

5.6 EXPLICIT STR

5.6.1 Pole-Placement Approach [18,20]

Explicit STRs are based on pole placement and do not require knowledge of time delay, as already mentioned in Sections 5.2.2 and 5.2.3. The standard plant model described by Eq. (5.31) is estimated. A range of time delays can be considered by overparameterizing the polynomial B. However, this leads to additional computation, as pointed out earlier.

Let us consider a range of values of k from k_1 to k_2. Let $B'(q^{-1})$ be a polynomial with $(n_b + k_2 - k_1)$ terms. Then the identified model is

$$A(q^{-1})y_p(t) = q^{-k_1}B'(q^{-1})u(t) + C(q^{-1})\xi(t) \tag{5.85}$$

If $k = k_1$, the first n_b coefficients of $\hat{B}'$ will be nonzero, and if $k = k_2$, the last n_b coefficients of $\hat{B}'$ will be nonzero. We can use $\hat{B}'$ in the algorithm, provided the control design is not sensitive to the resultant changes in the zeros of $\hat{B}'$. Of course, this resultant insensitivity will affect the performance.

The pole-placement approach of Wellstead et al. [18] is to consider only the closed-loop pole positions and not alter the zeros.

A standard feedback controller is

$$G(q^{-1})y_p(t) + E(q^{-1})u(t) = 0 \tag{5.86}$$

where the orders of G and E must be

$$n_g = n_a - 1$$
$$n_e = n_b + k - 1 \qquad (5.87)$$

It is to be noted that $g_0 = 1$ in the polynomial G in (5.86).

The model (5.85) is to be estimated. However, for simplicity, we use (5.31) along with (5.86) to yield the closed-loop equation.

$$(AE + q^{-k}BG)y_p(t) = CE\xi(t) \qquad (5.88)$$

The coefficients of G and E are chosen using the identity

$$AE + q^{-k}BG = CT \qquad (5.89)$$

where $T(q^{-1})$ is a polynomial of order n_t and is of the form

$$T(q^{-1}) = \pi_i(1 - \alpha_i q^{-1}) \qquad (5.90)$$

where α_i correspond to the stable poles, which are preselected by the engineer after due consideration of the concerned plant.

The closed loop is described by

$$y_p(t) = \frac{E}{T}\xi(t) \quad \text{and } u(t) = -\frac{G}{T}\xi(t) \qquad (5.91)$$

We see therefore that the engineer's specification of T affects the variance of $y_p(t)$. We note also that B is not a factor of the characteristic equation.

The objective of pole-assignment STR is to move the closed-loop poles from their open-loop locations to the prespecified values given by Eq. (5.90).

The number of closed-loop poles n_t is limited by

$$n_t < (n_a + n_b + k - n_c) \qquad (5.92)$$

This constraint ensures that the order of the right-hand side of Eq. (5.89) does not exceed the order of the left-hand side. If the system delay k is unknown, we can set k to be zero and Eq. (5.89) will still lead to a sensible regulator, provided that the order of the polynomial B is increased by k, to allow for k leading coefficients which may be zero. The assumption that $k = 0$ will lead to considerable simplification, though we should always note that time delay will be present, in general, and this is taken care of by a set of zero coefficients in the polynomial B. In many processes, significant changes occur in the transport delay, so that both B and k vary. This results in a fractional delay, leading to the system becoming non-minimum phase. In such cases, minimum-variance STRs will be unstable. The same difficulty arises if k increases beyond the value allowed for in the model. A pole-assignment STR can handle these phenomena by adapting its configuration, as during a parameter variation. This is a distinct advantage. Reference may be made to Wellstead et al. [18] for details of analysis and simulation results of two examples of varying time delay, and also experimental results with a hydraulic system which becomes non-minimum phase in its digitized form. It is shown how the pole-placement STR can successfully cope with difficult, but not uncommon, practical problems. Though the implicit STRs can also handle non-minimum-phase cases, with appropriate modifications [4,47], these attempt to cancel system dynamics, and this is much less robust than modifying them, as is done in pole assignment [18].

5.6.2 Explicit Self-Tuning Algorithm

Explicit self-tuning is achieved by an iteration at each sample interval through the following cycle: (1) identification, (2) control design strategy, and (3) control signal calculation, as shown in Fig. 5.4.

The pole-placement algorithm is developed as follows: Let $A_1(q^{-1})$ and $B_1(q^{-1})$ be the solutions of the identity:

$$A_1(q^{-1})E(q^{-1}) + q^{-k}B_1(q^{-1})G(q^{-1}) = T(q^{-1}) \tag{5.93}$$

where G and E are given by (5.89). It may be noted that C is omitted in (5.93).

Consider the sequence

$$\begin{aligned} A_1y_p(t) - q^{-k}B_1u(t) &= \frac{A_1E + q^{-k}B_1G}{T}\,\xi(t), \quad \text{using (5.92)} \\ &= \xi(t) \quad \text{using (5.93)} \end{aligned} \tag{5.94}$$

This is the model of a structure in which recursive least squares (RLS) can be used to get unbiased estimates of $A_1(q^{-1})$ and $B_1(q^{-1})$.

Self-tuning will then proceed as follows (Fig. 5.4):

Step 1: At each sampling interval, the parameters $\hat{A}_1(q^{-1})$ and $\hat{B}_1(q^{-1})$ are estimated by RLS.

Step 2: The polynomials $\hat{A}_1(q^{-1})$ and $\hat{B}_1(q^{-1})$ are next used to synthesize the regulator polynomials $\hat{E}(q^{-1})$ and $\hat{G}(q^{-1})$, using the identity (5.93) and prespecified $T(q^{-1})$.

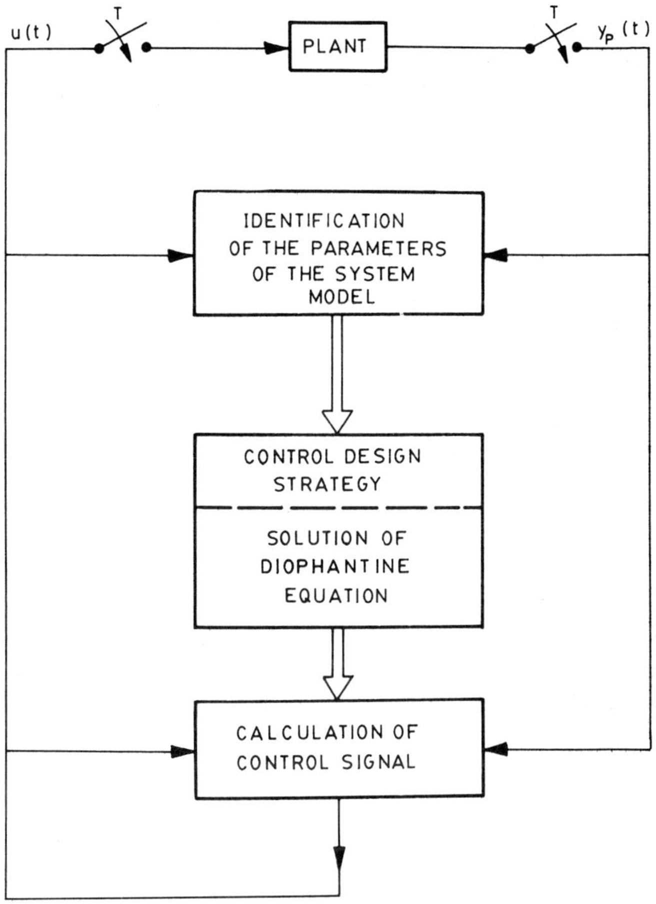

FIG. 5.4 Explicit self-tuning cycle. (From Ref. 86.)

Step 3: Next the control input is obtained by using $\hat{E}$ and $\hat{G}$ in the control law (5.86).

Repeat steps 1 to 3 at each sampling period until the estimated parameters $\hat{A}_1$ and $\hat{B}_1$ converge to steady values.

The pole-assignment self-tuner is basically intended for the regulator action to reduce the effect of disturbances on the output. However, it is possible to introduce reference input tracking into the algorithm. Wellstead et al. [21] discuss the servo case in which Eq. (5.94) is augmented by two additional signals, one including a precompensated set point and the other a feedforward signal which is added directly to the regulating control signal u(t). This leads to a self-tuning algorithm which tunes both the regulating and servo properties of the closed-loop system to the desired form. However, it must be pointed out that this will involve a significant increase in computational effort.

5.6.3 Pole-Zero-Placement Approach

Åström and Wittenmark [26] suggest a pole-zero-placement method in which only those zeros are canceled which are in some restricted stability region. They consider only the noise-free case.

Let us consider a general linear regulator described by

$$H(q^{-1})w(t) = G(q^{-1})y_P(t) + E(q^{-1})u(t) \tag{5.95}$$

The closed-loop transfer function is given by

$$y_P(t) = \frac{q^{-k}HB}{AE + q^{-k}BG}\, w(t) \tag{5.96}$$

In this, the precompensator is given by H/E and the feedback element by G/E. By a suitable choice of E, G, and H, the preceding transfer function is required to be made equal to a reference model described by

$$y_P(t) = \frac{q^{-k}B_M(q^{-1})}{A_M(q^{-1})}\, w(t) \tag{5.97}$$

The control signal is given by

$$u(t) = \frac{A(q^{-1})}{B(q^{-1})}\left(\frac{B_M(q^{-1})}{A_M(q^{-1})}\right) w(t) \tag{5.98}$$

We see that if any root of the denominator BA_M of (5.98) is outside some region, there are undesirable modes in this control, as, for instance, if B has unstable roots or poorly damped roots very near the unit circle. Since such zeros of B cannot be canceled, the polynomial B is factored as

$$B = B^+B^- \tag{5.99}$$

where B^+ has well-damped roots within the specified region and B^- has the remaining poorly damped or unstable roots which are outside the specified region.

From (5.98) we see that B_M must have a factor B^- so that

$$B_M = B'_M B^- \tag{5.100}$$

The specifications must be such that B^- is a factor of B_M; otherwise, we cannot find a solution to the design problem.

Next we note that the degree of A_M is normally less than the degree of the denominator of (5.96). Hence when comparing (5.96) and (5.97), we conclude that there are factors which cancel. In Refs. 26 and 107 it is shown that this canceling factor $A_0(q^{-1})$ is an observer polynomial. The introduction of the observer polynomial is guided by the state space solution to the problem, which is a combination of state feedback and an observer. From (5.89), we note that in the noisy case, the observer is, in fact, $C(q^{-1})$. In the noise-free case which we are concerned with here, A_0 can be so chosen that the roots are in the specified stability region.

We can therefore say that the polynomial $(AE + q^{-k}BG)$, which is the characteristic polynomial of the closed-loop system, may be designed to have three types of factors: the canceled process zeros, the desired model poles, and the desired observer poles.

$$AE + q^{-k}BG = B^+ A_M A_0 \tag{5.101}$$

Since B^+ divides B it follows from the preceding equation that B^+ divides E.

$$\therefore E = B^+ E_1 \tag{5.102}$$

We can rewrite (5.101) as

$$AE_1 + q^{-k}B^- G = A_M A_0 \tag{5.103}$$

The requirement that Eq. (5.96) should be equal to the desired relation (5.97) yields, with the use of (5.101), the following result.

$$H = A_0 B'_M \tag{5.104}$$

The following conditions guarantee that there exist solutions to (5.101) which give a causal control law [107]:

The feedback transfer function G/E is causal if

$$\deg G \leqslant \deg E \tag{5.105}$$

It is therefore desirable to find a solution G to (5.101) which has as low a degree as possible. There can always be a solution with

$$\deg G \leqslant (\deg A - 1) \tag{5.106}$$

From (5.101), we see therefore that

$$\deg E = (\deg B^+ + \deg A_M + \deg A_0 - \deg A) \tag{5.107}$$

Hence we obtain the following condition which guarantees that the feedback transfer function is causal:

$$\deg A_0 \geqslant (2 \deg A - \deg A_M - \deg B^+ - 1) \tag{5.108}$$

Similarly, we obtain the following condition which guarantees that the feedforward transfer function H/E is causal:

$$(\deg A_M - \deg B_M) \geqslant (\deg A - \deg B) \tag{5.109}$$

The following are the steps in the off-line design:

Step 1: Equation (5.103) is solved to obtain E_1 and G. (There are several solutions to the diophantine equation which satisfy conditions (5.108) and (5.109). All solutions give the same closed-loop transfer function. Further discussion is in Åström and Wittenmark [26]).

Step 2: The desired feedback is then given by (5.95) with E and H given by (5.102) and (5.104). The closed-loop transfer function (5.96) is now

$$\frac{q^{-k}HB}{AE + q^{-k}BG} = \frac{q^{-k}A_0B'_MB^+B^-}{AE_1B^+ + q^{-k}B^+B^-G}$$

$$= \frac{q^{-k}A_0B'_MB^+B^-}{A_MA_0B^+}$$

$$= q^{-k}\left(\frac{B_M}{A_M}\right), \quad \text{as required by Eq. (5.97)}$$

Step 3: The control law can be written from (5.95), using (5.97), (5.99), and (5.102)-(5.104).

$$u_0 = G_MG_P^{-1}w(t) - \frac{G}{E}(y_P - y_M) \tag{5.110}$$

where

$$G_P = \frac{B}{A}$$

$$G_M = \frac{B_M}{A_M}$$

This shows that this design can be interpreted as model following. This is important to establish the relations between the STR and the MRAC systems. However, Eq. (5.95) is preferable in realizations.

It may be noted that the set-point signal w(t) must be persistently exciting (as is the case in all servo problems) to get good estimates and so good control. However, this is not usually the case with industrial processes; in such cases this self-tuner serves as a design aid for fixed parameter controllers.

Note: The symbols G, E, and H are used in this section in place of S, R, and T, respectively, used by Åström [107].

Pole-Zero Placement STR Algorithm (Data: A_M, A_0, and B'_M are given.)

Step 1: The coefficients of the polynomials A and B are estimated, using RLS. Since it is assumed that there is no noise, there will be no bias. (Otherwise, we should use ELS or RML.) The polynomial B must be factorized on line at each step so that the decomposition B^+B^- can be made.

Step 2: Equation (5.103) is solved after substituting A and B by estimates from step 1, to obtain E_1 and G. Next E and H are calculated from (5.102) and (5.104).

Step 3: The control signal u_0 is calculated from (5.95). Repeat the steps 1 to 3 at each sampling period.

It has been suggested by Clarke [106] that it is possible to combine the two approaches of Wellstead et al. and Astrom et al. for combined regulation and servo cases by using ELS or RML to estimate all A, B, C parameters, by letting $\hat{C}$ be the observer polynomial A_0, and by using (5.89) or (5.103) to find the feedback parameters E and G. Another calculation would then give H, depending on the type and number of open-loop zeros which are to appear in the overall closed-loop transfer function.

5.6.4 Unification of Different Algorithms

Åström [107] gives an explicit algorithm considered in Section 5.6.3 and also two implicit algorithms based on the system equations discussed earlier and shows how from these algorithms, we can obtain, by appropriate substitutions, the following as special cases:

1. Minimum variance algorithm of Åström and Wittenmark [3]
2. Model reference adaptive algorithm of Monopoli [108]
3. Generalized minimum-variance algorithm of Clarke and Gawthrop [4, 5]
4. Pole-placement algorithm of Wellstead, Prager, and Zanker [18]
5. A simple version of the IDCOM algorithm of Richalet et al. [109]

It is worth referring to Åström [107] for further details showing the unification of different algorithms of STRs and MRAC.

Allidina and Hughes [25] show the link between implicit and explicit self-tuning controllers, and this relationship has been used to provide a generalized suboptimal controller, incorporating pole assignment, so that it possesses both the robustness of explicit self-tuning [18] and also the ease of reference signal tracking of the implicit self-tuning controller [4]. Such an extended controller can also be applied to non-minimum-phase plants.

5.6.5 Equivalence Between STR and MRAC

Self-tuning and MRAC are based on different design approaches. However, as pointed out in Section 4.6, Egardt [110] has shown the equivalence between the various adaptive schemes proposed for MRAC by Ionescu and Monopoli [111] and Bénéjean [112] as well as the different schemes proposed for self-tuning by Åström et al. [3, 26], Clarke and Gawthrop [4] and that based on input matching by Johnson and Tse [113].

Even more general than Egardt's algorithm is the generalized algorithm due to Landau and Lozano [114] which includes Egardt's algorithms as special cases. Section 4.6.6 showed how from this generalized algorithm it is possible to establish correspondence with the algorithms given by Ionescu and Monopoli [111] and Bénéjean [112], Egardt [110, 115], Goodwin et al. [116], and Åström and Wittenmark [3].

In a later paper Landau [117] discusses further the equivalence between MRAC and stochastic STRs. Other important papers showing the connections between MRAC and STR are by Ljung and Landau [118], Landau [119, 120], Schmid [121], Narendra and Lin [122], and Johnson [123].

5.7 LQG SELF-TUNING

5.7.1 Introduction

LQG self-tuning is most appropriate for complex systems subjected to noise inputs. Such LQG controllers can be used with plants which are either open-loop unstable or non-minimum phase. They can also be applied to multivariable systems with different delays in different loops [39]. Integral action may be introduced by an appropriate choice of the cost function. The time delay does not enter explicitly into the calculation of the controllers and so the magnitude of the delay need not be known a priori [124]. The controller has better robustness and stability characteristics and is easier to design than pole-placement controllers. However, as pointed out in Section 5.2.4, there is greater computational load. However, simplification can be introduced, making the amount of computation comparable to that required for pole-placement methods, as shown by Lam [32, 33]. Åström [125] surveys LQG self-tuners with special reference to such issues as convergence, persistent excitation, closed-loop identifiability, and robustness. The properties of the LQG self-tuners are illustrated by an application to concentration control.

5.7.2 System Description

The system structure is shown in Fig. 5.5 [124]. The reference, disturbance, and measurement noise are assumed to be stable and are denoted by r, n, and v, respectively. The noise-free output y has to be controlled by using the controller $C_0 = C_n/C_d$. The plant transfer function is $W_1 = B/A$. The plant may be unstable and non-minimum phase and may include a time delay of magnitude $k_1 \geqslant 1$.

The plant polynomials are assumed to be of the form:

$$B(q^{-1}) = q^{-k}(b_0 + b_1 q^{-1} + \cdots + b_{(n_b-k)} q^{-(n_b-k)}, \qquad b_{(n_b-k)} \neq 0 \tag{5.111}$$

$$A(q^{-1}) = 1 + a_1 q^{-1} + a_2 q^{-2} + \cdots + a_{n_a} q^{-n_a} \tag{5.112}$$

The orders of these polynomials are assumed to be known and n_a, $n_b \geqslant 1$. If the delay k_1 is known, then b_0 is nonzero and $k \triangleq k_1$. If the delay k_1 is unknown ($k_1 < n_b$), then k must be set to unity in (5.111), which is the minimum value for physical realizability. Then, if $k_1 > 1$, the estimated coefficients b_0, b_1, ..., b_{k_1-2} will be found to be equal to zero [124].

The disturbance and reference subsystems are denoted by $W_2 = C/A$ and $W_3 = E/A$, respectively. The polynomial A is the least common denominator for all the three subsystems. We assume that none of these subsystems includes unstable hidden modes. We assume that the signals v, ξ, and ω are stationary, zero mean sequences of uncorrelated random variables which are jointly independent and have the variances R, Q_2, and Q_3, respectively. The signal r can also represent an extra perturbation signal, so that closed-loop identifiability conditions can be met [95, 126].

We apply the generalized spectral factors introduced by Shaked [127] for the analysis. Let the performance criterion error and control weighting functions be denoted by Q_1 and R_1, respectively. Then the generalized spectral factors should satisfy:

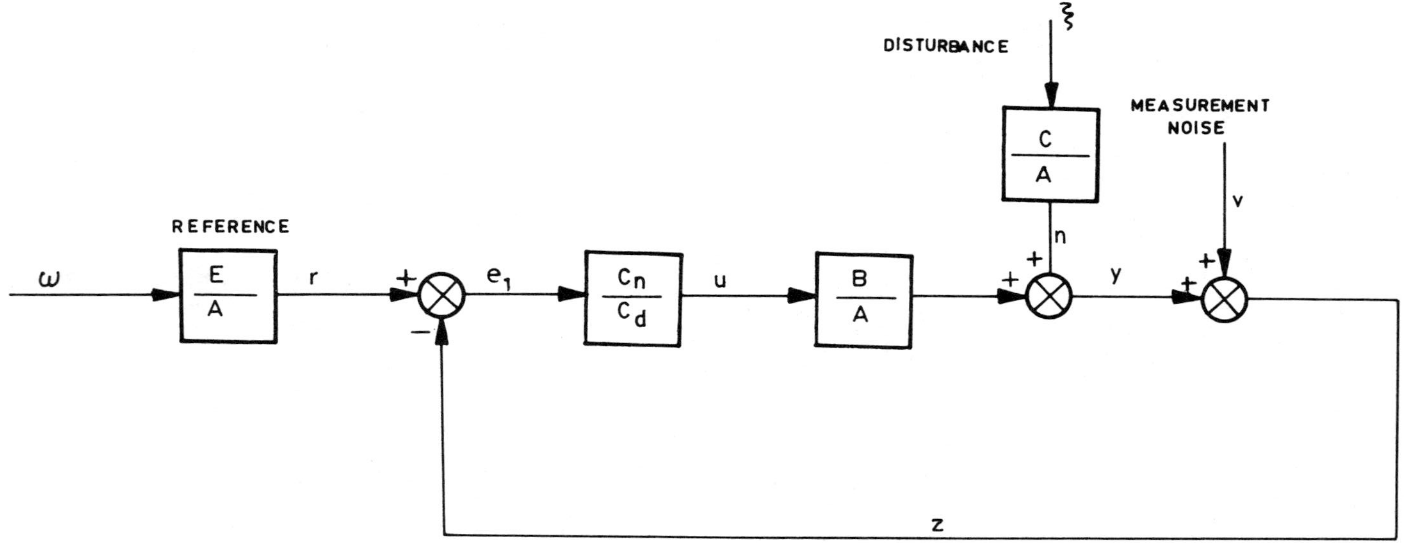

FIG. 5.5 Discrete-time feedback system. (From Ref. 124. Reprinted with permission from Pergamon Press, copyright 1984.)

$$Y\bar{Y} = (EQ_3\bar{E} + CQ_2\bar{C} + AR\bar{A})/(A\bar{A}) \tag{5.113}$$

$$\bar{Y}_1Y_1 = (\bar{B}Q_1B + \bar{A}R_1A)/(\bar{A}A) \tag{5.114}$$

The polynomials D and D_1 are stable (more correctly, $1/D$ and $1/D_1$ are stable) and satisfy

$$Y\bar{Y} = D\bar{D}/A\bar{A} \longrightarrow Y = D/A \tag{5.115}$$

$$\bar{Y}_1Y_1 = \bar{D}_1D_1/AA \longrightarrow Y_1 = D_1/A \tag{5.116}$$

where

$$D = d_0 + d_1q^{-1} + \cdots + d_{n_d}q^{-n_d} \tag{5.117}$$

$$\bar{D} = d_0 + d_1q + \cdots + d_{n_d}q^{n_d} \tag{5.118}$$

$$\tilde{D} = q^{-n_d}\bar{D} = d_{n_d} + d_{n_d-1}q^{-1} + \cdots + d_0q^{-n_d} \tag{5.119}$$

We have the following relations for the controller input,

$$e_1 = r - z \tag{5.120}$$

and the tracking error,

$$e \triangleq r - y \tag{5.121}$$

The plant can be represented by an "innovations model" (Fig. 5.6) in which we use the unit variance white noise ϵ_1. The plant equation is

$$e_1 = (D/A)\epsilon_1 - (B/A)u \tag{5.122}$$

5.7.3 Explicit LQG Algorithms [124]

The cost function to be minimized is defined by

$$J = E\{Q_1e^2(t) + R_1u^2(t)\} \tag{5.123}$$

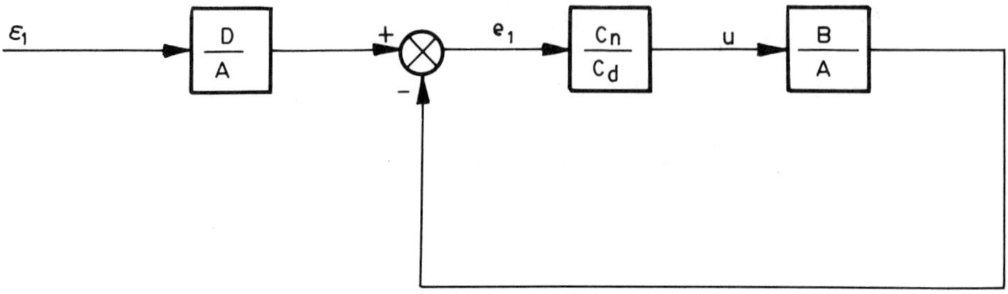

FIG. 5.6 Innovations model of the plant. (From Ref. 124. Reprinted with permission from Pergamon Press, copyright 1984.)

THEOREM 1: Optimal LQG Controller. Consider the cost function (5.123) and the system described in Section 5.7.2, which is in a steady-state condition. The optimal controller becomes

$$C_0 = G_0/H_0 \tag{5.124}$$

The following coupled diophantine equations, in terms of the unknown polynomials G, H, and F, will give the unique particular solution G_0, and H_0, with minimum degree with respect to F.

$$\begin{aligned} \bar{D}_1 q^{-g} G + FA &= \bar{B} q^{-g} QD \\ \bar{D}_1 q^{-2} H - FB &= \bar{A} q^{-g} R_1 D \end{aligned} \tag{5.125}$$

where $g \triangleq \max(n_{d_1}, n_b, n_a)$.

These equations can be combined to give the implied equation

$$AH + BG = D_1 D \tag{5.126}$$

and the closed-loop characteristic polynomial $\rho_c = D_1 D$. (For proof, see Grimble [124].)

Controller Computations

We assume that

1. A, B, and $\tilde{D}_1$ are relatively prime.
2. The orders of the polynomials satisfy

$$n_a > n_d - k \quad \text{and} \quad n_b > n_d$$

The minimal-degree solution with respect to F of the first two diophantine equations has $n_f < g$ and the orders of H and G become $n_h < n_b$ and $n_g < n_a$, respectively. The solution of the implied equation (5.126) is unique, as may be seen from Jezek [128]. We see therefore that the controller (H_0, G_0) can be computed from the solution of one diophantine equation (5.126).

The spectral factor D_1 leads to two types of control law, depending on the way it is defined.

1. Fixed criterion: We assume that Q_1 and R_1 are fixed. To calculate D_1 from (5.116), we require an on-line spectral factorization algorithm [129]. Because D_1 need not be computed at each sample instant, the computation load is low.
2. Fixed spectrum: We assume that D_1 is specified, so the controls of the system are fixed [18]. Then the controller is computed from (5.126). This approach is similar to that of Allidine and Hughes [25] and results in a stabilizing, nonoptimal control law.

The explicit LQG self-tuning strategy is based on the certainty equivalence principle. The algorithm involves a parameter identification and a control law calculation which uses the estimated parameters. The polynomials A, B, and D are estimated from the innovations form of the plant equation (5.122).

The algorithm is based on Theorem 1. The fixed criterion version is given. By omitting the spectral factorization stage (step 2) and prespecifying a fixed D_1, we obtain the fixed-spectrum version, which simplifies calculations and also leads to robust control.

Explicit Fixed-Criterion LQG Self-Tuning Algorithm

Data: Choose the cost function weighting polynomials.

Step 1: Estimate the parameters A, B, and D using the extended least-squares (ELS) method and the model

$$e_1 = \frac{D}{A}\epsilon_1 - Bu$$

Step 2: Calculate the stable spectral factor D_1 from

$$\bar{D}_1 D_1 = \bar{B}BQ_1 + \bar{A}AR_1$$

Step 3: Calculate the controller G_0/H_0 from

$$AH_0 + BG_0 = D_1 D$$

where $n_{h_0} \triangleq n_b - 1$ and $n_{g_0} \triangleq n_a - 1$.

Repeat steps 1 to 3.

5.7.4 Implicit LQG Algorithm [124]

As the controller parameters are directly calculated, the implicit controller is simpler to implement. We assume as earlier that A and B are relatively prime, so that a unique solution exists for (5.126). We use (5.122) also to obtain the implicit model:

$$Ae_1 = (AH_0 + BG_0)\epsilon_1/D_1 - Bu \tag{5.127}$$

$$D_1 e_1 = H_0 \epsilon_1 + \frac{BG_0}{AD}(Ae_1 + Bu) - \frac{D_1 B}{A} u$$

$$= H_0 \epsilon_1 + \frac{B}{D}(G_0 e_1 - H_0 u) \tag{5.128}$$

Let

$$\phi_1 \triangleq D_1 e_1$$

$$\phi_1(t) = H_0 \epsilon_1(t) + \frac{B}{D}[G_0 e_1(t) - H_0 u(t)] \tag{5.129}$$

This represents the desired implicit model from which G_0 and H_0 can be estimated. We note that $n_{h_d} = n_b - 1 \geq k - 1$ and if $n_b > k$, then $H_0 \epsilon_1(t)$ is correlated with the regressors $e_1(t - k)$, $e_1(t - k - 1)$, ..., $u(t - k)$, $u(t - k - 1)$, ... in (5.129). Therefore, we can write $H_0 = H_{0_1} + H_{0_2}$ where H_{0_1} includes all terms with powers of q^{-1} up to $q^{-(k-1)}$. We obtain the least-squares predictor:

$$\hat{\phi}(t|t - k) = H_{0_2} \epsilon_1(t) + \frac{B}{D}[G_0 e_1(t) - H_0 u(t)] \tag{5.130}$$

$$\text{Prediction error} = \phi(t|t - k) = H_{0_1} \epsilon_1(t) \tag{5.131}$$

The optimal control $u^o(t)$ is chosen to set the last term in (5.130) equal to zero. Then

$$u^o(t) = \left(\frac{G_0}{H_0}\right)e_1(t) \tag{5.132}$$

We can simplify the prediction equation by using the argument in [4].

$$\hat{\phi}(t|t-k) = H_{0_2}\epsilon(t) + B[G_0e_1(t) - H_0u(t)] \tag{5.133}$$

If the polynomials B and the prediction error are estimated separately, then we define $e_b = Be_1$ and $u_b = Bu$. In this case the parameters G_0 and H_0 are identified, by using

$$\begin{aligned}\hat{\phi}(t|t-k) &= X(t-k)\theta \\ &= G_0e_b - H_0u_b + H_{0_2}\epsilon\end{aligned} \tag{5.134}$$

If these quantities are unknown, then we can calculate G_0 and H_0 directly from (5.129), or by using the bilinear estimation technique proposed by Åström [49]. We use the former technique to obtain a two-stage estimation algorithm.

Implicit Fixed-Criterion LQG Self-Tuning Algorithm

Data: Choose the cost function weighting polynomials.

Step 1: Estimate the plant polynomials A, B, and D and the innovations ϵ_1 using ELS and the model

$$ae_1 = D\epsilon_1 - Bu$$

Step 2: Calculate the stable spectral factor D_1 using

$$\bar{D}_1D_1 = \bar{B}BQ_1 + \bar{A}AR_1$$

Step 3: Estimate G_0 and H_0 using least squares and the prediction model

$$\hat{\phi} = G_0e_b - H_0u_b + H_{0_2}\epsilon_1$$

Step 4: Calculate the optimal control

$$u^o = \left(\frac{G_0}{H_0}\right)e_1$$

Repeat steps 1 to 4.

The fixed-spectrum version can be obtained by avoiding step 2. If the assumptions made earlier in Section 5.7.3 following Theorem 1 are not satisfied, then the controller is not uniquely determined by (5.126). Then the coupled equations (5.125) should be solved. This does not give rise to any difficulty in the explicit algorithm, as step 3 can be altered to include (5.125). However, the implicit algorithm depends only on (5.126) and so results in suboptimal control.

Simulation results have been obtained for explicit and implicit algorithms, using an example of a non-minimum-phase plant considered by Åström and Wittenmark [48] and Clarke and Gawthrop [4]. Although improved performance was obtained for the implicit case compared to the explicit case, no general conclusion can be drawn. The results demonstrate the effectiveness of LQG self-tuners. These have been found to be robust in the presence of variations in both parameters and transport delay.

5.7.5 Discussion

The best form of LQG self-tuning controller will be determined by the computational complexity with respect to the number of arithmetic operations and storage requirements. Favier et al. [130] have proposed square root and fast algorithms for solving the LQG control problem for discrete-time systems.

Kumar [131] has considered the problem of adaptively controlling an unknown linear Gaussian system with a standard cost criterion, including control cost. A new adaptive control law, based on biasing the usual least-squares parameter estimation criterion with a term favoring parameters associated with lower optimal costs, is introduced. A salient feature of this adaptive control scheme is its imperviousness to the closed-loop identification problem. Properties such as closed-loop system identification, convergence of the adaptive control law to an optimal control law, overall stability of the controlled system, and optimality with respect to the long-term average cost of the adaptive controller are proved. Extensions to overcome the limitated formulation of the problem are necessary.

Grimble [132] provides further insight into LQG controllers by showing that these are equivalent to two forms of minimum-variance problems, involving either a minimum-phase or a non-minimum-phase equivalent plant model. Simulation trials show that a version of an LQG self-tuning controller for scalar systems operates well on non-minimum-phase unstable systems and can even deal with unknown time delays.

LQG self-tuning controllers have considerable potential in view of their desirable features and recent advances in hardware.

5.8 STRs FOR NON-MINIMUM-PHASE PLANTS

5.8.1 Introduction

Non-minimum-phase behavior occurs in many plants such as hydraulic power plants, hydraulic pumps, blast furnaces, dc motors with field regulation, and so on. If a minimum-phase continuous-time system is transformed into an equivalent discrete-time model, the latter is likely to exhibit non-minimum-phase behavior. Non-minimum-phase plants often occur because of time delays which are not integral multiples of the sample interval. This gives rise to a fractional time delay component, which when z-transformed, can result in non-minimum-phase discrete-time zeros. A table of z-transforms for systems involving fractional time delays is given in Wellstead et al. [18]. It is seen that fractional time delays usually occur in self-tuning, because of the finite time required to evaluate the next control. For slow-process systems, this is usually negligible, but for faster systems the computation delay is significant. Overcoming the problems caused by non-minimum-phase behavior has been the subject of many investigations mentioned in Section 5.2.6.

The relationship between the amplitude and phase of the frequency response for a minimum-phase system is governed by the classical Bode theorems. A plant with dead time e^{-sd} or with a right-half plane zero is called non-minimum phase, as these relationships fail. Such plants are not easy to control. In some respect all discrete-time systems are non-minimum phase, since for small sample intervals h, a zero-order hold is equivalent approximately to a dead time of h/2. In general, we term those discrete-time systems non-minimum phase whose zeros lie outside the unit-disk stability region.

Adaptive control of non-minimum-phase systems have proved difficult from both the theoretical and practical points of view. Early self-tuners and many MRAC algorithms involve pole-zero cancellations resulting in instability. However, some recently developed algorithms have overcome the problems posed by non-minimum-phase systems. In the following sections, two design procedures proposed recently will be considered.

5.8.2 A Design Procedure

The following design procedure is based on Unbehauen's work [54]. Consider a plant described by the following discrete transfer function:

$$G_P(z) = \frac{B(z)}{A(z)} z^{-k} = \frac{B^+(z)B^-(z)}{A(z)} z^{-k} \tag{5.135}$$

where $B^+(z)$ includes all zeros inside the unit circle in the z-domain and $B^-(z)$ all zeros outside and on the unit circle. The closed-loop transfer function is given by

$$G(z) = \frac{y_p(z)}{W(z)} = k_q \frac{B^-(z)}{P(z)} z^{-k} \tag{5.136}$$

where P(z) is the characteristic polynomial and

$$k_q = \frac{P(1)}{B^-(1)}$$

A compensating self-tuning controller is to be found. The controller transfer function is given by

$$\begin{aligned} G_c(z) &= \frac{1}{G_p(z)} \frac{G(z)}{1 - G(z)} \\ &= \frac{k_q A(z)B^-(z)}{B(z)[P(z) - k_q B^-(z)z^{-q}]} \end{aligned} \tag{5.137}$$

As $B^-(z)$ is included in G(z), the actuating signal will be bounded. As plant parameters vary, P(z) is to be calculated on the basis of non-minimum-phase behavior.

Let us consider the plant transfer function with one zero outside the right-hand side of the unit circle. Consider a second-order closed-loop transfer function which includes this non-minimum-phase zero, $z_0 = \beta > 1$:

$$G(z) = \frac{K(z - \beta)}{(z - p)(z - \bar{p})} \tag{5.138}$$

$$G(1) = 1$$

Then we can select the two poles $z_1 = p$ and $z_2 = \bar{p}$ arbitrarily.

Consider the step response of a second-order system with non-minimum-phase behavior as shown in Fig. 5.7 in which we have selected the undershoot e_u to be equal to e_{max}, the maximum allowable overshoot for the system. This is a reasonable choice.

The objective is to determine a suitable closed-loop transfer function which gives the desired step response as in Fig. 5.7 and with

$$e_u = e_{max} \tag{5.139}$$

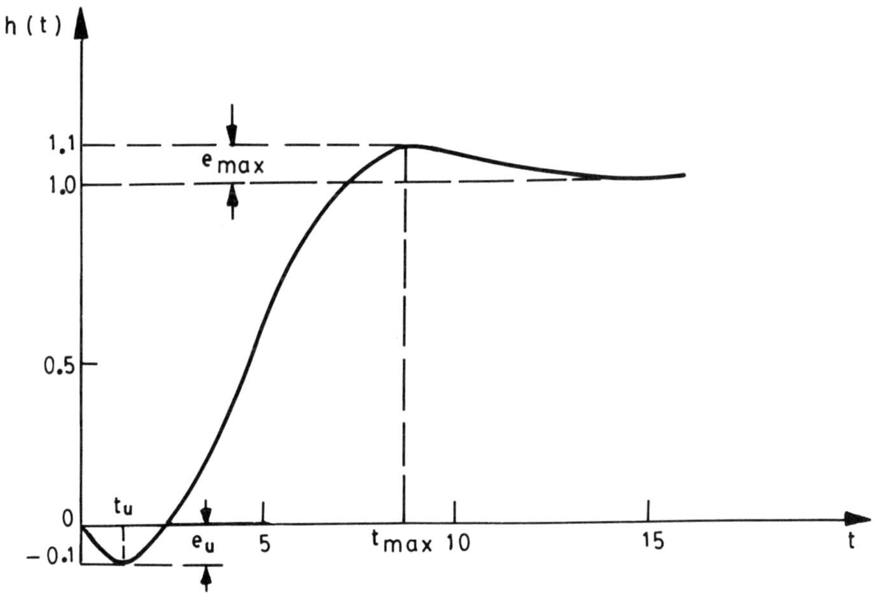

FIG. 5.7 Step response of a second-order non-minimum-phase closed-loop system with $e_u = e_{max}$. (From Ref. 54.)

We can assume that the discrete closed-loop transfer function (5.138) corresponds to the continuous closed-loop transfer function

$$G(s) = \frac{1 + s/f}{1 + (2d/\omega_0)s + (1/\omega_0^2)s^2} \tag{5.140}$$

The pole-zero configurations in the z- and s-planes are given in Fig. 5.8.

The closed-loop poles and zero in the s-domain are

$$\begin{aligned} s_{1,2} &= -\omega_0 d \pm j\omega_0\sqrt{1 - d^2} \\ s_0 &= -f \end{aligned} \tag{5.141}$$

For a given sampling time T_1, we can show that the above s-plane poles and zero given by (5.141) are related to the corresponding z-plane poles and zero by the following equations:

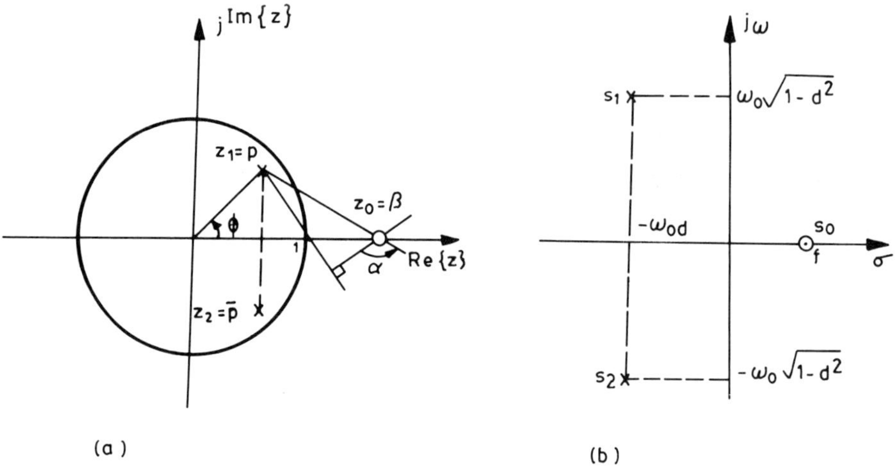

FIG. 5.8 Pole-zero configuration of the non-minimum-phase system in the (a) z- and (b) s-planes. (From Ref. 54.)

$$z_{1,2} = e^{-\omega_0 dT_1} \cos \omega_0 \sqrt{1 - d^2}\, T_1 \pm je^{-\omega_0 dT_1} \sin \omega_0 \sqrt{1 - d^2}\, T_1 \tag{5.142a}$$

$$z_0 = 1 + \frac{e^{-2\omega_0 dT_1} - 2e^{-\omega_0 dT_1} \cos \omega_0 \sqrt{1 - d^2}\, T_1 + 1}{e^{-\omega_0 dT_1} \cos \omega_0 \sqrt{1 - d^2}\, T_1 + \dfrac{e^{-\omega_0 dT_1}}{\sqrt{1 - d^2}} \dfrac{df - \omega_0}{f} \sin \omega_0 \sqrt{1 - d^2}\, T_1 - 1} \tag{5.142b}$$

The closed-loop step response can be described in discrete form, in terms of the angle α and ϕ of Fig. 5.8 by applying the sampled-data theory [133]

$$h(k') = 1 - \frac{|p|^{k'}}{\cos \alpha} \cos(k'\phi + \alpha) \tag{5.143}$$

where h(k') represents the sampling values of the corresponding time domain step response:

$$h(t) = 1 - \frac{|p|^{t}}{\cos \alpha} \cos(t\phi + \alpha) \tag{5.144}$$

Next we calculate t_u, t_{max}, e_u, and e_{max} of Fig. 5.7.

$$t_u = \left(\arctan \frac{\ln |p|}{\phi} - \alpha + \pi\right)\frac{1}{\phi} \tag{5.145}$$

$$t_{max} = \left(\arctan \frac{\ln |p|}{\phi} - \alpha + 2\pi\right)\frac{1}{\phi} \tag{5.146}$$

$$\frac{e_u}{100\%} = -1 - c|p|^{(\arctan \frac{\ln |p|}{\phi} - \alpha + \pi)\frac{1}{\phi}} \left[\frac{1}{\sqrt{1 + \left(\frac{\ln |p|}{\phi}\right)^2}}\right] \tag{5.147}$$

$$\frac{e_{max}}{100\%} = -c|p|^{(\arctan \frac{\ln |p|}{\phi} - \alpha + 2\pi)\frac{1}{\phi}} \left[\frac{1}{\sqrt{1 + \left(\frac{\ln |p|}{\phi}\right)^2}}\right] \tag{5.148}$$

where

$$c = \frac{1}{\cos \alpha}$$

From (5.139), (5.147), and (5.148), we get

$$\frac{\ln |p|}{\phi} = \frac{1}{\pi} \ln \frac{e_{max}}{e_{max} + 100\%} \tag{5.149}$$

Further, we can calculate α from the requirement of Eq. (5.139), $e_u = e_{max}$. From (5.147), we can obtain, using $|p| = e^{\ln |p|}$ and after some modification, the following relation:

$$F_1(\alpha) = \frac{e_u + 100\%}{100\%} + \cos \alpha \frac{e^{-\alpha \frac{\ln |p|}{\phi}} e^{(\arctan \frac{\ln |p|}{\phi} + \pi) \frac{\ln |p|}{\phi}}}{\sqrt{1 + \left(\frac{\ln |p|}{\phi}\right)^2}}$$

$$= 0 \tag{5.150}$$

Substituting $(\ln |p|)/\phi$ from (5.149) in (5.150), the parameter α can be calculated as an implicit solution of (5.150) using the Newton iteration method. But the iteration may be time-consuming because of small values of ϕ. Further, as the solution has to be performed during each sampling interval, it cannot be applied to a fast self-tuning algorithm. Hence an alternative procedure will be followed here.

Nöth [55] has derived the relationship between f, w_0, and d of (5.140) and the equal values of e_u and e_{max}. It has also been shown from (5.146) and (5.148) that the zero z_0 of the discrete transfer function can have any real value $z_0 > 1$ by variation of the sampling time T_1. Keeping this in view, the following method of calculations is suggested.

We replace the calculation of the unknown parameter ϕ by the calculation of the parameter T_1. The discrete zero z_0 is a function of T_1 as seen from (5.142). We calculate T_1 from the condition that z_0 must be identical to the zero β of the discrete transfer function of the plant. So we can transform the equation for z_0 in (5.142) into the form

$$F_2(T_1, \omega_0, d, f, \underset{\sigma}{\beta}) = 0 \tag{5.151}$$

The parameters ω_0 and d are determined by the given values of f = -1 and $e_{max} = e_u$, and the values of β are obtained from the on-line identification of the plant. Equation (5.151) can be solved for T_1 using the Newton iteration approach. We select as the initial value, say,

$$T_1(0) = 0.8\, T_{1z} \tag{5.152}$$

where T_{1z} is the value at which z_0 tends to infinity. It is found that the solution for T_1 is usually obtained within six iterations. This solution is used to find ϕ from

$$\phi = \omega_0 \sqrt{1 - d^2}\, T_1 \tag{5.153}$$

Using the parameters $|p|$, α, and ϕ, obtained from (5.149), (5.150), and (5.153) the desired closed-loop transfer function G(z) can be evaluated.

The open-loop transfer function is obtained from (5.135) and (5.137) as

$$G_0(z) = G_c(z)G_P(z) = \frac{k_q A(z)B^-(z)B(z)z^{-k}}{B(z)\left[P(z) - k_q B(z)z^{-k}\right]A(z)} \tag{5.154}$$

We see that there is cancellation of plant zeros by poles of the controller and vice versa (which is allowed only if all poles and zeros lie within the unit circle). We will therefore consider a modified structure as shown in Fig. 5.9.

The closed-loop transfer function for this system is

$$G(z) = \frac{B(z)T(z)z^{-k}}{A(z)R(z) + B(z)S(z)z^{-k}} \tag{5.155}$$

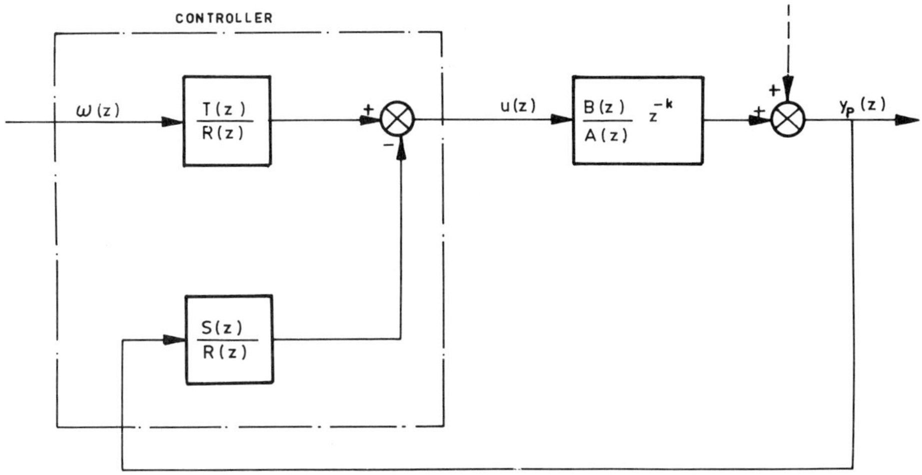

FIG. 5.9 Modified structure of the self-tuning control system. (From Ref. 54.)

This must be equal to the desired closed-loop transfer function given by (5.136), which can be rewritten as

$$G(z) = k_q \frac{B^-(z)}{P(z)} z^{-k} \frac{T_1(z)B^+(z)}{T_1(z)B^+(z)} \tag{5.156}$$

We can equate the numerators and denominators of (5.155) and (5.156) and thus obtain the equations for the synthesis of the controller:

(a) $B(z)T(z)z^{-k} = k_q B^-(z)B^+(z)T_1(z)z^{-k}$

from which we get

$$T(z) = k_q T_1(z) \tag{5.157}$$

(b) $P(z)T_1(z)B^+(z) = A(z)R(z) + B(z)S(z)z^{-k}$

we see that $B^+(z)$ must be a factor of R(z)

$$R(z) = B^+(z)R^*(z)$$
$$P(z)T_1(z) = A(z)R^*(z) + B^-(z)S(z)z^{-k} \tag{5.158}$$

Design Steps

Step 1: $T_1(z)$ can be selected as an asymptotic stable polynomial such that

$$\deg T_1 \leqslant \deg A$$

Step 2: The plant is identified continuously by RLS. Parameter changes are assumed to be slow.

Step 3: The results of the identification are used in (5.136) to calculate G(z) and then the factor $k_q = P(1)/B^-(1)$ is obtained. This is used in (5.157) to obtain the controller polynomial T(z).

Step 4: The other controller polynomials $R^*(z)$ and S(z) are obtained from (5.158). Nöth [55] has shown that a unique solution exists if the following conditions are satisfied

$$\deg R^*(z) = k - 1 + \deg B(z) \tag{5.159}$$

$$\deg S(z) = \deg A(z) - 1 \tag{5.160}$$

This design approach is systematic and has good convergence properties. The computer realization requires a minimum of storage and computing time. This method can be implemented on a microcomputer. The results of the computer realization of a controller for a non-minimum-phase plant at two operating points are given in Unbehauen [54].

5.8.3 Another Design Procedure

Kumar and Moore [59] propose an algorithm which links the adaptive pole-assignment and adaptive minimum-variance strategies, seeking the benefits of both. This work follows the stochastic global convergence theory of these authors [134] and also overcomes the minimum-phase restriction, as also reported in Kumar and Moore [135].

In this approach, weighting coefficients for the performance index, or equivalently the plant augmentations, are chosen adaptively based on plant parameter estimates. In this case we have minimum-variance control of a time-varying augmented plant which is intended to be an asymptotically time invariant system. The plant augmentations are updated as required by the solution of a Bezoutian polynomial equation so that the zeros of the estimated augmented plant are assigned to convenient "minimum-phase" locations. Such locations are assigned keeping in mind that when adaptive minimum-variance control is applied to a plant, the poles of the closed-loop system also converge to the plant zeros. Thus, in assigning the augmented plant zeros, the overall closed-loop poles are assigned to the same locations.

In this approach it is not necessary to have an a priori assumption that the signal model dimension is known, as in the well-known approach of Goodwin and Sin [136]. Another advantage is that this algorithm reverts to the simpler minimum-variance scheme when the control signals are suitably small, thus avoiding the solution of Bezoutian equations.

The control scheme is shown in Fig. 5.10. The plant is augmented with a series moving-average compensator $M(q^{-1})$ and a feedforward moving-average compensator $q^{-1}N(q^{-1})$. k is assumed to be known.

Note: The symbols N, M, L, I, and k are used here in place of E, F, H, k, and N, respectively, used in Kumar and Moore [59].

The augmented plant has zeros given by the zeros of $[q^{-k}B(q^{-1})M(q^{-1}) + q^{-1}N(q^{-1})A(q^{-1})]$. Let us specify the desired augmented plant zeros by means of a minimum-phase polynomial $\bar{L}(q^{-1})$ of appropriate dimensions. When $A(q^{-1})$ and $B(q^{-1})$ are coprime and are of degrees n_a and n_b, respectively, then $M(q^{-1})$ and $N(q^{-1})$ are of degree $r = \max[n_a^{-1}, n_b xk]$. This follows from the theory of Bezoutian equations, given in Wolovich [137].

We get the following relations (deleting q^{-1} arguments) for the augmented plant

$$\frac{\bar{y}_{P_i}}{u_i} = \frac{[q^{-k}BM + q^{-1}NA]}{A} = \frac{\bar{L}}{A}$$
$$y_{P_i} = \bar{L}^{-1}q^{-k}B\bar{y}_{P_i} \tag{5.161}$$

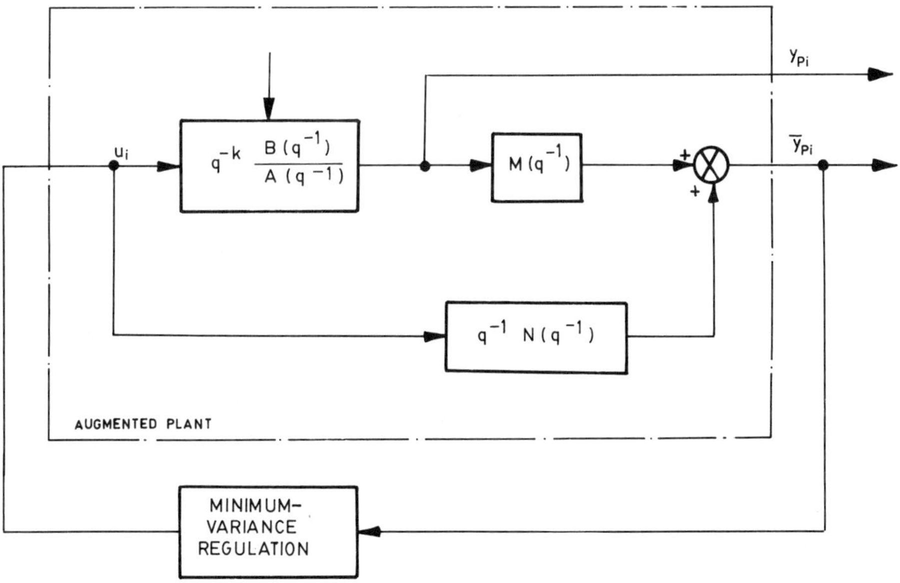

FIG. 5.10 Control scheme for non-minimum-phase plants. (From Ref. 59. Reprinted with permission from Pergamon Press, copyright 1983.)

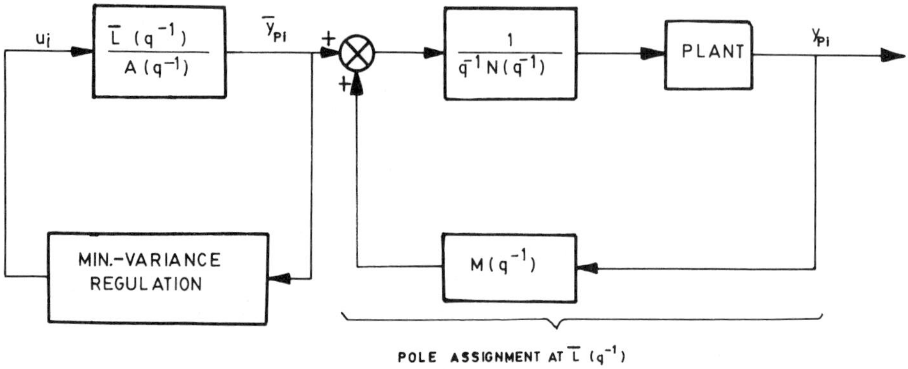

FIG. 5.11 Decomposition of the control scheme of Fig. 5.10. (From Ref. 59. Reprinted with permission from Pergamon Press, copyright 1983.)

Figure 5.10 can be decomposed to show the relationship to the pole-assignment scheme, as in Fig. 5.11.

In Fig. 5.11 a minimum-variance scheme for a minimum-phase plant, $\bar{L}(q^{-1})A^{-1}(q^{-1})$ generates $\bar{y}_{P_i}$, which then feeds into a pole-assignment scheme for assigning plant poles at $\bar{L}(q^{-1})$.

Consider now the case of an unknown plant, with only k, n_a, and n_b known. Then the standard weighted least-squares technique given by Kumar and Moore [134] can be used to estimate the polynomials $\hat{A}_i$ and $\hat{B}_i$ for each time i. Assuming that $\hat{A}_i$ and $\hat{B}_i$ are coprime, $\hat{M}_i$ and $\hat{N}_i$ are calculated to assign zeros for the estimated plant at some minimum-phase specification $\bar{L}(q^{-1})$. Next we have adaptive minimum-variance regulation of the time-varying augmented plant, by one of the two schemes proposed by Kumar and Moore [135]. Kumar and Moore [59] give further details and observations.

5.8.4 Discussion

As non-minimum-phase behavior is encountered in many engineering systems, it is important that such situations be taken care of by appropriate self-tuning strategies. Among the several approaches to this problem, two design procedures have been considered in sections 5.8.2 and 5.8.3, both of which have good convergence properties. The novelty of the method of Kumar and Moore [59] lies in the combination of both minimum-variance and pole-assignment approaches, so that simple and powerful minimum-variance control techniques which are globally convergent can be applied by means of adaptive compensators to control non-minimum-phase plants.

Clarke [60] discusses other approaches to the self-tuning control of non-minimum-phase plants. Simulations of some key methods are given in which a non-minimum-phase plant is subjected to a pattern of set-point changes. In applications the order of the process is not likely to be known. The pole-placement algorithm, which is particularly effective in the known-order case, is found to lack robustness for unspecified order. In the case of the generalized minimum-variance controller, the use of a large control weighting restores stability.

That adaptive control is viable for non-minimum-phase plants is confirmed by an increasing number of successful applications in which good control is achieved by suitable design and by using as much a priori information as possible [60]. However, a study of robustness against unmodeled dynamics is essential for a good adaptive algorithm.

5.9 HYBRID SELF-TUNING CONTROL

5.9.1 Introduction

A recent development of discrete-time adaptive control of continuous-time systems was discussed in Section 4.8. In a similar way Gawthrop [61] introduces hybrid self-tuners which combine both discrete-time and continuous-time features so that they have some advantages of both the methods.

Continuous-time algorithms cannot be applied with the present-day digital hardware, unlike discrete-time algorithms. However, usually the plants are described by continuous-time equations which are close to the real world. Further, a continuous-time plant with zeros in the left half-plane may have zeros outside the unit circle in the z-plane, when sampled [5,16,138]. With hybrid self-tuning it is also possible to avoid the undesirable intersample behavior associated with the discrete-time control of continuous-time systems with high pole-zero excess.

Self-tuning controllers consist basically of a model, a control law, and a parameter estimator. Gawthrop [61] uses a continuous-time model along with a discrete-time control law and estimator. It is implemented digitally by using a fast sample rate for the model and a relatively slow sample rate for the adaptive control law. Whereas Gawthrop has applied the hybrid approach to implicit adaptive controllers, it is also possible to extend this technique to explicit adaptive controllers and this is an area for further research.

Gawthrop has excluded time delays from the analysis to avoid obscuring the basic results. However, the analysis can be readily extended to systems with known time delays. Unknown delays remain a problem for further research.

5.9.2 Derivation of an Equivalent Closed-Loop System

The class of plants treated by Gawthrop is described by

$$A(s)y_P(t) = B(s)u(t) + C(s)\xi(t) \tag{5.162}$$

where A, B, and C are finite-degree polynomials in the differential operator s. $y_P(t)$, u(t), and $\xi(t)$ are the plant output, control input, and disturbance (white noise) respectively. C is assumed to have no roots in the right half-plane and on the imaginary axis.

The following pseudoprocess ϕ_y is defined as in (5.47)

$$\phi_y(t) = P(s)y_P(t) = \frac{P(s)B(s)}{A(s)} u(t) + \frac{P(s)C(s)}{A(s)} \xi(t) \tag{5.163}$$

where P(s) is taken to be a polynomial, but the results may be extended to include P(s) as a rational transfer function.

The process ϕ_y does not exist in any practical sense if CP/A is not strictly proper.

Consider the identity

$$\frac{CP}{A} = F + \frac{G}{A} \tag{5.164}$$

which is similar to the identity (5.49)

$$\deg G = \deg A - 1 \tag{5.165}$$

$$\deg F = \deg C + \deg P - \deg A \tag{5.166}$$

If CP and A have no common factors, Eqs. (5.165) and (5.166) will lead to a unique solution of the diophantine equation resulting from (5.164). The term G/A may be considered the strictly proper part of PC/A and F the improper remainder.

Note: The symbols F, G, ξ, and E are used here in place of E, F, ψ, and G, respectively, used by Gawthrop in [61].

By analogy with (5.50), consider the process ϕ_y^* defined by

$$\phi_y^*(t) = \frac{G(s)}{A(s)} \xi(t) + \frac{P(s)B(s)}{A(s)} u(t) \tag{5.167}$$

Substituting for $\xi(t)$ as in (5.51), we have

$$\phi_y^*(t) = \frac{G(s)}{C(s)} y_P(t) + \frac{E(s)}{C(s)} u(t) \tag{5.168}$$

where

$$E(s) = F(s)B(s) \tag{5.169}$$

The following constraint is found to be necessary, if ϕ_y^* is to be a well-defined (bounded) process with y_P given by (5.162) and u bounded:

$$\deg P \leqslant \deg A - \deg B \tag{5.170}$$

Hence we see that

$$\deg E \leqslant \deg C \tag{5.171}$$

The equality sign holds good simultaneously in (5.170) and (5.171). The "error" e(t) is obtained by using (5.163), (5.164), (5.167), and (5.169)

$$e(t) = \phi_y(t) - \phi_y^*(t) = F(s)\xi(t) \tag{5.172}$$

ϕ_y^* is the realizable approximation to the unrealizable process ϕ_y. We see from (5.172) that the error e(t) does not depend upon $y_P(t)$ or u(t) and depends on F(s) and $\xi(t)$.

We can consider the control law as in (5.57)

$$u(t) = \frac{1}{Q(s)}\left[w(t) - \phi_y^*(t)\right] \tag{5.173}$$

where w(t) is the desired value or the set point of $y_P(t)$.

An equivalent closed-loop system describing equations (5.162), (5.172), and (5.173) is given in Fig. 5.12.

From Fig. 5.12 and the equations obtained so far, we have

$$y_P(t) = \frac{FB + QC}{PB + QA}\,\xi(t) + \frac{B}{PB + QA}\,w(t) \tag{5.174}$$

$$u(t) = \frac{-G}{PB + QA}\,\xi(t) + \frac{A}{PB + QA}\,w(t) \tag{5.175}$$

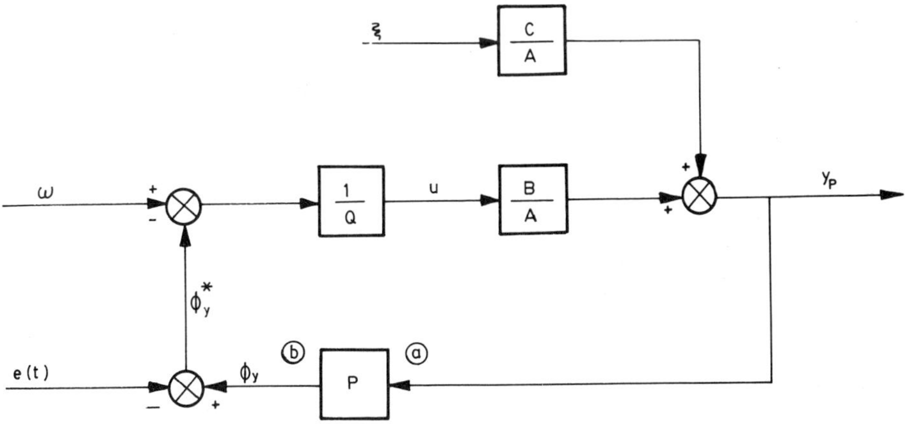

FIG. 5.12 Equivalent closed-loop system. (From Ref. 61.)

From (5.183), we see that Q^{-1} must be proper to avoid unfiltered white noise or set-point derivatives appearing in u(t). Realizable phase advance at all frequencies is obtained through P, subject to the condition (5.170). Therefore, using P and the associated predictor results in derivative action without excessive high frequency noise amplification.

From (5.164), we see that the predictor coefficients are dependent on the plant polynomials A, B, and C. As in Smith's method [139,140], the control law (5.173) is sensitive to polynomial uncertainties. Therefore, this method is most useful when combined with estimation to give a self-tuning algorithm.

5.9.3 Hybrid Predictor Design

Let us next introduce the following (finite variance) filtered version of ϕ_y

$$\tilde{\phi}_y(t) = \frac{1}{D(s)} \phi_y(t) \tag{5.176}$$

where D(s) is a filter polynomial with

$$\deg D > \deg P + \deg C - \deg A \tag{5.177}$$

The filter 1/D "smears out" the autocorrelation functions of the improper part of ϕ_y, that is, $F(s)\xi(t)$ from (5.172), over a width dependent on the bandwidth of the filter. Predicting a time ahead larger than this width makes this "smeared-out" signal once more unrealizable, as it occurs largely in the future. The predictions of ϕ_y and $\tilde{\phi}_y$ will thus be similar. A lemma with a proof is given in Gawthrop [61] in this connection.

Gawthrop gives two design methods, of which the first method has the disadvantage that the coefficients of the predictor have no limit as $T \longrightarrow 0$. The second method is outlined here. It is based on suitable approximation of the continuous predictor (5.167). This involves sampling of the sequence $y_P(t)$, and so samplers are used at both points a and b in Fig. 5.12. It has been shown that this does not affect stability, provided

$$\deg C = \deg A - 1 \tag{5.178}$$

These considerations lead to the following <u>design steps</u>:

<u>Step 1:</u> Choose the order of P as suggested by Edmunds [16]:

$$\deg P = \deg A - \deg B$$

<u>Step 2:</u> Derive the corresponding continuous-time predictor (5.167).

<u>Step 3:</u> Adjoin a sample-hold device to both E/C and G/C and derive the associated (exact) discrete-time transfer functions to give

$$\phi^*_{y'}(i + 1 | i) = \frac{G'(s')}{C'(s')} y'_P(i) + \frac{E'(s')}{C'(s')} u'(i) \tag{5.180}$$

where $s' = (1 - q^{-1})/T$ with T as the sample interval and q^{-1} as the backward shift operator. This is always possible, since

$$\deg G = \deg C = \deg E$$

It has been shown by a lemma in Gawthrop [61] that G', C', and E' are indeed G, C, and E, as $T \longrightarrow 0$. Equation (5.180) gives a T-step-ahead prediction of $\tilde{\phi}_y$ for suitable D.

Step 4: The discrete-version of (5.173) is given by

$$u'(i) = \frac{1}{Q'(s')} [\phi^*_{y'}(i) - w'(i)] \tag{5.181}$$

This gives the hybrid predictive control law. Gawthrop [61] examines the stability of this control law with reference to a theorem and shows that, unlike conventional predictive methods [3, 5, 6, 26], the preceding design procedure leads to a closed-loop system (and also stability properties) approximating the continuous-time equations (5.174) and (5.175). Reference may be made to the paper for further details.

5.9.4 Hybrid Self-Tuning Algorithm

The difference between discrete-time and hybrid self-tuners is small. As the former type has been discussed earlier, only the necessary modifications are given here.

Step 1: Choose
- (a) P(s) using continuous-time analysis,
- (b) A low-pass filter D(s) such that deg D > deg P, and
- (c) A sample interval T so that $\tilde{\phi}^*_y$ approximates to ϕ^*_y, while retaining finite variance.

Step 2: Generate $\tilde{\phi}_y = (P/D)y$ (using a digital filter with a sample interval $\ll$ T).

Step 3: Sample $\tilde{\phi}_y$ and use it in place of ϕ_y.

Step 4: Use the estimation model

$$\tilde{\phi}^*_y(i + 1 | i) = X^T(i)\theta$$

where

$$X^T(i) = \{u'(i), u'(i-1), \ldots, y'_P(i), y'_P(i-1), \ldots, \tilde{\phi}^*_y(i | i-1), \ldots\},$$

θ contains the corresponding coefficients from (5.180).

Gawthrop gives examples of simulation, illustrating the power of this method to deal with both regulation and set-point following and shows that there are no undesirable intersample effects.

Although the controller part of the algorithm can use analog components, in practice the hybrid self-tuner is implemented digitally [60]. By reasonable choices of the assumed plant model and the desired closed-loop model, we can formulate self-tuning PID laws [44].

5.9.5 Application in Multiloop Control

Gawthrop [140a] discusses the implementation of hybrid self-tuners in a modular form for the control of multiloop processes. Some problems which do not arise in the single-loop case are the following:

1. A tuning algorithm may have to tune controllers not designed for automatic tuning.
2. There must be some monitoring procedure to decide when loops need retuning. Such a facility is possible by a suitable choice of software modules.
3. There may be some interaction between the loops and this must be studied.
4. Software modules must be capable of being embedded in complex control systems.

In such applications, a number of questions occur such as

1. What should be the basic self-tuning module or modules?
2. How can such modules be interconnected and connected to more conventional modules such as a PID controller? In particular, the number of module types must be limited, and the data flow between modules must not be too great. A high-level language CONIC provides a natural framework for specifying and implementing such systems.

In the proposed multicontroller system with N (noninteracting) control loops, one possible configuration consists of (1) a one-tuner module, (2) N predictor modules and N controllers, and (3) a supervisory module which determines which of the N-loops is to be tuned at any instant of time. The predictor module contains those parts of the algorithms with related and regular sampling requirements. The tuner module contains those parts with related but irregular sampling requirements. The predictor module, having been tuned, can be disconnected from the tuning module, while still generating prediction errors. These prediction errors can be monitored on-line. Large errors indicate a need for retuning or an equipment failure. Such a configuration provides any loop with an automatic performance monitoring module without the overhead of one tuner per loop.

Gawthrop discusses [140b] two simple multiloop self-tuning configurations: a two-input/two-output interactive system and a two-loop cascade system. These two configurations show that a single-loop self-tuning controller can be used as a building block in a multiloop control system. It is possible to extend this approach to more complex systems. Reference may be made also to Gawthrop [140c-140e].

5.9.6 Discussion

This novel hybrid self-tuning algorithm improves on the discrete equivalent by yielding a control with properties associated with a continuous-time method. Further, continuous-time systems with high pole-zero excess may be controlled without the intersample phenomena resulting from the negative-real-axis zeros of the equivalent discrete-time systems. This method has an advantage over continuous-time MRAC since the model can be chosen on the basis of dynamics within the sampling bandwidth, so that with an appropriate sample interval nondominant poles and zeros may be ignored.

Further work is necessary to extend to (1) cases of explicit controllers, and (2) cases of unknown delays. It is also necessary to explore the implications of the values of sample rates and filter time constants.

5.10 CONVERGENCE ANALYSIS

Convergence analysis is important in self-tuning problems for the determination of convergence conditions, possible convergence points, and convergence rates.

Convergence analysis considers equations such as (5.22). Eykoff [141] discusses these problems thoroughly. In adaptive problems the following two situations arise. The plant input is generated by feedback, and so the excitation of the plant is affected by the disturbances. Hence it is difficult to show that the input is persistently exciting, which is necessary to prove the convergence of the estimate. Further, the input is correlated with the disturbance, with the result that the regression vector ϕ in (5.22) will depend on past estimates. Equation (5.22) will not therefore be a simple state equation. We will have to use results from ergodic theory and martingale theory to establish the convergence of adaptive systems. Martingale theorems are given in standard texts [142,143].

Sternby [144] gives a general proof for convergence of the least-squares algorithms using a martingale convergence theorem. An extension of this is applied to adaptive systems by Sternby and Rootzen [145], using a Bayesian approach.

Goodwin et al. [146] give a convergence theorem for a simple self-tuner based on estimation by modified stochastic approximation and minimum-variance control. They use a martingale convergence theorem.

Åström and Wittenmark [3] apply the method of averaging to find possible equilibrium points for the parameters in a self-tuning problem based on minimum-variance control and estimation by least-squares or stochastic approximation.

Ljung [147] has applied the method of averaging for the self-tuning problem. It was shown that asymptotically as time goes to infinity, the estimates converge to the solution of an ordinary differential equation. Hence this technique is called the ODE-approach. This method is useful for determination of possible equilibrium points and their local stability and for determination of convergence rates. This method is difficult to apply even for simple systems as the differential equation involved is complicated. Further, this method is based on the assumption that the signals are bounded. Ljung [148] applied the ODE-approach to a simple self-tuner based on stochastic approximation and minimum-variance control. A recent book by Ljung and Söderström [149] is a useful reference.

Holst [150] analyzes the local convergence problem for a self-tuner based on minimum variance and least-squares estimation. Kushner [151] considers convergence of a family of recursive adaptive and identification procedures via weak convergence theory. Some of the concepts of the theory of weak convergence of a sequence of probability measures allow relatively simple and natural proofs for many types of convergence and approximation problems and the results can be generalized in many directions. Reference may be made to a book by Kushner and Clark [152].

Solo [153] shows that provided a certain positive real condition is satisfied, the AML recursion for the parameters of a scalar ARMAX time series model converges with probability 1 without the need of monitoring. Previous proofs of convergence had effectively required that the recursion be monitored.

Gawthrop [154] analyzes the stability properties of a self-tuning controller using standard input/output concepts of stability criteria, the conic operator, and the small-gain theorem. Conditions for finite-gain input/output stability are derived. Using a martingale-based method, the stability results are shown to imply convergence (with probability 1) of the mean-square prediction error to the smallest value achievable with a causal control law. Further, a corresponding result is also obtained for the parameter estimates, using the concept of a storage function. The results in this paper are given in a unified manner and provide added insight into the self-tuning algorithm. However, the results are incomplete since at three points, extra assumptions are needed to enable the analysis to proceed. Furthermore, these assumptions are not (and possibly cannot) be proved for the self-tuning algorithm. However, the assumptions can be eliminated if the algorithm is suitably modified to bound the adaptive law coefficients, thus ensuring stability and convergence. In view of the close connections between self-tuning and MRAC methods, the methods developed in this paper are also applicable to some model reference systems.

Goodwin et al. [155] have analyzed discrete-time adaptive control and prediction algorithms for linear time-invariant systems having general delay and colored noise. It is shown that, under suitable conditions, the algorithms are globally convergent. By global convergence, it is meant that for all initial system and algorithm states, with probability 1, the (conditional) mean-square output tracking error is minimized and that this is achieved with a sample mean-square bounded input sequence.

Cordero and Mayne [156] consider the convergence properties of a modified self-tuning regulator with a variable forgetting factor developed by Fortescue et al. [99] and show that the tracking error, in response to a sudden change in the process, is bounded and converges to zero. They apply the convergence theory due to Goodwin et al. [116]. In order to establish convergence a slight modification has been made in the algorithm proposed by Fortescue et al. This study is done for a SISO, deterministic, discrete-time, time-invariant system. Although Fortescue's algorithm works well in practice and appears to prevent "blowing up" of the covariance matrix and subsequent unstable control, the analysis given in Cordero and Mayne [156] shows that in some cases the covariance does blow up. Hence there is need to introduce a further feature, that is, a mechanism to bound $||P(t)||$, in order to ensure convergence. This feature is invoked only infrequently, so the performance of the modified algorithm is substantially the same as that of Fortescue et al. This paper establishes the deterministic convergence of the modified algorithm (algorithm 1) when β_0 is known, and also the convergence of a further algorithm (algorithm 2) which estimates β_0 when β_0 is not known, where β_0 is a term involved in the expression for the B-parameter.

Lozano [156a] has also established convergence for an exponentially weighted least-squares algorithm in which the adjustment of the weighting depends on the covariance matrix. There are other algorithms for the time-varying problem such as that of Vogel and Edgar [28], who use an ordinary least-squares algorithm to which an additional term is added into the covariance matrix whenever parameter changes are detected. This approach also leads to the same result as the variable forgetting factor of Fortescue et al. [99], in view of the fact that a sudden drop in the weighting factor effectively resets the covariance.

Goodwin et al. [156b] also consider the use of an ordinary least-squares algorithm in which the covariance matrix is reset at various times. They establish deterministic convergence of a self-tuning regulator with covariance resetting. They also study an alternative scheme by adding an extra term into the covariance update at various times as in Vogel and Edgar [28]. In this case also, they show that global convergence results. In their simulation experience, it is found that this simple algorithm with the resetting of covariance has excellent practical features when applied to time-varying problems.

Global convergence results for stochastic adaptive control algorithms based on a minimum-variance controller using extended least-squares parameter estimates has been given by Kumar and Moore [134,157]. These results are based on the almost sure convergence of positive supermartingale and so ensure "almost sure convergence" and stability for adaptive control. These results are relevant for the case of a single run of the adaptive controller with a disturbance which is not known a priori. It is necessary to extend the results for many runs of the experiment in order to know the average behavior of the system.

Boel et al. [158] show how to guarantee convergence in mean for the adaptive control algorithm studied by Kumar and Moore [134]. By showing the existence of upper bounds on certain terms in the positive supermartingale, the dominated convergence theorem is applicable. This paper highlights some of the differences between "almost sure" and pth mean convergence of a stochastic adaptive system. They indicate that it is important to consider several different convergence concepts in order to obtain a more robust practical result. Mean convergence results are important if one wants to draw conclusions about the average behavior of a system over many runs of the experiment.

5.11 SELF-TUNING IMPLEMENTATION

Self-tuning implementation is possible in many ways. It can form part of a DDC package in a minicomputer control system; the self-tuner acts continuously on a few critical loops or can cycle around the loops in turn, adjusting the feedback parameters and leaving them fixed until the next cycle. Alternatively, a self-tuner works as an "out-station" in a microprocessor-based distributed control system. This is the latest trend in automation. We can also have dedicated microprocessor-based self-tuners, with special-purpose software. All this is possible with the increasing availability of low-cost powerful microprocessors.

Clarke et al. [104,105] present a feasibility study of the application of microprocessors to STR. Clarke discusses the implementation considerations of self-tuning controllers [106,159]. Wittenmark and his colleagues [160] and Cope [161] also consider the microprocessor implementation of self-tuning controllers. Zanker and Wellstead [162-164] discuss the practical features and the applications of self-tuning. Bengtsson [165] gives industrial experience with a microcomputer-based STR. Fjeld and Wilhelm [166] discuss the software aspects of self-tuning regulators. Glattfelder et al. [167] also deal with microcomputer-based self-tuning controllers. It is shown that microcomputers are well suited for hybrid implementation of the resulting control systems. This is in contrast to a completely digital design reported by Åström and Wittenmark [3]. A short survey of industrial applications of the methods developed is also included in Glattfelder et al. [167].

Clarke and Gawthrop [105] give the hardware and software details of a portable computer SESAME [168] with special reference to the use of a new high-level language, Control Basic [169]. SESAME has been used to test self-tuning on industrial processes such as the control of room-temperature, acid neutralization, and batch chemical reactors. The self-tuning program is in the form of a collection of modules for initial parameter entry, sequence control, self-tuning, parameter dumps, and error processing. It has been possible to have efficient on-site development and testing of self-tuning methods. It has also been shown how the self-tuning controller copes with the nonlinearities encountered in practical processes, though further theoretical analysis is called for.

Bonivento et al. [170] deal with the implementation of pole-zero placement STRs and generalized minimum-variance STRs using Intel-8085 microprocessors. The main features of the realized system are:

1. Use of simplified algorithms
2. Use of a time-varying forgetting factor for parameter estimation
3. Use of a UD-factorized recursive least-squares method
4. Use of a floating-point arithmetic realized via software
5. Significant use of high-level programming languages for noncritical parts of software

They point out the design features and also discuss the choices relating to memory and run-time requirements. They describe the controller prototype architecture. Experimental results on a laboratory plant agree with expected results. This paper demonstrates how rather sophisticated and advanced control algorithms can be implemented using quite simple and inexpensive components.

Since the self-tuning controller is used to obtain "superior" performance, its integrity is of considerable importance [106]. In some cases, absolute reliability is crucial. When designing a self-tuner, assumptions are made about the process and the environment. The usual assumptions are that the process is linearizable, that the signals are persistently exciting such that the estimator can produce a reasonable model, and that the parameters of the self-tuner have been correctly chosen. However, in practical application to a real process, these assumptions may not be valid, as the process may be nonlinear, signals may remain constant over long periods of time, and incorrect parameters may have been chosen. A well-designed self-tuner will have a robust control algorithm, will have jacketing software to ensure reliable estimation, and will be relatively insensitive to incorrect parameters. Clarke [106] gives many useful details about controller integrity, estimation integrity, and appropriate choice of system parameters, controller parameters, and estimator parameters so as to give the best performance. A practical example has also been given to verify the utility and robustness of self-tuning.

In practice, it is necessary to make sure that the self-tuner works satisfactorily under extreme conditions. Many ad hoc features have proved useful. It is often helpful to limit the control signal and its rate of change [171]. It is useful to use robust estimation [172,173] to avoid a single large measurement error which upsets the recursive estimation [107].

Some operational problems of self-tuning such as estimator wind-up, covariance wind-up, integral action, and parameterization are discussed at length by Åström [107,174]. Proper choices require insight and knowledge. There are cases where bad choices proved to be disastrous. The practical aspects of implementing self-tuning are treated by Zanker [164] and Oates [175].

Latawiec and Chyra [176] discuss long-term performance of self-tuning controllers with particular reference to the effects of drift. The related problem of "blow-up" protection for self-tuning controllers is also investigated and some effective measures are suggested.

Wittenmark et al. [177] have reported about the various practical issues in the implementation of self-tuning controllers. There are many disparities between academic and practical algorithms. In practice, self-tuning controllers should be able to handle nonlinearities, unmodeled dynamics, and unmodeled disturbances over a wide range of operating conditions. Practical algorithms must be formulated to take care of unusual and undesirable situations in a safe way. Much of practical work is done on an ad hoc basis depending on the specific application by extensive experimentation and simulation. It is important to have as much a priori knowledge about the process as possible, so as to choose appropriate specifications and select a design method. Many feasibility studies and applications indicate that STCs can be successfully used in many situations. However, the use of STCs is not simple and straightforward and many precautions must be taken. Some of the practical aspects of STCs such as robustness, signal conditioning, parameter tracking, estimator wind-up, reset actions, start-up procedures, anti-reset wind-up, and actuator nonlinearities are considered in detail in Wittenmark and Åström [177]. The advantages of STCs are that the controller can follow process variations, that more complex controllers can be used, and that dead-time and feedforward compensations can be easily handled.

5.12 CONCLUDING REMARKS

Self-tuning regulators have opened up several possibilities for application of adaptive control to diverse industrial processes because they are easily implemented with microelectronic devices. This chapter gives a comprehensive view of the subject, starting with a survey of different approaches to self-tuning. Basic concepts of stochastic control theory and some parameter estimation methods which are relevant to self-tuning are given. Algorithms for minimum-variance STR, the generalized minimum-variance

STR, pole placement and pole-zero placement, and LQG self-tuning are developed. Although MRAC systems and STRs are basically different approaches, the essential equivalence between these is also indicated. Design methods are given to cope with non-minimum-phase behavior which is usually present in many processes. An important development of hybrid self-tuning is also considered. Some aspects of convergence analysis and of self-tuning implementation are also outlined. While this chapter has focused attention on single-input/single-output self-tuners, the next chapter discusses multivariable self-tuning.

The self-tuner is rapidly maturing into a viable control method in industry. However, it is pertinent to point out that the application of self-tuning regulators in most fields is still in the stage of exploration and experimentation. There is considerable scope for further research and development, with greater interaction between researchers in academic institutions and in industry.

REFERENCES

1. Kalman, R. E., Design of a self-optimizing control system, Trans. ASME, 80: 468-478 (1958).
2. Peterka, V., Adaptive digital regulation of noisy systems, Proc. IFAC Symp. on Identification and Process Parameter Estimation, Prague (1970).
3. Åström, K. J., and Wittenmark, B., On self-tuning regulators, Automatica, 9: 185-199 (1973).
4. Clarke, D. W., and Gawthrop, P. J., Self-tuning controller, Proc. IEE, 122: 929-934 (1975).
5. Clarke, D. W., and Gawthrop, P. J., Self-tuning control, Proc. IEE, 126: 633-640 (1979).
6. Gawthrop, P. J., Some interpretations of the self-tuning controller, Proc. IEE, 124: 889-894 (1977).
7. Åström, K. J., Borison, U., Ljung, L., and Wittenmark, B., Theory and applications of self-tuning regulators, Automatica, 13: 457-476 (1977).
8. Grimble, M. J., Weighted minimum variance self-tuning control, Int. J. Contr., 36: 597-609 (1982).
9. Grimble, M. J., and Fung, P. T. K., Explicit minimum variance self-tuning controllers, Proc. IEEE-CDC, San Diego (1981).
10. Grimble, M. J., Weighted minimum variance control of nonlinear systems, Research Report No. ICU/15, Industrial Control Unit, Dept. of Electronic and Electrical Eng., Univ. of Strathclyde, Glasgow (1982).
11. Anbumani, K., Patnaik, L. M., and Sarma, I. G., Self-tuning minimum variance control of nonlinear systems of the Hammerstein model, IEEE Trans. Auto. Contr., AC-26: 959-961 (1981).

12. Peterka, V., Predictor based self-tuning control, Automatica, 20: 39-58 (1984).
13. Toivonen, H. T., Minimum variance control of first order systems with a constraint on the input amplitude, IEEE Trans. Auto. Contr., AC-26: 556-558 (1981).
14. Narendra, K. S., and Valavani, L. S., Direct and indirect adaptive control, Automatica, 15: 653-664 (1979).
15. Fuchs, J. J.-J., Explicit self-tuning methods, Proc. IEE, 127, Pt. D: 259-264 (1980).
16. Edmunds, J. M., Digital adaptive pole-shifting regulators, Doctoral Thesis, UMIST, Manchester, U.K. (1976).
17. Wouters, W. R. E., Adaptive pole-placement for linear stochastic systems with unknown parameters, Preprints IEEE-CDC, New Orleans, pp. 159-166 (1977).
18. Wellstead, P. E., Prager, D. L., and Zanker, P. M., Pole-assignment self-tuning regulator, Proc. IEE, 126: 781-787 (1979).
19. Åström, K. J., Westerberg, B., and Wittenmark, B., Self-tuning controllers based on pole-placement design, Report No. TFRT-3148, Dept. of Auto. Contr., Lund Inst. of Tech., Lund, Sweden (1978).
20. Wellstead, P. E., Edmunds, J. M., Prager, D. L., and Zanker, P. M., Self-tuning pole/zero assignment regulators, Int. J. Contr., 30: 1-26 (1979).
21. Wellstead, P. E., and Zanker, P. M., Servo self-tuners, Int. J. Contr., 30: 27-36 (1979).
22. Wellstead, P. E., Edmunds, J. M., Prager, D. L., and Zanker, P. M., Classical and optimal self-tuning control, Int. J. Contr., 31: 1003-1005 (1980).
23. Wellstead, P. E., and Sanoff, S. P., Extended self-tuning algorithm, Int. J. Contr., 34: 433-445 (1981).
24. Wittenmark, B., Self-tuning PID-controllers based on pole-placement, Report No. 7179, Dept. of Auto. Contr., Lund Inst. of Tech., Lund, Sweden (1979).
25. Allidina, A. Y., and Hughes, F. M., Generalized self-tuning controller with pole assignment, Proc. IEE, 127, Pt. D: 13-18 (1980).
26. Åström, K. J., and Wittenmark, B., Self-tuning controllers based on pole-zero placement, Proc. IEE, 127, Pt. D: 120-130 (1980).
27. Vogel, E. F., and Edgar, T. F., An adaptive dead-time compensator for process control, Proc. JACC, Paper No. TP 5-E (1980).
28. Vogel, E. F., and Edgar, T. F., Application of an adaptive pole-zero placement controller to chemical processes with variable dead time, Proc. JACC, Washington, D.C. (1982).
29. Peterka, V., and Åström, K. J., Control of multivariable systems with unknown but constant parameters, Proc. 3rd IFAC Symp. on Identification and Process Parameter Estimation, The Hague, pp. 535-544 (1973).

30. Zhao-Ying, Z., and Åström, K. J., A microcomputer implementation of LQG self-tuner, Report No. TFRT-7226, Dept. of Auto. Contr., Lund Inst. of Tech., Lund, Sweden (1981).
31. Åström, K. J., and Zhao-Ying, Z., A linear quadratic Gaussian self-tuner, Proc. Workshop on Adaptive Control, Florence (1982).
32. Lam, K. P., Implicit and explicit self-tuning controllers, Doctoral Thesis, Oxford University (1980).
33. Lam, K. P., Implicit and explicit self-tuning controllers, Report No. 1334/80, Eng. Lab., Oxford University (1980).
34. Salcudean, S., and Balanger, P. R., An indirect adaptive regulator: Computational algorithms and simulation results, Proc. 6th IFAC Symp. Identification and System Parameter Estimation, Washington, D.C., pp. 253-258 (1982).
35. Kumar, P. R., On the optimal adaptive control of linear systems, Proc. 6th IFAC Symp. Identification and System Parameter Estimation, Washington, D.C., pp. 627-630 (1982).
36. Grimble, M. J., Self-tuners based on LQG theory, in Encyclopedia of Systems and Control (M. Singh, ed.), Pergamon Press, Oxford, 1983.
37. Grimble, M. J., Robust LQG design of discrete systems using a dual criterion, Research Report No. ICU/24, Industrial Control Unit, Univ. of Strathclyde, Glasgow (1983).
38. Grimble, M. J., Observations weighted LQG controllers for self-tuning systems, Report No. EEE/72, Sheffield City Polytechnic, Sheffield, U.K. (1981).
39. Grimble, M. J., Recent trends in the design of multivariable control systems by optimal methods, Proc. IASTED Conf. on App. Control and Identification, Copenhagen (1983).
40. Åström, K. J., Ziegler-Nichols auto-tuners, Report No. TFRT-3167, Dept. of Auto. Contr., Lund Inst. of Tech., Lund, Sweden (1982).
41. Åström, K. J., and Hägglund, T., Automatic tuning of simple regulators with specifications on phase and amplitude margins, Automatica, 20: 645-651 (1984).
42. Ziegler, J. G., and Nichols, N. B., Optimum settings for automatic controllers, Trans. ASME, 65: 433-444 (1943).
43. Wittenmark, B., and Åström, K. J., Simple self-tuning controllers, in Methods and Applications of Adaptive Control (H. Unbehauen, ed.), Springer-Verlag, Berlin, pp. 21-30, 1980.
44. Gawthrop, P. J., Self-tuning PI and PID regulators, Proc. IEEE Conf. on Applications of Adaptive and Multivariable Control, Hull, U.K. (1982).
45. Kurz, H., and Goedecke, W., Digital parameter-adaptive control of processes with unknown dead time, Automatica, 17: 245-252 (1981).
46. Åström, K. J., and Zhao-Ying, Z., Self-tuners with automatic adjustment of the sampling period for processes with time delays, Report No. TFRT-7229, Dept. of Auto. Contr., Lund Inst. of Tech., Lund, Sweden (1981).

47. Åström, K. J., A self-tuning regulator for non-minimum phase systems, Report No. TFRT-7411, Dept. of Auto. Contr., Lund Inst. of Tech., Lund, Sweden (1974).
48. Åström, K. J., and Wittenmark, B., Analysis of a self-tuning regulator for non-minimum phase systems, Preprints IFAC Symp. Stochastic Control, Budapest (1974).
49. Åström, K. J., New implicit adaptive pole-zero placement algorithms for non-minimum phase system, Report No. TFRT 7172, Dept. of Auto. Contr., Lund Inst. of Tech., Lund, Sweden (1979).
50. Åström, K. J., Direct methods for non-minimum phase systems, Proc. 19th IEEE-CDC, Albuquerque, N.M., pp. 611-615 (1980).
51. Samson, C., and Fuchs, J. J.-J., Discrete adaptive regulation of not-necessarily minimum phase systems, Proc. IEE, <u>128</u>, Pt. D: 102-108 (1981).
52. Samson, C., An adaptive LQ controller for non-minimum phase systems, Int. J. Contr., <u>35</u>: 81-86 (1983).
53. Unbehauen, H., and Nöth, G., Einsatz von "Self-tuning" Reglern bei Regelstrecken mit nichtminimalen phasenverhalten, XXVII Int. Wissenschaftliches Kolloquium der TH Ilmenan (1982).
54. Unbehauen, H., Self-tuning controller design for non-minimum phase plants, Proc. 4th Polish-English Seminar on Real-Time Process Control, Jablonna, Poland (1982).
55. Nöth, G., Self-tuning—Strategien zur Reglang nichminimalphasiger Regelstrecken, Doctoral Thesis, Ruhr-University, Bochum, F.R.G. (1982).
56. Goodwin, G. C., and Sin, K. S., <u>Adaptive Filtering Prediction and Control</u>, Prentice-Hall, Englewood Cliffs, N.J., 1984
57. Grimble, K. J., A control weighted minimum variance controller for non-minimum phase systems, Int. J. Contr., <u>33</u>: 751-762 (1981).
58. Elliott, H., Direct adaptive pole-placement with application to non-minimum phase systems, IEEE Trans. Auto. Contr., <u>AC-27</u>: 720-722 (1982).
59. Kumar, R., and Moore, J. B., On adaptive minimum variance regulation for non-minimum phase plants, Automatica, <u>19</u>: 449-451 (1983).
60. Clarke, D. W., Self-tuning control of non-minimum phase systems, Automatica, <u>20</u>: 501-517 (1984).
61. Gawthrop, P. J., Hybrid self-tuning control, Proc. IEE, <u>127</u>, Pt. D: 229-236 (1980).
62. Gawthrop, P. J., and Clarke, D. W., Hybrid self-tuning control and its interpretation, Report No. 1131/80, Dept. of Eng. Science, Oxford University (1980).
63. Bar-Shalom, Y., and Tse, E., Dual effect, certainty equivalence and separation in stochastic control, IEEE Trans. Auto. Contr., <u>AC-19</u>: 494-500 (1974).
64. Fel'dbaum, A. A., <u>Optimal Control Systems</u>, Academic Press, New York, 1965.

65. Wittenmark, B., Stochastic adaptive control methods—A survey, Int. J. Contr., 21: 705-730 (1975).
66. Joseph, P. D., and Tou, J. T., On linear control theory, AIEE Trans. (Appl. Ind.), 80: 193-196 (1961).
67. Saridis, G. M., Self-Organizing Control of Stochastic Systems, Marcel Dekker, New York, 1977.
68. Jacobs, O. L. R., and Patchell, J. W., Caution and probing in stochastic control, Int. J. Contr., 16: 189-199 (1972).
69. Tse, E., and Bar-Shalom, Y., Generalized certainty equivalence and dual effect in stochastic control, IEEE Trans. Auto. Contr., AC-20: 817-819 (1975).
70. Simon, H., Dynamic programming under uncertainty with a quadratic function, Econometrica, 24: 74-81 (1956).
71. Theil, H., A note on certainty equivalence in dynamic planning, Econometrica, 25: 346-349 (1956).
72. Jacobs, O. L. R., Introduction to adaptive control, in Self-Tuning and Adaptive Control (C. J. Harris and S. A. Billings, eds.), Peter Peregrinus, London, 1981.
73. Harris, C. J., and Billings, S. A. (eds.), Self-Tuning and Adaptive Control: Theory and Applications, Peter Peregrinus, London, 1981.
74. Bar-Shalom, Y., and Tse, E., Caution, probing and the value of information in the control of uncertain systems, Ann. Econ. Soc. Meas., 5: 323 (1976).
75. Patchell, J. W., and Jacobs, O. L. R., Separability, neutrality, and certainty equivalence, Int. J. Contr., 13: 337-342 (1971).
76. Striebel, G., Sufficient statistics in the control of stochastic systems, J. Math. An. App., 12: 576-592 (1965).
77. Wonham, W. M., On the separation theorem of stochastic control, SIAM J. Contr., 6: 312-326 (1968).
78. Akashi, H., and Nose, H., On certainty-equivalence of stochastic optimal control problem, Int. J. Contr., 21: 857-863 (1975).
79. Alspach, D. L., and Sorensen, H., Stochastic optimal control for linear but non-Gaussian systems, Int. J. Contr., 13: 1169-1181 (1971).
80. Root, J. G., Optimum control of non-Gaussian linear systems with inaccessible state variables, SIAM J. Contr., 7: 317-323 (1969).
81. Eykoff, P., System Identification, Wiley, New York, 1974.
82. Sage, A. P., and Melsa, J. L., Estimation Theory, with Applications to Communications and Control, McGraw-Hill, New York, 1971.
83. Bierman, G. J., Factorization Methods for Discrete System Estimation, Academic Press, New York, 1977.
84. Åström, K. J., and Eykoff, P., System identification—A Survey, Automatica, 7: 123-162 (1971).
85. Ljung, L., and Söderström, T., Theory and Practice of Recursive Identification, MIT Press, Cambridge, Mass., 1983.

86. Grimble, M. J., Fung, P. T. K., and Johnson, M. A., Optimal self-tuning control systems: Theory and application. Pt. 2: Identification and self-tuning, Trans. Inst. Meas. Contr., 4: 25-36 (1982).
87. Panuska, V., An adaptive recursive least squares identification algorithm, Proc. IEEE Symp. Adaptive Processes, Decision, and Control (1969).
88. Young, P. C., An instrumental variable method for real time identification of a noisy process, Automatica, 6: 271-288 (1970).
89. Ljung, L., Convergence analysis of parametric identification methods, IEEE Trans. Auto. Contr., AC-23: 770-783 (1978).
90. Ljung, L., Söderström, T., and Gustavsson, I., Counter examples to general convergence of a commonly used recursive identification method, IEEE Trans. Auto. Contr., AC-20: 643-652 (1975).
91. Söderström, T., An on-line algorithm for approximate maximum likelihood identification of linear dynamic systems, Report No. TFRT-3052, Dept. of Auto. Contr., Lund Inst. of Tech., Lund, Sweden (1973).
92. Ljung, L., Convergence of recursive estimators, Proc. IFAC Symp. Identification and System Parameter Estimation, Darmstadt, F.R.G. (1979).
93. Söderström, T., Ljung, L., and Gustavsson, I., A comparative study of recursive identification methods, Report No. TFRT-7427 Dept. of Auto. Contr., Lund Inst. of Tech., Lund, Sweden (1974).
94. Isermann, R., Parameter adaptive control algorithms—A tutorial, Automatica, 18: 513-528 (1982).
95. Kurz, H., Isermann, R., and Schumann, R., Experimental comparison and application of various parameter adaptive control algorithms, Automatica, 16: 117-133 (1980).
96. Peterka, V., A square root filter for real-time multivariable regression, Kybernetika, 11: 53-67 (1975).
97. Unton, F. Z., Recursive estimator of solutions of linear equation sequence, IEEE Trans. Auto. Contr., AC-29: 177-179 (1984).
97a. Unbehauen, H., System identification methods using parameter estimation—A survey, Proc. 2nd Int. Symp. on Systems Analysis and Simulation, Berlin, DDR (1985).
98. Clarke, D. W., Introduction to self-tuning controllers, in Self-Tuning and Adaptive Control (C. J. Harris and S. A. Billings, eds.), Peter Perigrinus, London, 1981.
99. Fortescue, T. R., Kershenbaum, L. S., and Ydstie, B. E., Implementation of self-tuning regulators with variable forgetting factors, Automatica, 17: 831-835 (1981).
100. Xianya, X., and Evans, R. J., Discrete-time stochastic adaptive control for time-varying systems, IEEE Trans. Auto. Contr., AC-29: 638-640 (1984).
101. Xianya, X., and Evans, R. J., Discrete-time adaptive control for deterministic time-varying systems, Automatica, 20: 309-319 (1984).

102. Grimble, M. J., and Johnson, M. A., Optimal self-tuning control systems: Theory and application: Part 1—Introduction and controller design, Trans. Inst. Meas. Contr., 2: 115-120 (1980).
103. Ljung, L., and Wittenmark, B., Analysis of a class of adaptive regulators, Proc. IFAC Symp. Stoch. Control, Budapest (1974).
104. Clarke, D. W., Cope, S. N., and Gawthrop, P. J., Feasibility study of the application of microprocessors to self-tuning regulators, Report No. 1137/75, Oxford Univ. Eng. Lab. (1975).
105. Clarke, D. W., and Gawthrop, P. J., Implementation and application of microprocessor-based self-tuners, Automatica, 17: 233-244 (1981).
106. Clarke, D. W., Implementation of self-tuning controllers, in Self-Tuning and Adaptive Control (C. J. Harris and S. A. Billings, eds.), Peter Peregrinus, London, 1981.
107. Åström, K. J., Theory and applications of adaptive control—A survey, Automatica, 19: 471-486 (1983).
108. Monopoli, R. V., Model reference adaptive control with an augmented error signal, IEEE Trans. Auto. Contr., AC-19: 474-484 (1974).
109. Richalet, J., Rault, A., Testud, J. L., and Papon, J., Model predictive heuristic control: Application to industrial processes, Automatica, 14: 413-428 (1978).
110. Egardt, B., Unification of some discrete-time adaptive control schemes, IEEE Trans. Auto. Contr., AC-25: 693-697 (1980).
111. Ionescu, T., and Monopoli, R. V., Discrete model reference adaptive control with an augmented error signal, Automatica, 13: 507-518 (1977).
112. Bénéjean, R., La commande adaptive à modèle de référence evolutif (application à la commande sans accés aux variables d'état), Doctoral Thesis, 3me cycle, Institut National Polytechnique de Grenoble (1977).
113. Johnson, C. R., Jr., and Tse, E., Adaptive implementation of one-step-ahead optimal control via input matching, IEEE Trans. Auto. Contr., AC-25: 865-872 (1978).
114. Landau, I. D., and Lozano, R., Unification of discrete-time explicit model reference adaptive control designs, Automatica, 17: 593-611 (1981).
115. Egardt, B., Stability analysis of discrete-data adaptive control schemes, IEEE Trans. Auto. Contr., AC-25: 710-716 (1980).
116. Goodwin, G. C., Ramadge, P. J., and Caines, P. E., Discrete-time multivariable adaptive control, IEEE Trans. Auto. Contr., AC-25: 449-456 (1980).
117. Landau, I. D., Combining model reference adaptive control systems and stochastic self-tuning regulators, Automatica, 18: 77-84 (1982).
118. Ljung, L., and Landau, I. D., Model reference adaptive systems and self-tuning regulators—Some connections, Proc. 7th IFAC Congress, Helsinki, pp. 1973-1979 (1978).

119. Landau, I. D., Model reference adaptive control and stochastic self-tuning regulators, AFOSR Workshop on Adaptive Control, Univ. of Illinois, Urbana-Champaign (1979).
120. Landau, I. D., Adaptive controllers with explicit and implicit reference models and stochastic self-tuning regulators—Equivalence and duality aspects, Proc. IEEE-CDC, San Diego (1979).
121. Schmid, Ch., Ein Beitrag zur Realiserung adaptiver Regelungssysteme mit dem Prozeβrechner, Doctoral Thesis, Ruhr-Universitat, Bochum, F.R.G. (1979).
122. Narendra, K. S., and Lin, Y.-H., Stable discrete adaptive control, IEEE Trans. Auto. Contr., AC-25: 456-461 (1980).
123. Johnson, C. R., Jr., Input matching, error augmentation, self-tuning, and output error identification, IEEE Trans. Auto. Contr., AC-25: 697-703 (1980).
124. Grimble, M. J., Implicit and explicit LQG self-tuning controllers, Automatica, 20: 661-669 (1984).
125. Åström, K. J., LQG self-tuners, Proc. IFAC Workshop on Adaptive Systems in Control and Signal Processing, San Francisco, pp. 137-146 (1983).
126. Gustavson, I., Ljung, L., and Söderström, T., Identification of processes in closed-loop—Identifiability and accuracy aspects, Automatica, 13: 59-75 (1977).
127. Shaked, U., A general transfer function approach to linear stationary filtering and steady-state optimal control problems, Int. J. Contr., 24: 741-770 (1976).
128. Jezek, J., New algorithm for minimal solution of linear polynomial equations, Kybernetika, 18: 505-516 (1982).
129. Kučera, V., Discrete Linear Control: The Polynomial Equation Approach, Wiley, New York, 1979.
130. Favier, G., Alangrin, G., and Orsini, M., Square root and fast algorithms for solving the LQG problem in discrete-time systems, Proc. IEEE-CDC, pp. 340-345 (1981).
131. Kumar, P. R., Optimal adaptive control of LQG systems, SIAM J. Contr. Optim., 21: 163-178 (1983).
132. Grimble, M. J., LQG multivariable controllers—Minimum variance interpretation for use in self-tuning systems, Int. J. Contr., 40: 831-842 (1984).
133. Lindorff, D., Theory of Sampled-Data Control Systems, Wiley, New York, 1965.
134. Kumar, R., and Moore, J. B., Convergence of adaptive minimum variance algorithms via weighting coefficient selection, IEEE Trans. Auto. Contr., AC-27: 146-153 (1982).
135. Kumar, R., and Moore, J. B., Adaptive minimum variance control harnessed for non-minimum phase plants, Proc. IFAC 8th World Congress (1981).

136. Goodwin, G. C., and Sin, K. S., Adaptive control of non-minimum phase systems, IEEE Trans. Auto. Contr., AC-26: 478-483 (1981).
137. Wolovich, W. A., Linear Multivariable Systems, Springer, New York, 1974.
138. Åström, K. J., Hagander, P., and Sternby, J., Zeros of sampled systems, Proc. IEEE-CDC (1980).
139. Marshall, J. E., Control of Time-Delay Systems, Peter Peregrinus, London, 1979.
140. Smith, O. J. M., A controller to overcome dead-time, ISA Trans., 6: 28-33 (1959).
140a. Gawthrop, P. J., Implementation of distributed self-tuning controllers, Proc. EUROCON, Brighton, U.K. (1984).
140b. Gawthrop, P. J., Self-tuning multiloop process control: An outline, Proc. 4th Polish-English Seminar on Real-Time Process Control, Warsaw, pp. 89-97 (1983).
140c. Gawthrop, P. J., Hybrid self-tuning control, in Encyclopedia of Systems and Control (M. G. Singh, ed.), Pergamon Press, Oxford, 1983.
140d. Gawthrop, P. J., Multiloop self-tuning control: Cascade system, Proc. 9th IFAC World Congress, Budapest (1984).
140e. Gawthrop, P. J., Multiloop self-tuning control: Two-input two-output systems, Report No. CE/T/4, University of Sussex, School of Eng. and Appl. Sci. (1983).
141. Eykoff, P. (ed.), Trends and Progress in System Identification, Pergamon Press, Oxford, 1981.
142. Chung, K. L., A Course in Probability Theory, 2nd ed., Academic Press, New York, 1975.
143. Doob, J. L., Stochastic Processes, Wiley, New York, 1953.
144. Sternby, J., On consistency for the method of least squares using martingale theory, IEEE Trans. Auto. Contr., AC-22: 346-352 (1977).
145. Sternby, J., and Rootzen, H., Martingale theory in Bayesian least squares estimation, Proc. 6th IFAC Symp. on Identification and System Parameter Estimation, Washington, D.C. (1982).
146. Goodwin, G. C., Ramadge, P. J., and Caines, P. E., Discrete-time stochastic adaptive control, SIAM J. Contr. Optim., 19: 829-853 (1981).
147. Ljung, L., Analysis of recursive stochastic algorithms, IEEE Trans. Auto. Contr., AC-22: 551-575 (1977).
148. Ljung, L., On positive real transfer functions and the convergence of some recursive schemes, IEEE Trans. Auto. Contr., AC-22: 539-550 (1977).
149. Ljung, L., and Söderström, T., Theory and Practice of Recursive Identification, MIT Press, Cambridge, Mass., 1983.

150. Holst, J., Local convergence of some recursive stochastic algorithms, Proc. IFAC Symp. on Identification and System Parameter Estimation, Darmstadt, F.R.G. (1979).
151. Kushner, H. J., Convergence of recursive adaptive and identification procedures via weak convergence theory, IEEE Trans. Auto. Contr., AC-22: 921-930 (1977).
152. Kushner, H. J., and Clark, D. S., Stochastic Approximation Methods for Constrained and Unconstrained Systems, Springer, New York, 1978.
153. Solo, V., The convergence of AML, IEEE Trans. Auto. Contr., AC-24: 958-962 (1979).
154. Gawthrop, P. J., On the stability and convergence of self-tuning controller, Int. J. Contr., 31: 973-998 (1980).
155. Goodwin, G. C., Sin, K. S., and Saluja, K. K., Stochastic adaptive control and prediction—The general delay-colored noise case, IEEE Trans. Auto. Contr., AC-25: 946-950 (1980).
156. Cordero, A. O., and Mayne, D. Q., Deterministic convergence of a self-tuning regulator with variable forgetting factor, IEE Proc., 128, Pt. D: 19-23 (1981).
156a. Lozano, R., Adaptive control with forgetting factor, Proc. IFAC World Congress, Kyoto, Japan (1981).
156b. Goodwin, G. C., Elliott, H., and Teoh, E. K., Deterministic convergence of a self-tuning regulator with covariance resetting, IEE Proc., 130, Pt. D: 6-8 (1983).
157. Kumar, R., Simultaneous adaptive control and identification via the weighted least square algorithm, IEEE Trans. Auto. Contr., AC-29: 259-263 (1984).
158. Boel, R. K., Teo, K. L., and Moore, J. B., Convergence in mean for ELS-based adaptive control, Syst. Contr. Lett., 5: 389-396 (1985).
159. Clarke, D. W., Some implementation considerations of self-tuning controllers, in Numerical Techniques in Stochastic Systems (F. Archetti and M. Cugiani, eds.), North-Holland, Amsterdam, 1980.
160. Wittenmark, B., Hagander, P., and Gustavsson, I., STUPID—Implementation of a self-tuning PID-controller, Report, Dept. of Auto. Contr., Lund Inst. of Tech., Lund, Sweden (1980).
161. Cope, S. N., Microprocessor implementation of a self-tuning controller, IEE Colloquium on Self-Tuning and Adaptive Systems (1976).
162. Wellstead, P. E., and Zanker, P., Techniques of self-tuning, Report No. 432, Control Systems Centre, UMIST, Manchester, U.K. (1978).
163. Zanker, P. M., and Wellstead, P. E., Practical features of self-tuning, Proc. IEE Conf. Trends in On-line Computer Control Systems, Sheffield, U.K., pp. 160-166 (1979).

164. Zanker, P. M., Application of self-tuning, Doctoral Thesis, Control Systems Centre, UMIST, Manchester, U.K. (1980).
165. Bengtsson, G., Industrial experience with a microcomputer-based self-tuning regulator, Preprints, 18th IEEE-CDC, Fort Lauderdale, Fla. (1979).
166. Fjeld, M., and Wilhelm, R. G., Jr., Self-tuning regulators—The software way, Contr. Eng., p. 99 (1981).
167. Glattfelder, A. H., Huguenin, F., and Schaufelberger, W., Microcomputer-based self-tuning and self-tuning controllers, Automatica, 16: 1-8 (1980).
168. Evans, S. E., Application of microprocessors to adaptive control, M.Sc. Thesis, Brunel University (1978).
169. Clarke, D. W., and Frost, P. J., Control BASIC for microcomputers, IEE Conf. on Trends in On-line Computer Control Systems, Sheffield, U.K. (1979).
170. Bonivento, C., and Spisni, F., Realization of a microprocessor-based adaptive controller, Control and Computers (J. IASTED), 11: 1-5 (1983).
171. Wittenmark, B., A self-tuning regulator, Doctoral Thesis, Dept. of Auto. Contr., Lund Inst. of Tech., Lund, Sweden (1973).
172. Huber, P. J., Robust Statistics, Wiley, New York, 1981.
173. Poljak, B. T., and Tsypkin, Ya. Z., Robust identification, Preprints, 4th IFAC Symp. Identification and System Parameter Estimation, Tbilisi, U.S.S.R. (1976).
174. Åström, K. J., Design principles for self-tuning regulators, in Methods and Applications in Adaptive Control (H. Unbehauen, ed.), Springer-Verlag, Berlin, pp. 1-20, 1980.
175. Oates, C., Doctoral Thesis, UMIST, Manchester, U.K. (1980).
176. Latawiec, K., and Chyra, M., On low-frequency and long run effects in self-tuning control, Automatica, 19: 419-424 (1983).
177. Wittenmark, B., and Åström, K. J., Practical issues in the implementation of self-tuning control, Automatica, 20: 595-605 (1984).

BIBLIOGRAPHY

Allidina, A. Y., and Hughes, F. M., Self-tuning controller with integral action, Opt. Contr. Appl. Math., 3: 4 (1982).

Allidina, A. Y., and Hughes, F. M., Self-tuning regulators for deterministic systems, Int. J. Contr., 37: 831-842 (1983).

Allidina, A. Y., Hughes, F. M., and Tye, C., Self-tuning control for systems employing feedforward, IEE Proc., 128: 283-291 (1981).

Andriev, N., A new dimension: A self-tuning controller that continually optimizes PID constants, Contr. Eng., 84 (1981).

Åström, K. J., Lectures on the identification problem—The least squares method, Report No. 6806, Dept. of Auto. Contr., Lund Inst. of Tech., Lund, Sweden (1968).

Åström, K. J., Introduction to Stochastic Control Theory, Academic Press, New York, 1970.

Åström, K. J., Stochastic control problems, in Mathematical Control Theory (W. A. Coppel, ed.), Springer-Verlag, Berlin (1978).

Åström, K. J., Simple self-tuner, Report No. TFRT-7184, Dept. of Auto. Contr., Lund Inst. of Tech., Lund, Sweden (1979).

Åström, K. J., and Gustavsson, I., Analysis of a self-tuning regulator in a servo loop, Report No. TFRT-3150, Dept. of Auto. Contr., Lund Inst. of Tech., Lund, Sweden (1978).

Åström, K. J., Ljung, Borisson, U., and Wittenmark, B., Theory and applications of adaptive regulators based on recursive parameter estimation, Paper No. 50.1, Proc. IFAC Congress, Boston, Mass. (1974).

Banyasz, Cs., and Keviczky, L., Direct methods for self-tuning PID regulators, Proc. IFAC Symp. Identification and System Parameter Estimation, Washington, D.C., pp. 1395-1400 (1982).

Barabanov, A. E., Use of the method of least squares for designing an adaptive optimal control for a linear dynamic plant, Auto. Rem. Contr., 44: 1574-1581 (1984).

Berger, C. S., New pole-placement design method for adaptive controllers, IEE Proc., 129, Pt. D: 13-14 (1982).

Bethoux, G., Approche unitaire des méthodes d'identification et de commande adaptive des procédés dynamiques, Doctoral Thesis, Inst. Polytechnique de Grenoble, France (1976).

Cameron, F., and Seborg, D. E., A self-tuning controller for PID structure, Proc. IFAC/IFIP Symp. on Real Time Digital Control Applications, Guadalajara, Mexico, pp. 613-622 (1983).

Chan, Y. M., and Cruz, J. B., Jr., Self-tuning leader-follower games, Automatica, 19: 237-245 (1983).

Clarke, D. W., Hodgson, A. J. F., and Tuffs, P. S., Offset problem and k-incremental predictors in self-tuning control, IEE Proc., 130, Pt. D: 217-225 (1983).

De Keyser, R. M. C., Design and evaluation of self-tuning controllers, Doctoral Thesis, State University of Ghent, Belgium (1980).

Fegley, K. A., Baldvinsson, G. I., and Chen, H., Design of self-adjusting controllers, Proc. IFAC Symp. Identification and System Parameter Estimation, Washington, D.C., pp. 1377-1382 (1982).

Fogel, E., On self-tuning regulators and the least squares procedure, Proc. IFAC Symp. Identification and System Parameter Estimation, Washington, D.C., pp. 1199-1203 (1982).

Fogel, E., and Graupe, D., Convergence of least squares identification algorithms applied to unstable stochastic processes, Int. J. Syst. Sci., 8: 611-618 (1977).

Gawthrop, P. J., Developments in optimal and self-tuning control theory, Doctoral Thesis, Univ. of Oxford (1977).

Gawthrop, P. J., A continuous-time approach to discrete-time self-tuning control, Opt. Contr. Appl. Meth., 3: 399-414 (1982).

Godfrey, K. R., and Brown, R. F., Practical aspects of identification of process dynamics, Trans. Inst. M.C., 2: 85-95 (1979).

Goodwin, G. C., and Payne, R. L., Dynamic System Identification: Experiment Design and Data Analysis, Academic Press, New York, 1977.

Goodwin, G. C., and Sin, K. S., Stochastic adaptive control using a modified least squares algorithm, Tech. Report No. EE 7907, Dept. of Elec. Eng., University of Newcastle, N.S.W., Australia (1979).

Hägglund, T., Recursive least squares identification with forgetting of old data, Report No. TFRT-7254, Dept. of Auto. Contr., Lund Inst. of Tech., Lund, Sweden (1983).

Hahn, V., and Unbehauen, H., Direct adaptive control of non-minimum phase systems, Proc. IEEE Conf. Applns. of Adaptive and Multivariable Control, Hull, U.K., pp. 170-175 (1982).

Hijab, O. B., The adaptive LQG problem—Part I, IEEE Trans. Auto. Contr., AC-28: 171-178 (1983).

Hrusak, J., Magdy, M. A., and Simandl, M., A new self-tuning controller for plants disturbed by correlated noise, Proc. IFAC Symp. Identification and System Parameter Estimation, Washington, D.C., pp. 1187-1192 (1982).

Hughes, F. M., and Allidina, A. Y., Self-tuning controller design incorporating frequency response criteria, Proc. IFAC Symp. Identification and System Parameter Estimation, Washington, D.C., pp. 1383-1388 (1982).

Johnson, C. R., Jr., and Elliott, H., On three similar (but different) approaches to direct adaptive pole-placement, Proc. 15th Conf. Inform. Syst. Sci., Baltimore, Md., pp. 257-262 (1981).

Johnstone, R. M., Johnson, C. R., Jr., Bitmead, R. R., and Anderson, B. D. O., Exponential convergence of recursive least squares with exponential forgetting factor, Syst. Contr. Lett., 2: 77-81 (1982).

Karpela, K., and Koivo, H. N., On self-tuning control of a class of nonlinear scalar processes, Proc. 6th IFAC Conf., Washington, D.C. (1979).

Keviczky, L., and Vajk, T., A self-tuning extremum regulator, Proc. 12th ASILOMAR Conf. on Circuits, Systems and Computers, pp. 69-72 (1978).

Kul'chitski, O. Yu., Stochastic approximation algorithms operating in the adaptation loop of a discrete stochastic linear dynamic system, Auto. Rem. Contr., 44: 1189-1203 (1984).

Kumar, R., and Moore, J. B., Detection techniques in least squares identification, Automatica, 17: 805-809 (1981).

Lammers, H. C., and Verbruggen, H. B., Simple self-tuning control, Proc. IFAC Symp. Identification and System Parameter Estimation, Washington, D.C., pp. 1181-1186 (1982).

Lawson, C. L., and Hanson, R. J., Solving Least Squares Problems, Prentice-Hall, Englewood Cliffs, N.J. (1974).

Ljung, L., and Wittenmark, B., Asymptotic properties of self-tuning regulators, Report No. TFRT-3071, Dept. of Auto. Contr., Lund Inst. of Tech., Lund, Sweden (1974).

Ljung, L., and Wittenmark, B., On a stabilizing property of adaptive regulators, Proc. IFAC Symp. Identification and System Parameter Estimation, Tbilisi, U.S.S.R. (1976).

Mäkilä, P. M., Self-tuning control with linear input constraints, Opt. Contr. Appln. Meth., 3: 4 (1982).

Morris, A. J., Nazer, Y., and Chisholm, K., A comparison of identification techniques for robust self-tuning control, Proc. 5th IFAC Symp. Identification and System Parameter Estimation, Darmstadt, F.R.G. (1979).

Peterka, V., On the steady-state minimum variance control strategy, Kybernetica, 8: 219-231 (1972).

Sanoff, S. P., and Wellstead, P. E., Extended self-tuning—Practical aspects, Proc. IFAC Symp. on Identification and System Parameter Estimation, Washington, D.C., pp. 1175-1180 (1982).

Solo, V., Some aspects of recursive parameter estimation, Int. J. Contr., 32: 395-410 (1980).

Toivonen, H. T., Self-tuning regulators with explicit variance restrictions, Int. J. Contr., 38: 229-235 (1983).

Tuffs, P. S., Self-tuning control: Algorithms and applications, Doctoral Thesis, Univ. of Oxford (1985).

Unbehauen, H., System identification methods using parameter estimation—A survey, Preprints, 2nd Int. Symp. on Systems Analysis and Simulation, Berlin, DDR (1985).

Wellstead, P. E., Self-tuning digital control systems: The pole-zero assignment approach, Report No. 490, Control Systems Center, UMIST, Manchester, U.K. (1980).

Wellstead, P. E., and Zanker, P., Techniques of self-tuning, Opt. Contr. Appln. Meth., 3: 305 (1982).

Wellstead, P. E., Zanker, P., and Edmunds, J. M., Self-tuning pole/zero assignment regulators, Report No. 404, Control System Center, UMIST, Manchester, U.K. (1978).

Wellstead, P. E., and Zarrop, M. B., Self-tuning regulators: Nonparametric algorithms, Int. J. Contr., 37: 787-808 (1983).

Wittenmark, B., A self-tuning predictor, Automatica, 17: 848 (1981).

Wong, K. Y., and Bayoumi, M. M., A self-tuning control algorithm for systems with unknown time delay, Proc. IFAC Symp. Identification and System Parameter Estimation, Washington, D.C., pp. 1193-1198 (1982).

Ydstie, B. E., and Co, T., Recursive estimation with adaptive divergence control, IEE Proc., 132, Pt. D: 124-130 (1985).

Ydstie, B. E., and Sargent, R. W. H., Deterministic convergence of an adaptive regulator with variable forgetting factor, Proc. IFAC Symp. Identification and System Parameter Estimation, Washington, D.C., pp. 259-263 (1982).

Zhang, Y.-H., Stochastic adaptive control and prediction based on a modified least squares—The general delay: Colored noise case, IEEE Trans. Auto. Contr., AC-27: 1257-1260 (1982).

Zhao-Ying, Z., and Åström, K. J., Polynomial factorization using recursive formulas for complex integrals, Report No. TFRT-7223, Lund Inst. of Tech., Lund, Sweden (1981).

Zhao-Ying, Z., and Åström, K. J., Regulator synthesis based on polynomial manipulation, Report No. TFRT-7225, Lund Inst. of Tech., Lund, Sweden (1981).

6

Multivariable Self-Tuning Regulators

6.1 INTRODUCTION

Multivariable self-tuning regulators were first considered by Peterka and Åström [1]. The problem was converted into LQG formulation with iterative solution of the discrete Riccati equation in between the sampling instants. The computations required were rather large. Borison [2,3] proposes a multivariable STR based on the minimum-variance strategy, using a recursive least-squares estimator and a linear controller obtained directly from the current estimates. The class of systems is restricted to include only minimum-phase systems with the same number of inputs and outputs. This analysis gives an insight into the case when several SISO self-tuning regulators are used in cascade. Some further improvements are suggested by Keviczky and Hetthessy [4]. They apply such a regulator to the control of raw material blending in a cement plant [5]. The approach of Borison and Keviczky et al. is based on the estimation of the controller parameters instead of the process parameters. This approach requires a larger number of parameters than is needed.

Bayoumi and El-Bagoury [6,7] discuss the disadvantages of identifying controller parameters instead of process parameters. They consider a controller based on minimizing a quadratic function with penalties on the amplitude of the input and on the expected output deviation from the reference vector. They introduce a multivariable self-tuning augmented regulator [8]. The process model parameters are first estimated by the least-squares method and then the estimated parameters are used to find the control signal, which is then augmented with a random signal. The transient properties of this combination are improved over the basic multivariable STR. The peak excursions of inputs and outputs are reduced. This is also reflected in the reduction of the cost function. Further investigation is possible to find the optimal character of the random signal.

Bayoumi et al. [9] give a design method for multivariable plants with constant, unknown parameters. The algorithm operates on-line with a priori information about the time delay. If the order of the system is unknown, it can be determined by a generalized likelihood-ratio statistical test, which is discussed in the paper. Estimation of the input/output system model parameters is based on the least-squares principle. It avoids estimation of excess parameters. The control is computed to minimize the combined cost of output deviation and control energy. The controller is therefore more conservative than the minimum-variance controller because of the penalty on the control magnitude.

Minimum-variance decoupling control has been proposed by Sinha [10]. Cascaded sets of SISO self-tuners are used with decoupled subsystems of the overall multivariable system. The "decoupler" can be designed by any standard technique, as given by Rosenbrock [11]. Morris et al. [12] use the approach of Clarke and Gawthrop to develop a controller which can deal with non-minimum-phase systems and which does not have the limitation that all loop time delays are the same. A recursive factorization parameter estimator based on the least-squares method is used. This, no doubt, increases the computation load. This multivariable controller is applied to such a complex process as product composition control in a distillation column. Simulation and experimental results show that robust control is possible.

Koivo [13] extends the method of Clarke and Gawthrop [14] to a multivariable self-tuning controller, having the same number of inputs and outputs. The controller structure is obtained by penalizing the control terms in the cost function and then obtaining the control law. A k-step-ahead predictor which is slightly different from the one derived by Borison [2] is utilized. A recursive least-squares algorithm in square-root form [15] identifies the controller parameters. Time-varying reference signals can also be tracked and the plant may be non-minimum phase. Thus, this method broadens the scope of Borison's minimum-variance controller [3].

Minimum-variance regulators can sometimes introduce large feedback loop gains, resulting in excessive control signals. Prager [16,17] proposes "detuned" minimum-variance regulators which introduce "moderating poles" into the closed-loop equations, so that the control signal excursions are not large.

Prager and Wellstead [18] extend the pole assignment idea to the multivariable case, to avoid the disadvantages of minimum-variance controllers. The latter due to Borison [2] cannot be applied to non-minimum-phase plants. However, Koivo's approach overcomes this difficulty, as stated earlier. Minimum-variance controllers are also sensitive to variations in system time delays. Further, the minimum-variance design requires that all loops have the same time delay, which is usually not possible in real systems. In pole assignment regulators, these difficulties do not arise. However, the main disadvantage lies in the complexity of the computation required, which restricts the maximum allowable sampling frequency. This is not a problem now with the availability of faster and more powerful microprocessors. Further, robustness is attained at the cost of optimality of the closed-loop configuration; but it must be noted that in many practical applications, optimal control is not the chief requirement. Examples illustrate that a pole assignment self-tuner can provide good results when operating on a real process.

Wellstead and Prager [17] discuss the various methods available for multivariable self-tuning with special reference to their relative merits. They deal also with the detuned minimum-variance regulators, due to Prager [16]. They consider two examples, including a non-minimum-phase plant in which loop time delays also differ and show how the pole-shifting method can cope with both these problems. Bezanson and Harris [19] discuss state variable design of multivariable STRs based on eigenvalue assignment. State variable feedback is used. This approach is applicable for non-minimum-phase plants and for open-loop unstable systems. Grimble [20] reports about minimum-variance interpretation of LQG multivariable controllers for self-tuning.

This chapter is laid out as follows: Section 6.2 discusses the minimum-variance approach to multivariable self-tuning as proposed by Koivo [13]. In Section 6.3 the pole assignment method of Prager and Wellstead [18] is dealt with. Section 6.4 comprises the concluding remarks.

6.2 MINIMUM-VARIANCE APPROACH

6.2.1 System Representation

A multivariable self-tuning controller as proposed by Koivo [13] will be considered here. This is an extension of the method of Clarke and Gawthrop [14].

Consider the system described by the linear vector difference equation:

$$A(q^{-1})\underline{y}_P(t) = B(q^{-1})\underline{u}(t - k) + \underline{d} + C(q^{-1})\underline{\xi}(t) \tag{6.1}$$

where $\underline{y}_P$ is the output, $\underline{u}$ is the control, $\underline{d}$ is the constant steady-state output response for a zero input signal, $\underline{\xi}$ is a sequence of independent equally distributed random vectors with zero mean value and covariance $E\{\xi(t)\xi^T(t)\} = r_\xi$, k is the time delay, and q^{-1} is the backward shift operator, $q^{-1}\underline{y}(t) = \underline{y}(t - 1)$. A, B, and C are polynomial m × m matrices given by:

$$A(q^{-1}) = I + A_1 q^{-1} + \cdots + A_n q^{-n}$$

$$B(q^{-1}) = B_0 + B_1 q^{-1} + \cdots + B_n q^{-n} \qquad (B_0 \text{ is nonsingular})$$

$$C(q^{-1}) = I + C_1 q^{-1} + \cdots + C_n q^{-n}$$

The following are the basic assumptions pertaining to the plant (6.1):

1. The number of outputs is equal to the number of inputs.
2. B_0 is nonsingular.
3. Roots of det $C(\alpha)$ lie outside unit disk.

Whereas in Borison's approach [2], only minimum-phase plants can be considered, the present approach can handle certain non-minimum-phase plants also.

6.2.2 Cost Function

The cost function is of the form

$$J = E\left\{ \left\| \sum_{i=0}^{k-1} p_i \underline{y}_P(t + k - i) - \sum_{i=0}^{N} R_i \underline{w}(t - i) \right\|^2 + \left\| \sum_{i=0}^{M} Q_i' \underline{u}(t - i) \right\|^2 \right\} \tag{6.2}$$

where $\underline{w}(t)$ is the known reference signal, p_i are real numbers, $p_0 = 1$, R_i and Q_i' are $m \times m$ matrices. Notation $\|\underline{x}\|^2 = \underline{x}^T\underline{x}$ has been used. N and M are the upper bounds in the sums of $\underline{w}$ and $\underline{u}$. The values of N and M are, in most cases, $N = 0$ and $M = 0$ or $M = 1$.

Note: The following changes in the symbols have been made:

[13]	E'	F'	e	ϵ	I	I_1	I_2	J	G'
Present work	F'	G'	ϵ	ϵ'	J	J_1	J_2	J'	E'

The cost function (6.2) is minimized over all the admissible strategies. A control strategy is admissible if the control signal at time t, $\underline{u}(t)$ is a function of all the observed outputs up to time t, $\underline{y}_P(t)$, $\underline{y}_P(t - 1)$, ..., and all previously applied control signals, $\underline{u}(t - 1)$, $\underline{u}(t - 2)$, ...

Equation (6.2) can be rewritten as:

$$J = E\{\|P(q^{-1})\underline{y}_P(t+k) - R(q^{-1})\underline{w}(t)\|^2 + \|Q'(q^{-1})\underline{u}(t)\|^2\} \tag{6.3}$$

where P, Q', and R are polynomial matrices obtained from (6.2).

The cost function (6.3) includes the following two cases:

1. $J_1 = E\{\|\underline{y}_P(t+k) - \underline{w}(t)\|^2 + \underline{u}^T(t)\,\mathrm{diag}\{\lambda_1,\ldots,\lambda_m\}\underline{u}(t)\}$
2. $J_2 = E\{\|\underline{y}_P(t+k) - \underline{w}(t)\|^2 + [\underline{u}(t) - \underline{u}(t-1)]^T \times \mathrm{diag}\{\lambda_1,\ldots,\lambda_m\}[\underline{u}(t) - \underline{u}(t-1)]\}$

J_1 penalizes deviations of the output from the reference signal and variations in control. It is well known [14] that J_1 does not guarantee in general that the mean value of $\underline{y}_P(t)$, namely $\overline{\underline{y}}_P(t)$, would equal the mean values of $\underline{w}(t)$, namely, $\overline{\underline{w}}(t)$, unless $\lambda_i = 0$, $i = 1, 2, \ldots, m$.

J_2 introduces additional integrators into the loop and assures the equality of $\overline{\underline{y}}_P(t)$ and $\overline{\underline{w}}(t)$, but the dynamic response will be sluggish.

6.2.3 Derivation of Control Law

The optimal control law is deduced by using the least-squares predictor $\hat{\underline{y}}_P(t+k|t)$, which predicts, at time t, the output k steps ahead using the previous data:

$$\{\underline{y}_P(t), \underline{y}_P(t-1), \ldots, \underline{u}(t), \underline{u}(t-1), \ldots\}$$

The optimal k-step-ahead predictor for (6.1) has been derived in the appendix of Koivo [13] and is given by

$$\hat{\underline{y}}_P(t+k|t) = \tilde{C}^{-1}(q^{-1})[\tilde{G}'(q^{-1})\underline{y}_P(t) + \tilde{F}'(q^{-1})B(q^{-1})\underline{u}(t) + \gamma] \tag{6.4}$$

where

(i) $\gamma = \tilde{G}(1)\underline{d}$.

(ii) $\tilde{F}'$ and $\tilde{G}'$ are as defined in Section 5.5, and F' and G' are, as in Borison [3], nonunique matrices such that the following "pseudocommutivity" relation holds good:

$$\tilde{F}(q^{-1})G'(q^{-1}) = \tilde{G}'(q^{-1})F'(q^{-1})$$

where det $\widetilde{F}'(q^{-1})$ = det $F'(q^{-1})$ and $\widetilde{F}'(0) = 1$. The polynomial matrices $\widetilde{F}'$ and $\widetilde{G}'$ always exist, but they are not unique [21].

(iii) $\widetilde{C}(q^{-1}) = \widetilde{F}'(q^{-1})A(q^{-1}) + q^{-k}\widetilde{F}'(q^{-1})$

$$\widetilde{C}(q^{-1})F'(q^{-1}) = \widetilde{F}'(q^{-1})C(q^{-1})$$

On the basis of these, it is also shown in the appendix of Koivo [13] that

$$\underline{y}_P(t + k) = \{\widetilde{C}^{-1}(q^{-1})[\widetilde{G}'(q^{-1})\underline{y}_P(t) + \widetilde{F}'(q^{-1})B(q^{-1})\underline{u}(t) + \gamma] + \xi(t + k) + \cdots + F'_{k-1}\xi(t + 1)\} \tag{6.5}$$

Equations (6.4) and (6.5) are similar in form to Eqs. (5.52) and (5.51) already obtained for SISO self-tuners.

From (6.4) and (6.5), we get the prediction error, as in Section 5.5.,

$$\begin{aligned}\underline{\epsilon}(t + k) &= \underline{y}_P(t + k) - \hat{\underline{y}}_P(t + k|t)\\ &= \xi(t + k) + \cdots + F'_{k-1}\xi(t + 1)\end{aligned} \tag{6.6}$$

This is uncorrelated with $\underline{y}_P(t)$, $\underline{y}_P(t - 1)$, ..., and $\underline{u}(t)$, $\underline{u}(t - 1)$, ... and hence with $\hat{\underline{y}}_P(t + k|t)$.

Substituting (6.6) in (6.3), we get

$$J = E\{||P(q^{-1})[\hat{\underline{y}}_P(t + k|t) + \underline{\epsilon}(t + k)] - R(q^{-1})\underline{w}(t)||^2 + ||Q'(q^{-1})\underline{u}(t)||^2\} \tag{6.7}$$

$P(q^{-1})\underline{\epsilon}(t + k)$ is uncorrelated with $\underline{u}(t - i)$, $\underline{w}(t - i)$, and $\underline{y}_P(t - i)$ and with $P(q^{-1})\hat{\underline{y}}_P(t + i|t)$, $0 \leq i \leq t$ as shown in the appendix of Koivo [13]. Hence (6.7) can be written as

$$J = ||P(q^{-1})\hat{\underline{y}}_P(t + k|t) - R(q^{-1})\underline{w}(t)||^2 + ||Q'(q^{-1})\underline{u}(t)||^2 + E\{||P(q^{-1})\underline{\epsilon}(t + k)||^2\} \tag{6.8}$$

The problem has been reduced to a deterministic optimization problem and the control law can be determined. However, for self-tuning purposes this form has not been found successful and so we introduce an auxiliary output by defining a matrix polynomial,

$$Q(q^{-1}) \triangleq (B_0^T)^{-1}(Q'(0)^T Q'(q^{-1})) \tag{6.9}$$

and a vector function (as in Section 5.5),

$$\underline{\hat{\phi}}(t+k|t) \triangleq P(q^{-1})\underline{\hat{y}}_P(t+k|t) - R(q^{-1})\underline{w}(t) + Q(q^{-1})\underline{u}(t) \tag{6.10}$$

The optimal control is found by setting

$$\underline{\hat{\phi}}(t+k|t) = 0$$

Another closely related vector function $\underline{\phi}$ is defined by

$$\underline{\phi}(t+k|t) \triangleq P(q^{-1})\underline{y}_P(t+k) - R(q^{-1})\underline{w}(t) + Q(q^{-1})\underline{u}(t) \tag{6.11}$$

Subtracting (6.10) from (6.11) and using (6.6), we have

$$\underline{\phi}(t+k) = \underline{\hat{\phi}}(t+k|t) + \underline{\epsilon}'(t+k) \tag{6.12}$$

where

$$\underline{\epsilon}'(t+k) \triangleq \sum_{i=0}^{k-1} p_i \underline{\epsilon}(t+k-i) \tag{6.13}$$

and is uncorrelated with $\underline{\hat{\phi}}_y(t+k|t)$. So $\underline{\hat{\phi}}_y(t+k|t)$ is the optimal predictor of $\underline{\phi}(t+k)$ at time t.

A minimization problem equivalent to (6.3), producing the same control law, is to minimize

$$J' = E\{\underline{\phi}^2(t+k)\} = \underline{\hat{\phi}}(t+k|t)^2 + E\{\underline{\epsilon}'^2(t+k)\} \tag{6.14}$$

Substituting the predictor (6.4) into (6.10), we get

$$\underline{\hat{\phi}}(t+k|t) = \left\{ \sum_{j=0}^{k-1} p_j \tilde{C}^{-1}(q^{-1})[\tilde{G}'(q^{-1})\underline{y}_P(t) + \tilde{E}'_{k-j}(q^{-1})\underline{u}(t-j) + \gamma_{k-j}] \right.$$
$$\left. - R(q^{-1})\underline{w}(t) + Q(q^{-1})\underline{u}(t) \right\} = 0 \tag{6.15}$$

where

$$\tilde{E}'_{k-j} = \tilde{F}'(q^{-1})B(q^{-1})$$

Let us next define the following matrix polynomials $\tilde{E}$, $\tilde{G}$, and $\tilde{H}$ as

$$\begin{aligned}\tilde{E}(q^{-1}) &\triangleq \Sigma\, p_j q^{-j} \tilde{E}_{k-j} + \tilde{C}(q^{-1}) Q(q^{-1}) \\ \tilde{G}(q^{-1}) &\triangleq \Sigma\, p_j \tilde{G}'_{k-j} \\ \tilde{H}(q^{-1}) &\triangleq -\tilde{C}(q^{-1}) R(q^{-1})\end{aligned} \tag{6.16}$$

and a constant vector

$$\underline{\delta} \triangleq \Sigma\, p_j \underline{\gamma}_{k-j} \tag{6.17}$$

Using the preceding equations in (6.15) and premultiplying by $\tilde{C}$, we get

$$\tilde{C}(q^{-1})\, \underline{\hat{\phi}}(t + k|t) = \tilde{G}(q^{-1})\, \underline{y}_P(t) + \tilde{E}(q^{-1})\, \underline{u}(t) + \tilde{H}(q^{-1})\, \underline{w}(t) + \underline{\delta} = 0 \tag{6.18}$$

We get the optimal control law by solving (6.18) for $\underline{u}(t)$.

It is to be noted that the premultiplication requires that P and C commute, and so P is taken as a polynomial and not a matrix polynomial.

The following observations may be made:

1. If $P = 1$, and $Q = 0$ in (6.18), we get minimum-variance control. The overall system is unstable if B is non-minimum phase.
2. If $Q \neq 0$, we can stabilize the overall system by an appropriate choice of Q in certain cases. Let $Q = \lambda I$, where λ is a constant tuning parameter. Then
   a. If A is open-loop unstable but B is minimum phase, we can get a stable closed-loop system if λ is small enough.
   b. If A is open-loop stable and B is non-minimum phase, we get a stable closed-loop system if λ is large enough.

6.2.4 Self-Tuning Algorithm

Next, to obtain the self-tuning controller algorithm we consider the equivalent system of (6.18) and (6.12).

Consider the case when $C(q^{-1}) = I$. It follows that $\tilde{C}(q^{-1}) = I$ as in Borison [2]. Then Eqs. (6.18) and (6.12) can be written as

$$\underline{\hat{\phi}}(t + k|t) = \tilde{G}(q^{-1})\, \underline{y}_P(t) + \tilde{E}(q^{-1})\, \underline{u}(t) + \tilde{H}(q^{-1})\, \underline{w}(t) + \underline{\delta} = 0 \tag{6.19}$$

$$\underline{\phi}(t + k) = \left\{\tilde{G}(q^{-1})\, \underline{y}_P(t) + \tilde{E}(q^{-1})\, \underline{u}(t) + \tilde{H}(q^{-1})\, \underline{w}(t) + \underline{\delta} + \underline{\epsilon}'(t + k)\right\} \tag{6.20}$$

Let us define the data vector $\underline{x}(t)$ by

$$\underline{x}(t) \triangleq [\underline{y}_P^T(t), \underline{y}_P^T(t-1), \ldots; \underline{u}^T(t), \underline{u}^T(t-1), \ldots; \underline{w}^T(t), \underline{w}^T(t-1), \ldots; 1] \tag{6.21}$$

and the parameter matrix $\underline{\theta}$ (dropping ~):

$$\underline{\theta} = [\underline{\theta}_1, \ldots, \underline{\theta}_m] \triangleq [G_0, G_1, \ldots, E_0, E_1, \ldots, H_0, H_1, \ldots, \gamma]^T \tag{6.22}$$

where the column vector $\underline{\theta}_i$ is

$$\underline{\theta}_i = [g_{i1}^0, \ldots, g_{im}^0, g_{i1}^1, \ldots, g_{im}^1, \ldots; e_{i1}^0, \ldots, e_{im}^0, e_{i1}^1, \ldots, e_{im}^1, \ldots;$$

$$h_{i1}^0, \ldots, h_{im}^0, h_{i1}^1, \ldots, h_{im}^1, \ldots; \gamma_i]^T, \quad i = 1, \ldots, m \tag{6.23}$$

Hence (6.20) can be rewritten component-wise as

$$\phi_i(t+k) = \underline{x}(t)\underline{\theta}_i + \underline{\epsilon}_i'(t+k) \tag{6.24}$$

In (6.24) the components of $\underline{x}(t)$ are uncorrelated with $\underline{\epsilon}_i'(t+k)$. We can obtain the control $\underline{u}(t)$ from (6.19):

$$E_0\underline{u}(t) = -\left[\sum_{i\geq 0} G_i\underline{y}_P(t-i) + \sum_{i\geq 1} E_i\underline{u}(t-i) + \sum_{i\geq 0} H_i\underline{w}(t-i) + \underline{\gamma}\right] \tag{6.25}$$

The control parameters $\underline{\theta}$ can be estimated by a standard recursive least-squares method as in Borison [2]. But computationally it is more advantageous to use Peterka's square-root algorithm [15] for updating the parameters. If the dimension of $\underline{x}(t)$ is ρ, then the recursive algorithm in square-rcot form is given as

$$\hat{\theta}_i(t+1) = \hat{\theta}_i(t) + K(t)[\underline{\phi}_i(t) - \underline{x}(t-k)\hat{\theta}_i(t)], \quad i = 1, \ldots, m \tag{6.26}$$

$$K(t) = e^{(\rho)}/\sigma_\rho^2 \tag{6.27}$$

$$P(t+1) = S(t+1)S^T(t+1) \tag{6.28}$$

where

$$S(t+1)_{ij} = \frac{\sigma_{j-1}[S(t)_{ij} - g_i s_i^{(j-1)}/\sigma_{j-1}^2]}{\sigma_j\sqrt{\beta}}, \quad i, j = 1, \ldots, \rho$$

$$\sigma_0 = \sqrt{\beta}$$

$$\sigma_j = (\sigma_{j-1}^2 + g_j^2)^{1/2}, \qquad j = 1, \ldots, \rho$$

$$g_j = \sum_{i=1}^{j} S(t)_{ij} \underline{x}(t-k)_i, \qquad i = 1, \ldots, \rho$$

$$s_i^{(1)} = \begin{cases} \displaystyle\sum_{k_1=i}^{l} S(t)_{ik_1} g_{k_1}, & i = 1, 2, \ldots, l \\ 0, & i > l \end{cases}$$

$(\cdot)_{ij}$ indicates the (ij)th element of the matrix.

The following is the self-tuning algorithm with $\rho = 1$, $Q = \lambda I$, where λ is a constant tuning parameter.

Algorithm

Step 1: Read new output $\underline{y}_P(t)$ and set point $\underline{w}(t)$.
Step 2: Compute $\underline{\phi}(t) = \underline{y}_P(t) - R\underline{w}(t-k) + \lambda \underline{u}(t-k)$.
Step 3: Form the data vector (6.21).
Step 4: Update $\hat{\theta}_i$ by recursive square-root algorithm (6.26)-(6.28).
Step 5: Obtain new control $\underline{u}(t)$ from (6.25)

$$\hat{E}_0 \underline{u}(t) = -\left[\sum_{i\geq 0} \hat{G}_i \underline{y}_P(t-i) + \sum_{i\geq 1} \hat{E}_i \underline{u}(t-i) + \sum_{i\geq 0} \hat{H}_i \underline{w}(t-i) + \hat{\underline{\gamma}}\right]$$

Step 6: Set $t = t + 1$ and repeat.

We had assumed earlier that $C(q^{-1}) = I$. For cases in which this is not true, we have, by using (6.18) in (6.12):

$$\phi_i(t+k) = \{\underline{x}(t)\underline{\theta}_i + \epsilon_i'(t+k) + [1 - C_{ii}(q^{-1})]\hat{\phi}_i(t+k\,|\,t) - \sum_{j\neq 1} C_{ij}(q^{-1})\hat{\phi}_i(t+k\,|\,t)\} \tag{6.29}$$

This indicates that $\hat{\underline{\phi}}$ is correlated with $\underline{x}$. Then the least-squares are biased. However, since the control law sets $\hat{\underline{\phi}}$ equal to zero, the last two terms in (6.29) vanish.

Set-Point Changes

In process control it is usual to change the system from one set point to another. The controller structure (6.25) can easily handle this situation. If we have to eliminate the steady-state error, an integrator should be introduced by using the cost function J_2. Then $Q = \lambda(1 - q^{-1})I$. Then the preceding algorithm (with $\rho = 1$) has to be modified only in step 2 as follows:

Step 2: Compute

$$\underline{\phi}(t) = \underline{y}_P(t) - R\underline{w}(t - k) + \lambda[\underline{u}(t - k) - \underline{u}(t - k - 1)] \tag{6.30}$$

Another alternative way of introducing an integrator is, as suggested by Wittenmark [22], to use $[\underline{y}_P(t) - \underline{w}(t)]$ wherever $\underline{y}_P(t)$ occurs, and $[\underline{u}(t) - \underline{u}(t - 1)]$ wherever $\underline{u}(t)$ occurs.

6.2.5 Discussion

Koivo [13] gives the simulation results for three examples, two of which were studied by Borison [3] and one of a non-minimum-phase plant and shows how this approach is an improvement over Borison's approach, as non-minimum-phase plants as well as time-varying reference signals can be handled. Further, this controller has the desired advantage of limiting the control signal significantly, compared with the one based on Borison's approach.

Hahn et al. [23] apply the scheme of Koivo to a non-minimum-phase plant which is characterized by rather fast dynamics and strong nonlinearities and show experimentally that satisfactory control is possible for such a system consisting of a small synchronous generator coupled to a small turbine. A brief discussion is given in Section 9.6.3. The modified scheme is applicable also to systems with varying time delays. Hahn et al. suggest that stochastic approximation technique, which is simpler, could be applied for estimation, instead of the least-squares method, which requires relatively more time and storage.

6.3 POLE-ASSIGNMENT APPROACH

6.3.1 System Representation

The pole-assignment approach of Prager and Wellstead [18] eliminates some of the disadvantages of minimum-variance controllers, as already mentioned in Section 6.1. The objective is to design a regulator which assigns the closed-loop poles to the values specified by the designer. The plant, which is both controllable and observable, may be represented by the difference equation

$$[I + A(q^{-1})]\underline{y}_P = q^{-k}B(q^{-1})\underline{u} + [I + C(q^{-1})]\underline{\xi} \tag{6.31}$$

where A, B, and C are polynomial matrices in the backward shift operator q^{-1} and are of the form

$$X(q^{-1}) = X_1 q^{-1} + \cdots + X_{n_x} q^{-n_x} \tag{6.32}$$

where X_i, $i = 1, 2, \ldots, n$, are $p \times p$ matrix coefficients and k is the system time delay, which is an integer multiple of the sampling time. Other time delays (integer or noninteger multiples of the sampling time) are included in B. Generally, $n_a = n_b = n$.

Let us introduce a control law of the form

$$\underline{u} = G(q^{-1})(I + F(q^{-1}))^{-1}\underline{y}_P \tag{6.33}$$

where

$$G(q^{-1}) = G_0 + G_1 q^{-1} + \cdots + G_{n_g} q^{-n_g}$$

and $F(q^{-1})$ is given by (6.32).

The coefficient matrices G_i, $i = 0, 1, \ldots, n_g$ and F_i, $i = 1, 2, \ldots, n_f$, are of dimension $p \times p$. Generally, $n_g = n_a - 1$ and $n_f = n_b + k - 1$.

Substituting (6.33) in (6.31), we get the closed-loop equation

$$\underline{y}_P = [I + F(q^{-1})][I + P(q^{-1})]^{-1}[I + C(q^{-1})]\underline{\xi} \tag{6.34}$$

where

$$[I + P(q^{-1})] = [I + A(q^{-1})][I + F(q^{-1})] - q^{-k}B(q^{-1})G(q^{-1}) \tag{6.35}$$

Let us choose the coefficients of the polynomials F and G such that

$$[I + P(q^{-1})] = [I + C(q^{-1})][I + T(q^{-1})] \tag{6.36}$$

where the order n_t of the polynomial $T(q^{-1})$ is determined by

$$n_t \leqslant (n_a + n_b + k - 1 - n_c) \tag{6.37}$$

Then the closed-loop system is described by

$$\underline{y}_p = [I + F(q^{-1})][I + T(q^{-1})]^{-1}\underline{\xi} \tag{6.38}$$

We assume that $[I + F(q^{-1})]$ and $[I + T(q^{-1})]$ are relatively prime. This is true in general.

The poles of (6.38) are given by $|I + T(q^{-1})|$ and these are to be chosen by the designer.

The solution of Eqs. (6.35) and (6.36) requires the solution of a set of simultaneous linear equations

$$[(n_a + n_b + k - 1)p]\left\{\begin{bmatrix} \underbrace{\begin{matrix} I & & \mathbf{0} \\ A_1 & \ddots & \\ \vdots & & I \\ \vdots & & \vdots \\ A_{n_a} & & \vdots \\ & \ddots & A_{n_a} \\ \mathbf{0} & & \end{matrix}}_{(n_b + k - 1)p} & \underbrace{\begin{matrix} (pk)\left\{\begin{matrix} 0 \\ \vdots \\ 0 \end{matrix}\right. & \mathbf{0} \\ -B_1 & \\ \vdots & -B_1 \\ -B_{n_b} & \vdots \\ \mathbf{0} & -B_{n_b} \end{matrix}}_{(n_a p)} \end{bmatrix}\right. \begin{bmatrix} F_1 \\ \vdots \\ F_{n_f} \\ G_0 \\ \vdots \\ G_{n_g} \end{bmatrix}$$

$$= \begin{bmatrix} P_1 \\ \vdots \\ P_{n_p} \end{bmatrix} - \begin{bmatrix} A_1 \\ \vdots \\ A_{n_a} \\ 0 \\ \vdots \\ 0 \end{bmatrix} \tag{6.39}$$

The matrix on the left-hand side must be nonsingular for the solution to exist. This is almost always true, with $[I + A(q^{-1})]$, and $[B(q^{-1})]$ relatively prime and the plant observable and controllable.

6.3.2 Derivation of Control Law

The control law may be implemented in two ways. The first method uses

$$[I + F(q^{-1})]^{-1} = \{\det[I + F(q^{-1})]\}^{-1} \operatorname{adj}[I + F(q^{-1})] \tag{6.40}$$

so that from (6.33) we have

$$\{\det[I + F(q^{-1})]\} I\underline{u} = G(q^{-1})\{\mathrm{adj}[I + F(q^{-1})]\}\underline{y}_P \tag{6.41}$$

from which the control signal $\underline{u}$ is computed.

In the second method another assumption on the form of the regulator [Eq. (6.33)] must be made. Let

$$F^*(q) = q^{n_f}(I + F(q^{-1})) \tag{6.42}$$

$$G^*(q) = q^{n_g}G(q^{-1}) \tag{6.43}$$

We assume that $F^*(q)$ and $G^*(q)$ are relatively prime, with all observability and controllability indices equal to n_f, and $n_a = n_b = n$.

Then the regulator equation (6.33) may be written in the form [2,21]:

$$\underline{u} = [I + \tilde{F}(q^{-1})]^{-1}\tilde{G}(q^{-1})\underline{y}_P \tag{6.44}$$

where

$$n_{\tilde{f}} = n_f; \quad n_{\tilde{g}} = n_g \tag{6.45}$$

The assumption ensures that the transformation matrix required in the transformation from $[I + F(q^{-1}), G(q^{-1})]$ to $[I + \tilde{F}(q^{-1}), \tilde{G}(q^{-1})]$ has full ranks and the equations (6.44) and (6.45) hold. The following is the transformation matrix:

$$(n_f + n_g)\left\{\left[\begin{array}{cccc|cccc}
G_0^T & & \mathbf{0} & & -I & & & \mathbf{0} \\
G_1^T & G_0^T & & & -F_1^T & \ddots & & \\
\vdots & \vdots & \ddots & & \vdots & & -I & \\
\vdots & \vdots & & G_0^T & -F_{n_f}^T & & \vdots & \\
G_{n_g}^T & \vdots & & \vdots & & \ddots & \vdots & \\
 & G_{n_g}^T & & \vdots & & & \ddots & \\
\mathbf{0} & & \ddots & G_{n_g}^T & \mathbf{0} & & & -F_{n_f}^T
\end{array}\right]\right. \tag{6.46}$$

$$\underbrace{\qquad\qquad\qquad}_{n_f}\qquad\underbrace{\qquad\qquad\qquad}_{n_g}$$

We get the control law, from (6.44)

$$\underline{u} = -\tilde{F}(q^{-1})\underline{u} + \tilde{G}(q^{-1})\underline{y}_P \tag{6.47}$$

This can be easily implemented.

It is preferable to use the second method, as it uses routines already required in order to compute the pole-shifting law parameters $F(q^{-1})$ and $G(q^{-1})$.

6.3.3 Self-Tuning Algorithm

An on-line model of the system is formed

$$[I + \hat{A}(q^{-1})]\underline{y}_P = \hat{B}(q^{-1})\underline{u} + \underline{\xi} \tag{6.48}$$

The transport delay k has been absorbed in the estimated polynomial matrix $\hat{B}(q^{-1})$, with the order given by

$$n_{\hat{b}} = n_b + k \tag{6.49}$$

$\hat{A}$ and $\hat{B}$ are estimated using recursive least-squares method.

The control law is defined by Eq. (6.33). F and G are evaluated by solving the matrix polynomial identity:

$$[I + \hat{A}(q^{-1})][I + \hat{F}(q^{-1})] - \hat{B}(q^{-1})\hat{G}(q^{-1}) = [I + T(q^{-1})] \tag{6.50}$$

where the order of T, that is, n_t, is determined by (6.37). It may be noted that explicit knowledge of $C(q^{-1})$ is not required in (6.50). Nevertheless, the algorithm has been shown in the appendix of Prager and Wellstead [18] to converge to the closed-loop configuration defined by (6.38).

The determinant $|I + T(q^{-1})|$ gives the closed-loop system poles, as mentioned earlier.

<u>Algorithm</u>

<u>Step 1:</u> Estimate the parameters of the matrix polynomials $\hat{A}(q^{-1})$ and $\hat{B}(q^{-1})$ in (6.48) by using recursive least-squares method. Then $\{\xi\}$ becomes the sequence of fitting errors.

<u>Step 2:</u> Solve the pole-shifting equation (6.50). This involves the solution of a set of linear simultaneous equations like (6.39) with A^S, B^S, F^S, and G^S replaced by $\hat{A}^S$, $\hat{B}^S$, $\hat{F}^S$ and $\hat{G}^S$, respectively.

Step 3: Solve for $\tilde{F}(q^{-1})$ and $\tilde{G}(q^{-1})$, where

$$[I + \tilde{F}(q^{-1})]G(q^{-1}) = \tilde{G}(q^{-1})[I + F(q^{-1})] \quad (6.51)$$

and

$$n_{\tilde{f}} = n_f, \quad n_{\tilde{g}} = n_g$$

Then the control law may be expressed in the form given by (6.44). This also is done by the solution of a set of linear simultaneous equations.

Step 4: Apply the control $\underline{u}$ calculated from the law derived in step 3.

Proceed to the next iteration.

The preceding algorithm is a relatively simple, reliable, and robust algorithm for the regulator problem.

6.3.4 Discussion

As stated earlier, the pole-shifting approach has advantages over the minimum variance method in dealing with non-minimum-phase plants and also systems in which the loop pure time delays are not all the same. The closed-loop poles may be determined by the designer, by specifying the zeros of $\det[I + T(q^{-1})]$. However, this is at the cost of the optimality of regulation. Variations in time delay are treated in the same way as parameter variations, by taking the order of the $\hat{B}(q^{-1})$ polynomial as $(n_b + k_{max})$ where k_{max} is the maximum expected value of the time delay.

The pole placement approach requires considerably more computation than the minimum-variance law, as stated earlier. However, if a fast processor is used with floating-point hardware, computation times are not excessive. The pole-shifting law requires the solution of $(n_f + n_g + 1)$ simultaneous linear equations (6.50) in order to find $\hat{F}(q^{-1})$ and $\hat{G}(q^{-1})$. In addition, $(n_f + n_g)p$ simultaneous equations must be solved to find $\tilde{F}(q^{-1})$ and $\tilde{G}(q^{-1})$.

The preceding algorithm is meant for the regulator problem. Although it is possible to adapt it for constant-reference input tracking, there will be detuning during a change in reference input, and so if there are frequent set-point changes, the performance will not be satisfactory; thus this problem needs further investigation.

Prager and Wellstead [18] deal, in the paper, with two examples: one involving simulation of a non-minimum-phase plant with different time delays in the two loops and the other involving laboratory experiments for the control of a hydraulic system comprising two water tanks coupled by an interconnecting orifice. The objective in the second example is to independently control the level of water in each tank of this coupled multivariable system.

Results demonstrate the capability of the pole-placement controller when a real process is being operated. In this setup the DDC technique of rate-limiting should be introduced to avoid too much self-tuner detuning due to set-point changes.

Wellstead and Sanoff [24] discuss an extended self-tuning algorithm which can be applied to pole-assignment servo-following problems.

6.4 CONCLUDING REMARKS

This chapter has discussed multi-input/multi-output self-tuning. Because many industrial processes are multivariable, there has been significant interest in multivariable self-tuning in recent years. This chapter has considered two approaches, namely, the minimum-variance approach and the pole-assignment method. There is a need for further investigation of multivariable self-tuners.

REFERENCES

1. Peterka, V., and Åström, K. J., Control of multivariable systems with unknown but constant parameters, Proc. 3rd IFAC Symp. on Identification and System Parameter Estimation, The Hague, pp. 535-544 (1973).
2. Borison, U., Self-tuning regulators—Industrial application and multivariable theory, Report No. 7513, Dept. of Auto. Contr., Lund Inst. of Tech., Lund, Sweden (1975).
3. Borison, U., Self-tuning regulator for a class of multivariable systems, Automatica, 15: 209-215 (1979).
4. Keviczky, L., and Hetthessy, J., Self-tuning minimum variance control of MIMO discrete time systems, Auto. Contr. Theor. Appl., 14: 525-532 (1977).
5. Keviczky, L., Hetthessy, J., Hilger, M., and Kolostori, J., Self-tuning adaptive control of cement raw material blending, Automatica, 14: 525-532 (1978).
6. Bayoumi, M. M., and El-Bagoury, M. A., Comments on "Self-tuning adaptive control of cement raw material blending," Automatica, 15: 693-694 (1979).
7. Bayoumi, M. M., and El-Bagoury, M. A., A self-tuning regulator for multivariable systems, Research Report, Queen's University, Canada (1979).
8. El-Bagoury, M. A., and Bayoumi, M. M., Multivariable self-tuning augmented regulator, in Methods and Applications in Adaptive Control (H. Unbehauen, ed.), Springer-Verlag, Berlin, 1980.
9. Bayoumi, M. M., Wong, K. Y., and El-Bagoury, M. A., A self-tuning regulator for multivariable systems, Automatica, 17: 575-592 (1981).

10. Sinha, A. K., Minimum interaction minimum variance controller for multivariable systems, IEEE Trans. Auto. Contr., AC-22: 274-275 (1977).
11. Rosenbrock, H. H., Computer-Aided Control System Design, Academic Press, New York, 1974.
12. Morris, A., Nazer, Y., and Wood, R. K., Single and multivariable applications of self-tuning controllers, in Self-Tuning and Adaptive Control (C. J. Harris and S. A. Billings, eds.), Peter Peregrinus, London, 1981.
13. Koivo, A. J., A multivariable self-tuning controller, Automatica, 16: 351-366 (1980).
14. Clarke, D. W., and Gawthrop, P. J., Self-tuning controller, Proc. IEE, 122: 929-934 (1975).
15. Peterka, V., A square root filter for real-time multivariable regression, Kybernetika, 11: 53-67 (1975).
16. Prager, D. L., Doctoral Thesis, Control Systems Centre, UMIST, Manchester, U.K. (1980).
17. Wellstead, P. E., and Prager, D. L., Self-tuning multivariable regulators, in Self-Tuning and Adaptive Control (C. J. Harris and S. A. Billings, eds.), Peter Perigrinus, London, 1981.
18. Prager, D. L., and Wellstead, P. E., Multivariable pole-assignment regulator, Proc. IEE, 128, Pt. D: 9-18 (1981).
19. Bezanson, L. W., and Harris, S. L., State space design of multivariable self-tuning regulator, Int. J. Contr., 39: 395-412 (1984).
20. Grimble, M. J., LQG multivariable controllers—Minimum variance interpretation for use in self-tuning systems, Int. J. Contr., 40: 831-842 (1984).
21. Wolovich, W. A., Linear Multivariable Systems, Springer, New York, 1974.
22. Wittenmark, B., A self-tuning regulator, Doctoral Thesis, Dept. of Auto. Contr., Lund Inst. of Tech., Lund, Sweden (1973).
23. Hahn, V., Unbehauen, H., and Stoffer, R., Self-tuning control of a multivariable non-minimum phase pilot plant, Workshop on Adaptive Control, Florenz (1982).
24. Wellstead, P. E., and Sanoff, S. P., Extended self-tuning algorithm, Int. J. Contr., 34: 433-445 (1981).

BIBLIOGRAPHY

Davison, E. J., Multivariable tuning regulators: The feedforward and robust control of a general servomechanism problem, IEEE Trans. Auto. Contr., AC-21: 35-47 (1976).

Diekmann, K., and Unbehauen, H., Recursive identification of multi-input multi-output systems, Proc. 5th IFAC Symp. Identification and System Parameter Estimation, Darmstadt, F.R.G. (1979).

El-Bagoury, M. A., and Bayoumi, M. M., A two-level multivariable self-tuning regulator, Proc. IFAC Symp. Identification and System Parameter Estimation, Washington, D.C., pp. 1407-1412 (1982).

Elliott, H., and Wolovich, W. A., Multivariable adaptive pole-placement, Proc. IEEE Conf. Decision and Control, San Diego, Calif., pp. 975-977 (1981).

Gauthier, A., and Landau, I. D., On the recursive identification of multi-input multi-output systems, Automatica, 14: 609-614 (1978).

Grimble, M. J., and Moir, T. J., Multivariable weighted minimum variable STC, Proc. IFAC Workshop Adaptive Systems in Control and Signal Processing, San Francisco, pp. 341-345 (1983).

Nazer, Y., Single and multivariable self-tuning controllers, Doctoral Thesis, University of Newcastle-upon-Tyne, U.K. (1981).

Pentinen, J., and Koivo, H. M., Multivariable tuning regulators for unknown systems, Automatica, 16: 393-398 (1980).

7

Other Adaptive Control Strategies

7.1 INTRODUCTION

Adaptive control has been an area of intense research in recent years. Apart from extensive studies made on MRAC systems and STRs which have been discussed in the preceding chapters, many researchers have proposed novel adaptive control strategies.

The purpose of this chapter is to introduce some of these interesting schemes which appear to have many potential applications and need further investigation. Section 7.2 deals with variable-structure schemes. Reduced-order schemes are discussed in Section 7.3, and adaptive predictive control systems are taken up in Section 7.4. Section 7.5 introduces fuzzy logic adaptive control methods, applicable to very complex processes which have been difficult to control automatically. Section 7.6 considers adaptation using the internal model principle. Section 7.7 discusses adaptive control by inverse modeling. Concluding remarks are made in Section 7.8.

7.2 VARIABLE-STRUCTURE SCHEMES

7.2.1 Introduction

In variable-structure systems (VSS) changes can occur in the system structure during the transient periods, unlike conventional control systems. The current state of the system determines the times at which these changes occur and the nature of the structure thus formed. Variable-structure controllers are characterized by a discontinuous control action which changes structure upon reaching a set of switching surfaces. The operation depends on the values of the state variables. The chief advantages of VSS controllers are (1) high speed of response, (2) good transient response, (3) simplicity of physical realization, and (4) insensitivity to changes in system parameters. Variable-structure controllers are therefore usefully employed in systems with uncertain and time-varying parameters.

Variable structure schemes for control systems were developed mostly in the USSR by Petrov [1], Emel'yanov [2, 3], Bakakin et al. [4, 5], Utkin [6, 7], and Itkis [8]. In a survey paper, Utkin [9] discusses the design and analysis aspects of VSS. These schemes were applied for adaptive control by Young [10, 11], Zinober [12-16], and by Ambrosino et al. [17]. This strategy has also been applied for power system problems by Bengiamin and Chan [18], Chan et al. [19-21] and Kumar et al. [22]. A variable structure load-frequency controller has been designed by Sivaramakrishnan et al. [23]. Hsu and Chan [24] report about a VSS controller for dc motor speed control.

Zinober [13] applies VSS strategy for adaptive control of second-order plants and shows that the resulting system response is close to the desired response. This technique has been extended to third-order systems using nonlinear switching surface theory [14]. Ortega [25] examines the VSS as an alternative to robust adaptive control. Ambrosino et al. [17] report about the design of variable-structure model reference adaptive control systems. This design procedure avoids the use of differentiators and makes it possible to prescribe the transient response of the error in advance.

7.2.2 Basic Principle of VSS [23]

Consider a second-order system described by the state equations:

$$\begin{aligned} \dot{x}_1 &= x_2 \\ \dot{x}_2 &= -a_2 x_2 - a_1 x_1 + b_1 u, \qquad \text{where } b_1 > 0 \end{aligned} \tag{7.1}$$

The control law is described by

$$u = -\psi x_1 \tag{7.2}$$

where ψ is a piecewise constant feedback gain, such that

$$\psi = \begin{cases} \alpha, & \text{if } x_1\sigma > 0 \\ \beta, & \text{if } x_1\sigma < 0 \end{cases} \tag{7.3}$$

and

$$\sigma = C_1 x_1 + C_2 x_2 = 0 \tag{7.4}$$

Equation (7.4) gives the "switching line" in the state space.

The value of ψ can be adjusted according to the location of the system operating point in the state space as shown in Fig. 7.1.

In general, a switching hyperplane is described by

$$\sigma = C^T X \tag{7.5}$$

where σ is a constant vector and X is the state vector.

The objective of control is to bring the operating point from any initial position in the state space to the switching line, by appropriate choice of α, β, C_1, and C_2. When the operating point reaches the switching line, the gains will switch between α and β and the control will force the operating point to move along the switching line. The controller forces the point to come back to the switching line, whenever the point leaves it, and this is possible by varying the feedback structure. The motion of the system along

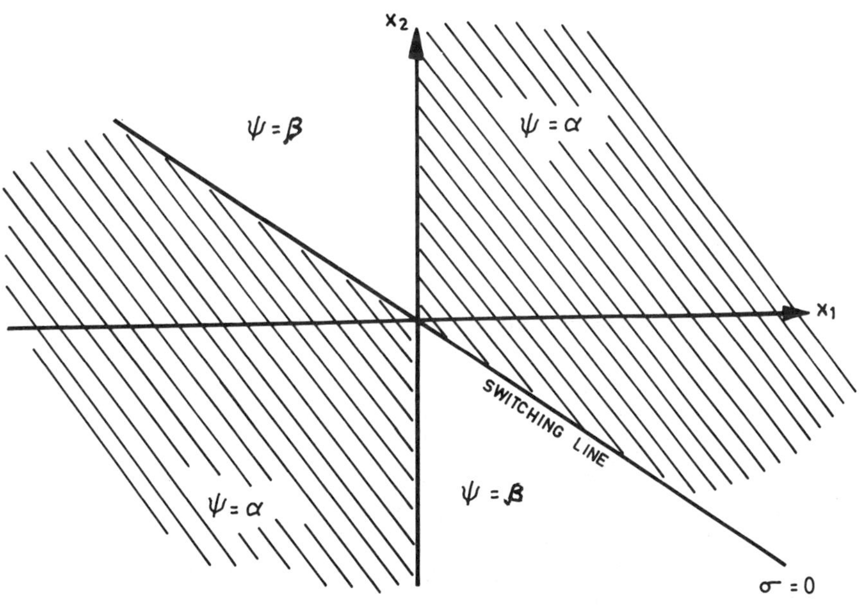

FIG. 7.1 Switching line in state space. (From Ref. 23.)

the switching line $\sigma = 0$ is called the sliding mode. The necessary and sufficient condition [8] for the existence of a sliding mode on $\sigma = 0$ is given by

$$\lim_{\sigma \to 0} \sigma\dot{\sigma} \leq 0 \tag{7.6}$$

The system will have invariance properties yielding motion which is independent of certain system parameters and disturbances.

7.2.3 Adaptive Strategy [13]

We are concerned with the case of parameter uncertainty. It is assumed that the plant parameters are constant but unknown. The VSS controller dispenses with the identification phase of a conventional adaptive controller and does not do the adaptive correction. The controller does not require knowledge of the plant dynamics for its implementation. It identifies a certain surface in the state space. This surface is associated with sliding motion and is found by rotating the switching hyperplane in the state space whenever the system is in a sliding mode. The controller is easily implemented and yields satisfactory system response.

7.2.4 Design of a Variable-Structure Controller [23]

A schematic diagram of the variable-structure controller is shown in Fig. 7.2.

In Fig. 7.2 the switch is operated according to the change of signs of the signal $X\sigma$. σ is defined by Eq. (7.5) and C is the switching vector.

The control law is, in general, given by

$$u = \sum_{i=1}^{n} \psi_i X_i \tag{7.7}$$

where the piecewise constant gains ψ are given by

$$\psi_i = \begin{cases} \alpha_i, & \text{if } X_i\sigma > 0 \\ \beta_i, & \text{if } X_i\sigma < 0 \end{cases} \tag{7.8}$$

The objective is to find the elements of the switching vector C and the feedback gains ψ such that the control u satisfies the specified dynamic performance measures while maintaining the plant output at the desired value; in addition, the system response should be insensitive to plant parameter variations when it is operated in the sliding mode.

Following the work of Utkin and Yang [26], a method is developed below for specifying the elements of the switching vector, using a pole-assignment

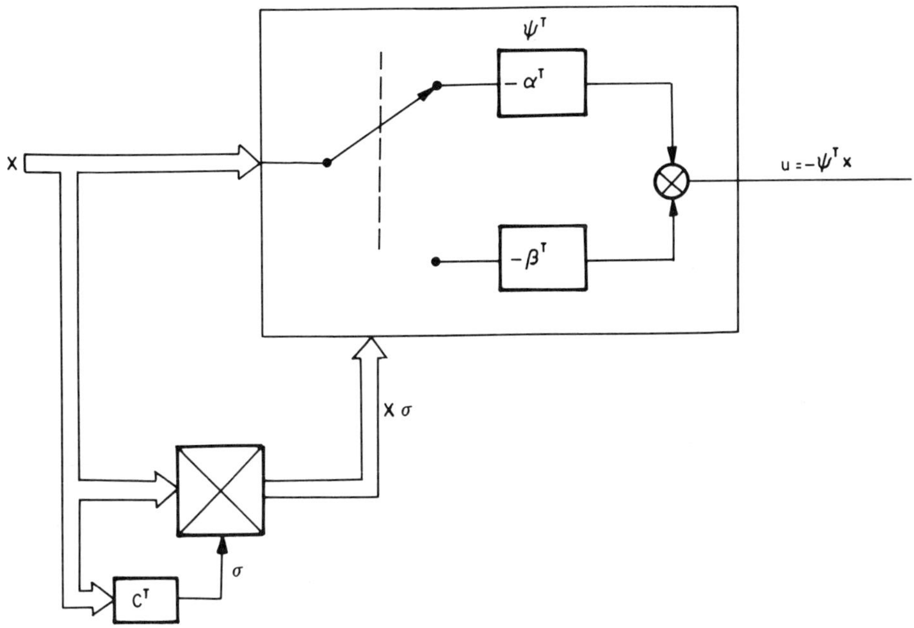

FIG. 7.2 Variable-structure controller. (From Ref. 23.)

approach. The VSS controller changes the structure of the system according to a law of structural change. This enables utilization of the useful properties of each of the structures, resulting in improved dynamic performance of the system.

Design Steps

Consider a linear system described by

$$\dot{X} = AX + BU \tag{7.9}$$

where X is the state vector of dimension n × 1, U is the control vector of dimension m × 1, and A and B are constant matrices of dimensions n × n and n × m, respectively.

Step 1: Define the coordinate transformation

$$Y = MX \tag{7.10}$$

such that

$$MB = \begin{bmatrix} 0 \\ \hline B_2 \end{bmatrix} \tag{7.11}$$

where M is a nonsingular n × m matrix and B_2 is a singular m × m matrix.

From Eqs. (7.9) and (7.10), we have

$$\dot{Y} = MAM^{-1}Y + MBU \tag{7.12}$$

Using (7.11), we can rewrite (7.12) as

$$\begin{bmatrix} \dot{Y}_1 \\ \hline \dot{Y}_2 \end{bmatrix} = \begin{bmatrix} A_{11} & A_{12} \\ \hline A_{21} & A_{22} \end{bmatrix} \begin{bmatrix} Y_1 \\ \hline Y_2 \end{bmatrix} + \begin{bmatrix} 0 \\ \hline B_2 \end{bmatrix} U \tag{7.13}$$

where A_{11}, A_{12}, A_{21}, and A_{22} are (n - m) × (n - m), (n - m) × m, m × (n - m), and (m × m) submatrices, respectively.

The motion of the system in sliding mode is described by the first equation of (7.13), and

$$\sigma_j(X) = C_j^T X = 0, \qquad j = 1, 2, \ldots, m \tag{7.14}$$

that is,

$$\dot{Y}_1 = A_{11}Y_1 + A_{12}Y_2 \tag{7.15}$$

$$\sigma(Y) = C_{11}Y_1 + C_{12}Y_2 = 0 \tag{7.16}$$

where C_{11} and C_{12} are m × (n - m), and (m × m) matrices respectively. These satisfy the equation

$$[C_{11} \;\; C_{12}] = C^T M^{-1} \tag{7.17}$$

Equations (7.15) and (7.16) describe the dynamics in the sliding mode over the intersection of the switching hyperplanes $\sigma(X) = 0$.

Equation (7.15) represents an open-loop control system with the state vector Y_1 and the control vector Y_2. The form of control is obtained from (7.16):

$$Y_2 = -C_{12}^{-1}C_{11}Y_1 \tag{7.18}$$

We see therefore that the problem of designing a system with desirable properties in a sliding mode is a linear feedback design problem. We can assume, without loss of generality, that

$$C_{12} = I \tag{7.19}$$

Step 2: We next combine (7.15), (7.18), and (7.19) to get

$$\dot{Y}_1 = [A_{11} - A_{12}C_{11}]Y_1 \tag{7.20}$$

Utikin and Yang [26] have shown that if the pair (A, B) is controllable, the pair (A_{11}, A_{12}) is controllable, the eigenvalues of $[A_{11} - A_{12}C_{11}]$ in the sliding mode can be arbitrarily placed by a suitable choice of C_{11}.

Step 3: With $C_{12} = I$, Eq. (7.17) leads to

$$C^T = [C_{11} \mid I]M \tag{7.21}$$

The switching hyperplanes are given by

$$\sigma_j(X) = C_j^T X = 0, \quad j = 1, 2, \ldots, m \tag{7.22}$$

7.2.5 Discussion

Appropriate selection of the switching vector is very important for providing improvements in system performance. The pole-assignment approach gives a systematic procedure for the proper choice of the switching vector. It has been shown by simulation of the load-frequency control problem [23] that the VSS controller exhibits insensitivity to plant parameter variations when operated in the sliding mode. The switching logic of the VSS controller is simple to implement.

7.2.6 Variable-Structure MRAC Systems

Ambrosino et al. [17] gives a design procedure for variable-structure MRAC systems.

Consider a single-input/single-output plant which is completely described by the linear time-invariant differential equation

$$A_p(p)y_p(t) = B_p(p)u(t) \tag{7.23}$$

where

$$A_P(p) = p^n + \sum_{i=0}^{n-1} a_i p^i \tag{7.24}$$

$$B_P(p) = \sum_{i=0}^{m} b_i p^i \tag{7.25}$$

and

$$p = d/dt$$

The following symbols are used here:

[17]	y	$\hat{y}$	ϵ	A	B	$\hat{A}$	$\hat{B}$	e	s	γ
Present work	y_P	y_M	e	A_P	B_P	A_M	B_M	ϵ	σ	C

The following assumptions are made:

1. $B_P(p)$ is Hurwitz.
2. Degrees n and m ($m < n$) are known.
3. Upper and lower bounds of the unknown parameters are known; that is,

$$a_i^0 - \Delta a_i < a_i < a_i^0 + \Delta a_i, \qquad i = 0, \ldots, n-1$$

$$b_i^0 - \Delta b_i < b_i < b_i^0 + \Delta b_i, \qquad i = 0, \ldots, m$$

The reference model is described by

$$A_M(p)y_M(t) = B_M(p)r(t) \tag{7.26}$$

where $A_M(p)$ is a monic Hurwitz polynomial of degree n, $B_M(p)$ is a polynomial of degree $m_1 < m$, and r(t) is a uniformly bounded piecewise continuous signal. The control signal u(t) which makes the plant output track the model output is given by

$$u(t) = \frac{B_M(p)}{F(p)} r(t) - w(t) \tag{7.27}$$

where

$$F(p) = \sum_{i=0}^{n-1} f_i p^i \tag{7.28}$$

F(p) is an arbitrary Hurwitz polynomial and w(t) is a signal to be adaptively synthesized. Such a feedforward structure has been suggested by Hang and Tang [27].

Let us define

$$e(t) = y_M(t) - y_P(t)$$

Subtracting (7.23) from (7.26) and using (7.27), we get the error equation

$$A_M(p)e(t) = [A_P(p) - A_M(p)]y_P(t) + [F(p) - B_P(p)]u(t) + F(p)w(t) \tag{7.29}$$

Let us define the following "filtered" variables, e_F, u_F, and y_{P_F}:

$$F(p)e_F(t) = e(t) \tag{7.30}$$

$$F(p)u_F(t) = u(t) \tag{7.31}$$

$$F(p)y_{P_F}(t) = y_P(t) \tag{7.32}$$

Dividing (7.29) by F(p) gives the filtered error equation

$$A_M(p)e_F(t) = [A_P(p) - A_M(p)]y_{P_F}(t) + [F(p) - B_P(p)]u_F(t) + w(t) \tag{7.33}$$

We can rewrite this equation in a state form:

$$\dot{\epsilon} = \bar{A}\epsilon + \bar{D}\phi + \bar{E}\theta_P + \bar{b}w \tag{7.34}$$

where

$$\epsilon^T = (e_F, pe_F, \ldots, p^{n-1}e_F)$$

$$\phi^T = (u_F, pu_F, \ldots, p^{n-1}u_F)$$

$$\theta_P^T = (y_{P_F}, py_{P_F}, \ldots, p^{n-1}y_{P_F})$$

$$\bar{A} = \begin{bmatrix} 0 & 1 & 0 \cdots 0 \\ 0 & 0 & 1 \cdots 0 \\ \vdots & \vdots & \vdots \quad \vdots \\ 0 & 0 & 0 \cdots 1 \\ -a_{M_0} & & \cdots -a_{M_{n-1}} \end{bmatrix}$$

$$\bar{D} = \begin{bmatrix} \mathbf{0} \\ \hline (f_0 - b_0) \cdots (f_m - b_m) f_{m+1} \cdots f_{n-1} \end{bmatrix}$$

$$\bar{E} = \begin{bmatrix} \mathbf{0} \\ \hline (a_0 - a_{M_0}) \cdots (a_{n-1} - a_{M_{n-1}}) \end{bmatrix}$$

$$\bar{b} = \begin{bmatrix} 0 \\ \vdots \\ 0 \\ 1 \end{bmatrix}$$

It may be noted that the vector signals ϵ, ϕ, θ are completely available if the filtered signals e_F, u_F, and y_{P_F} are obtained by the state variable filters (SVF), which are low-pass filters having (n - 1) integrators, allowing us to get the filtered derivatives up to the (n - 1)th order [28]. This allows us to apply VSS theory [6] in order to synthesize the adaptive signal w(t) assuming the asymptotic convergence of the error to zero.

Following this approach, w(t) must be chosen as a function of $\epsilon(t)$, $\phi(t)$, $\theta(t)$, discontinuous on some surface in the error state space

$$\sigma(\epsilon) = C^T \epsilon = 0, \quad C^T = (C_1, \ldots, C_{n-1}, 1) \tag{7.35}$$

In the present design procedure the VSS control has the form

$$w(t) = -[\beta_\phi^T \phi(t) + \beta_{\theta_P}^T \theta_P(t) + \beta_{\theta_M}^T \theta_M(t)] - [\alpha_\phi^T |\phi(t)| + \alpha_{\theta_P}^T |\theta_P(t)| + \alpha_{\theta_M}^T |\theta_M(t)|] \operatorname{sgn}(\sigma) \tag{7.36}$$

where

$$\theta_M(t) = e(t) + \theta_P(t) = (y_{M_F}, py_{M_F}, \ldots, p^{n-1} y_{M_F})^T$$

$$|\phi| = (|\phi_1|, \ldots, |\phi_n|)^T$$

and so on, and β_ϕ, β_{θ_P}, β_{θ_M}, α_ϕ, α_{θ_P}, α_{θ_M} are n-vectors of design parameters. As we have seen in (7.6), the VSS theory assures that if these vectors are chosen such that

$$\sigma\dot{\sigma} < 0 \tag{7.37}$$

then the error trajectories move toward the surface given by (7.35) and a sliding mode appears on it. During sliding, $\sigma = 0$. Then the resulting error dynamics are given by

$$C_1 e + C_2 pe + \cdots + C_{n-2} p^{n-2} e + p^{n-1} e = 0 \tag{7.38}$$

We can specify this arbitrarily, by a suitable choice of the parameters C_i.

When in sliding mode, w(t) chatters at infinite frequency, according to theory. However, in practice, the actual signal w(t) can be considered to be of finite frequency, consisting of an average- and a high-frequency component. Utkin [29] has shown that the error behavior during sliding can be obtained by replacing w(t) in (7.34) with an equivalent control $w_{eq}(t)$ and it can be obtained by solving the albegraic equations $\dot{\sigma} = 0$

$$\begin{aligned} w_{eq}(t) &= -C^T\bar{A}\epsilon - C^T\bar{D}\phi - C^T\bar{E}\theta_P \\ &= \{[A_M(p) - p\Gamma(p)]e_F(t) - [F(p) - B_M(p)]u_F(t) \\ &\quad - [A_P(p) - A_M(p)]y_{P_F}(t)\} \end{aligned} \tag{7.39}$$

where

$$\Gamma(p) = p^{n-1} + \sum_{i=0}^{n-2} C_{i+1} p^i \tag{7.40}$$

To guarantee asymptotic stability, it is important to check the boundedness of the plant input signal u(t) during sliding [10]. This signal can be obtained from (7.27) by substituting $w_{eq}(t)$ for w(t). Then

$$B_P(p)u(t) = \{B_M(p)r(t) - [A_M(p) - p\Gamma(p)]e(t) + [A_P(p) - A_M(p)]y_P(t)\} \tag{7.41}$$

As assumption 1 holds and $m_1 < m$, asymptotic stability is guaranteed.

We can express condition (7.37) using (7.34) and (7.36) as

$$\begin{aligned}\sigma\{&(f_0 - b_0, \ldots, f_m - b_m, f_{m+1}, \ldots, f_{n-1})\phi \\ &+ (a_0, a_1 - C_1, \ldots, a_{n-1} - C_{n-1})\theta_P \\ &+ (-a_{M_0}, C_1 - a_{M_1}, \ldots, C_{n-1} - a_{M_{n-1}})\theta_M \\ &- \beta_\phi^T\phi - \beta_{\theta_P}^T\theta_P - \beta_{\theta_M}^T\theta_M \\ &- (\alpha_\phi|\phi| + \alpha_{\theta_P}|\theta_P| + \alpha_{\theta_M}|\theta_M|)\ \mathrm{sgn}\ (\sigma)\} < 0\end{aligned} \tag{7.42}$$

Let us choose

$$\beta_\phi^T = (f_0 - b_0^0, \ldots, f_m - b_m^0, f_{m+1}, \ldots, f_{n-1} + \eta) \tag{7.43}$$

$$\beta_{\theta_P}^T = (a_0^0, a_1^0 - C_1, \ldots, a_{n-1}^0 - C_{n-1}) \tag{7.44}$$

$$\beta_{\theta_M}^T = (-a_{M_0}, C_1 - a_{M_1}, \ldots, C_{n-1} - a_{M_{n-1}}) \tag{7.45}$$

Then we have

$$\alpha_\phi^T = (\Delta b_0, \ldots, \Delta b_m, 0, \ldots, 0, \eta) \tag{7.46}$$

$$\alpha_{\theta_P}^T > (\Delta a_0, \ldots, \Delta a_{n-1}) \tag{7.47}$$

$$\alpha_{\theta_M} = 0 \tag{7.48}$$

where η is an arbitrarily small positive constant assuring the physical feasibility of the controller.

Equations (7.41)-(7.48) can be used to mechanize the configuration of a variable-structure MRAC system (Fig. 7.3).

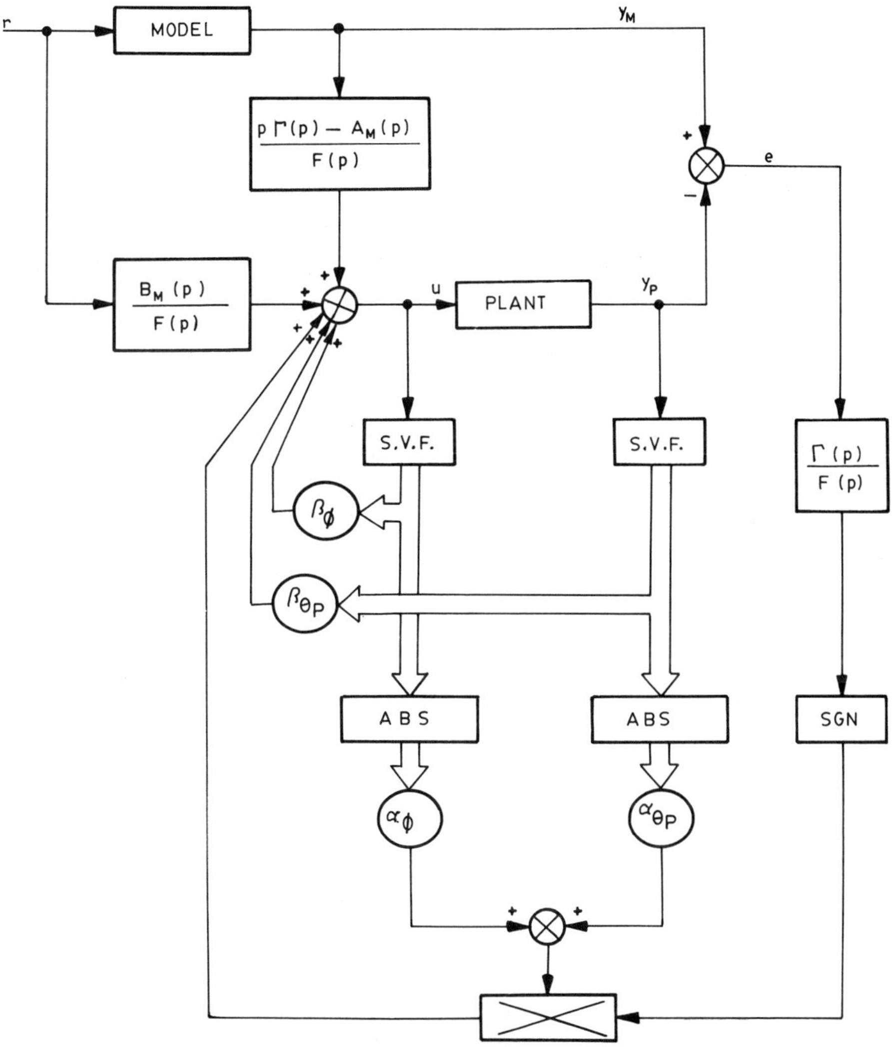

FIG. 7.3 Variable-structure MRAC system. (From Ref. 17.)

7.2.7 Discussion

Ambrosino et al. [17] give simulation results for two examples and show that there is high-frequency chattering of the control signal, even though it is possible to specify arbitrarily the rate of convergence of the error. Such a control signal cannot be directly applied to the plant. The chattering of the control signal can be reduced by modification of the discontinuous adaptive law. An alternative scheme is given in Ambrosino et al. [17] using continuous adaptation laws with nonideal sliding motion which result in similar performance of the error. Computer simulations show the effectiveness of this scheme in obtaining a smooth input signal.

Zinober et al. [30] consider the design of multivariable adaptive model-following control systems using VSS theory. Both the plant transient and the error response transient can be specified in advance. The control is discontinuous on a number of switching hyperplanes. The dynamic response is insensitive to certain parameter variations and disturbances. The design technique is straightforward and requires little computational effort. This approach is applied to the design of adaptive controllers for a fourth-order multivariable system and a single-input system. The implementation of the controller can be carried out easily with microprocessors. Further investigations are possible in (1) the nonperfect model-following case, (2) sensitivity to inaccuracy in switching structure implementation, and (3) problems due to the nonavailability of some of the states.

In view of the advantages of the VSS approach to adaptive control, there is a considerable scope for forther exploration.

7.3 REDUCED-ORDER SCHEMES

7.3.1 Introduction

Higher-order models are theoretically feasible; however, they present serious practical and numerical difficulties. Furthermore, it must be borne in mind that all industrial processes can be modeled only approximately. It is therefore necessary to consider the study of adaptive controllers designed with reduced-order models from the point of view of successful practical applications.

Some attempts have been made to analyze reduced-order adaptive schemes. Specific results such as error bounds have been reported for adaptive observers and identifiers [31-35]. Ioannou [36] deals with adaptive systems with reduced models. Kreisselmeier [37] proves local stability for a reduced-order indirect adaptive regulator. Rohrs et al. [38] and Johnson and Balas [39] study reduced-order direct adaptive control. They have demonstrated that unmodeled parasitics can lead to an unstable closed-loop system. Ortega [40] considers assessment of stability robustness of reduced-order adaptive controllers. Ioannou and Kokotovic [41] give an estimate of the region of attraction for adaptive regulation and also a modified adaptive law to guarantee boundedness in the case of tracking.

7.3.2 Scalar Reduced-Order Problem

Let us consider a simple example of a reduced-order adaptive control in which the output y_p of a second-order plant is given by [41]:

$$\dot{y}_P = a_P y_P + 2z - u \tag{7.49}$$

$$\mu\dot{z} = -z + u \tag{7.50}$$

where a_P and μ are the unknown constant parameters and u is the control input.

The preceding system is required to track the state y_M of a first-order model:

$$\dot{y}_M = -a_M y_M + r, \quad a_M > 0 \tag{7.51}$$

where $r = r(t)$ is a reference input, a uniformly bounded function of time.

The model-plant mismatch is due to some "parasitic" time constants which appear as multiples of a singular perturbation parameter μ and introduce the "parasitic" state η.

In (7.49) and (7.50) the parasitic state is defined as

$$\eta \triangleq z - u$$

Using this expression, we get the following equations

$$\dot{y}_P = a_P y_P + 2\eta + u \tag{7.52}$$

$$\mu\dot{\eta} = -\eta - \mu\dot{u} \tag{7.53}$$

where the "dominant" part (7.52) and the "parasitic" part (7.53) of the plant appear explicitly.

Let us apply to the preceding plant with parasitics the same adaptive law we would have applied to the plant without parasitics. That is to say, we apply the control

$$u = -K(t)y_P + r(t) \tag{7.54}$$

and the adaptive law

$$\dot{K} = \gamma e y_P, \quad \gamma > 0 \tag{7.55}$$

Then we have

$$\dot{e} = -a_M e - [K(t) - K^*](e + y_M) + 2\eta \tag{7.56}$$

$$\mu\dot{\eta} = -\eta + \mu[\gamma e(e + y_M)^2 - K(K - a_P)(e + y_M) + 2K\eta + Kr - \dot{r}] \tag{7.57}$$

$$\dot{K} = \gamma e(e + y_M) \tag{7.58}$$

where

$$e = (y_P - y_M); \quad K^* = a_M + a_P \tag{7.59}$$

The existing theory of adaptive control [42] guarantees stability for the case without parasitics, $\mu = 0$, when (7.56)-(7.58) reduce to

$$\dot{\bar{e}} = -a_M\bar{e} - [\bar{K}(t) - K^*][\bar{e} + y_M] \tag{7.60}$$

$$\dot{\bar{K}} = \gamma\bar{e}(\bar{e} + y_M) \tag{7.61}$$

LEMMA 1. For any bounded initial conditions $\bar{e}(0)$, $\bar{K}(0)$, the solution $\bar{e}(t)$, $K(t)$ of (7.60) and (7.61) is bounded and

$$\lim_{t\to\infty} \bar{e}(t) = 0, \qquad \lim_{t\to\infty} \bar{K}(t) = K_s$$

where K_s is, in general, a function of $\bar{e}(0)$, $\bar{K}(0)$. Furthermore, if $r(t)$ is sufficiently rich, then $\lim_{t\to\infty} K(t) = K^*$, independent of $\bar{e}(0)$, $\bar{K}(0)$.

Next the question arises: Given that the adaptive system without parasitics, in this case (7.60), (7.61), possesses the properties given in Lemma 1, how will these properties be modified by the parasitics? In other words, what are the stability properties of (7.56)-(7.58)? Which modification of the adaptive law would help to preserve some of the desirable properties?

7.3.3 Regulation Problem

For the regulation problem, (7.56)-(7.58) become

$$r(t) = 0, \quad y_M(t) = 0, \quad e(t) = y_P(t) \tag{7.62}$$

$$\dot{y}_P = a_P y_P + u + 2\eta \tag{7.63}$$

$$\mu\dot{\eta} = -\eta - \mu\dot{u} \tag{7.64}$$

$$u = -K(t)y_P \tag{7.65}$$

$$\dot{K} = \gamma y_P^2 \tag{7.66}$$

The objective is to drive y_P to zero despite the presence of parasitics, while assuring that all signals in the closed-loop system (7.63)-(7.66) remain bounded.

We should note that the open-loop system (7.63)-(7.64) might not be stabilizable by constant-gain output feedback for a given value of μ. If this is the case, the adaptive controller (7.65)-(7.66) cannot stabilize the equilibrium of (7.63)-(7.64).

The following lemma characterizes the parasitics for which a linear output stabilizing feedback law exists.

LEMMA 2. There exists a $\mu_1 > 0$ and a constant K_0 such that for all $\mu \in (0, \mu_1]$ the system (7.63)-(7.64) with the feedback law

$$u = -K_0 y_P \tag{7.67}$$

is an asymptotically stable closed-loop system. Furthermore,

$$\mu_1 < \frac{1}{2a_P} \tag{7.68}$$

and

$$\left(\frac{1}{\mu} - a_P\right) > K_0 > a_P \tag{7.69}$$

The following theorem establishes the stability properties of the adaptive control system (7.63)-(7.66) for $\mu < \mu_1$.

THEOREM 1. There exists $\mu^* < \mu_1$ and positive numbers $\alpha < 1/2$, C_1, C_2 such that for each $\mu \in (0, \mu^*]$, any solution $y_P(t)$, $\eta(t)$, $K(t)$ of (7.63)-(7.66) starting from the set

$$\mathcal{D}(\mu) = \{y_P, \eta, K: |y_P| + |K| < C_1\mu^{-\alpha}, |\eta| < C_2\mu^{-\alpha-(1/2)}\} \tag{7.70}$$

is bounded and $y_P(t) \longrightarrow 0$, $\eta(t) \longrightarrow 0$, $K(t) \longrightarrow$ constant, as $t \longrightarrow \infty$.

For proof, and also some observations on this theorem, one may refer to Ioannou and Kokotovic [41], which also gives simulation results showing the effect of different values of μ, γ, and initial conditions on the stability properties of the system (7.63)-(7.66) for $a_P = 4$.

7.3.4 Tracking Problem

In the tracking problem, when $r(t) \neq 0$, the adaptive system with parasitics such as (7.56)-(7.58) may not converge to or may not even possess an equilibrium. A practical goal is to guarantee some boundedness properties. That is, the solutions starting in $\mathfrak{D}(\mu)$ remain bounded and converge to a residual set $\mathfrak{D}_0(\mu)$. The design objective is to make $\mathfrak{D}(\mu)$ as large as possible and $\mathfrak{D}_0(\mu)$ as small as possible.

To obtain such a result, we modify the adaptive law (7.58) as

$$\dot{K} = -\sigma K + \gamma e(e + y_M) \tag{7.71}$$

where σ is a positive design parameter.

We can therefore write the equations for the tracking problem in the presence of parasitics as

$$\dot{e} = -a_M e - [K(t) - K^*](e + y_M) + 2\eta \tag{7.72}$$

$$\mu\dot{\eta} = -\eta + \mu[\gamma e(e + y_M)^2 - K(K - a_P - \sigma)(e + y_M) + 2K\eta + Kr - \dot{r}] \tag{7.73}$$

$$\dot{K} = -\sigma K + \gamma e(e + y_M) \tag{7.74}$$

The following theorem establishes the stability properties of the preceding system.

THEOREM 2. Let the reference input, r(t), satisfy

$$|r(t)| < r_1; \quad |\dot{r}(t)| < r_2, \quad \forall t < 0 \tag{7.75}$$

where r_1, r_2 are given positive constants. There exist positive constants t_1, μ^*, σ, $\alpha < 1/2$, C_1, C_2, C_3, and $C_4 \geq 1$ such that for each $\mu \in (0, \mu^*]$ every solution of (7.72)-(7.74) starting from $t = 0$ from the set

$$\mathfrak{D}(\mu) = \{e, K, \eta: |e| + |K| < C_1\mu^{-\alpha}, |\eta| > C_2\mu^{-\alpha-(1/2)}\} \tag{7.76}$$

enters the residual set

$$\mathfrak{D}_0(\mu) = \{e, K, \eta: \frac{a_M}{4}|e|^2 + \frac{|\eta|^2}{8} + \frac{\sigma}{2\gamma}|K - K^*|^2$$

$$< C_4\left[\frac{\sigma|K^*|^2}{2\gamma}\right] + \mu^2\left(1 + \frac{4}{a_M}\right)|2K^* y_M - \dot{r}|^2\} \tag{7.77}$$

at $t = t_1$ and remains in $\mathcal{D}_0(\mu)$ for all $t > t_1$. Furthermore, $\sigma \geq C_3\mu^{2(1-\alpha)}$.

For proof, and some observations on this theorem, see Ioannou and Kokotovic [41], who also give simulation results showing the effect of different values of μ, σ, initial conditions, and reference input characteristics on the stability properties of the system with $a_P = 4$, $a_M = 3$ and $\gamma = 5$.

7.3.5 Discussion

Ioannou and Kokotovic [41] also obtain results for a continuous-time SISO adaptive control problem where the transfer function of the dominant part of the plant has a relative degree of 1. These results are expressed in the form of a theorem similar to Theorem 2 stated above. The same methodology can be extended to more complicated continuous- and discrete-time adaptive control systems.

Djaferis et al. [43] suggest a method based on reduced-order adaptive pole placement for multivariable systems. They show that the resulting design significantly reduces the complexity in comparison to any other multivariable scheme, since the number of parameters that need to be estimated is considerably reduced. We should know the order of the system and an upper bound on the observability index to implement this method. There is scope for further study of this approach, especially with regard to global stability analysis.

The reduced-order problem is a very important theoretical question in the area of adaptive control. The effects of this problem are felt in many open questions in this field and a solution of this problem will have an important repercussions in both theory and practice. A considerable amount of research is currently in progress; however, even a proper formulation of this problem is a formidable one.

7.4 ADAPTIVE PREDICTIVE CONTROL SYSTEMS

7.4.1 Introduction

Adaptive predictive control systems (APCS) have been developed according to a methodology which is different from that of MRAC systems and STRs. This approach was originated in 1976 by Martin-Sanchez [44,45].

APCS have been applied successfully to the control of many processes such as a nonlinear SISO process in a F-8 aircraft [46], a multivariable nonlinear system with time delays as in a distillation column [47], a multivariable pilot plant evaporator [48]. In all cases it has been found that the performance of APCS is excellent and at least as good as that of alternative control systems [49].

7.4.2 Principle of APCS [50]

APCS are characterized by the following basic principles:

1. At each step, a future desired process output is generated, and the control signal is computed to make the predicted process output equal to the desired process output.
2. The model parameters are updated to solve the prediction problem.
3. The above considered desired process output belongs to a desired process output trajectory which verifies a certain performance criterion. For example, this trajectory can start from the current "state" of the plant and evolve according to some chosen dynamics to the final desired set point.

These steps are illustrated in Fig. 7.4 [49].

Driver Block

The driver block can be interpreted as an extension of the conventional "reference model." At each sampling instant, this block generates a desired process output vector y_d for a future sampling instant; y_{SP} is the set-point vector specified by the operator.

Control Block

The concept of predictive control [51] is the basis of APCS methodology. At each sampling instant k, the control block generates a control vector u, by use of an adaptive predictive (AP) model. This control signal, u, will make the predicted process output vector equal to the desired output vector y_d. The AP model predicts the process output using a matrix $\hat{\theta}$ of the estimated process parameters and the process input/output vector x. The vector x includes measured or filtered values of the process output vector y, the control vector u, other measured process outputs and/or intermediate variables z, and other measured disturbances or other external variables w.

$$x^T = [y^T, \ldots, u^T, \ldots, z^T, \ldots, w^T, \ldots]$$

Estimation Block

The estimation block updates the parameters of the AP model in such a way that the prediction error, that is, the error between the predicted process output and the measured process output, tends to a minimum value.

Process

The process is subject to unmeasured external disturbances ξ, plus additive measurement and process noise (n_y, n_u, n_z, n_w) in all the variables. We assure that these are bounded.

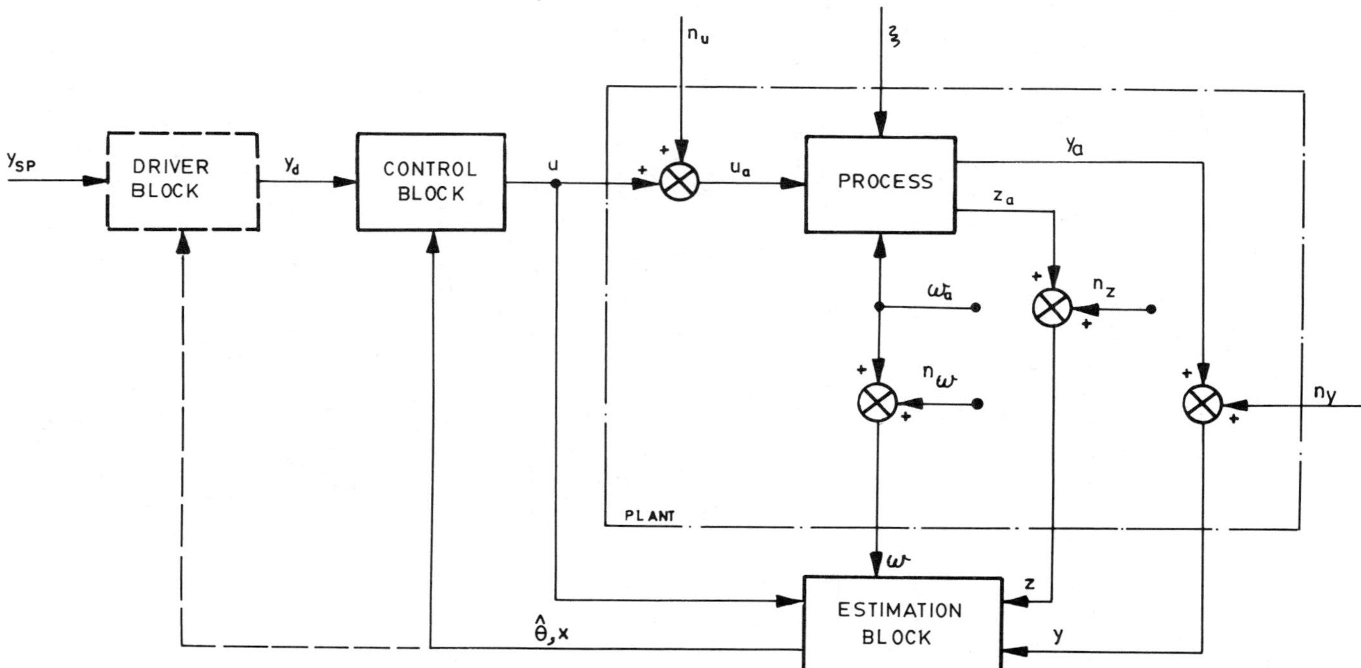

FIG. 7.4 Schematic diagram of APCS. (From Ref. 49.)

The APCS algorithm has been studied by Martin-Sanchez et al. [49] for an unknown, MIMO, ARMA, time-invariant, stable-inverse process with bounded noise and unmeasured disturbances, and a proof of global stability has been given for the same. Asymptotic stability for the "deterministic case" has been shown to be a special case of the general result. It may be mentioned that the APCS algorithm first given by Martin-Sanchez [44] is identical to the "projection algorithm I" proposed by Goodwin et al. [52] for a delay-free, discrete, SISO or MIMO system. Martin-Sanchez [50] works out the proof for global stability of APCS for discrete linear processes with pure time delays in the presence of bounded noise and unmeasured disturbances. In both these cases the proof is based on minimization of the a posteriori estimation error and convergence of the estimated parameters. Earlier work on similar studies of stability by Goodwin et al. [53], Egardt [54], Peterson and Narendra [42], and Samson [55] is based on conditions which are generally very restrictive. Unlike these results, Martin-Sanchez et al. [49] do not consider any previous estimation of the process parameters nor use the condition of boundedness of the rate of growth of the system signals, and so this approach is stronger than the earlier studies.

Martin-Sanchez and Shah [47] discuss the practical issues in APCS implementation and also a successful experimental evaluation of single-loop and multivariable adaptive predictive control of a binary distillation column. This adaptive control system was tested under conditions typical of an industrial plant, thus demonstrating the potential scope of applications of APCS to industrial processes. The extensive set of experiments was carried out in the presence of step and stochastic feed-flow disturbances, set-point changes, and measurement noise, in the control of a nonlinear, multivariable plant with time delays. The experimental results for both the SISO and MIMO cases confirm that APCS schemes outperform conventional techniques, showing that adaptive predictive control is now well beyond the theoretical stage.

7.4.3 Mathematical Description [49]

The process shown in Fig. 7.4 is described by the following discrete, multivariable ARMA model:

$$y_a(k) = \theta_0 \phi_a(k-1) + \theta_1 u_a(k-1) + \xi(k) \tag{7.78}$$

where

$$\phi_a(k-1)^T = [y_a(k-1)^T, \ldots, u_a(k-2)^T, \ldots, z_a(k-1)^T, \ldots, w_a(k-1)^T, \ldots] \tag{7.79}$$

that is, a vector of past actual (noise-free) values of the process output vector y_a, control vector u_a, other process variables z_a, and other measured inputs in the vector w_a.

The input and output vectors, u_a and y_a, are assumed to be of dimension n. The dimension of ϕ_a depends on the order of the assumed process model. $\xi(k)$ represents the effect of unmeasured disturbances on the process output y_a at time k. θ_0 and θ_1 are unknown matrices of process parameters of appropriate dimensions.

The available measured process variables differ from the actual values because of measurement errors plus noise and so forth.

$$\begin{aligned} y(k) &= y_a(k) + n_y(k) \\ u(k) &= u_a(k) + n_u(k) \\ w(k) &= w_a(k) + n_w(k) \\ z(k) &= z_a(k) + n_z(k) \end{aligned} \tag{7.80}$$

Then the corresponding measured vector ϕ becomes

$$\phi(k) = \phi_a(k) + n_\phi(k) \tag{7.81}$$

Substituting (7.80) and (7.81) in (7.78), we get

$$y(k) = \theta x(k-1) + \Delta(k) \tag{7.82}$$

where the process parameter matrix,

$$\theta = [\theta_0, \theta_1] \tag{7.83}$$

the process input/output vector,

$$x(k-1)^T = [\phi(k-1)^T, u(k-1)^T] \tag{7.84}$$

the noise vector,

$$n_x(k-1)^T = [n_\phi(k-1)^T, n_u(k-1)^T] \tag{7.85}$$

$$\Delta(k) = n_y(k) - \theta n_x(k-1) + \xi(k) \tag{7.86}$$

$\Delta(k)$ is termed the perturbation vector.

7.4.4 APCS Algorithm

The a priori estimation or one-step-ahead prediction, $\hat{y}(k|k-1)$ of the process output y(k) at time k is based on the estimation at time (k - 1), $\hat{\theta}(k-1)$ of the process parameter matrix . That is,

$$\hat{y}(k|k-1) = \hat{\theta}(k-1)x(k-1) \tag{7.87}$$

The corresponding a priori estimation error is

$$\begin{aligned} e(k|k-1) &= y(k) - \hat{y}(k|k-1) \\ &= y(k) - \hat{\theta}(k-1)x(k-1) \end{aligned} \tag{7.88}$$

The estimated parameter matrix is generated from the following recursive relationship:

$$\hat{\theta}_i(k) = \hat{\theta}_i(k-1) + \frac{\psi_i(k)^2 e_i(k|k-1)x(k-1)}{1 + \psi_i(k)^2 x(k-1)^T x(k-1)}, \quad i = 1, \ldots, n \tag{7.89}$$

where $\hat{\theta}_i(k)^T$ is the ith row of matrix $\hat{\theta}(k)$ and $e_i(k|k-1)$ is the ith component of the prediction error $e(k|k-1)$. $\psi_i(k)^2$, $i = 1, \ldots, n$, are scalars, as defined below; these determine a criterion for stopping parameter adaptation which is essential for the proof of stability.

DEFINITION

(a) $\psi_i(k)^2 = 0$

if

$$|e_i(k|k-1)| \leq \Delta'_{ib}(\psi_{il}, \Delta_{ib}, k) \leq 2\Delta_{ib} < \phi \tag{7.90}$$

and

$$\Delta'_{ib}(\psi_{il}, \Delta_{ib}, k) = \frac{2 + 2\psi_{il}^2 x(k-1)^T x(k-1)}{2 + \psi_{il}^2 x(k-1)^T x(k-1)} \Delta_{ib} \tag{7.91}$$

with

$$0 < \psi_{il}^2 < \phi \quad \text{and} \quad \Delta_{ib} \geq \Delta_{im} = \max_{0<k\leq\phi} |\Delta_i(k)| \tag{7.92}$$

where $\Delta_i(k)$, Δ_{ib}, and Δ_{im} are respectively the ith components, $i = 1, \ldots,$ n, of the vectors $\Delta(k)$, Δ_b, and Δ_m, where Δ_b is an estimate of a constant, upper bound on the absolute value of the perturbation vector $\Delta(k)$ for all k and Δ_m is the minimum value of this upper bound.

$$\text{(b)} \quad \psi_{il}^2 < \psi_i(k)^2 \leq \psi_{ib}(k)^2 \leq \psi_{iu}^2 < \phi \tag{7.93}$$

if

$$|e_i(k|k-1)| > \Delta'_{ib}(\psi_{il}, \Delta_{ib}, k) \geq \Delta_{ib}$$

where $\psi_{ib}(k)^2$ is defined as

$$\psi_{ib}(k)^2 = \psi_{iu}^2 \tag{7.94}$$

if

$$|e_i(k|k-1)| > \Delta'_{ib}(\psi_{iu}, \Delta_{ib}, k)$$

where, according to (7.91),

$$\Delta'_{ib}(\psi_{iu}, \Delta_{ib}, k) = \frac{2 + 2\psi_{iu}^2 x(k-1)^T x(k-1)}{2 + \psi_{iu}^2 x(k-1)^T x(k-1)} \Delta_{ib} \tag{7.95}$$

and

$$\psi_{ib}(k)^2 = \frac{2(|e_i(k|k-1)| - \Delta_{ib})}{(2\Delta_{ib} - |e_i(k|k-1)|x(k-1)^T x(k-1)} \tag{7.96}$$

if

$$\Delta'_{ib}(\psi_{il}, \Delta_{ib}, k) < |e_i(k|k-1)| \leq \Delta'_{ib}(\psi_{iu}, \Delta_{ib}, k)$$

For all $\psi_i(k)^2$ such that $\psi_{il}^2 < \psi_i(k)^2 \leq \psi_{ib}(k)^2$, the following holds good:

$$|e_i(k|k-1)| \geq \Delta'_{ib}(\psi_i(k), \Delta_{ib}, k) = \frac{2 + 2\psi_i(k)^2 x(k-1)^T x(k-1)}{2 + \psi_i(k)^2 x(k-1)^T x(k-1)} \Delta_{ib} \tag{7.97}$$

Consequently, along the adaptive algorithm (7.87)-(7.96), the adaptation of $\hat{\theta}_i(k)$ $(i = 1, \ldots, n)$ will be stopped at time k, that is, $\hat{\theta}_i(k) = \hat{\theta}_i(k-1)$, if the absolute value of the ith component of the prediction error, $|e_i(k|k-1)|$, is less than or equal to $\Delta'_{ib}(\psi_{il}, \Delta_{ib}, k)$. Otherwise, if the adaptation is not stopped, the value of $\psi_i(k)^2$ can be chosen in an interval greater than a selected value ψ^2_{il} and less than or equal to a value ψ^2_{ib}, which has been defined in such a way that condition (7.97) is satisfied.

The prediction $\hat{y}(k+1|k)$ of the process output at $(k+1)$ is given by

$$\hat{y}(k+1|k) = \hat{\theta}(k)x(k) = \hat{\theta}_0(k)\phi(k) + \hat{\theta}_1(k)u(k) \tag{7.98}$$

where $\hat{\theta}_0(k)$ and $\hat{\theta}_1(k)$ are estimates of the actual process matrices θ_0 and θ_1, and

$$\hat{\theta}(k) = [\hat{\theta}_0(k), \hat{\theta}_1(k)] \tag{7.99}$$

The control vector u(k) is computed to make the predicted output $\hat{y}(k+1|k)$ equal to the desired output $y_d(k+1)$ and is given by the control law:

$$u(k) = \hat{\theta}_1(k)^{-1}[y_d(k+1) - \hat{\theta}_0(k)\phi(k)] \tag{7.100}$$

In the preceding control law, $\hat{\theta}_1(k)$ can always be made nonzero by an appropriate choice of $\psi_i(k)$ $(i = 1, \ldots, n)$ at each time k.

The control or tracking error $\epsilon(k)$, which is equal to the prediction error $e(k|k-1)$, provided that $\hat{\theta}_i(k)$ is nonsingular, is defined by

$$\epsilon(k) = y(k) - y_d(k) \tag{7.101}$$

The APCS algorithm is given by the Eqs. (7.87)-(7.100), which are all that is required for implementation.

7.4.5 Assumptions to Prove Stability

Global stability of APCS can be proved on the basis of the following assumptions:

1. An upper bound on the dimension of the process parameter matrix is known.
2. The perturbation vector $\Delta(k)$ in (7.86) is bounded. Two cases are considered:
   a. The "general stochastic" case where a constant upper bound Δ_b on the absolute value of $\Delta(k)$ for all k is known and

$$\Delta_b - \Delta_m = \delta \tag{7.102}$$

where

$$\delta_i > 0 \quad (i = 1, \ldots, n) \quad \text{and} \quad \Delta_m = \max_{0<k\leq\infty} |\Delta(k)|$$

b. The deterministic (noise-free) case where

$$\Delta(k) = \Delta_b = \Delta_m = \quad = 0 \tag{7.103}$$

3. The desired process output $y_d(k+1)$ is known at time k and is bounded:

$$||y_d(k+1)|| \leq \lambda^2 < \infty, \quad \forall k \tag{7.104}$$

4. The sequence $\{||x(k)||\}$ is unbounded only if there is a subsequence $\{k_s\}$ such that

a. $$\lim_{k_s \to \infty} ||x(k_s - 1)|| = \infty \tag{7.105}$$

b. $$||y(k_s)|| > \alpha_1 ||x(k_s - 1)|| - \lambda_2, \quad \forall k_s \tag{7.106}$$

where

$$0 < \alpha_1 < \infty \quad \text{and} \quad 0 \leq \alpha_2 < \infty$$

This assumption is always verified for MIMO, ARMA, stable-inverse processes of the form (7.82), where θ_1 is nonsingular and vectors z and w are not included.

7.4.6 Global Stability of APCS

The proof of global stability of APCS is based on these four assumptions, as well as four lemmas, which are given along with proofs in [49] to which reference may be made. The main results are given in the following two theorems:

THEOREM 1: General Stochastic Case. Subject to assumptions 1, 2a, 3, and 4, the following properties are true if the APCS algorithm equations (7.87)-(7.100) are applied to a process described by (7.78)-(7.81):

(i) $||x(k)|| < \Omega < \infty, \quad \forall k$

(ii) $\exists k_1 < \infty$ such that $\forall k > k_1, \quad \hat{\theta}(k) = \hat{\theta}(k-1)$, and

$$|\epsilon_i(k)| \leq \Delta'_{ib}(\psi_{i1}, \Delta_{ib}, k) < 2\Delta_{ib} \quad (i = 1, \ldots, n)$$

(iii) $\lim_{\psi_{il}^2 \to 0} \Delta'_{ib}(\psi_{il}, \Delta_{ib}, k) = \Delta_{ib}$, and

$\lim_{\delta_i \to 0} \Delta_{ib} = \Delta_{im} \quad (i = 1, \ldots, n)$

THEOREM 2: Deterministic Case. Subject to assumptions 1, 2b, 3, and 4, the following properties are true if APCS equations (7.87)-(7.100) are applied to a process described by (7.78)-(7.81):

(i) $||x(k)|| < \Omega < \infty, \quad \forall k$

(ii) $\lim_{k \to \infty} [\hat{\theta}(k) - \hat{\theta}(k-1)] = 0$

(iii) $\lim_{k \to \infty} \epsilon(k) = 0$

For proofs of the preceding theorems, and also certain interpretations and other details, interested readers may see Martin-Sanchez et al. [49].

7.4.7 Discussion

The APCS strategy is an important alternative approach to adaptive control with entirely different concepts. The perturbation signal concept as defined in the ACPS process description makes the formulation very general in scope, and further practical extensions, interpretations and applications can be explored.

7.5 FUZZY LOGIC ADAPTIVE CONTROL

7.5.1 Introduction

Among the various adaptive strategies the fuzzy logic approach has a very bright future and many potential applications in sophisticated and complex processes which are difficult to control or which have, in the past, relied on the experience and judgment of a human operator.

The simple fuzzy logic controller was originally proposed in 1974 by Assilian and Mamdani [56] to regulate the outputs of a process around a specified set point. Mamdani et al. [57-60] reported on other advances in this field. Other contributors using this method are Braae and Rutherford [61, 62], Gupta [62a], Rutherford et al. [63, 64], and Tong [65].

Fuzzy controllers are designed by what is termed the rule-based approach. The conventional approach depends on a complete analysis of the process to be controlled. If the quantitative model is available, control can be effected by strictly numerical algorithms. However, in the rule-based technique the main emphasis is not on the actual process, but on the existing human decision-maker. Such a technique is referred to as an "expert system." Such an "expert" tackles the problem linguistically by "If... then" rules. This technique is therefore qualitative and the rules obtained are usually somewhat vague. However, such rules are implementable on a computer, using fuzzy set theory. In an adaptive fuzzy controller, we do not require a complete set of rules at the starting point. The objective, as in all control problems, is to minimize errors which may arise as quickly as possible even in the presence of disturbances and this objective is used to cause the controller to organize itself and learn new control rules.

In recent years there has been a significant development in the application of fuzzy controllers to a number of industrial processes, as is evidenced in the following list:

Cement plants: Egelov [66], Holmblad and Østergaard [67, 68], Sheridan and Skøth [69]
Sinter plants: Bloore [70], Carter et al. [71, 72], Rutherford et al. [63, 64]
Heat exchanger plants: Østergaard [73]
Warm water plants: Kickert and Van Nauta Lemke [74]
Various industrial processes: Mamdani [75], Jensen [76], King and Mamdani [77], Larsen [78]

7.5.2 Fuzzy Adaptive Controllers

This novel approach has been extended to adaptive control by Procyk [79] and Procyk and Mamdani [80]. They label such controllers as self-organizing controllers (SOC). They have shown by a series of experiments that fuzzy logic SOC have the following important advantages:

1. The controller is relatively insensitive to the variations in the parameters and in the environment.
2. The controller can be applied to a variety of processes including multivariable and nonlinear plants.
3. The control actions are not confined to only the bang-bang type as in some conventional adaptive controllers.
4. The controller can be primed with a rough initial control policy which would guard against violent changes during adaptation.
5. The structure, which is rule-based, can lend itself to a simple method of modifying the rules.
6. Fast convergence rates can be obtained. Even if convergence does not take place, the response is still found to be satisfactory.
7. The controller is robust.

We can say that a fuzzy adaptive controller belongs to the domain of "artificial intelligence" (AI). In the design of such sophisticated controllers, we have to utilize some aspects of human intelligence. A human being is capable of taking decisions even when working with vague or imprecise concepts which are expressed linguistically.

Zadeh [81] proposed one method of modeling this decision-making process, using the theory of approximate reasoning (AR), by which some classes of linguistic statements can be translated into mathematical terms. Zadeh has presented other important basic papers in this field [82-84] and these concepts are useful in the design of fuzzy controllers.

The following are the steps in the design of fuzzy adaptive controllers:

Step 1: The input is in the form of linguistic input rules.
Step 2: These linguistic values are converted into numerical form, utilizing fuzzy set theory.
Step 3: All numerical computations are executed inside the adaptive controller.
Step 4: Next the numerical subsets are "translated" to their closest linguistic equivalents.
Step 5: The output is in the form of linguistic output rules.

Such a fuzzy adaptive controller will have the following functional blocks:

1. The performance measure
2. The process model
3. The controller modification algorithm

These are considered in some detail in Section 7.5.4.

7.5.3 A Simple Fuzzy Controller

Before considering the adaptive controller in some detail, we shall deal briefly with a simple fuzzy controller [56]. It comprises a program running on a digital computer and is arranged in a feedforward configuration as shown in Fig. 7.5.

Some knowledge of basic fuzzy set theory is assumed. The fuzzy controller consists of a set of linguistic statements or rules which define individual control situations. A part of such an algorithm may read as follows:

IF ERROR IS POSITIVE SMALL THEN IF CHANGE IN ERROR IS ZERO THEN CHANGE IN PROCESS INPUT IS NEGATIVE SMALL ELSE

IF ERROR IS NEGATIVE LARGE THEN IF CHANGE IN ERROR IS POSITIVE SMALL THEN CHANGE IN PROCESS INPUT IS NEGATIVE MEDIUM ELSE ...

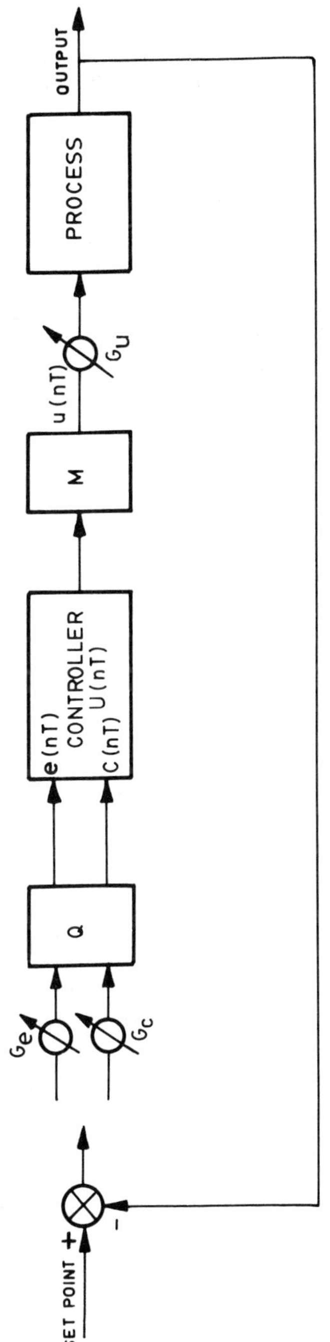

FIG. 7.5 Simple fuzzy controller. (From Ref. 80. Reprinted with permission from Pergamon Press, copyright 1979.)

Each rule relates the process ERROR and CHANGE IN ERROR (which is the difference between the present and last outputs) with CHANGE IN PROCESS INPUT which has to be implemented. These are three fuzzy variables (E, C, U, respectively), which are defined by fixed universes of discourse. They give the range of measurements these variables can take. In the present case we assume that the universes are discrete and finite. Each universe is a set of elements

$$E = \{e\}, \quad C = \{c\}, \quad U = \{u\}$$

In each control the fuzzy variables are given fuzzy values which represent their linguistic meaning. These fuzzy values are expressed as fuzzy subsets of the corresponding universe of discourse. Thus, a control rule k can be written in a short statement:

If E is E_k, then if C is C_k then U is U_k

The fuzzy subsets E_k, C_k, U_k are defined by a set of ordered pairs:

$$E_k = \{(e, \mu_{E_k}(e))\} \subset E$$
$$C_k = \{(c, \mu_{C_k}(c))\} \subset C$$
$$U_k = \{(u, \mu_{U_k}(u))\} \subset U$$

where e, c, and u are elements of the discrete universes and $\mu_{E_k}(e)$, $\mu_{C_k}(c)$, and $\mu_{U_k}(u)$ the corresponding membership values which give the degree to which the element is a member of the subset.

A control rule k can be regarded as an implication

$$E_k \rightarrow C_k \rightarrow U_k$$

which produces a relation matrix in three dimensions on the space $E \times C \times U$ given by the outer product

$$R_k = E_k \times C_k \times U_k$$

For the controller made up of several such implications or rules, the overall relation R is calculated as the union of the individual relations, that is,

$$R = R_1 V R_2 V R_3 \cdots R_k V = V_k R_k \tag{7.107}$$

R is therefore a matrix of membership values, $\mu_R(e, c, u)$.

Thus, the entire fuzzy algorithm or control strategy is represented by a single matrix R.

The objective of the controller is to infer, from the error and change in error, the change in control action to be applied. The values of error and change in error used by the controller are elements of the corresponding universes of discourse which can be represented in time by e(nT) and c(nT) where T is the sampling period and n the sampling number. These values are calculated from the actual process error and change in error. We should first scale the actual values by a suitable scaling factor and then quantize the result to the closest element of the universe of discourse.

If the process output at the nth sample is x(nT) and S is the set point, then

$$e(nT) = Q[\{S - x(nT)\} x G_e] \tag{7.108}$$

$$c(nT) = Q[\{x(nT - x(nT - T))\} x G_c] \tag{7.109}$$

where G_e and G_c are the scaling factors for the error and change in error, respectively, and Q represents the quantization process.

The output from the controller can therefore be obtained from the inputs by using Zadeh's compositional rule of inference. In this implementation the controller inputs are nonfuzzy. Hence the compositional rule of inference yields a fuzzy subset, U(nT), whose membership values are defined by

$$\mu_{U(nT)}(u) = \mu_k(e(nT), c(nT), u) \tag{7.110}$$

The result is therefore a fuzzy subset and for practical reasons a decision is required concerning what deterministic control action to take.

One solution is to choose the element of the universe, u(nT), corresponding to the maximum membership value of the calculated subset, U(nT). If this value is not unique, the mean of these elements is taken. This is represented by

$$u(nT) = M\{U(nT)\} \tag{7.111}$$

This is then multiplied by the scaling factor G_u to calculate the actual change in process input. The new process input is given by

$$i(nT + T) = i(nT) + G_u \times u(nT) \tag{7.112}$$

The scaling factors G_e, G_c, and G_u are chosen by the user, similar to the gain settings of a PID controller.

It may be pointed out that a large computer memory will be required for the relation matrix, even for single-input/single-output controllers. Simplification is possible by the use of a look-up table which is precalculated from the control rules before the controller is run. An example of such a look-up table is given by Procyk and Mamdani [80].

7.5.4 Design of a Fuzzy Adaptive Controller

The simple fuzzy controller considered above is applicable to processes which are not very complex. In the case of more difficult processes which exhibit nonlinear characteristics and which are of higher dimensions, we should have the control rules generated automatically. In other words, we should design a controller in an adaptive fashion.

Performance Measure

An adaptive controller must be able to assess its own performance in order to improve its control strategy. We can have either (1) a global criterion which measures the overall performance or (2) a local criterion which measures the performance over a small set of plant states. The second option is relatively simpler. The desired closed-loop response is the goal the controller should seek. It represents a sequence of "per sample" performance criteria which constitutes the global one, that is, the desired response.

The performance of the controller in relation to each process input is measured by the deviation of the actual response from the desired one. This is represented by a variable which gives a rough indication of the amount of the corrections required at the output. The response can be monitored by e(nT) and c(nT). The performance measure will then assume the role of a decision maker which finds the output correction required from a knowledge of e(nT) and c(nT). This is expressed as a decision table, an example of which is given by Procyk and Mamdani [80]. The entries in this table not only take into account the distance from the set point, but also the rate of approach to it. Thus a small error and small change in error is considered more satisfactory than a small error and a large change in error, because of the possibility of overshoot. This performance decision table is invoked at every sampling instant to get a measure of deviation of each output from the trajectory.

Process Model

The next step is to translate these output variations into input corrections. The following questions arise:

1. How are these input corrections to be computed from a knowledge of output variations?

2. In a MIMO plant, which of the inputs should be corrected and by how much?
3. Which previous control actions have resulted in the present poor performance?

To find the answers to these questions, we must have knowledge of the relationship between the process inputs and outputs and also the time lags in the process. In other words, we should have some sort of a model of the process. Whereas an accurate model is required for a nonadaptive controller, we can have a less accurate model for an adaptive controller, since identification and control take place simultaneously and any information that is inadequate can be found during the adaptation process.

The model is essentially an incremental one, as it has to relate the input changes to the output changes. Let us consider, for example, a two-input/two-output plant described by the state space equations:

$$\dot{X} = F(X, U, V) \tag{7.113}$$

$$\dot{Y} = G(X, U, V) \tag{7.114}$$

For small input changes, the resultant changes in the first output derivate are given by

$$\begin{bmatrix} \delta\dot{X} \\ \delta\dot{Y} \end{bmatrix} = \begin{bmatrix} \partial F/\partial U & \partial F/\partial V \\ \partial G/\partial U & \partial G/\partial V \end{bmatrix} \begin{bmatrix} \delta U \\ \delta V \end{bmatrix} \tag{7.115}$$

Input changes ΔU and ΔV will produce output changes ΔX and ΔY after one sample of time T, and these can be approximated by

$$\begin{bmatrix} \Delta X \\ \Delta Y \end{bmatrix} \approx \begin{bmatrix} T\delta\dot{X} \\ T\delta\dot{Y} \end{bmatrix} = TJ \begin{bmatrix} \Delta U \\ \Delta V \end{bmatrix} = M \begin{bmatrix} \Delta U \\ \Delta V \end{bmatrix} \tag{7.116}$$

where J is the Jacobian matrix and M is the incremental model of the process. In general, if p(nT) is a vector of output corrections and r(nT) is a vector of input corrections, then from (7.116), we have

$$r(nT) = M^{-1} p(nT) \tag{7.117}$$

This model is to be scaled, using scaling factors already indicated, and then normalized. More details are given in [79, 80] about the model.

A single-input/single-output plant incorporating the performance measure and the model are shown in Fig. 7.6. A simple controller forms the bottom level and the performance measure and the model constitute the upper level of the fuzzy adaptive controller.

The dynamics of the process determines the control action regarding to which samples the input corrections are to be applied. Higher-order processes with large time lags will require control actions further back from the present to overcome these delays. Low-order processes with short lags will require more recent control actions. Hence, control action is set by the user according to the judgment about the process to be controlled. This also implies a model of the dynamics of the process, from the time-lag considerations.

Controller Modification Algorithm

The next step is to change the control strategy in such a way that future control actions result in an improvement. The fuzzy controller of the bottom level (in Fig. 7.6) consists of a set of linguistic rules which can be modified simply. The controller can be primed with an initial rough set of control rules obtained from experience. The modification algorithm improves on the initial set of rules.

Let us assume that the present poor performance is due to a control action, m samples in the past. Then

$$\text{Error at that time} = e(nT - mT)$$

$$\text{Change in error at that time} = c(nT - mT)$$

$$\text{Controller output at that time} = u(nT - mT)$$

As seen earlier, the controller output should have been $[u(nT - mT) + r(nT)]$ and not $u(nT - mT)$, where $r(nT)$ is the input correction.

Let us construct fuzzy subsets around the elements $e(nT - mT)$, $c(nT - mT)$, $u(nT - mT)$, and $[u(nT - mT) + r(nT)]$ of the corresponding universes of discourse as follows:

$$E(nT - mT) = F_z\{e(nT - mT)\} \tag{7.118}$$

$$C(nT - mT) = F_z\{c(nT - mT)\} \tag{7.119}$$

$$U(nT - mT) = F_z\{u(nT - mT)\} \tag{7.120}$$

$$V(nT - mT) = F_z\{u(nT - mT) + r(nT)\} \tag{7.121}$$

where F_Z represents the process of fuzzification around a single element.

The controller modification is involved because we have to <u>replace</u> the implication

$$E(nT - mT) \longrightarrow C(nT - mT) \longrightarrow U(nT - mT) \tag{7.122}$$

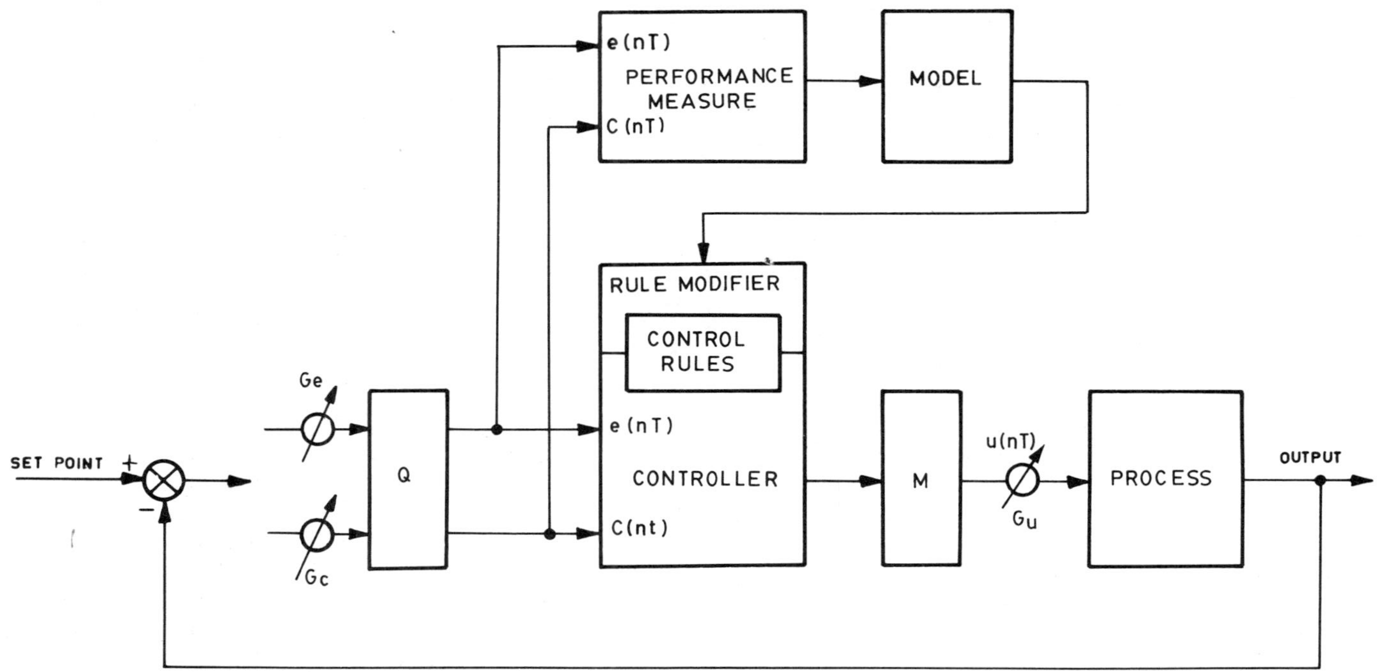

FIG. 7.6 Fuzzy adaptive controller. (From Ref. 80. Reprinted with permission from Pergamon Press, copyright 1979.)

by the new implication

$$E(nT - mT) \longrightarrow C(nT - mT) \longrightarrow V(nT - mT) \tag{7.123}$$

These two implications form relation matrices R'(nT) and R"(nT) given by the outer products

$$R'(nT) = E(nT - mT) \times C(nT - mT) \times U(nT - mT) \tag{7.124}$$

$$R''(nT) = E(nT - mT) \times C(nT - mT) \times V(nT - mT) \tag{7.125}$$

One way of effecting this replacement is by a linguistic statement

$$R(nT + T) = \{R(nT) \text{ but not } R'(nT)\} \text{ else } R''(nT) \tag{7.126}$$

where R(nT) is the current controller relation matrix and R(nT + T) is the new modified one. It must be noted that the method of modification (7.126) is not unique, but one of several alternatives possible. Further, the preceding theory is based on relation matrices and in practice this has certain disadvantages. It is possible to consider another approach of storing the rules instead of the relation matrices, as outlined in the appendix of Procyk and Mamdani [80].

7.5.5 Discussion

In the preceding discussion we have considered only SISO systems and we have assumed that only one control action is responsible for the poor performance. In MIMO systems the number of variables and also the size of the matrices increase. When more than one action is involved, we have to repeat the preceding procedure for each of the control actions.

Procyk and Mamdani [80] report about a series of experiments to examine the performance of such a controller with different processes and with different internal parameter settings. Strong "adaptive" properties were observed even when dealing with a highly nonlinear process. They also consider systems with time lags and two-input/two-output processes and show how convergence can be achieved in all cases.

There appears to be considerable scope in the future for this new approach to adaptive control, and further investigations may lead to a wide variety of applications.

7.6 ADAPTATION USING THE INTERNAL MODEL PRINCIPLE [85]

7.6.1 Introduction

Elliott and Goodwin [85] have proposed a new algorithm for discrete-time adaptive pole placement using the internal model principle [86-88]. This allows rejection of deterministically modeled disturbances and tracking of simple reference signals. Such disturbances can be modeled as uncontrollable modes of the system to be controlled. This leads to an algorithm which identifies a higher-order system model containing these modes. The model is then used to adaptively implement an error feedback compensator which places all closed-loop poles and also includes models of the disturbance and any reference signal to be tracked. After convergence, the tuned controller renders the disturbances unobservable in the output and ensures tracking of the desired reference signal. Use is made of sequential least squares with covariance reset for identification.

In the implementation of the adaptive controller, two time scales are employed [89,90]. The parameters in the adjustable controller are held constant for blocks of N samples, by choosing an appropriate value of N. Global stability depends on the assumption that the reference signal is persistently exciting. Thus, the reference signal should be supplemented by additional sinusoidal components.

7.6.2 Modeling of Disturbances

Consider an unknown linear time-invariant plant subject to deterministic disturbances as shown in Fig. 7.7. The signals $\tilde{u}(t)$ and $\tilde{y}_P(t)$ are the true plant input and output signals. The input and output disturbances are represented by $d_i(t)$ and $d_o(t)$.

The plant input/output data are modeled as the solution to the ARMA difference equation

$$\tilde{A}(q^{-1})\tilde{y}_P(t) = \tilde{B}(q^{-1})\tilde{u}(t) \tag{7.127}$$

where $\tilde{A}(0) = 1$, $\tilde{B}(0) = 0$, and $\max(\deg(\tilde{A}), \deg(\tilde{B})) = \tilde{n}$. The system order $\tilde{n}$ is assumed to be known, but the coefficients $\tilde{A}$ and $\tilde{B}$ are unknown.

The modeled disturbances are represented as solutions to the homogeneous difference equations.

$$D_2(q^{-1})d_i(t) = 0 \tag{7.128}$$

$$D_3(q^{-1})d_o(t) = 0 \tag{7.129}$$

$$D(q^{-1}) \triangleq D_2(q^{-1})D_3(q^{-1}) \tag{7.130}$$

where

$$D_2(0) = D_3(0) = 1, \quad \deg(D_2) = l_2, \quad \deg(D_3) = l_3$$

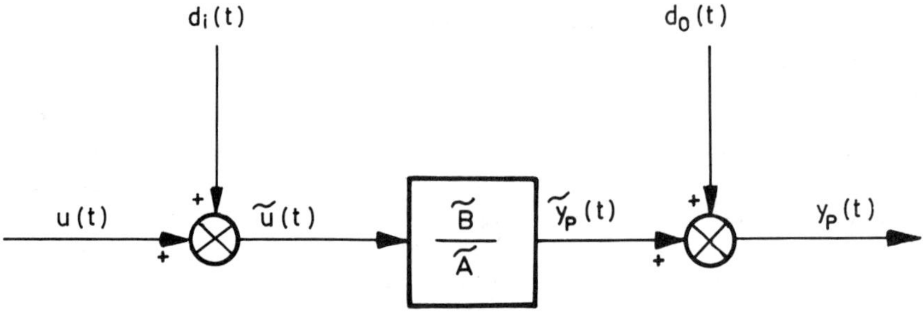

FIG. 7.7 Unknown linear time-invariant plant subject to deterministic disturbances. (From Ref. 85.)

The following assumptions are made:

1. The degrees of D_2 and D_3 are known, but their coefficients are unknown.
2. The disturbance is bounded and nondiminishing.
3. The roots of $D_2(q^{-1})$ and $D_3(q^{-1})$ lie on the unit circle and are distinct.
4. The polynomials $\tilde{A}D_2$ and $\tilde{B}D_3$ are relatively prime and so are $\tilde{B}$ and D_3. Physically, the first requirement is explained thus: If D_2 has a common factor with $\tilde{B}$, then the corresponding mode of disturbance will be annihilated as it enters the plant and hence cannot be identified. Similarly if D_3 and $\tilde{A}$ have a common factor, it cannot be determined whether the corresponding disturbance mode is actually present or whether that mode corresponds to an initial condition response of the plant. There, this strong condition of $\tilde{A}D_2$ and $\tilde{B}D_3$ being prime is needed to guarantee proper performance of the system identification algorithm. In terms of the actual analysis, it is necessary to assure persistency of excitation.

Using (7.127)-(7.130) we can show that

$$A(q^{-1})y_p(t) = B(q^{-1})u(t) \tag{7.131}$$

$$A(q^{-1}) = \tilde{A}(q^{-1})D(q^{-1}) \tag{7.132}$$

$$B(q^{-1}) = \tilde{B}(q^{-1})D(q^{-1}) \tag{7.133}$$

Let $n = \max(\deg(A), \deg(B))$. Then

$$n = \tilde{n} + l_2 + l_3 \tag{7.134}$$

This inclusion of modeled disturbances results in a higher-order ARMA representation with common factors in the A and B polynomials.

Consider the following partial state representation for this system:

$$A(q^{-1})z_1(t) - C_2z_2(t) = u(t) \tag{7.135}$$

$$D_2(q^{-1})z_2(t) = 0 \tag{7.136}$$

$$D_3(q^{-1})z_3(t) = 0 \tag{7.137}$$

$$y_P(t) = \tilde{B}(q^{-1})z_1(t) + C_3z_3(t) \tag{7.138}$$

where $z_1(t)$ is the internal state of the plant and $z_2(t)$ and $z_3(t)$ are the uncontrollable states associated with the modeled disturbances. The constants C_2, C_3 relate to the modeled disturbances as

$$d_i(t) = C_2z_2(t)$$

$$d_o(t) = C_3z_3(t)$$

and help in representing the cases of presence or absence of the disturbances.

7.6.3 Internal Model Principle

The implementation of the internal model principle is shown in Fig. 7.8.

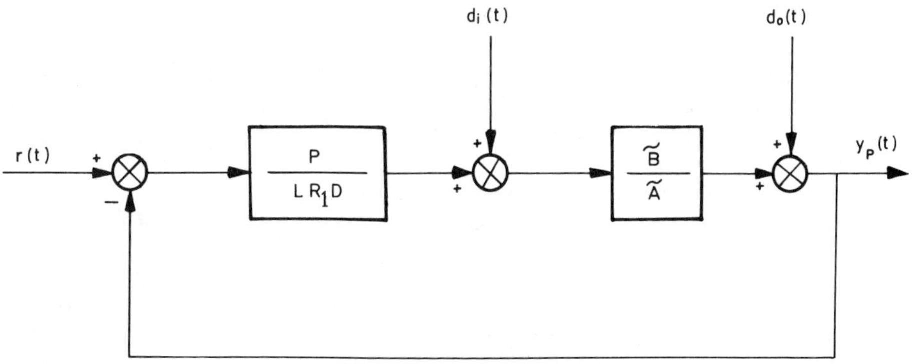

FIG. 7.8 Implementation of internal model principle. (From Ref. 85.)

In Fig. 7.8, we have

$$L(q^{-1})R_1(q^{-1})D(q^{-1})u(t) = P(q^{-1})[r(t) - y_p(t)] \tag{7.139}$$

where $L(q^{-1})$ and $P(q^{-1})$ satisfy the diophantine equation

$$L(q^{-1})R_1(q^{-1})D(q^{-1})\tilde{A}(q^{-1}) + P(q^{-1})\tilde{B}(q^{-1}) = A^*(q^{-1}) \tag{7.140}$$

The polynomial $A^*(q^{-1})$ is an arbitrary Hurwitz polynomial representing desired closed-loop pole locations. The reference signal r(t) has two components, $r_1(t)$ and $r_2(t)$, which satisfy

$$R_1(q^{-1})r_1(t) = 0$$

$$R_2(q^{-1})r_2(t) = 0$$

where

$$R_1(0) = R_2(0) = 1 \quad \text{and} \quad \deg(R_1) = m_1, \ \deg(R_2) = m_2$$

The signal $r_1(t)$ is a signal which has to be tracked and $r_2(t)$ is a signal which has been introduced to assure persistency of excitation. Elliott and Goodwin [85] show that the reference signal r(t) must contain $(m_1 + m_2) \geqslant 3n$ complex sinusoidal components. We assume that $R_1(q^{-1})$ and $R_2(q^{-1})$ also have roots on the unit circle and that $R_1(q^{-1})$ is known. However, if $R_1(q^{-1})$ is unknown, then Fig. 7.8 must be modified by taking $r_1(t)$ as an output disturbance included in the unknown polynomial $D(q^{-1})$.

In addition to the constraints already imposed on $\tilde{A}$, $\tilde{B}$, D_1, and D_2 in Section 7.6.2, we require that R_1, D, $\tilde{A}$, and $\tilde{B}$ be relatively prime if a solution should exist for (7.139) for arbitrary A^*. This is equivalent to requiring each of the polynomial pair, R_1 and $\tilde{B}$, D_2 and $\tilde{B}$, D_3 and $\tilde{B}$, and $\tilde{A}$ and $\tilde{B}$ to be prime.

We obtain the closed-loop equation from Eqs. (7.139) and (7.140):

$$A^*(q^{-1})y_p(t) = P(q^{-1})\tilde{B}(q^{-1})r(t) \tag{7.141}$$

This control law is the direct application of the internal model principle. It renders the uncontrollable disturbances and the signal $r_1(t)$ unobservable through the error signal $e(t) = y_p(t) - r(t)$.

Thus in the absence of the unmodeled disturbances and the supplemental excitation $r_2(t)$, $y_p(t)$ will track r(t) asymptotically.

7.6.4 Adaptive Control

Let us use the idea of a two-time-scale operation as mentioned earlier to obtain an adaptive version of (7.139):

$$L_k(q^{-1})R_1(q^{-1})D_k(q^{-1})u(t) = P_k(q^{-1})[r(t) - y_P(t)] \tag{7.142}$$

where (i) $t \in [kN, (k + 1)N - 1]$, (ii) $k \in [0, 1, 2, \ldots)$, (iii) L_k and P_k satisfy

$$L_k(q^{-1})R_1(q^{-1})D_k(q^{-1})\tilde{A}_k(q^{-1}) + P_k(q^{-1})\tilde{B}_k(q^{-1}) = A^*(q^{-1}) \tag{7.143}$$

and (iv) $\tilde{A}_k$, $\tilde{B}_k$, D_k are estimates of $\tilde{A}$, $\tilde{B}$, and D obtained at time $t = kN$.

The idea of two time scales results in the estimates of L, P, and D used in (7.142) being frozen over the interval $t \in [kN, (k + 1)N - 1]$. The interval length N should be chosen appropriately. Elliott and Goodwin [85] show that N should satisfy the condition: $N \geqslant (3n + 2m_1)$.

Computational procedures for calculating $\tilde{A}_k$, $\tilde{B}_k$, D_k, L_k, and P_k will be considered next.

The system model (7.131)-(7.134) can be written in regression form as

$$y_P(t) = \phi(t - 1)^T \theta_0 \tag{7.144}$$

$$\theta_0 = [-a_1, \ldots, -a_n, b_1, \ldots, b_n]^T \tag{7.145}$$

$$\phi(t - 1) = [y_P(t - 1), \ldots, y_P(t - n), u(t - 1), \ldots, u(t - n)]^T \tag{7.146}$$

The sequential least-squares algorithm with covariance reset will be used for parameter identification:

$$\hat{\theta}(t) = \hat{\theta}(t - 1) + \frac{P(t - 2)\phi(t - 1)\epsilon(t)}{1 + \phi(t - 1)^T P(t - 2)\phi(t - 1)} \tag{7.147}$$

$$P(t - 1) = \begin{cases} P(t - 2) + \dfrac{P(t - 2)\phi(t - 1)\phi(t - 1)^T P(t - 2)}{1 + \phi(t - 1)^T P(t - 2)\phi(t - 1)}, & t \neq kN \quad (7.148) \\[2ex] \sigma_0 I, & t = kN \quad (7.149) \end{cases}$$

$$\epsilon(t) = y_P(t) - \phi(t - 1)^T \hat{\theta}(t - 1) \tag{7.150}$$

Let us define

$$\hat{\theta}(t) \triangleq [-a_1(t), \ldots, -a_n(t), b_1(t), \ldots, b_n(t)] \tag{7.151}$$

$$A_k(q^{-1}) \triangleq 1 + \sum_{i=1}^{n} a_i(kN)q^{-i} \tag{7.152}$$

$$B_k(q^{-1}) \triangleq \sum_{i=1}^{n} b_i(kN)q^{-i} \tag{7.153}$$

To calculate L_k, and P_k in (7.142) and (7.143), we must generate the estimates of $\tilde{A}_k$, $\tilde{B}_k$, and D_k of $\tilde{A}$, $\tilde{B}$, and D, using the estimate A_k, and B_k of A, and B, and the additional fact that $A = \tilde{A}D$ and $B = \tilde{B}D$. We have therefore

$$\tilde{B}\tilde{A}D = \tilde{B}A = B\tilde{A}$$

Let us define $\tilde{A}_k$ and $\tilde{B}_k$ as solutions to

$$\tilde{A}_k(q^{-1})B_k(q^{-1}) - \tilde{B}_k(q^{-1})A_k(q^{-1}) = 0$$

Since $\tilde{A}_k(0) = 1$, we can write the preceding equation as

$$[\tilde{A}_k(q^{-1}) - 1]B_k(q^{-1}) + \tilde{B}_k(q^{-1})[-A_k(q^{-1})] = -B_k(q^{-1}) \tag{7.154}$$

By equating the coefficients of q^{-1} on both sides of (7.154), we get 2n equations for $2\tilde{n}$ unknown coefficients of $\tilde{A}_k$ and $\tilde{B}_k$. These are solved using a pseudo left inverse.

Next we use a similar approach to generate $D_k(q^{-1})$. An appropriate estimate of D_k would be the solution of equations:

$$D_k(q^{-1})\tilde{B}_k(q^{-1}) = B_k(q^{-1}) \tag{7.155}$$

$$D_k(q^{-1})\tilde{A}_k(q^{-1}) = A_k(q^{-1}) \tag{7.156}$$

By equating the coefficients of q^{-1} on both sides, we get 2n equations for the $(l_2 + l_3)$ unknowns which can be solved using a pseudo left inverse.

Next we solve (7.143) for L_k and P_k, by equating the coefficients of q^{-1}. A sufficient condition for solutions to exist when the true parameters are known is for $A^*(q^{-1})$ to be of degree $(2n + m_1 - 1)$ or less, and $L_k(q^{-1})$ and $P_k(q^{-1})$ to be of degree $(n + m_1 - 1)$.

REMARK 1. In actual practice, $R_1 D_k \tilde{A}_k$ and $\tilde{B}_k$ may have a common factor instantaneously resulting in the ill-conditioning of the matrix involved in solving the linear equations associated with (7.143). To avoid this we assume a lower bound on the magnitude of the determinant of the matrix obtained from the true parameters. If, at the instant kN, the determinant of the matrix of the estimated parameters is below this bound, then we set $L_k = L_k^{-1}$ and $P_k = P_k^{-1}$. With this modification, bounded estimates of L_k and P_k are ensured, provided $\tilde{A}_k$, $\tilde{B}_k$, and D_k remain bounded. The latter are bounded if the estimates A_k and B_k remain bounded, which is so since the least-squares algorithm does yield bounded estimates of A_k and B_k.

REMARK 2. We are using an error feedback compensator to implement the internal model principle (instead of the usual observer). We should see that the numerator $P_k(q^{-1})$ of the compensator does not converge to zero. So we assume a lower bound on $||P||$, associated with the vector of coefficients of $P(q^{-1})$. If $||P_k||$ falls below this bound, we set

$$L_k = L_{k-1} \quad \text{and} \quad P_k = P_{k-1}$$

7.6.5 Stability

A detailed global stability analysis in the presence and absence of modeled disturbances is given in Elliott and Goodwin [85] to which reference may be made. The results can be summarized as follows: If the reference one wishes to track is supplemented by additional sinusoidal components, the block length N is long enough, and the disturbances are present, then the algorithm is globally stable and parameters converge exponentially. If there are no disturbances or fewer than expected, then global stability is still assured, although exponential convergence of the parameters is not guaranteed.

7.6.6 Discussion

This is an interesting application of the internal model principle to allow tracking in the presence of modeled deterministic disturbances. Although there is need to have additional sinusoidal components in the reference signal to assure persistence of excitation, it should be noted that if set points change frequently, the DFT of such a signal over each block will contain the necessary frequency content.

There is scope to extend this approach to cases of unmodeled disturbances, unmodeled dynamics, and time-varying parameters.

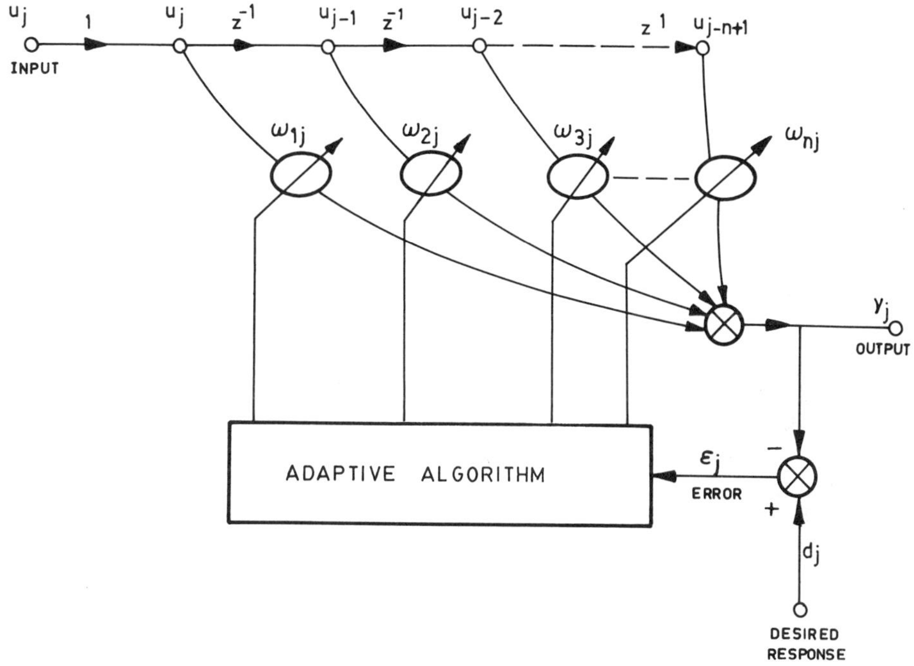

FIG. 7.9 An adaptive filter. (From Ref. 92. Reprinted with permission from Pergamon Press, copyright 1983.)

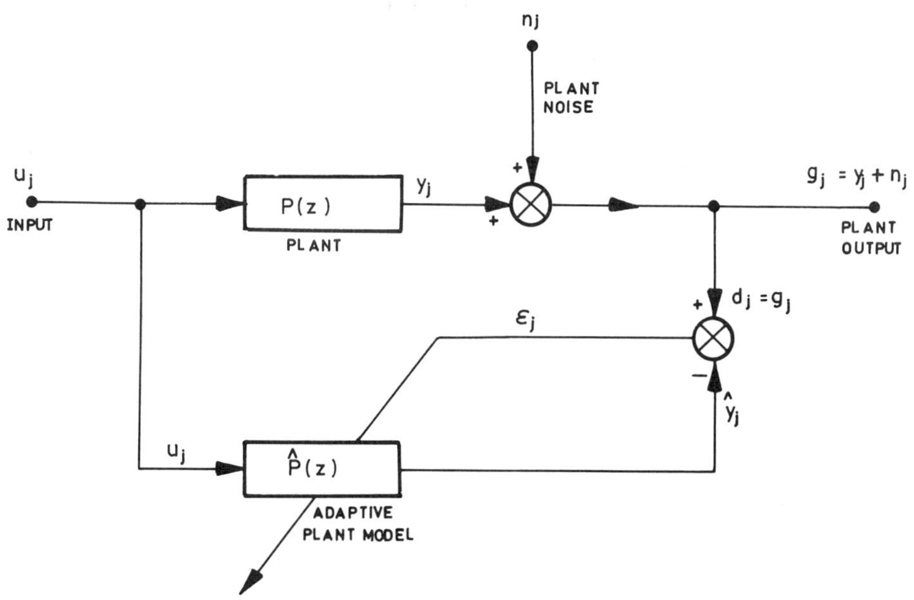

FIG. 7.10 Adaptive modeling. (From Ref. 92. Reprinted with permission from Pergamon Press, copyright 1983.)

the adaptive transversal filter can achieve a close fit to an unknown plant having many poles and zeros.

7.7.4 Plant Inverse Modeling

Figure 7.11 shows the inverse model of the unknown plant. The adaptive filter input is the plant output. The filter is adapted to cause its output to be a best least-squares fit to the plant input. This implies that the unknown plant with the LMS filter is cascade will have a unit "transfer function." As mentioned earlier, it is possible to obtain close fits even when the unknown plant has many poles and zeros.

If the plant is stable, all the poles lie in the left half-plane of the s-plane. If the plant is non-minimum phase, some of its zeros may lie in the right half-plane. A minimum-phase plant has no problem with stability of the inverse, but for a non-minimum-phase plant, the stability of the inverse is a problem. However, we can show that we can construct stable inverses if we are permitted noncausal two-sided impulse responses. By allowing a delay in the modeling process, as shown in Fig. 7.11, we can obtain approximate inverse models for minimum-phase and non-minimum-phase plants. We need not know a priori whether the plant is minimum phase or not. However, some knowledge of plant characteristics is helpful in choosing the delay Δ and the length of the transversal filter.

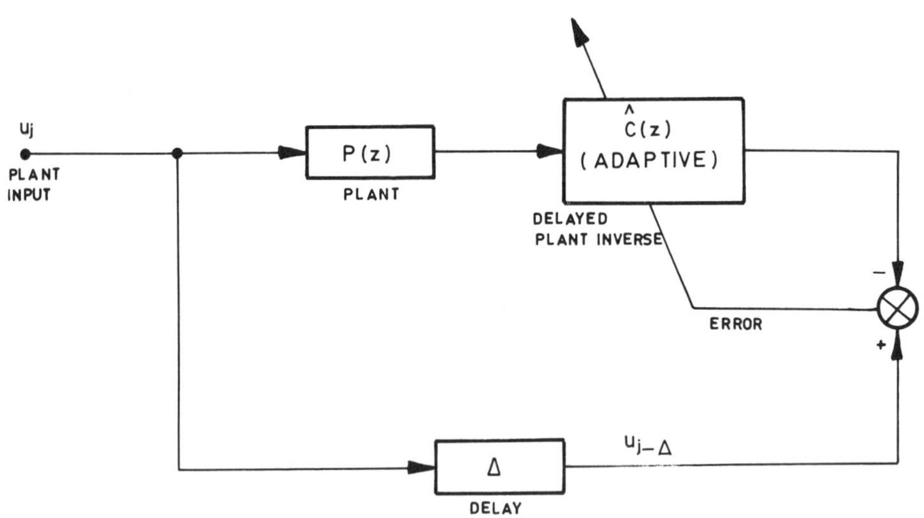

FIGURE 7.11 Delayed inverse modeling. (From Ref. 92. Reprinted with permission from Pergamon Press, copyright 1983.)

7.7.5 Adaptive Inverse Control Scheme

An adaptive inverse control system is shown in Fig. 7.12. The controller is a copy of the inverse model. The command input i_j (which is the desired output for the plant) is applied to the controller. The controller output is the driving function for the plant. Then the plant output, assuming no noise, will be an exact copy of the input reference command, but delayed, that is, $y_j = I_{j-\Delta}$. Computer simulations show that very good plant inverses can be obtained even for non-minimum-phase plants.

7.7.6 Model Reference Adaptive Inverse Control System

A reference model can be readily incorporated into the adaptive inverse control system as shown in Fig. 7.13. The reference model is a smoothing model. Such a system is used whenever we desire that the plant output track not the command input itself, but a delayed or smoothed version of the command input. The system of Fig. 7.12 becomes a special case of the system of Fig. 7.13 when the reference model equals the delay Δ. If the plant inverse model has enough weights and if the reference model contains enough delay, this approach results in excellent control of plant dynamics, as shown in simulation studies.

7.7.7 MRAC for a Noisy Plant

If any noise is present in the output of the plant, performance is severely affected. This problem is tackled by the system shown in Fig. 7.14. A model $\hat{p}(z)$ of the plant P(z) is a Wiener solution that is not biased by plant noise. An "off-line" process can then be used to obtain a controller $\hat{C}(z)$ from $\hat{p}(z)$ and the reference model M(z). This process adjusts $\hat{C}(z)$ to cause the output of the cascade of $\hat{p}(z)$ and $\hat{C}(z)$ to be a best least-squares match to the output of the reference model M(z) when both the cascade and the reference model are driven simultaneously by a "modeling signal" having an appropriate spectral character. An exact digital copy of $\hat{C}(z)$ is then used as a controller, as shown in Fig. 7.14. This results in the dynamic response of the cascade of controller and plant to closely follow the dynamic response of the reference model.

7.7.8 Discussion

The technique proposed by Widrow et al. is easy to implement and exhibits robust, predictable behavior. Additional work (to be reported) has been done to enhance the potential capabilities of this approach. The following issues have been studied:

1. Cancelling the plant noise by feeding appropriate signal at the plant input
2. Using dither signals for the modeling process when sufficient ambient signal activity is not present

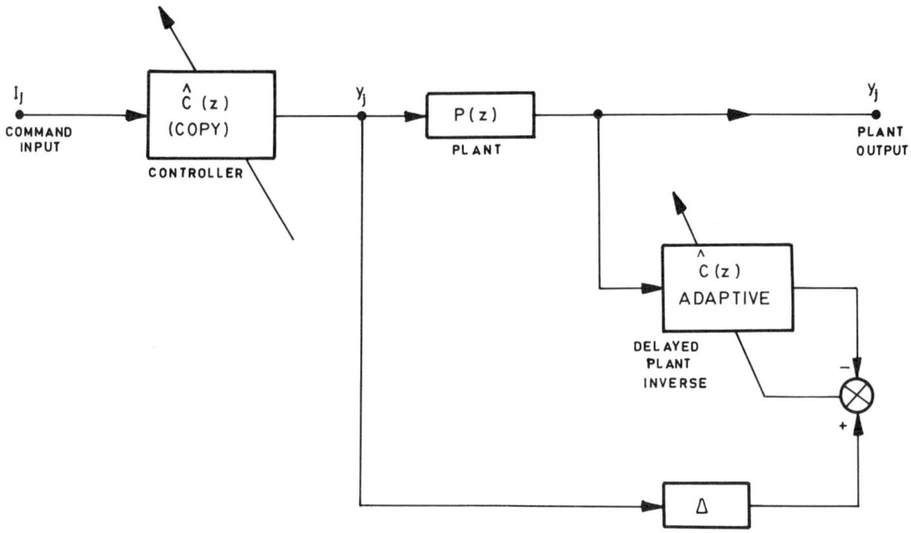

FIG. 7.12 An adaptive inverse control system. (From Ref. 92. Reprinted with permission from Pergamon Press, copyright 1983.)

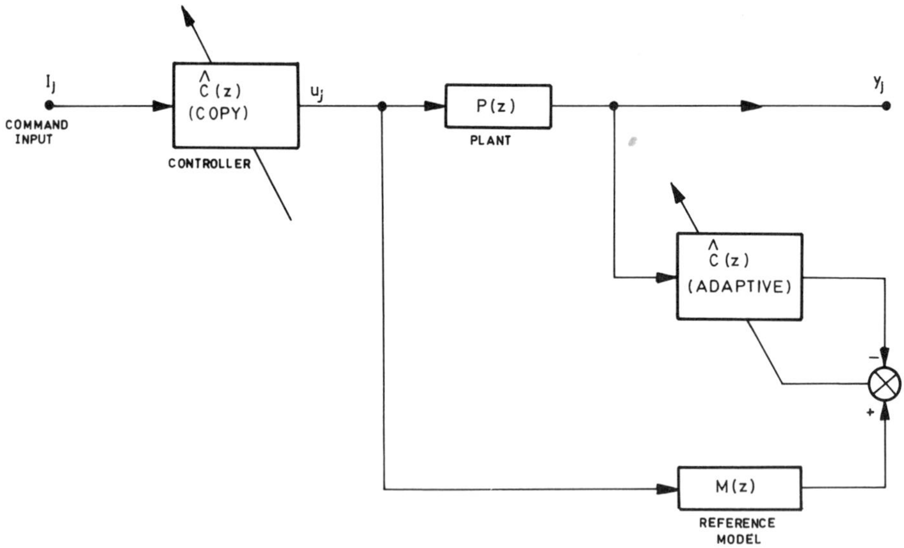

FIG. 7.13 A model reference adaptive inverse control system. (From Ref. 92. Reprinted with permission from Pergamon Press, copyright 1983.)

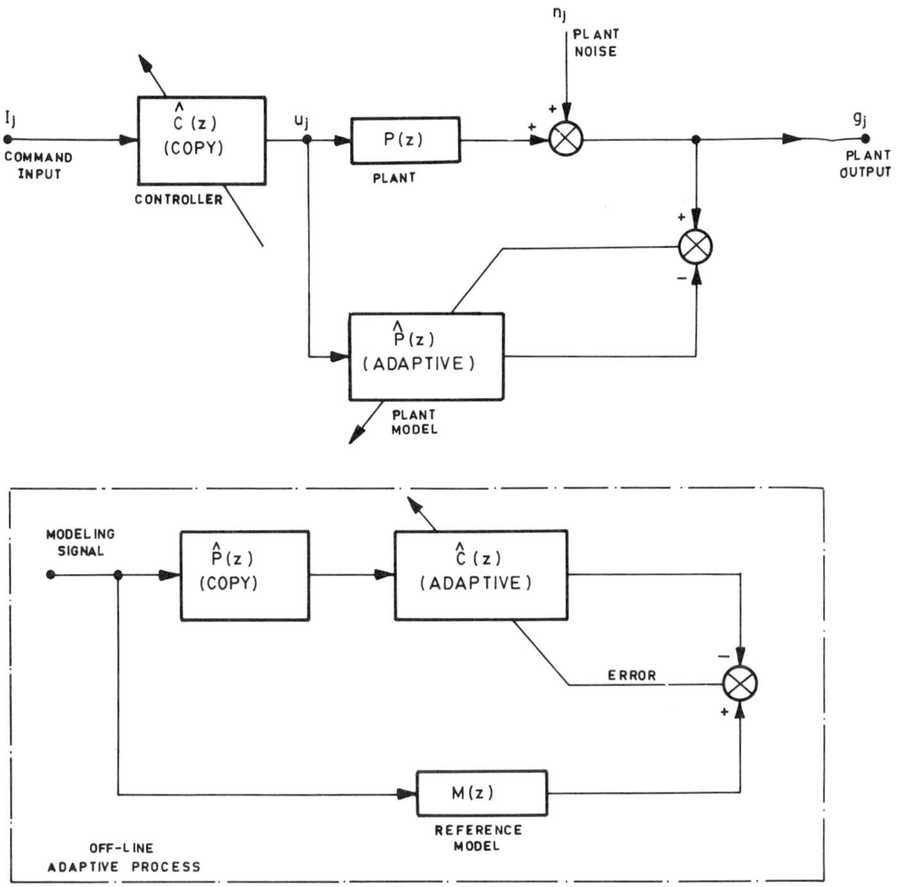

FIG. 7.14 A model reference adaptive inverse control system for a noisy plant. (From Ref. 92. Reprinted with permission from Pergamon Press, copyright 1983.)

3. Adaptive control of MIMO plants
4. Detailed quantitative performance and stability analysis

Further work is necessary to eliminate the constraint that the plant be stable, while retaining the basic Widrow approach.

7.8 CONCLUDING REMARKS

The preceding chapters considered the most commonly applied schemes, namely, the model reference adaptive control systems and self-tuning regulators. In this chapter we have discussed certain other recently proposed adaptive control strategies which have many possibilities for applications. The various methods discussed are (1) variable structure schemes (VSS), (2) reduced-order schemes, (3) adaptive predictive control systems (APCS), (4) fuzzy logic adaptive control systems, (5) adaptation using the internal model principle, and (6) adaptive control by inverse modeling. There is considerable scope for further studies of all these interesting and promising techniques.

REFERENCES

1. Petrov, B. N., Ulanov, G. M., and Emel'yanov, S. V., Invariant nost i optimizacia sistem avtomalicheskogo regulirovania s zsestoko i peremennoi stukturoi, Proc. IFAC Congress (1963).
2. Emel'yanov, S. V., Design of variable structure control systems with discontinuous switching functions, Eng. Cybern., 1: 156-160 (1964).
3. Emel'yanov, S. V., Automatic Control Systems of Variable Structure, Nauka, Moscow, 1967.
4. Bakakin, A. A., and Utkin, V. I., Variable structure systems with pure delay in switching devices, in Variable Structure Systems and their Application to Flight Automation, Nauka, Moscow, 1968.
5. Bakakin, A. A., Bermant, M. A., and Ezerov, V. B., Application of variable structure systems to stabilization of time-varying plants in the presence of restrictions on the movement of the controlling device, Autom. Rem. Contr., 7: 1016-1021 (1964).
6. Utkin, V. I., Sliding Modes and their Application in Variable Structure Systems (English translation), MIR, Moscow, 1978.
7. Utkin, V. I., Variable structure systems with sliding modes, IEEE Trans. Auto. Contr., AC-22: 212-222 (1977).
8. Itkis, U., Control Systems of Variable Structure, Keter Publishing House, Jerusalem, 1976.
9. Utkin, V. I., Variable structure systems: Present and future, Autom. Rem. Contr., 44: 1105-1120 (1984).
10. Young, K.-K. D., Asymptotic stability of model reference systems with variable structure control, IEEE Trans. Auto. Contr., AC-22: 279-281 (1977).
11. Young, K.-K. D., Design of variable structure model-following control systems, IEEE Trans. Auto. Contr., AC-23: 1079-1085 (1978).
12. Zinober, A. S. I., Relay control of plants subject to parameter uncertainty, Doctoral Thesis, University of Cambridge, U.K. (1974).
13. Zinober, A. S. I., Adaptive relay control of second-order systems, Int. J. Contr., 21: 81-98 (1975).

14. Zinober, A. S. I., Analysis of an adaptive third-order relay control system using nonlinear switching surface theory, Proc. R. Soc., Edinburgh, 76A: 239-254 (1977).
15. Zinober, A. S. I., Adaptive variable structure systems, Proc. 3rd IMA Conf. on Control Theory, Sheffield, U.K. (1980).
16. Zinober, A. S. I., El-Ghezawi, O. M. E., and Billings, S. A., Variable structure control of adaptive model following systems, Proc. IEE Conf. on Control and Its Applications, Warwick, U.K., pp. 123-127 (1981).
17. Ambrosino, G., Celentano, G., and Garofalo, F., Variable structure model reference adaptive control systems, Int. J. Contr., 39: 1339-1349 (1984).
18. Bengiamin, N. N., and Chan, W. C., Variable structure control of electric power generation, IEEE Trans. Power App. and Systems, PAS-101: 376-380 (1982).
19. Chan, W. C., and Hsu, Y. Y., Automatic generation control of interconnected power systems using variable structure controllers, Proc. IEE, 128, Pt. C: 269-279 (1981).
20. Chan, W. C., and Hsu, Y. Y., An optimal variable structure stabilizer for power system stabilization, IEEE Trans. Power App. and Systems, PAS-102: 1738-1746 (1983).
21. Hsu, Y. Y., and Chan, W. C., Stabilization of power systems using a variable structure stabilizer, Elec. Power System Res., 6: 129-139 (1983).
22. Kumar, A., Malik, O. P., and Hope, G. S., Variable structure system control applied to AGC of an interconnected power system, Proc. IEE, 132, Pt. C: 23-29 (1985).
23. Sivaramakrishnan, A. Y., Hariharan, M. V., and Srisailam, M. C., Design of variable structure load-frequency controller using pole assignment technique, Int. J. Contr., 40: 487-498 (1984).
24. Hsu, Y. Y., and Chan, W. C., Optimal variable structure controller for d.c. motor speed control, Proc. IEE, 131, Pt. D: 233-237 (1984).
25. Ortega, R., Variable structure systems—An alternative to robust adaptive control?, Coll. Nat. Belle Ile, France (1982).
26. Utkin, V. I., and Yang, K. D., Methods for constructing discontinuous planes in multidimensional variable structure systems, Auto. Rem. Contr., 39: 1466-1470 (1978).
27. Hang, C. C., and Tang, H. S., Model reference adaptive control using only input and output measurements, Int. J. Contr., 34: 853-868 (1981).
28. Monopoli, R. V., Model reference adaptive control with an augmented error signal, IEEE Trans. Auto. Contr., AC-19: 474-484 (1974).
29. Utkin, V. I., Equations of sliding mode in discontinuous systems—I, Autom. Rem. Contr., 21: 1897-1907 (1971).

30. Zinober, A. S. I., El-Ghezawi, O. M. E., and Billings, S. A., Multivariable variable-structure adaptive model-following control systems, Proc. IEE, 129, Pt. D: 6-12 (1982).
31. Ioannou, P. A., and Johnson, C. R., Jr., Reduced-order performance of parallel and series-parallel identifiers of two-time-scale systems, Proc. Workshop on Applns. Adapt. Syst. Theory, Yale University, New Haven, Conn. (1981).
32. Kokotovic, P. V., and Ioannou, P. A., Robustness re-design of continuous-time adaptive schemes, Proc. IEEE-CDC, San Diego, Calif., pp. 522-527 (1981).
33. Ioannou, P. A., and Kokotovic, P. V., An asymptotic error analysis of identifiers and adaptive observers in the presence of parasitics, IEEE Trans. Auto. Contr., AC-27: 921-927 (1982).
34. Anderson, B. D. O., and Johnson, C. R., Jr., On reduced order adaptive output error identification and adaptive IIR filtering, IEEE Trans. Auto. Contr., AC-27: 927-933 (1982).
35. Balas, M. S., and Johnson, C. R., Jr., Adaptive identification and control of large-scale or distributed parameter systems using reduced-order models, Proc. Workshop on Applns. Adapt. Syst. Theory, Yale University, New Haven, Conn. (1981).
36. Ioannou, P. A., and Kokotovic, P. V., Adaptive Systems with Reduced Models, Springer-Verlag, Berlin, 1983.
37. Kreisselmeier, G., On adaptive state regulation, IEEE Trans. Auto. Contr., AC-27: 3-17 (1982).
38. Rohrs, C. E., Valavani, L., Athans, M., and Stein, G., Analytical verification of undesirable properties of direct MRAC algorithms, Proc. 20th IEEE-CDC, pp. 1272-1284 (1981).
39. Johnson, C. R., Jr., and Balas, M. S., Reduced-order adaptive controller studies, Proc. JACC (1980).
40. Ortega, R., Assessment of stability robustness for adaptive controllers, IEEE Trans. Auto. Contr., AC-28: 1106-1109 (1983).
41. Ioannou, P. A., and Kokotovic, P. V., Robust re-design of adaptive control, IEEE Trans. Auto. Contr., AC-29: 202-211 (1984).
42. Peterson, B. B., and Narendra, K. S., Bounded error adaptive control, IEEE Trans. Auto. Contr., AC-27: 1161-1169 (1982).
43. Djaferis, T., Das, M., and Elliott, H., Reduced order adaptive pole-placement for multivariable systems, IEEE Trans. Auto. Contr., AC-29: 637-638 (1984).
44. Martin-Sanchez, J. M., A new solution to adaptive control, Proc. IEEE, 64: 1209-1218 (1976).
45. Martin-Sanchez, J. M., Adaptive Predictive Control System, U.S.A. Patent No. 4, 197, 576 (1976).
46. Martin-Sanchez, J. M., Implementation of an adaptive auto-pilot scheme for the F-8 aircraft using the adaptive predictive control system, Proc. 13th ASILOMAR Conf. on Circuits, Systems, and Computers, Pacific Grove, Calif. (1978).

47. Martin-Sanchez, J. M., and Shah, S. L., Multivariable adaptive predictive control of a binary distillation column, Automatica, 20: 607-620 (1984).
48. Martin-Sanchez, J. M., Fisher, D. G., and Shah, S. L., Application of a multivariable adaptive predictive control system, Proc. Workshop on Applns. of Adapt. Syst. Theory, Yale University, New Haven, Conn., pp. 138-148 (1981).
49. Martin-Sanchez, J. M., Shah, S. L., and Fisher, D. G., A stable adaptive predictive control system, Int. J. Contr., 39: 215-234 (1984).
50. Martin-Sanchez, J. M., A globally stable APCS in the presence of bounded noises and disturbances, IEEE Trans. Auto. Contr., AC-29: 461-464 (1984).
51. Martin-Sanchez, J. M., Contributions to model reference adaptive systems with hyperstability theory, Doctoral Thesis, Universidad Polytecnica de Barcelona (1974).
52. Goodwin, G. C., Ramadge, P. J., and Caines, P. E., Discrete-time multivariable adaptive control, IEEE Trans. Auto. Contr., AC-25: 449-456 (1980).
53. Goodwin, G. C., Ramadge, P. J., and Caines, P. E., Discrete-time stochastic adaptive control, SIAM J. Contr. Optim., 19: 829-853 (1981).
54. Egardt, B., Stability of Adaptive Controllers, Springer-Verlag, Berlin, 1979.
55. Samson, C., Stability analysis of adaptively controlled systems subject to bounded disturbances, Automatica, 19: 81-86 (1983).
56. Assilian, S., and Mamdani, E. H., An experiment in liguistic synthesis with a fuzzy logic controller, Int. J. Man-Mach. Stud., 7: 1-13 (1974).
57. Mamdani, E. H., and Baaklini, Prescriptive method for deriving control policy in a fuzzy logic controller, Electron. Lett., 11: 625-626 (1975).
58. Mamdani, E. H., Advances in the linguistic synthesis of fuzzy controllers, Int. J. Man-Mach. Stud., 8: 669-678 (1976).
59. Mamdani, E. H., Application of fuzzy logic to approximate reasoning using linguistic synthesis, IEEE Trans. Computers, C-26: 1182-1191 (1977).
60. Kickert, W. J. M., and Mamdani, E. H., Analysis of a fuzzy logic controller, Fuzzy Sets and Systems, 1: 29-44 (1978).
61. Braae, M., and Rutherford, D. A., Fuzzy relations in control setting, Kybernetics (1977).
62. Braae, M., and Rutherford, D. A., Theoretical and linguistic aspects of the fuzzy logic controller, Automatica, 15: 553-577 (1979).
62a. Gupta, M. M., Fuzzy logic controllers for feedback control systems, Proc. IEEE Int. Conf. on Systems, Man and Cybernetics, Bombay, India, p. 523 (1983).

63. Rutherford, D. A., and Bloore, C. G., The implementation of fuzzy algorithms for control, Proc. IEEE, 64: 572-573 (1976).
64. Rutherford, D. A., and Carter, G. A., A heuristic adaptive controller for a sinter plant, Proc. 2nd IFAC Symp. Automation in Mining, Mineral, and Metal Processing (1976).
65. Tong, R. M., Analysis of fuzzy control algorithms using relations matrix, Report No. CUED/F-CAMS/TR 133, Cambridge University (1976).
66. Egelöv, A. H., Experience with NO measurements in connection with fuzzy logic control of rotary kilns, Zement-Kalk-Gips, 3: 122-126 (1982).
67. Holmblad, L. P., and Østergaard, J.-J., Fuzzy logic control—Operator experience applied in automatic process control, Zement-Kalk-Gips, 3: 127-133 (1982).
68. Holmblad, L. P., and Østergaard, J.-J., Control of a cement kiln by fuzzy logic, FLS Rev., 67 (1982) (F. L. Smidth & Co., Copenhagen, Denmark).
69. Sheridan, S. E., and Skjoth, P., Automatic kiln control at Oregon Portland Cement Company's Durkee plant, utilizing fuzzy logic, Proc. 25th IEEE Cement Industry Tech. Conf., San Antonio, Tex. (1983).
70. Bloore, C. G., An heuristic adaptive controller for a sinter plant, M.Sc. Dissertation, UMIST, Manchester, U.K. (1974).
71. Carter, G. A., and Hague, M. J., Fuzzy control of raw mix permeability at a sinter plant, Proc. Fuzzy Workshop Conf., Queen Mary College, London (1976).
72. Carter, G. A., and Rutherford, D. A., A heuristic adaptive controller for a sinter plant, IFAC Symp. on Automation in Mining, Mineral, and Metal Processing, Johannesburg (1976).
73. Østergaard, J.-J., Fuzzy logic control of a heat exchanger process, Report No. 7601, Elec. Power Eng. Dept., Tech. Univ. of Denmark, Lyngby, Denmark (1976).
74. Kickert, W. J. M., and Van Nauta Lemke, H. R., Applications of a fuzzy controller to a warm water plant, Automatica, 12: 307-308 (1976).
75. Mamdani, E. H., Application of fuzzy algorithms for control of simple dynamic plant, Proc. IEE, 121: 1585-1588 (1974).
76. Jensen, J. H., Application of fuzzy logic control, Report No. 7607, Elec. Power Eng. Dept., Tech. Univ. of Denmark, Lyngby, Denmark (1976).
77. King, P. J., and Mamdani, E. H., Application of fuzzy control systems to industrial processes, Automatica, 13: 235-242 (1977).
78. Larsen, P. M., Industrial applications of fuzzy logic control, Int. J. Man-Mach. Studies, 12: (1980).
79. Procyk, T. J., A self-organizing controller for dynamic processes, Doctoral Thesis, Queen Mary College, London (1977).
80. Procyk, T. J., and Mamdani, E. H., A linguistic self-organizing process controller, Automatica, 15: 15-30 (1979).

81. Zadeh, L. A., Outline of a new approach to the analysis of complex systems and decision processes, IEEE Trans. Syst. Man & Cybern., SMC-3: 28-44 (1973).
82. Zadeh, L. A., Fuzzy algorithms, Inf. Contr., 12: 94-102 (1968).
83. Zadeh, L. A., Quantitative fuzzy semantics, Inf. Sci. (1971).
84. Zadeh, L. A., Fuzzy sets, Inf. Contr., 8: 338-353 (1963).
85. Elliott, H., and Goodwin, G. C., Adaptive implementation of the internal model principle, Tech. Report No. UMASS-ECE-MA 84-1, Department of Electrical and Computer Eng., University of Massachusetts, Amherst, Mass. (1984).
86. Wonham, W. M., and Pearson, J. B., Regulation and internal stabilization in linear multivariable systems, SIAM J. Contr., 12: 1-15 (1974).
87. Wolovich, W. A., and Ferreira, P., Output regulation and tracking in linear multivariable systems, IEEE Trans. Auto. Contr., AC-24: 460-465 (1979).
88. Francis, B. A., and Wonham, W. M., The internal model principle of control theory, Automatica, 12: 457-465 (1976).
89. Elliott, H., Cristi, R., and Das, M., Global stability of adaptive pole-placement algorithms, Tech. Report No. UMASS-ECE-SE 83-1, Department of Electrical and Computer Eng., University of Massachusetts, Amherst, Mass. (1983).
90. Goodwin, G. C., and Teoh, E. K., Persistency of excitation in the presence of possibly unbounded signals, Tech. Report No. EE 8320, Department of Electrical and Computer Eng., University of Newcastle, Newcastle, N.S.W., Australia (1983).
91. Widrow, B., McCool, J. M., and Medoff, B. P., Adaptive control by inverse modeling, Proc. ASILOMAR Conf. (1978).
92. Widrow, B., and Walach, E., Adaptive signal processing for adaptive control, Proc. IFAC Workshop on Adaptive Systems in Control and Signal Processing, San Francisco, pp. 7-12 (1983).
93. Anderson, B. D. O., and Johnstone, R. M., Convergence results for Widrow's adaptive controller, Proc. 6th IFAC Symp. Identification and System Parameter Estimation, Washington, D.C., pp. 247-251 (1982).
94. Widrow, B., and Hoff, M. E., Adaptive switching circuits, Tech. Report No. 1533-1, Stanford University (1960); also Wescon Conv. Record, Pt. 4, pp. 96-140 (1960).

BIBLIOGRAPHY

Averkin, A. N., and Dulin, S. K., Fuzzy modeling relation for robots, Proc. IFAC Symp. on Artificial Intelligence, Leningrad, USSR, pp. 331-336 (1983).

Dubois, D., and Prada, H., Fuzzy Sets and Systems: Theory and Applications, Academic Press, New York, 1980.

El-Ghezawi, O. M. E., Billings, S. A., and Zinober, A. S. I., Variable structure systems and system zeros, IEE Proc., 130, Pt. D: 1-5 (1983).

Erschler, J., Roubellat, F., and Vernhes, J. A., Automation of hydroelectric power station using variable structure controllers, Automatica, 10: 31-36 (1974).

Gupta, M. M., Ragade, R. K., and Yager, R. R., Advances in Fuzzy Set Theory and Applications, North-Holland, Amsterdam, 1979.

Gupta, M. M., Saridis, G. N., and Gaines, B. R. (eds.), Fuzzy Automata and Decision Processes, North-Holland, Amsterdam, 1977.

Hsu, Y. Y., and Chan, W. C., Optimal variable structure controllers for the load-frequency control of interconnected hydrothermal power systems, Elec. Power and Energy Systems, 6: 221-229 (1984).

Krishnan, A. Y. S., Hariharan, M. V., and Srisailam, M. C., Variable structure control applied to automatic generation control of power systems, Proc. Int. Conf. on Computers, Systems and Signal Processing, Bangalore, India, pp. 659-661 (1984).

Kumar, A., Malik, O. P., and Hope, G. S., Application of VSS concept to power system control problems, Proc. 8th Power Computation Conf., Helsinki, pp. 1119-1123 (1984).

Kumar, A., Malik, O. P., and Hope, G. S., Variable structure concept applied to a load frequency control problem, Proc. Int. Conf. on Computers, Systems and Signal Processing, Bangalore, India, pp. 655-658 (1984).

Mamdani, E. H., and Gaines, B. R. (eds.), Fuzzy Reasoning and Its Applications, Academic Press, London, 1981.

Mamdani, E. H., Østergaard, J. J., and Lembessis, E., Use of fuzzy logic for implementing rule-based control of industrial processes, in Fuzzy Sets and Decision Analysis (J. Zimmermann, L. A. Zadeh, and B. R. Gaines, eds.), North-Holland, Amsterdam, pp. 429-445, 1984.

Pappis, C. P., and Mamdani, E. H., A fuzzy logic controller for a traffic junction, IEEE Trans. Syst., Man and Cybern., SMC-7: 707-712 (1977).

Prade, H., Data bases with fuzzy information and approximate reasoning in expert systems, Proc. IFAC Symp. on Artificial Intelligence, Leningrad, USSR, pp. 113-120 (1983).

Sheridan, S. E., and Skjoth, P., Automatic kiln control at Oregon Portland Cement Company's Durke plant utilizing fuzzy logic, IEEE Trans. Ind. Appl., IA-20: 562-568 (1984).

Tong, R. M., A control engineering review of fuzzy systems, Automatica, 13: 559-569 (1983).

White, B. A., Reduced order switching functions in variable structure control systems, IEE Proc., 130, Pt. D: 33-40 (1983).

White, B. A., and Silson, P. M., Reachability in variable structure control systems, IEE Proc., 131, Pt. D: 85-91 (1984).

Yahunobu, S., Miyamoto, S., and Ihara, H., Fuzzy control for automatic train operation systems, Proc. 4th IFAC Conf. on Control in Transportation Systems, Baden-Baden, F.R.G. (1983).

Young, K.-K. D., Controller design for a manipulator using theory of variable structure systems, IEEE Trans. Syst. Man and Cybern., SMC-8: 101-109 (1978).

Zadeh, L. A., Fu, K. S., Tanaka, K., and Shimira, M. (eds.), Fuzzy Sets and Their Applications to Cognitive and Decision Processes, Academic Press, New York, 1975.

Zimmermann, H. J., Zadeh, L. A., and Gaines, B. R. (eds.), Fuzzy Sets and Decision Analysis, North-Holland, Amsterdam, 1984.

Zinober, A. S. I., Controller design using the theory of variable structure systems, in Self-Tuning and Adaptive Control (C. J. Harris and S. A. Billings, eds.), Peter Peregrinus, London, 1981.

8

Recent Trends in Adaptive Control Theory

8.1 INTRODUCTION

The field of adaptive control has been a very rapidly developing area. There has been intensive research directed toward formulating new algorithms, improving existing algorithms, bridging some of the gaps prevalent in the technical literature, and also taking up new studies. The purpose of this chapter is to give the state-of-the-art of this fast-evolving subject by identifying certain significant issues, presenting some recent trends, and pointing out the new directions in which more work is called for.

The chapter is organized as follows: Section 8.2 deals with some novel ideas in self-tuning control. Robustness studies are discussed in Section 8.3. In Section 8.4 are considered some methods of improvement of control behavior during the adaptation phase. The important problem of parameter convergence is taken up in Section 8.5. Nonlinear systems and stochastic systems form the subject matter of Sections 8.6 and 8.7, respectively. Multivariable systems are dealt with in Section 8.8, while the still unexplored area of distributed parameter systems is considered in Section 8.9. Section 8.10 discusses model updating, which seems to have a lot of promise. Section 8.11 examines the important issue of persistent excitation and concluding remarks are given in Section 8.12.

8.2 RECENT TRENDS IN SELF-TUNING

8.2.1 State Space Self-Tuning Controllers

State space self-tuning controllers were proposed by Tsay and Shieh [1] and Warwick [2], who in both cases used a pole-placing controller design. The advantage of the state space method is that the SISO case is readily extended to deal with MIMO systems, while the order of the closed-loop system is maintained to be the same as that of the open-loop system. However, this means that the state vector is to be estimated. Tsay and Shieh [1] estimate the states by using an extended least-squares estimator and deriving the state estimator gains from the noise model terms. Warwick [2] obtains the state estimates using the normal least-squares approach. It is possible to avoid state estimation and still utilize the state space design techniques, operating on measurements of the plant inputs and outputs. Hesketh [3] shows how to formulate a suitable multivariable STC with input and output values in the state vector. The resulting pole-placing design is simple and exhibits good control with very rapid convergence of estimated parameters. Warwick [4] obtains a general rule for the self-tuning property of state space self-tuners based on pole placement and shows that the previous strategies are special cases of this general rule.

Warwick and Westcott [5] have suggested a self-tuner which employs a control input dependent on state feedback in combination with linear output feedback to provide a pole-placement scheme with additional features. They consider the possibilities of reduction in output variance by the correct choice of the parameter associated with the extra feedback. In this scheme, in addition to the selection of closed-loop poles, the output feedback parameter must also be specified, prior to system operation, to minimize the problem of noise on the output signal. This has the effect of moving the closed-loop zeros to alternative positions. They discuss an algorithm with two tuning properties: (1) state feedback for pole placement, and (2) linear feedback for optimization. Simulations are done using a variable forgetting factor, with an initial startup value of 0.94 and a window of 100 samples. Simulation studies show the effect of the linear output feedback parameter on output variance, under different conditions. A much improved value of output variance is obtained than when only state feedback is used, although there is an increase in computing time.

Shieh et al. [6] propose a state space approach for self-tuning control of a class of multivariable stochastic systems having the same number of inputs as outputs. A multivariable state feedback controller, based on the pole-assignment concept, has been developed. This multivariable STC can be applied to stable/unstable, and minimum/non-minimum-phase linear time-invariant multivariable systems. They also present a multivariable reduced-order self-tuner and also a state feedback minimum-variance self-tuner, extending the state space approach of Tsay and Shieh [1]. The proposed approach has the advantages of simplicity and flexibility, which facilitate the practical applications of STCs to real systems.

7.7 ADAPTIVE CONTROL BY INVERSE MODELING

7.7.1 Introduction

Widrow et al. [91, 92] have proposed a novel adaptive control scheme for the task of following a prescribed reference trajectory, which relies on adaptively identifying a finite impulse response model of the inverse of the unknown plant. The structure of this controller is unlike any other type of adaptive controller.

The basic idea is to create a good transversal filter model of the plant, then utilize it to obtain an inverse (or delayed inverse) of the plant. This inverse is used as an open-loop controller of the system. Since such a controller is realized as a transversal filter, the stability of the system is assured. Simulations indicate that this algorithm is robust and is capable of handling non-minimum-phase plants. Although this approach seems to be promising, a key restriction is that the algorithm applies only to stable plants. Further, when applied to plants other than all-pole plants, some approximations are involved. Anderson and Johnstone [93] establish convergence conditions for Widrow's adaptive controller. This demands persistence of the excitation condition on the reference trajectory. This condition ensures exponentially fast decay of errors to zero. Without this condition, convergence may not occur (in contrast to some other adaptive control algorithms).

7.7.2 Adaptive Filtering

A schematic representation of an adaptive filter is shown in Fig. 7.9. The filter has an input u_j and an output y_j. The desired response is d_j. The error ϵ_j is the difference between the desired and actual output responses. The filter is assumed to be transversal. Its weights $\omega_{1j}, \omega_{2j}, \ldots, \omega_{nj}$ are adapted so as to minimize the expected square of the error ϵ_j. We can apply several adaptation algorithms for this purpose. In this approach, the LMS steepest descent algorithm of Widrow and Hoff [94] is employed.

7.7.3 Plant Modeling

Consider the direct modeling of an unknown plant as shown in Fig. 7.10, to illustrate the application of the LMS adaptive filter and to show how one obtains an input and a desired response in a control environment.

When given the same input signal as that of an unknown plant, the adaptive model adjusts itself to cause its output to be a best least-squares fit to the actual plant output. The unknown plant may have both poles and zeros, but the adaptive transversal filter can realize only "zeros." It must, however, be borne in mind that the adaptive filter is time varying and does not strictly have a transfer function. We may therefore think of "instantaneous zeros," corresponding to the zeros that would exist if the weights were frozen at their instantaneous values. With a sufficient number of weights,

8.2.2 Variance-Constrained STC

The minimum-variance STR has been widely applied to a variety of industrial processes. However, in practice there are situations which require large control signals. Such cases are formulated as LQG control problems. In such designs the weights in the quadratic loss function are used as design variables, which affect the closed-loop response. It may be difficult for a designer to select proper values for the weights. In practice, it is easier to select a constraint on the variance of the input.

Toivonen [7] has proposed such a variance-constrained STC, in which the weights are adjusted on-line so that the desired control performance is obtained. In this method an explicit restriction on the input variance is used. We then have to minimize the stationary variance of the output subject to the restriction that the input variance is less than or equal to the predetermined constraint. This becomes a linear quadratic control problem, where the weight on the input can be looked upon as a Lagrange multiplier.

The variance-constrained STC uses on-line identification to estimate the parameters of a linear, stochastic process model. At each sampling instant the control strategy is found by solving a linear quadratic control problem. The weight on the input in the quadratic loss function is adjusted at each sampling interval using the Robbins-Monro scheme. The solution to the variance-constrained problem implies a solution to a steady-state Riccati equation. However, a much simpler algorithm is obtained by using a single-step loss function. Then this algorithm is related to that of Clarke and Gawthrop, with the difference that the adjustment of the input weight is done on-line. This strategy gives good performance in many cases. The results of Modén and Söderström [8] show that this approach can also be applied to systems with a positive real transfer function. However, this approach is not optimal and it may not give satisfactory performance for non-minimum-phase plants. In such cases a more complex algorithm, which solves a steady-state Riccati equation, will have to be employed.

A simulated example shows that this method has good convergence properties. The algorithm given in Toivonen [7] is for SISO systems. This can be generalized to the multivariable systems in a straightforward manner.

In a related paper, Toivonen [9] presents a self-tuning algorithm for on-line adaptation of the design weight in a single-step optimal control law. The algorithm is based on a simple relationship between the closed-loop variances and the design variable. This approach is useful in practical applications, as it is often more straightforward to use a design criterion based on closed-loop variances than to select a design weight in a quadratic loss function.

8.2.3 STC for Single-Chip Implementation

Dexter [10] has proposed a low-cost single-chip implementation for consumer control applications such as domestic heating systems. This is made possible by using single-chip microcomputers which have the microprocessor, the program and data memories, and both analog and digital interfaces all in one integrated-circuit package [11]. Although controllers based on single-chip microcomputers have limitations due to their restricted processing power, limited on-chip program storage, and limited on-chip data storage capability, they can have applications in certain low-order, stable consumer plants which have a known constant transport delay and set point and which have parameters that are only slowly varying or even constant. In such situations we can simplify the self-tuner and reduce the required computation without sacrificing the overall performance of the control system. Dexter develops such a control scheme by simplification of the generalized minimum-variance STC of Clark and Gawthrop. Practical results are given to illustrate the behavior of the control scheme when implemented on a commercial single-chip microcomputer and controlling a real system. This simplified self-tuner cannot deal with unknown or varying time delays. A careful choice of sampling rate is required. Further modifications are necessary to allow estimation of more controller parameters, to improve its robustness, and to incorporate time-varying set points.

8.2.4 Extended Self-Tuning Controller

Latawiec et al. [12] report on an extended self-tuning controller (ESTC) which is conceptually new. It provides effective self-tuning control of drifting systems, without the need for external tuning, or fixing any parameter, including a weighting coefficient. It can also control open-loop stable non-minimum-phase plants. It has high control accuracy and robustness. It is possible to switch from a low value of the weighting coefficient providing the appropriate control accuracy in normal operating conditions, to a higher value which will ensure robustness. It is also feasible to switch to another optional service in case of any process emergency. Such an extension of self-adjusting capabilities is not possible for direct performance index adaptive control schemes, unless the models are reconverted into an indirect scheme model. This may be numerically burdensome.

The generalized self-tuning controller of Allidina et al. [13] somewhat resembles the extended self-tuning controller of Latwiec et al. It has slower convergence because of control law estimation problems.

Simulation examples demonstrate the superiority of extended ESTC over conventional STC from the point of view of both drift rejection and blow-up protection. A brief report is also given of the application of the extended self-tuning controller to a full-size cement clinker rotary kiln which displays non-minimum-phase behavior. This control system incorporating two coupled extended self-tuners and fitted with special blow-up countermeasures, drift-compensating/disregarding means, nonlinear effects, and emergency measures has been operating on-line in the cement plant. This shows that ESTC is promising for efficient adaptive control of complex industrial processes.

8.2.5 Self-Tuning Controller with Deadbeat Convergence

Yan et al. [14] have developed an adaptive algorithm based on a deadbeat controller. They use the polynomial linear controller discussed by Kučera [15], and the finite time setting adaptive observer reported by Yan et al. [16]. Then system is designed so that the output of a non-minimum-phase plant can exactly follow the reference input in finite time. The stability of the system can be assured. However, this scheme relies heavily on the absence of any exogenous input to the system other than this particular reference; otherwise the deadbeat performance is lost.

Minamide [17] has also considered the design of a deadbeat adaptive tracking system. Using an orthogonal decomposition technique, a matrix pseudoinversion lemma is applied to obtain recursive algorithms for computing the best approximate solution to the sequential estimation problem. A deadbeat convergence property of the algorithm is demonstrated. Simulation results show perfect tracking of system output to the periodic reference trajectory in finite steps, even in the case of an unstable, non-minimum-phase third-order plant which was subject to slow parameter variation and an abrupt parameter variation at an unforeseen time.

8.2.6 Predictor-Based STC

A new approach to self-tuning control has been proposed by Peterka [18], using a linear finite memory output predictor, updated in real time, as an internal representation of the uncertain process to be controlled. The control strategy is designed to minimize, for the most recent predictor available, the expected value of a suitably chosen quadratic criterion embracing a given (finite or asymptotically infinite) control horizon. In such a self-tuning controller the control law has to be repeatedly updated in each control step. Hence the algorithm should be numerically stable, free of singularities, uniform for all situations that may arise, and at the same time as simple as possible. These requirements and the need to tune the controller with respect to the preprogrammed sequence of set points are met by this new method proposed by Peterka for a linear quadratic (LQ) optimum control synthesis which forms the basis for the predictor-based STC. The synthesis covers both the servo problem and the regulation problem, including program control (with preprogrammed command signal) and feedforward from measurable external disturbances. Unlike the Riccati equation approach, this method results in algorithms (or explicit formulae in low-order cases) which are numerically robust and therefore suitable for real-time computation using microprocessors with reduced word length.

The predictor-based STC has definite advantages over the conventional STC based on the certainty equivalence hypothesis and exponential forgetting of old data. The latter can result in undesirable effects such as "bursts" [19] and "covariance wind-up" [20] if the closed control loop is not excited externally (for instance by set-point changes) and if no precautions are taken [21]. The main cause of this difficulty is that the fixed linear control law, to which the self-tuning algorithm finally converges, leads to linear dependence on the data in the regressor used to update the estimates of the predictor parameters. In the present method this linear dependence does not arise. Further work on these simple robust STCs and practical applications are possible.

8.2.7 STC for Rapidly Varying Parameters

There exist few STC algorithms which will work for other than slow time-varying systems. Simulation results show that in the fast time-varying case the performance of existing algorithms is limited. This is due to the fact that these algorithms are based on a time-invariant model. Xianya and Evans [22] have developed a new algorithm which is able to successfully control unknown systems with quite rapidly varying parameters. They present a series of improvements to the standard least-squares estimator and one-step-ahead controller structure which ensure good adaptive controller performance for rapidly varying system parameters. The first improvement concerns the speed of convergence of the parameter estimation algorithm, since the speed of parameter estimator convergence and the rate of time variation allowed are directly related. This modified algorithm is shown to converge faster than the standard least-squares algorithm. Secondly, a resetting procedure is employed to ensure that the estimation algorithm stays "awake." Finally, and most important of all, the unknown parameter variations are modeled by the use of a restricted complexity time-varying model (offset linear ramp). This allows good control performance even when the unknown parameters are varying sinusoidally and with step changes. Several simulation studies clearly demonstrate the ability of this algorithm to control unknown, time-varying systems. In particular, step, ramp, and sinusoidally varying unknown parameters are considered. The effects of different resetting procedures and of different parameter models are studied. It is important to note that in all the cases, a standard deterministic adaptive controller is unable to control the system and rapidly blows up, whereas the modified algorithm performs well. Although the convergence of this algorithm in the general case where the parameters are varying in an arbitrary fashion is not proved, a convergence proof is given for the linearly time-varying parameter case, which includes the time-invariant case as a special one. A strong assumption of persistent excitation is required. If the real plant parameters do not vary linearly but vary smoothly and can be reasonably approximated by a piecewise linear function, then good control is achieved.

8.2.8 Dual-Rate Sampling STR

Certain applications necessitate that the control interval be large because of the dynamics of the process, as in slowly responding systems. Load frequency control in a power system is an example of such a system. There may also be constraints such as wear and tear on the control equipment, which may limit the frequency at which the control should be reset. A conventional STR cannot be used in such cases. It is desirable to use different data sampling and control intervals, so that the sampling frequency is maintained at an appropriate value. Kanniah and Malik [23] have proposed such a dual-rate sampling regulator (DSSTR) in which a smaller sampling interval and a larger control interval are used, resulting in a periodically constant control.

The advantages of using DSSTR are as follows:

1. In cases where a reduction in control interval can help move the zeros inside the unit circle, dual-rate sampling can be used to achieve this. This has been demonstrated by an example. If the plant is non-minimum phase for the major sample selected, minor samples are governed by unstable roots in a closed loop, even when controlled by a DSSTR. If the computation time cannot be reduced any further, we have to resort to a suboptimal strategy. However, for plants of order 3 or more, there will be one less parameter at least and this can reduce the computation time for the same computing power. For a class of plants, this will move the zeros inside the unit circle.

2. Near non-minimum-phase plants can be readily controlled by DSSTR.

3. Since the controller is a digital filter, the well-known quantization problems may show up if a fixed-word-length microcomputer is used as a controller. However, if a DSSTR is used, problems due to quantization do not arise and we can have fixed-word-length processor implementation. This does not require increased computational effort. Stable and optimal response is obtained.

4. For an arbitrary plant it has been demonstrated that variance in terms of the major samples can be reduced by resorting to dual-rate sampling. It has also been shown that the output of a DSSTR is superior in terms of minor sample autocorrelation. The DSSTR will perform better than a conventional STR for a given application.

5. For a third- or higher-order plant it is possible to represent the plant only by two B-parameters, without any approximation. In cases where major samples are governed by computation time, advantage can be taken of this reduction in parameters to have reduced control interval.

6. DSSTR does not require state estimation, but does require additional samples to be taken.

7. In interrupt-controlled sampling implemented with a M6800 microprocessor, an interrupt servicing takes a maximum of 250 μs to sample, to convert, and to read data. This is insignificant compared to the time taken for identification and control computation, which may be of the order of 100 ms.

A successful application of DSSTR for the excitation control of a synchronous generator is reported in Kanniah et al. [24] and this is discussed in Section 9.4. The authors have also reported another version of the paper in Kanniah et al. [25].

In Kanniah et al. [25a] the authors discuss a microprocessor-based universal regulator using dual-rate sampling. Relevant implementation details are given with emphasis on data acquisition, interrupt-controlled sampling, time sufficiency test, and scaling. The software has been developed using a Motorola EXORciser and has been written in Motorola Assembly Language. Experimental details are given to show that satisfactory performance is possible with this technique for the excitation control of a synchronous generator. For these tests the synchronous machine identification is done as third-order model. On account of dual-rate sampling only two B-parameters are sufficient. Major sampling and minor sampling intervals are fixed at 100 ms and 16.67 ms, respectively. A step change was introduced to the generator input to obtain a control which maintains the voltage error to a minimum and avoids excessive oscillations in the load angle. A study was made also of the response following a fault and subsequent opening of one circuit of a double-circuit transmission line. Experimental results show that even after a major disturbance in the controlled variable, control is fairly stable and stays within tolerable limits. Such a smooth control would not be possible with a conventional STR, since at least one root of the B-polynomial would be very close to the unit circle. Dual-rate sampling causes the zero to move inside.

There is great scope for exploiting the DSSTR in a variety of applications.

8.2.9 Parametric LQ Self-Tuners

Mäkilä [26] has developed STRs using optimal output feedback theory for applications such as auto-tuning of PID regulators and adaptive decentralized control. The underlying problem is a parametric linear quadratic (LQ) problem for fixed-structure controllers. The regulator design is based on the idea of minimizing a steady-state quadratic criterion with respect to the regulator parameters. The steady-state quadratic criterion is justifiable in many process control problems. In this approach the order or structure of the regulator is specified independently of the estimated process model. This results in some flexibility as, for instance, in auto-tuning of simple regulators. Further, the problem of zero cancellation, restricting the use of some adaptive schemes to systems with stable zeros only [27] can be avoided by this method. An explicit parametric LQ self-tuner has been developed using the computationally attractice linear descent mapping (LDM) method [28] to solve the optimal output feedback problem [29]. The proposed STR may be interpreted as a generalization of state space STRs based on LQG theory to parametric LQ problems. An example of a multivariable model of a cement kiln has been given to illustrate some of the properties of this self-tuner. The parametric LQ self-tuner can be generalized to arbitrarily fixed linear regulator structures using an appropriate form of the LDM method. This makes this STR algorithm useful in adaptive decentralized control.

8.2.10 MUSMAR Self-Tuners

The acronym MUSMAR refers to MUlti Step Multivariable Adaptive Regulator. The MUSMAR algorithm was introduced by Menga and Mosca [30]. Further work was done by Mosca et al. [31-36] in respect to analysis and application for plants with ill-defined or time-varying i/o transport delay.

In this approach, reported in detail in Greco et al. [37], an attempt has been made to retain the advantages of computational simplicity and numerical robustness of direct adaptive schemes, without losing the greater flexibility and better performance of indirect adaptive schemes. This is possible by adjusting the feedback parameters according to a multistep quadratic cost functional of plant future i/o samples. The weighting coefficients of the cost functional allow design flexibility. Computational simplicity is retained by using least-squares estimates of a "multistep prediction model" of the closed-loop system. The resulting regulator yields an approximate LQG solution and so has the intrinsic robustness of LQG design. Extensive simulations have confirmed several attractive features of MUSMAR self-tuners when applied to plants for which other adaptive controllers fail, for example, non-minimum-phase plants, plants whose order is underestimated, and plants with nonpositive real transfer functions. A case study has also been given for the adaptive control of a robot arm.

8.2.11 Auto-Pole-Placement Self-Tuning Controller

An interesting idea of auto-pole-placement STC has been presented by McDermott et al. [38]. This new technique allows the closed-loop poles to be autotuned on-line to optimize the set-point response, leaving the designer to specify only internal model orders and noise characteristics on which to initiate retuning. The sampling period is also determined on-line. The resulting controller is quite robust and can easily handle unstable, non-minimum-phase systems in a single-loop configuration. This allows a much broader range of applications, which few of the existing methods can deal with. This explicit STC incorporates closed-loop pole location as the principal design criterion for the objective function of Clark and Gawthrop, and also uses the approach of Allidina and Hughes [13]. The algorithm includes several extensions to improve robustness: a variable dead-time compensator to deal with unknown or time-varying delay, and a method of treating steady-state offset that avoids retaining an integrator in the loop continuously. None of the pole-placement methods can deal efficiently with unknown load disturbances, that is, by estimating the disturbance and compensating for it directly. Simulation studies show the effectiveness of this technique. A nonlinear model of a fixed-bed autothermal reactor has also been investigated over such a large range that auto-tuning of the poles is a necessity. This application is briefly discussed in Section 9.8. With the technique of pole auto-tuning, this algorithm is more truly self-tuning, since the algorithm has the ability of choosing on-line its own values for the optimum closed-loop poles as well as the optimum sampling period. This strategy has the following merits:

1. The controller can reoptimize its pole locations as required, without having to perturb the process.
2. The controller will not be sensitive to the choice of sampling period—a difficulty with other self-tuners. Once a sampling period is selected, the technique yields directly the optimal closed-loop pole for set-point changes.
3. An on-line method of auto-tuning the sampling period can maintain the closed-loop pole at an appropriate value despite process changes.

There is scope for further exploration of this promising technique.

8.2.12 Distributed Control Using STRs

A recent trend in industrial process control is the application of distributed control because of increased reliability and capability of digital computers along with significant decreases in the cost of the hardware. However, control theory for decentralized control of time-varying or unknown systems has not kept pace and there is considerable scope for research.

Costin and Buchner [39] have proposed a modified self-tuning algorithm for the case of decentralized controllers. In completely decentralized control of a multivariable system of order n, the controllers are restricted to a set of n SISO controllers. Since each controller has only local information available, the set of n SISO controllers cannot have the same structure as the system's minimum-variance controller (except when the system is fully decoupled). Hence, such a use of SISO controllers cannot, in general, reduce the variance of the measurements to the minimum value. It is desirable to minimize subsystem interactions by coordinating the individual local controllers. This reduces the possibility of individual controllers regulating their local measurements at the expense of causing upsets to other subsystems.

Costin and Buchner propose that the coordination be done by modifying the updating equation used to calculate the controller parameters of the STR. This modification involves the use of measurements of the other subsystems. In typical decentralized digital control applications, this information is usually available as each loop measurement is often sent to the centralized operator interface. Information about the other subsystems is not required for the controller portion of the algorithm.

Simulation results show that the proposed algorithm results in significant improvements over a completely decentralized set of STRs for controlling a coupled multivariable process. There is reduction both in the control variances and in the weighted sum of the output variances of the controlled plant. Further, this algorithm can be implemented naturally in a distributed computer control system.

8.2.13 Elimination of Offset in Self-Tuning Control

The problem of offset arises in many process control applications on account of load disturbances, which can be modeled by drift (Brownian motion) or as random steps occurring at random instants of time. This offset should be eliminated to make the self-tuning controller industrially acceptable.

One of the ad hoc methods adopted is to insert an integrator in the forward path. But this causes difficulty in the analysis of the adaptive part of the controller, since the corresponding noise model is also affected.

Tuffs and Clarke [39a] suggest an appropriate model for the disturbed process from which the integrating aspect of the controller is derived in a natural way.

In self-tuning control we generally assume that the disturbance is stationary with rational spectral density. This leads to the controlled autoregressive moving average or CARMA model. Such a CARMA model results in an offset between the measurement and the desired value when regulating a plant subjected to constant load disturbances about a nonzero set point. In view of this we should employ another model, a controlled autoregressive integrated moving average or CARIMA model in which the noise term is nonstationary. Because of its properties and of the fact that integrating control laws arise naturally, this CARIMA model is a more realistic basis for obtaining the adaptive algorithms and also leads to a unified approach for the elimination of offset. It has been shown that good adaptive behavior is possible both for generalized minimum variance and pole-placement self-tuning controllers. This approach generates integral control terms in a natural manner and leads directly to a general PID algorithm for simplified plant models.

In this paper two approaches to self-tuning control of the CARIMA plant model are given:

1. k-incremental and integrating "model-following" control, analogous to the Clarke and Gawthrop approach
2. The suboptimal approach of k-incremental and integrating pole-placement techniques extended for set-point following, analogous to those proposed by Clarke [39b] and Wellstead and Sanoff [39c].

Four different algorithms are obtained from these approaches and are analyzed to compare their relative merits with special reference to the expected response to a step load disturbance. It has been shown that the Clarke-Gawthrop self-tuning controller outperforms the equivalent pole-placement controller as regards time of rejection of disturbance, at the expense of restrictions on admissible B-polynomials and a priori knowledge of the time delay.

Extensive simulations show that the "integrating" form of the Clarke-Gawthrop self-tuner is particularly effective in terms of rejection of load disturbances and good parameter estimation. However, this is applicable only to minimum-phase plants. For non-minimum-phase plants we have to use either detuning (by use of control weighting) or a pole-placement controller again based on the CARIMA model.

There can be a convergence difficulty if the plant is described by the CARIMA model and this can be circumvented by using the observer (or data filter) polynomial $T(z^{-1})$. Reference may be made to Tuffs and Clarke [39a] for details; the main point of this paper is that the CARIMA model is a more realistic representation of an industrial plant. From this model it is possible to obtain control laws with naturally integrating behavior. Further, it is easy to produce algorithms which are extensions to classical PID regulators, but which can deal with plants that have significant dead time.

8.3 ROBUSTNESS STUDIES

Until recently, researchers had considered global stability as the design objective of most adaptive controllers and no provisions were included to consider robustness for types of uncertainties other than parametric uncertainties. It was tacitly assumed that a set of parameters exists for a given regulator structure which satisfies the control objective. On this basis, parammeter adaptation algorithms were developed to ensure overall stability. However, even mild violations of the assumptions may result in instability. In practice, such violations are unavoidable, and questions arose about the ability of the controller to maintain suitable performance in the presence of other types of uncertainties, such as fast parasitics, bounded output disturbances, unmodeled dynamics, or unstructured uncertainties. Athans and Valavani [40] raise some critical questions about the existing adaptive control algorithms with particular reference to certain robustness issues pertaining to their closed-loop stability properties in the presence of unmodeled high-frequency dynamics. An adaptive loop must remain stable in the presence of unstructured modeling uncertainty, which always exists and cannot be adequately modeled in any physical system.

Robustness study is therefore an essential part of design as it deals with preservation of stability or boundedness when ideal conditions are not met. This is a fairly new and active area of investigation, a brief survey of which is given by Ortega [41].

Anderson and Johnstone [42] study the behavior of four adaptive algorithms in the presence of plant parameter variations. Of the various conclusions drawn, the most general is that the various adaptive algorithms will behave robustly, provided a persistency of excitation condition is satisfied on inputs, since it is just such a condition which guarantees exponential stability of the idealized algorithm. These robust Lyapunov stability results can also be extended to consider the effects of measurement noise, plant nonlinearity, and undermodeling of the plant order. This is possible by obtaining the true state variable equations for an idealized plant, assuming excitation conditions which guarantee uniform or exponential stability, and finally showing that the nonideal character of the plant corresponds to one of the standard variations to the equations considered by the robust Lyapunov stability results. An important idea is that, in general, the less well initialized an adaptive scheme is, the less tolerance there will be of variations from the idealizations.

Ortega [43] proposes a method of improving the stability characteristics of discrete-time SISO adaptive controllers. In this approach, adjustment is made of the observer poles of an indirect adaptive scheme to ensure stability of the closed-loop system in spite of possible unstructured uncertainties, such as process-model order mismatch. This method uses the results given by Åström [44], who considers the robustness of a design method based on assignment of poles and zeros. Assuming that the identified model transfer function module converges within a previously defined uncertainty range, the frequency response-like compensation method is shown to lead to a stable design, provided that the reduced-order estimated model is stable and stably invertible and system is open-loop stable. A systematic procedure is given for robustness enhancement and has been illustrated by simulation results. There is ample scope for further studies in this area. Ortega addresses himself [45, 46] to the problem of robustness enhancement of adaptive controllers by incorporation, into the design, of the a priori knowledge about the process transfer function. An analytical framework is given to design a fixed outer-loop compensator for adaptive controllers. The outer loop is a regulation loop around the overall adaptive system with the tracking error as the output. The main result is the establishment of a global stability condition which makes it possible directly to incorporate the prior knowledge about the process into the design of a suitable outer-loop compensator to improve the stability robustness properties with no deterioration of performance. The key method used is the characterization of uncertain systems with conic-bounded transfer functions and the corresponding conic sector stability theorem [47]. This approach is applicable to almost all existing adaptive schemes. In contrast to the other modified adaptive schemes, this method fully exploits the capabilities of the adaptive controller to tune to the unknown system because of the outer-loop nature of the compensation scheme. Thus it is possible to obtain improvement in the robustness of the adaptive controllers without modification of the parameter adaptation algorithm or performance degradation. Also see other relevant papers by Ortega and Landau [48, 49].

The problem of robustness of MRAC systems has been studied by Kokotovic and Ioannou [50, 51], who showed that several adaptive schemes are robust to modeling errors when the fast parasitics are "weakly observable." In a 1982 paper [52] they consider singular perturbations for robust redesign in the presence of unmodeled dynamics. In a later paper [53] they obtain the analytical conditions for robustness of direct adaptive control with respect to high-frequency dynamics. They show that the stability of the adaptive control schemes are crucially affected by (1) the frequency content and magnitude of the reference input signal, (2) the speed ratio of slow versus fast phenomena, (3) the adaptive gain, and (4) the initial conditions. This methodology can be extended to more complicated continuous and discrete-time adaptive controllers. In a 1984 paper [54] they examine the effects of high-frequency dynamics and bounded disturbances on stability and performance of adaptive control schemes and analyze five possible types of instability mechanisms, using simple examples. They obtain sufficient conditions under which none of these five types of instability can occur. They construct Lyapunov-like functions for a modified adaptive controller applied to a plant of relative degree 1. This modified scheme is robust in the sense that it guarantees the existence of a large region of attraction from which all the trajectories remain bounded and the state errors converge exponentially to a much smaller residual set. The size of the region of attraction depends on the speed of parasitics, and if the parasitics become infinitely fast, the region of attraction becomes the whole space. If the parasitics are small, the region of attraction is sufficiently large for practical operation of the adaptive systems. The region of attraction is also larger for smaller magnitudes of disturbance and reference inputs. The proposed modification in this paper [54] indicates that there is need for further research in robustness redesign of adaptive control schemes.

Anderson and Johnstone [55] present a set of robust Lyapunov stability results and show that some MRAC schemes possess a certain degree of robustness to mismodeling effects, provided the input signal is sufficiently rich. Rohrs et al. [56, 57] study first-order systems with unmodeled dynamics and conclude that the systems remain stable with unmodeled pole in the plant, and also for sufficiently small inputs with two unmodeled poles. Rohrs [58] suggests the need for a frequency-selective adaptive algorithm. Cook and Chen [59] show that some simple MRAC systems with unmodeled dynamics and output disturbances possess some robustness under certain conditions. These results can be extended to higher-order MRAC systems. Chen [60] gives an elaborate study of the robustness of MRAC systems with respect to disturbances, measurement noise, time-varying parameters, and unmodeled dynamics. However, Rohrs et al. [61] raise a controversy by stating that the existing adaptive algorithms cannot be used with confidence, since they contain a infinite-gain operator in the dynamic system that defines command reference errors and parameter errors. This drastic conclusion has been contested by Chen and Cook [62]. They give a robustness analysis and simulation studies of the same numerical example as in Rohrs et al. [61]. They conclude that the BIBO property can be expected within a certain domain, provided the input signal is sufficiently rich and the modeling error is small enough. In addition, the system will also be stable if a sufficiently small output disturbance is present in the system. In view of the existence of a certain limited robustness of most MRAC systems with mismodeling, they opine that before a practical system is implemented the robust margin of the scheme should be investigated.

The study of robustness of the existing MRAC algorithms has therefore assumed importance since, in reality, the structural properties of the plant are uncertain and a real system is always subjected to output disturbances, measurement noise, or time-varying parameters. Furthermore, as we have seen earlier in Chapter 3, certain restrictive assumptions have to be made to ensure global asymptotic stability. For instance, the upper bound of the number of poles and the relative degree of the plant should be known for continuous-time MRAC systems and there should be no other external signals, other than the system inputs. Similar restrictive assumptions are necessary for discrete-time MRAC systems. However, in actual practice these assumptions can never be satisfied. In view of these discrepancies, the behavior of adaptive controllers will be drastically different in practice. Hence there is a need for further extensive investigations regarding the robustness of adaptive controllers.

Rohrs et al. [63] report further investigations to demonstrate that some stability problems can be reduced by following some design guidelines:

1. To maintain stability in the presence of unmodeled dynamics, the adaptive gain should be kept small and the adaptation proceed slowly.
2. The nominal control loop is designed so that it is robust and approximate model matching can be obtained even in the presence of unmodeled dynamics.
3. The system should be sampled slowly enough to remove the effects of unmodeled dynamics.

These guidelines are demonstrated by performing a linearization analysis of the nonlinear adaptive control system. These guidelines hold far more generally and are very helpful in practical implementation of adaptive algorithms. Although some of the problems can be alleviated by sticking to these design guidelines, further research is necessary to resolve some serious problems of adaptive controllers in the presence of unmodeled dynamics, in order to design robust, easily usable adaptive controllers.

Kosut and Johnson [64] raise the question whether one can abandon the requirement of "global" stability—a requirement that is beyond the needs of most actual systems—and develop conditions for "local" stability. They use the term "local" in the sense that the plant uncertainty and external inputs are limited in a defined way, for example, by restricting the magnitude and spectrum of the reference commands and disturbances, as well as the initial adaptive parameter error. They present an input/output view of robustness in adaptive control, with particular reference to unmodeled plant dynamics and to uncertain, but bounded disturbances. The generic adaptive controller is converted into an equivalent general error system. The input/output properties of the error system relate the performance of the nominal control system to that of the adaptive control system. Applying passivity theory, the error system gives the conditions for global stability where bound (in norm) is required on the external inputs (e.g., disturbance, reference, and initial conditions). Small-gain theory is used to develop local stability results, where the magnitudes of the external inputs are restricted. In the global case a particular system operator (not the plant) is required to be strictly passive; this condition will not be satisfied, in practice, because of unmodeled dynamics. However, the local results are not so restrictive and allow for unmodeled dynamics. In this case, an estimate of the stability margin is given under the condition of persistent excitation.

Silveria and Doraiswami [65] have proposed a new structure for an adaptive controller based on robust servomechanism theory; it consists of a tracking error driven servocompensator and a stabilizer. Such a robust structure has been widely used in practice in nonadaptive problems. Incorporating this structure with an adaptive controller ensures asymptotic tracking with disturbance rejection. The controller structure tends to the robust servomechanism structure, when the adaptation is over. The external signals may be unbounded, but satisfy a known linear difference equation. A tracking error driven internal model whose parameters are fixed is included and the parameters of the controller are recursively updated so that the augmented system is asymptotically stable. The parameter adaptation algorithm ensures global convergence of the tracking error under suitable conditions. The control input synthesis is based on a pole-positioning algorithm; that is, the roots of the characteristic equation of the closed-loop system obtained by the process parameter estimates are the desired roots. The proposed scheme has a very simple structure and is immune to the presence of deterministic perturbations. The reference and the disturbance signals are not restricted to be bounded. This structure has also the following advantage: the parameter update mechanism need not be turned on for any changes in the reference and/or disturbance signals. The controller parameter adaptation is required only when the process parameter variations lead to deterioration of the closed-loop system performance, caused by the drift of the closed-loop poles away from those specified. If the modes of the external signals are unknown, then both the parameters of the servocompensator and the stabilizer must be updated. Otherwise, the parameters of the servocompensator are fixed and only the stabilizer parameters are updated, unlike the case of adaptive controllers based on feedforward servomechanism theory, in which the parameters of both the feedforward blocks and the stabilizer must be updated.

In the analysis of self-tuning regulators the main assumptions usually made are that the controlled system is linear with a known upper bound to its order and that the disturbance is a stationary stochastic process with rational spectral density. In a real industrial process neither of these assumptions holds good. It is therefore necessary to examine the consequences of operating a self-tuning algorithm when these assumptions are not valid. It is crucial that self-tuning algorithms are robust against unmodeled dynamics and nonlinearities. Analysis of robustness of self-tuning methods is yet in its early stage [66]. Lim [67] considers some methods using input/output and passivity arguments to show how certain design polynomials can be chosen in order to improve the robustness of the algorithms. More work needs to be done, particularly in the area of explicit algorithms [66].

Gawthrop and Lim [68] examine the robustness of STC to find the extent to which the assumptions may be violated while retaining the stability of the self-tuning algorithm. They demonstrate that a self-tuning algorithm based on a generalized cost function may be more robust than the one based on minimizing output variance. Some frequency-domain conditions are given for the robustness of STC in the presence of certain modeling errors. Some conditions are rather restrictive; however, the results give some guidelines as to how certain design parameters may be selected to improve robustness and also indicate what forms of unmodeled dynamics and nonlinearity are likely to give problems regarding stability. The main results are illustrated by means of some examples which include nonlinearities. The examples also show that the use of a model with a time constant somewhat larger than the sample interval gives better results than a deadbeat or minimum-variance strategy. Further studies are necessary for a comprehensive understanding of the robustness problem.

8.4 IMPROVEMENT OF CONTROL BEHAVIOR DURING THE ADAPTATION PHASE

There has been significant research effort, in recent years, to solve the problem of global asymptotic convergence of adaptive controllers. However, these properties do not ensure an appropriate adaptation transient. In spite of this, there has not been much work in improving the adaptation transients, though mention is made of this problem by Jacobs and Saratchandran [69] and Wittenmark [70]. Heuristic rules to improve the transients are also given, such as to use a forgetting factor near unity for systems with time-varying parameters, during the estimation. It was also found through experimentation that self-tuning (ST) and optimal k-step-ahead (OK) control have good transient responses and open-loop feedback optimal control (OLFO) often has poor response [69-71]. In rapidly varying parameters, it was found [69] that OK control may have better response than ST control. There have not been mathematical formulations for such problems, which are, no doubt, important. This is, perhaps, due to the complicated nonlinear structure of the resulting adaptive schemes.

8.4.1 Optimization Approach

de la Sen [72] has proposed a method for improving the adaptation transients in discrete-time adaptive control systems. This method is based on the well-known fact that we can usually choose some of the parameters entering the adaptive gain, within certain bounds, while maintaining the stability of the whole system. The key idea of the method is to interpret these parameters as additional inputs to the adaptive control system to be computed on-line in order to minimize a quadratic criterion of the tracking error. As an intermediate step, an equivalent time-varying, near-linear system is obtained as an approximate model of the entire system. This simplified equivalent system is used for optimization of the plant model error. A bang-bang (or eventually linear) mode suboptimal control is computed using classical optimization techniques. The resulting control is finally translated in terms of modifications of the free parameters entering the adaptation gain updating equation. The optimization-correction scheme implementation is made by using a modified adaptation mechanism, which alternately works either in a priori or a posteriori mode directly related, respectively, to the actions of fixing a priori equivalent system parameters for the new optimization horizon and computation of the corrected estimates on this one based on the modification rules for the algorithm parameters. Reference may be made to the paper for full details of the sequence of actions associated with the a priori and a posteriori modes. Analytical proof of the maintenance of standard stability properties is also given. Simulation results show the influence of the various choices made in the design, such as prediction method used, optimization horizon size, and repeated a posteriori mode of the adaptation mechanism. This technique can be extended to other adaptive control algorithms also by appropriate modification of the adaptive controller. It may be noted that this method does not require extra input signals or richness of the reference signal.

8.4.2 Adaptive Sampling Approach

In a later paper de la Sen [73] discusses another approach for improving the behavior during the adaptation transient. The method is based on the use of adaptive sampling, taking the tracking (or regulation) error between the reference model and the controlled system outputs as the signal to be adapted. No auxiliary input to the adaptive scheme is considered and no extra hypothesis on the input to the plant or reference model output is made. The adaptation algorithms used are of the same type as those commonly applied for constant sampling. The adaptation mechanism consists of (1) the usual feedforward-feedback controller (first control level), (2) the adaptation algorithms (second control level), and, in addition, (3) a sampling controller, which supplies an on-line adaptation of the sampling period to the error signal between the reference model and the controlled plant, and (4) an induced sampling controller acting on the output (while the sampling controller acts on the input). Results of simulation show that adaptive sampling yields better performance than periodic sampling. Although consideration is given to response during the adaptation transient, the stability of the overall scheme is preserved. de la Sen has summarized [73a] the guidelines used for updating the free parameters and the sampling rate in adaptive systems to improve the adaptation transients and has also focused on practical implementation by using appropriate algorithms. Another paper of interest is de la Sen [73b].

8.4.3 Use of a Third Level for Coordination

When parameter-adaptive control is applied in closed loop from the very beginning, an undefined control action is obtained in the first part of the adaptation phase. This depends on the type of adaptation algorithm used, the initial state of the system and the external disturbances. To avoid large and unacceptable process input and output variations during this phase, Schumann et al. [74] suggest that a probing signal be used to estimate the parameters for a reasonable starting model in open loop, before the loop is closed with the parameter-adaptive controller after an a priori fixed number of samples. Experimental results indicate that this procedure results in a significant reduction in signal variations.

To improve the control behavior especially during the first adaptation phase, the following additional measures are proposed by Schumann et al.:

1. Modification of the control algorithm, by changing the specification parameters, or by increasing the sampling time period for the control action, so as to limit the control signal variations
2. Automatic on-line determination of appropriate system model structure parameters
3. Automatic on-line switch-off of the parameter estimation to avoid the "bursting" of parameter estimates, observed sometimes in long-time behavior.

To achieve the above, a new feedback level is introduced into the parameter-adaptive scheme which provides several supervisory control functions in order to ensure satisfactory control. In fact, the introduction of the third level of coordination in the control scheme results in a considerable improvement in overall control behavior and a significant reduction of the tuning time for the "self-tuning" controller. Simulation and experimental results for an air-conditioning pilot plant are given to demonstrate the improvements in the overall performance of the extended parameter-adaptive control scheme.

This extended scheme incorporating the third coordination level can simplify and act as a safeguard for satisfactory applications. The third feedback level can carry out the following tasks:

1. On-line search for model order and dead time
2. On-line search for the control sampling time
3. Coordination of the first adaptation phase (e.g., starting with a defined probing signal)
4. Supervision of the parameter estimation
5. Supervision of the control performance

Tasks 1 to 3 are very helpful for the commissioning phase. Tasks 4 and 5 are useful in ensuring a good control performance, under all circumstances, which may arise in practice.

8.5 PARAMETER CONVERGENCE

Recent studies on the problems of stability and convergence have been considered earlier in Sections 3.9 and 5.10. Parameter convergence is an important problem and is still an open issue. One advantage of having parameter convergence is that this can improve the robustness of the algorithm to unmodeled dynamics and slowly time-varying plants.

For continuous-time MRAC systems, it has been shown [75-77] that under a suitable adaptive control law, the output of the controlled plant tracks the output of the stable reference model asymptotically, despite the fact that the parameter error may not converge to zero (or may not converge at all). Some investigations on parameter error convergence [78-81] have established the uniform asymptotic stability of the adaptive schemes under a sufficient richness condition. The main difficulty arises because this condition is imposed on certain signals which appear inside the time-varying feedback loop around the unknown plant [82]. It is therefore not possible to determine beforehand if a given reference input will lead to sufficient rich signals, thus ensuring the convergence of parameter error to zero.

Anderson and Johnson [82] have overcome this difficulty. The conditions for persistent excitation involving input and output quantities are converted into the ones involving, for the identification problem, input quantities alone, and for the adaptive control problem, output quantities alone. The first projection algorithm of Goodwin et al. [83], which is an underlying basis for several approaches to adaptive control, has been considered for a SISO system. It has been shown that when the output of the unknown plant is required to follow a sufficiently rich reference trajectory—in the sense of the output only conditions referred to above—then the estimate of the plant parameters will converge exponentially fast to the true value, as will the plant output to the reference trajectory. This line of approach with minor modifications is also applicable to other adaptive control algorithms. The main conclusion regarding the adaptive control problem may be stated: If we model the plant order correctly, if the plant has zeros inside $|z| = 1$, or we otherwise ensure the control input remains provably bounded, and if the reference trajectory is persistently exciting (or spanning) we get exponential convergence. These results can also be extended to multiinput/multioutput systems with appropriate parameterizations. It should be noted that in changing over from input/output conditions to input-only conditions, it is necessary to demand that the inputs not become faster and faster; otherwise the contributions to the output become smaller and smaller. This can be taken care of by a technique due to Yuan and Wonham [81] and also used by Anderson [84]. We may note that in Anderson and Johnson [82], we do not have a formula for the exponential convergence rate, even though the paper proves simply the existence of such a rate. This fact should spur investigations to find such a rate which is important in a variety of problems in adaptive systems.

Boyd and Sastry [85] have suggested a remedy to the difficulty, which arises because conditions have to be imposed on an internal signal. They introduce the idea of spectral lines and show that when the reference input (which is the exogenous input to the adaptive system) has as many spectral lines as there are unknown parameters, the output error and parameter error both converge to zero exponentially. Their results can be stated as follows:

(1) A function $u(t)\colon \mathcal{R}^+ \longrightarrow \mathcal{R}^n$ is said to have a spectral line at frequency ν of amplitude $\hat{u}(\nu) \in \mathcal{C}^n$ iff

$$\left\{ \frac{1}{T} \int_s^{s+T} u(t)e^{-j\nu t}\, dt \right\}$$

converges to $\hat{u}(\nu)$, as $T \longrightarrow \infty$, uniformly in s. When $\hat{u}(\nu) \neq 0$, then we say that u has a spectral line at ν.

(2) Let $x(t) \in \mathcal{R}^N$ have spectral lines at frequencies $\nu_1, \nu_2, \ldots, \nu_N$ Further, let

$$\{\hat{x}(\nu_1), \hat{x}(\nu_2), \ldots \hat{x}(\nu_N)\}$$

be linearly dependent in $\mathcal{C}^N$. Then x(t) is sufficiently rich, that is, $\exists\ \alpha, \delta > 0$, such that

$$\forall s \in \mathcal{R}^+, \quad \int_s^{s+\delta} xx^T\, dt \geqslant \alpha I$$

(3) We say that r(t) is persistently exciting at frequencies $\nu_1, \nu_2, \ldots,$ ν_N, if it has spectral lines at these frequencies.

The results of Boyd and Sastry are also applicable to the discrete-time algorithm given by Narendra and Lin [85a], provided we modify the definition of spectral lines for discrete-time signals. We can also apply this approach to the SISO Algorithm 1 of Goodwin et al. [83] for which parameter convergence has been shown by Anderson and Johnson [82]. Boyd and Sastry also sketch how prior bounds on parameters and plant-model state errors can be used to obtain an estimate of the rate of exponential convergence.

Becker et al. [86] examine the geometric properties exhibited by the sequence of parameter estimates, which yield valuable insight into the behavior of the stochastic approximation based algorithm, as it is used in minimum-variance adaptive control. Sufficient conditions are also given for some of these results to hold for parameter estimation schemes other than stochastic approximation. If "self-tuning" is the purpose of using an adaptive controller, then convergence of parameter estimates is of great importance. They note that there are no rigorous results available on the convergence of the parameter estimates themselves and so consider the following questions:

1. Do the parameter estimates converge to the true parameter?
2. Do the parameter estimates converge to some parameter which leads to an optimal control law (e.g., a multiple of the true parameter)?
3. Does the adaptive control law converge to an optimal control law?
4. Is the cost optimal?

They show that question 1 does not hold, even though it is the most desirable property. If the system does not possess reduced-order minimum-variance controllers, then they prove that question 2 holds. But if the system does possess reduced-order minimum-variance controllers, then it is shown that the limit set of parameter estimate is a hypersphere of dimension strictly less than that of the space. Further, they show that question 3 holds in a Cesaro sense. It should be noted that overmodeling per se is not a problem. By an example, they show that convergence of the parameters can occur, even if we overmodel the orders of the system.

There is another important departure from the usual approach to the convergence of parameter estimates based on the idea of persistent excitation. In adaptive control, such conditions for persistent excitation are not verifiable a priori. In this work, persistency of excitation does <u>not</u> hold (i.e., it holds with zero probability), because of the linear relationship between the outputs and inputs induced by the convergence of the adaptive control law. One way to ensure persistency of excitation is to "inject" noise into the system.

It should be mentioned that stochastic approximation-based adaptive control algorithm is not usually implemented in practice due to its poor convergence rate. However, since it is easier to analyze, results are given for this scheme and then generalized to its least-squares like counterparts. Further work is needed for analyzing the least-square scheme and also the "tracking" problem, since only the "regulation" problem has been considered.

In pole-assignment adaptive control algorithms, we should first show the boundedness of inputs and outputs before establishing parameter convergence. However, Goodwin et al. [87] have reported results, without requiring a priori bounds on the system variables.

Anderson et al. [88] examine the model reference adaptive scheme proposed by Morse [77] for exponential convergence of parameter estimates. This algorithm has been shown to be globally stable. However, even though the tracking error converges to zero, the parameter estimates do not necessarily converge to their desired values. It is found that when the high-frequency gain is known a priori, there will be exponential convergence of parameter estimates, whenever the reference input is persistently exciting. But Anderson et al. [88] consider the situation in which the high-frequency gain is not known and show that the algorithm does not result in exponential convergence, even with sufficiently rich inputs. The problem appears to lie in the identification of the gain parameter because the signal involved in the identification converges asymptotically to zero and, therefore, loses persistence of excitation. This conclusion is expected to be applicable in general to plants with relative degree greater than 1.

Elliott et al. [89] present a proof of global stability for direct and indirect adaptive control schemes which arbitrarily assign the closed-loop poles of a SISO system. The results are applicable to the control of either minimum or non-minimum-phase systems. Both hybrid continuous-time and discrete-time formulations are given. They verify the persistent excitation condition for potentially unbounded data resulting from closed-loop adaptive control using a sequential estimation procedure. Hence the convergence of a sequential least-squares identification algorithm is proved. The results are applicable to standard sequential least squares, and least squares with a covariance test. The main limitation of the results is that they are applicable only to the case of sinusoidal excitation. An important extension would be for other persistently exciting input signals. Some suggestions are given on the lines on which future work can be done. Goodwin and Teoh [90] establish a general convergence result for parameter estimates in the presence of possibly unbounded signals. Trulsson [91] has analyzed an instrumental variable scheme to establish parameter convergence with possibly unbounded noisy signals.

The stochastic case is more difficult to analyze with respect to asymptotic properties because of the highly nonlinear manner in which the variables are coupled to each other. Chen [92] establishes parameter convergence for the modified least-squares algorithm, and that for the stochastic gradient algorithm is shown by Chen and Caines [93]. The case of adaptive control problems where the system parameters evolve randomly is discussed by Chen and Caines [94, 94a, 94b].

The issue of persistent excitation, which is relevant to parameter convergence as well as other problems in adaptive systems, is considered in Section 8.11.

8.6 NONLINEAR SYSTEMS

The study of adaptive control algorithms for nonlinear and time-varying systems has not been extensively reported in the literature. Some specific problems have been discussed. Goodwin et al. [95, 96] deal with adaptive control of bilinear systems. Evans et al. [97] establish global convergence for classes of nonlinear systems and also systems whose output is corrupted by unknown but bounded noise. Anbumani et al. [98] discuss the self-tuning minimum-variance control of nonlinear systems of the Hammerstein model. They have reported in a later paper [99] about self-tuning cascade control of linear and nonlinear systems. This is of interest in some industrial processes which consist of two subsystems in cascade—one comparatively faster subsystem driving a slower one, with the disturbances acting on both the subsystems. In such a scheme, the output (y_2) of the faster subsystem is measured and employed for feedback regulation against the disturbance, before this disturbance can affect the primary output (y_1). Self-tuning control is applied to such cascade control, by characterizing the nonlinear subsystem by the Hammerstein model. Simulation studies show that self-tuning can result in the successful cascade control of unknown systems subjected to unknown disturbances.

Korpela and Koivo [100] consider the self-tuning control of a class of nonlinear scalar processes. The process input nonlinearity is assumed to be separable from the linear part. In the first algorithm the nonlinearity is assumed to be known or its output to be measurable. Knowledge of the characteristics of the nonlinearity is used for algorithmic linearization. The nonlinear model is not updated. In the second control algorithm the nonlinearity need not be known, because the model is updated based on measurements. The nonlinearity is described by a Volterra series in a Hammerstein model. In both the cases a reduction in the variance of the output signal error and the input signal is obtained. However, in some cases a steady-state error may be observed.

Grimble [101] proposes a weighted minimum-variance controller for nonlinear plants, which can have a mathematical description that is quite general. If the system is represented by a Hammerstein model, that is, by a single polynomial nonlinearity in cascade with a dynamic linear block, the resulting controller is quite simple. These controllers are useful when the exact form of the nonlinearities and the dynamics are not known. Amplitude constraints can be applied on the optimal control signal variations. Such a controller can also be applied to non-minimum-phase plants. This approach can also be of value in adaptive extremum control problems [102-105].

Kung and Womack [106] consider stability and convergence of an adaptive controller for a system consisting of a static nonlinearity followed by a linear dynamic plant. The static nonlinearity is described by a finite odd order polynomial. This is a very restricted class of nonlinear systems. In a later paper [107] they discuss a method for discrete-time adaptive control of a nonlinear system which consists of a cascade connection of a preload nonlinearity followed by a linear system. It has been shown that the algorithm will ensure that the closed-loop system is globally convergent with a specific residue and stable for both deterministic and stochastic cases.

Urwin and Swanick [108] develop an adaptive controller with stability assurance for a highly nonlinear plant. A quadratic model is used to investigate the nonlinear structure of the system. This recognition of the structure may be deemed as identification, since this gives an indication of the position of nonlinearities in the state space structure of the physical process. Although the quadratic model may not represent exactly the order or structure of the process, this technique assumes each system to have a fast and continuous updating of the quadratic approximation. The objective of the control law is to attain a preset operating level and to maintain that level even when there are disturbances. The stability of the nonlinear control has also been investigated using an extension of Lyapunov's indirect method. Simulation studies show that stable as well as unstable nonlinear systems can be controlled. The superiority of the nonlinear approach with respect to computing time, accuracy, and stability is established in comparison with the linear model method of adaptive control such as pole-placement technique. This nonlinear approach has been applied to a two-output/one-input system; however, further work is possible using several of these nonlinear model adaptive controllers in a hierarchical regime to control and stabilize large-scale systems. Investigations can also be made using other polynomial approximations to give even more generalized estimates of the nonlinear elements.

Balestrino et al. [109] describe three approaches to the synthesis of adaptive model-following controllers for nonlinear time-varying plants with disturbances. The first technique is based on the theory of multivariable variable structure systems (already considered in Section 7.2), which have desirable properties such as parameter invariance and disturbance rejection. The second approach is based on hyperstability theory and the third one is a suitable combination of the previous approaches. These techniques lead to control laws which are discontinuous on linear switching hyperplanes. These schemes do not include any parameter identification algorithms. Adaptation is accomplished by signal synthesis in a fixed-parameter structure. Since they do not require any global convergence properties, these approaches are well suited to some classes of uncertain time-varying plants. The design procedures given in Balestrino [109] are theoretically well founded for nonlinear time-varying or uncertain plants and all give robust discontinuous control. These techniques are especially applicable whenever the plant dynamics are so fast that identification or on-line computations make other approaches not feasible. The main areas of applications are in electrical power systems, electrical drives, robotics, and aerospace.

Ambrosino et al. [110] give a new design method, using the hyperstability approach, for the adaptive control of a class of nonlinear systems with nonlinearities of known form. The earlier methods, due to Young [111] based on VSS theory, Leitmann and Gutman [112], and Gutman [113] based on the Lyapunov approach, and Balestrino et al. [114], based on the hyperstability approach, lead to adaptive control laws which are discontinuous on an attractive hyperplane in the error state space. These methods result in the "chattering" of the control input at a theoretically infinite frequency. Such control is not only difficult to achieve in practice, but is also not directly applicable to the plant. Hence Ambrosino et al. suggest an alternative approach which gives a continuous adaptation law. A systematic procedure is given for the design of multivariable plants. The structure of the controller, as well as the adaptation law, is a modified version of the standard one used in adaptive model-following control for linear plants. The stability proof of the error equation in this case is based on hyperstability theory. The proposed technique leads to adaptation laws of the integral type. These laws give rise to smooth control signals, and hence avoid high-frequency "chattering" of the plant input. Computer simulation of a standard nonlinear control problem demonstrates the capability of the adaptive system to quickly nullify the state error resulting from an initial misalignment between the plant and the model.

Similar work is reported by Ambrosino et al. [114a] who propose a continuous control law that guarantees not only that the error remains bounded, but also that it tends to an arbitrarily small neighborhood of the origin with a prescribed rate of convergence. A procedure has been suggested by which the rate of convergence can be arbitrarily increased simply by solving a pole-placement problem. The possibility of arbitrarily prescribing the rate of convergence of the error is advantageous in solving tracking problems. The stability analysis makes use of some results of VSS theory. This design technique guarantees the attractiveness of a certain neighborhood of a hyperplane. When the error enters this neighborhood, the system becomes less sensitive to plant parameter variations and disturbance inputs. The design procedure and the performance of the resulting control system are illustrated by a simple example, demonstrating that the proposed scheme may be very useful for industrial applications.

Raju and Riad [114b] present a procedure for the design of signal synthesis model reference adaptive controllers for linear and nonlinear plants which can also be time-varying. The plants considered are those which can be represented by a set of multiple-input/single-output subplants and which have at least as many inputs as outputs. Each subplant output is used to derive one input signal so that its output will asymptotically approach the corresponding output of a reference model. The models are linear, stable, and of lower order. The design of the controller is based on the Lyapunov method. However, no derivatives of the error signals are required. Filtered signals using state variable filters are employed. No prior information of the plant in terms of plant parameters or the form of the nonlinearity is required. Simulation results including those for a two-input/two-output nonlinear system illustrate the effectiveness of the design procedure.

8.7 STOCHASTIC SYSTEMS

8.7.1 MRAC Systems with Bounded Disturbances

In Chapter 3 we considered noise-free MRAC systems and have shown how they can be designed so as to be globally asymptotically stable. This is, no doubt, a major advance in recent years. However, the results obtained in the disturbance-free case are not entirely satisfactory. From the practical point of view, it is more realistic to investigate the performance of adaptive schemes in the presence of external disturbances. The adverse effects of measurement noise in MRAC systems were first considered by Monopoli and Kessler [115] and Gutman et al. [116], who suggested some practical methods to ensure system stability. Egardt [117,118] has shown that the introduction of even small disturbances can lead to instability. Hence it is of utmost necessity to examine how the adaptive systems perform in the presence of disturbances. Only recently have such studies been undertaken.

Peterson and Narendra [119] have shown that when bounded output disturbances are present with no modification of the adaptive laws, the parameter error vector $\phi(t)$ can grow without bound even though the state error vector between the plant and the model is bounded. It is therefore essential that adaptive schemes be appropriately modified when external disturbances are also present.

Peterson and Narendra [119] also show that when exact model matching is not possible owing to the presence of disturbances, it is necessary to revise the ideas about the acceptable behavior of the overall system. A relaxation in performance objective will give greater freedom in the development of the algorithms. The minimal requirement of any such algorithm is that the adaptive controller assures the uniform boundedness of all the signals in the adaptive loop. Conditions are also imposed on the asymptotic behavior of the control parameter vector, so that it tends to a constant value. It is also desirable that as the disturbance level tends to zero, the performance of the adaptive system approaches that of the noise-free case. In this work a nonlinear adaptive law using a dead zone is suggested which assures the boundedness of all the signals in the system. The presence of the dead zone, however, results in finite parameter and output errors even when no external disturbance is present. In this approach there is also no direct way of taking into account changes in the disturbance levels. Instability can be caused by a larger disturbance than that used in the design of the dead zone. Further, a smaller disturbance does not result in a smaller error.

An alternative approach simultaneously developed by Kreisselmeir and Narendra [120] has a significant advantage over the approach in Peterson and Narendra [119] in that the changes in the disturbance level are directly reflected in the magnitude of the output error. In particular, if the disturbance tends to zero, the output error also tends to zero. The structure of the controller used is identical to that used in Narendra et al. [76] for the disturbance-free case. This assures the existence of a constant control parameter vector θ^* for which the transfer function of the plant, along with the controller, matches exactly that of the model. As part of prior information, it is assumed that θ^* has a norm less than a specified value $\|\theta^*\|_{max}$ and the disturbance is bounded, although it is not necessary to know the explicit value of a bound. Since $\|\theta^*\|_{max}$ is known, it is necessary to search only within a sphere S of radius $\|\theta^*\|_{max}$ in parameter space. This assures the boundedness of the parameter error vector $\phi(t)$. The adaptive law used to update $\theta(t)$ is the same as that in the disturbance-free case when θ lies in the interior of S and is modified when it reaches the boundary of S or lies outside it. It has been shown in this paper that such a scheme yields bounded signals in the adaptive system, even in the presence of disturbances. In the absence of disturbances the modification is of no consequence and the output error e(t) goes to zero as t tends to infinity. This clearly demonstrates that this scheme retains the properties of earlier schemes and is, in addition, robust with respect to bounded disturbances, which makes it suitable for practical applications.

It may be pointed out that both the approaches [119, 120] operate in an identical manner when the output error is larger than the dead zone and the control parameter vector $\theta(t)$ lies in the interior of S. Further, since both schemes are globally stable, it is possible to estimate the plant parameters on-line in the two cases. It will be interesting if these two approaches can be combined judiciously taking advantage of the individual merits of each. There is considerable scope for further research in this important area.

8.7.2 Stochastic Adaptive Controllers

A stochastic adaptive controller is an algorithm which combines on-line parameter estimation with on-line control to generate a control law which can be applied to systems having unknown parameters and random disturbances [121]. Although control laws for such cases have been studied for quite some time, only recently has rigorous convergence analysis been done.

In the case of deterministic disturbances such as periodic signals, Goodwin and Chan [122] establish global convergence for a scheme having a simple stochastic gradient parameter estimator with a minimum variance control law. Simulations have confirmed the good performance of their algorithm. Global convergence has been proved, provided the disturbance is bounded. It is also necessary to consider disturbances which are of quite arbitrary wave form.

Samson [123] has studied the stability of adaptively controlled systems subject to bounded disturbances with unknown statistics. This approach is an extension of that used for the disturbance-free case [124] and is based on the idea that the stability of the adaptively controlled system can be established quickly as soon as the identification algorithm associated with adaptive control verifies certain properties. One of the important aspects of this approach is that stability results are established not only for minimum-phase systems, but also for non-minimum-phase systems. In each case the uniform boundedness of all signals is proved, provided a specified condition is fulfilled. This approach also has the advantage of being effective for a large set of control methods.

Samson considers "indirect" adaptive control schemes only, since a result of stability obtained for an indirect scheme can be extended to a corresponding direct scheme, without much difficulty. Although only the regulator problem has been discussed, the results of the stability analysis remain unchanged for any uniformly bounded reference signal. The disturbance sequence is assumed to be any uniformly bounded sequence of real numbers. No assumption has been made on the statistical properties of the disturbance.

Extension of this approach is possible by making the properties to be verified by the identification algorithm less restrictive than those considered. Then it is possible to increase the number of identification algorithms which will be admissible in adaptive control. It is also possible to extend this approach by exploiting the statistical properties of the disturbances, without the results of stability depending on the properties of these disturbances.

Noise can also be modeled as a stochastic process. The convergence analysis in this case is more difficult, as we have to do combined parameter and state estimation. As in the deterministic case, convergence is dependent on the boundedness of system variables. Goodwin et al. [125] establish global convergence of a stochastic adaptive control algorithm for discrete-time linear systems. For all initial system and algorithm states, the (conditional) mean square output tracking error is minimized, with probability 1, and this is achieved with a sample mean square bounded input sequence. The algorithms have a simple structure and are applicable to both single-input/single-output systems and multi-input/multi-output systems. These results are the first rigorous analysis of any stochastic adaptive control algorithm of this type.

Sin and Goodwin [126] establish global convergence for a slightly modified least-squares stochastic adaptive control algorithm for linear, time-invariant, discrete-time systems. In Goodwin et al. [125], the algorithm is based on a stochastic approximation iteration, whereas in practically all applications of stochastic adaptive control, a least-squares iteration is used. The motivation for using the least-squares method is that it has superior rates of convergence compared to stochastic approximation. The modifications suggested in Sin and Goodwin [126] to the least-squares algorithm have both theoretical and practical advantages over the other approaches. In the modified algorithm they use a posteriori predictions in the regression vector. Secondly, a full least-square step is taken only when this is consistent with convergence criteria developed by them. In fact, the need for modifying the algorithm to ensure stability and convergence was also indicated earlier by Gawthrop [127]. In Sin and Goodwin [126] both the white-noise and colored-noise cases are treated for SISO systems with unit delay. It is shown that, with probability 1, this algorithm will ensure that the system inputs and outputs are sample mean-square bounded and the mean-square output tracking error achieves its global minimum possible value for linear feedback control. Chen [92] also deals with convergence results for least-squares-based adaptive control algorithms.

Caines [128] has analyzed the performance of the stochastic gradient algorithm of Goodwin et al. [125], applied to systems with (converging) Martingale parameters. Such work on time-varying parameters is relevant, because, in practice, we do meet with such systems and we should gain insight into the convergence properties of such systems. Further results for systems having random parameters are discussed by Chen and Caines [94], who establish global convergence by modeling parameter variations as a bounded Martingale difference sequence. It must be noted that analysis based on Martingale theorems applies only if the disturbance acting on the system satisfies the statistical properties of a moving average of independent zero mean random variables. In practice, such conditions are not satisfied and so the analysis may be said to lack robustness.

Fuchs [129] has examined the general delay, white-noise case for indirect stochastic adaptive control. In the earlier reported work, for delays k greater than unity, k interlaced adaptation algorithms were required. Fuchs has shown that a single adaptation algorithm suffices to establish that, with probability 1, the system output and input and the output tracking error are sample mean-square bounded. In this indirect adaptive control approach, we identify the system, compute at each step the current prediction form of the system, and apply the control equating the so-obtained system output with the desired output. The mean-square output tracking error is also shown to converge to its global minimum value, that is, the value achievable in the known parameter case. The same indirect approach has been applied by Fuchs [130] to the general delay-colored noise case and convergence properties are achieved by a single adaptation algorithm. These results are readily applicable to the adaptive predictor problems of the type suggested in [131,132], thus allowing us to reduce the memory space requirement. This approach can also be extended to other indirect control schemes.

Salyga et al. [133] deal with dynamic stochastic systems which are described by a Markovian decision process. The true value of the parameters of the transition function which describes the system are not known. A sequential identification method and a procedure for adaptive control have been developed within the framework of Markovian decision processes. The recurrent procedure ensures convergence to the optimal true control strategy within a finite number of steps. This recurrent procedure has some useful properties. It is sequential so that each new observation is used for updating the estimates and adaptive control. These can be made good enough with a few observations. This strategy is applicable to design of optimal models and control methods for industrial, management, ecological, and other systems.

Caines and Lafortune [134] propose a stochastic adaptive control algorithm which asymptotically optimizes the behavior of the controlled system and, at the same time, generates consistent estimates of its parameters. None of the earlier developed algorithms has been shown to yield, either directly or indirectly, consistent parameter estimates. This approach, based on Goodwin et al. [125], can be viewed as a means to identify unstable stochastic systems by, first, stabilizing them via adaptive control and ensuring that persistent excitation condition is satisfied, and second, applying an on-line identification method which is known to work under these conditions. To be specific, the algorithm has the following properties when applied to a possibly unstable, inverse stable, linear stochastic system with unknown parameters, whenever the system satisfies a certain positive real condition on its (moving average) noise dynamics:

1. The adaptive control part of the algorithm stabilizes and asymptotically optimizes the behavior of the system, in the sense that the (limit of the) sample mean-square variation of the output around a given demand level equals that of a minimum-variance control strategy implemented with known parameters (this value being the variance of the linear least-squares prediction error process). The optimal behavior is subject to an offset μ^2 where μ^2 is the variance of a dither signal added to the control action in order to produce a "continually disturbed control." For $\mu^2 > 0$, it has been shown that the input/output process satisfies a persistent excitation property, and hence, subject to a simple identifiability condition, the next property holds.
2. The observed input and output of the controlled system may be considered as input to an approximate maximum-likelihood algorithm (AML) [135], which generates strongly consistent estimates of the parameters of the system.

This paper [134] considers both scalar and multivariable cases, and also the multiple-delay case. The techniques of this paper, especially the interesting idea of "continually disturbed controls" can be applied to the analysis and synthesis of adaptive control laws for systems with randomly varying parameters.

Chen [136] has presented a modified least-squares (MLS) algorithm. This is in contrast to the preceding paper [134], which uses two algorithms. The MLS algorithm can be applied both for tracking and estimation purposes and this is done with generally weaker conditions than the persistent excitation condition.

Goodwin et al. [121] have reported that standard stochastic adaptive control algorithms for time-invariant systems have an inherent robustness property which renders them applicable, without modification, to time-varying systems whose parameters converge exponentially. This paper has a similar approach to that discussed earlier [94,128]. However, since the parameter variations are deterministic, the method of analysis is different. It is shown that the convergence properties are retained if the steady-state model is replaced by the time-varying Kalman filter model. The key observation is that the Kalman filter converges exponentially fast to the steady-state solution [87]. Thus the results of Goodwin et al. [121] allow global convergence to be established for the standard adaptive control algorithms when applied to these systems. This represents a relaxation of the usual modeling assumption employed by earlier workers that the system is described by an ARMAX model or equivalently a steady-state Kalman filter model.

Xianya and Evans [137] have presented an algorithm for adaptive control of plants with an unknown periodically or nonperiodically time-varying deterministic disturbance. When the system parameters are also unknown, a persistent excitation condition is required for convergence. When the system parameters are known and only the disturbance is unknown, the persistent excitation condition is automatically satisfied. Another paper [138] by them extends the results of discrete-time adaptive control of time-varying systems to the stochastic case. Simulation results show that this scheme is able to track time-varying parameters more effectively than some previous schemes.

Dugard et al. [139] present a derivation of a stochastic controller using the "general interactor matrix" for multiinput/multioutput systems. The concept of the "system interaction matrix" was introduced earlier in Section 4.7 in connection with the design of multivariable discrete-time adaptive control systems. If the interactor matrix is assumed known, it is possible to develop a stochastic adaptive control algorithm which is globally convergent. However, the requirement of the knowledge of the interactor matrix imposes certain difficulties, in practice. This matrix contains noninteger-valued real variables, a knowledge of which will imply that we know the complete system transfer function. Hence a method has been given which eliminates the need to know the interactor matrix in multivariable stochastic adaptive control. This algorithm appears to be promising from a practical point of view, and some extension and further studies are possible. The idea is based on a scheme due to Ydstie [140] for SISO systems. A multistep optimization horizon is used. In the MIMO case this avoids the requirement to know the interactor matrix, while retaining the simplicity of the self-tuning algorithm. Based on this idea, an indirect adaptive control strategy has been developed. The convergence of the resulting algorithm has to be studied further to get an insight into the method.

It is pertinent to mention that the problem of dealing with multivariable systems having an unknown interactor matrix was also studied by Johannson [141] and Dugard et al. [142], who estimte the noninteger-valued variables in the interactor matrix along with the other system parameters. An alternative idea, given by Singh and Narendra [143], is to apply a suitably chosen precompensator to the system so that the interactor matrix is transformed to a diagonal matrix having only integer-valued parameters. This method works for a large class of multivariable systems. This is an area where further work is possible.

8.7.3 Multiple-Model Adaptive Control

The control of linear stochastic systems with unknown parameters is a problem of major importance. The parameter uncertainty renders an optimal control solution not possible. Several parameter adaptive suboptimal control strategies have been proposed. Partitioning adaptive control strategy, based on Lainiotis' partition theorem [144], is one of the methods. This method is also known as multiple-model adaptive control (MMAC). This approach is computationally implementable and also effective as a practical stochastic regulator for both the longitudinal and lateral dynamics of the NASA F-8C aircraft [145].

Watanabe [146] has presented a passive-type MMAC algorithm for linear discrete-time stochastic systems with observation subsystems which involve unknown but constant parameters. These unknown parameters occur only in the observation matrix. It is also assumed that they are members of a finite set with known probability distribution and are constant during the period of interest. A passive-type (i.e., taking no account for the fact that future observations will be made) MMAC algorithm is derived. In this method a suboptimal control policy is adopted. Instead of searching for an adaptive control which minimizes the "original quadratic cost," an elemental control which approximately minimizes the "elemental cost" is used. This scheme is slightly different from the method of Lainiotis et al. [147] in that the elemental control is fed back to the elemental filter. However, the solution of the elemental filter gain depends on both the elemental and adaptive control policies. It has been shown, through some Monte Carlo simulations, that for unstable systems, this method yields performance improvements over the method of Lainiotis et al. [147]. Further work for the case of a switching parameter such as the Markov jump is in progress.

8.8 MULTIVARIABLE SYSTEMS

There has been considerable progress in the development of the theory of SISO adaptive control, including such complex situations as effects of disturbances, nonlinearities, and unmodeled dynamics. Only recently has such work been extended to MIMO systems.

Kidd and Munro [148] give a new design approach for multivariable MRAC systems, using frequency-domain methods. This approach is to first design a robust feedback system, and then to augment it with nonlinear model reference adaptation laws. MRAC designs can easily be converted to robust feedback systems, by removing the adaptation mechanism and replacing it with a linear compensator, which acts on the error signal between the reference model and the plant. This kind of system has been referred to as a conditional feedback system and is ideally suited to the proposed design technique because of the ease with which the linear compensator can be augmented by nonlinear adaptation laws, whenever the linear compensation is inadequate for the task on hand. The paper gives the design for the required linear compensation and also a multivariable frequency-domain performance criterion. This criterion is then related to other scalar frequency-domain performance criteria. A new design method is carried out using Rosenbrock's direct Nyquist array [149]. A number of plant frequency-response arrays are displayed on a computer graphical-display unit, along with the frequency-response array of the reference model. The designer then proceeds to develop the appropriate compensation to reduce the error between the moduli of the plant's frequency-response diagonal terms and the moduli of the reference model's frequency-response diagonal terms.

This design approach is illustrated by application to a two-input/two-output plant with parameter uncertainty in both the diagonal and off-diagonal elements. This approach is simpler and utilizes recent developments in linear multivariable frequency-domain design methods.

Further work is possible for the design of a more general multivariable system, including a suitable disturbance rejection criterion. The question of robustness under conditions of plant-model-order uncertainty needs investigation.

Elliott and Wolovich [150] present a strategy of model matching, based on direct zero canceling. That is, the control parameters are directly estimated from input/output data, and the designs are based on fixed control strategies which assign the open-loop zeros as closed-loop poles. In Elliott et al. [151] a direct strategy which does not involve zero cancellation is given. It is based on arbitrary pole placement. This method is applicable to the class of systems characterized by proper full rank transfer matrices and can be implemented with only a priori knowledge of the system controllability indexes and an upper bound on the observability index. No knowledge of the system interactor matrix is required. The adaptive scheme is based on a fixed observer structure and arbitrarily assigns the closed-loop system poles. This methodology is applicable to both continuous-time and discrete-time linear multivariable systems. The scheme is direct in nature in that no parameterized model of the unknown plant needs to be explicitly identified. The control parameters, and an "equivalent" plant parameterization are simultaneously estimated from input/output data, using linear parameter estimation procedures. Simulation results are given for three cases of two-input/two-output unstable non-minimum-phase systems. As stated by Elliott et al. [151], one potential limitation of this procedure is that when the controllability indexes are unequal, a unique closed-loop structure cannot be guaranteed. Although some local stability results are available, further work is necessary on global stability for unstable open-loop systems.

Elliott and Wolovich [152] give an overview of multivariable adaptive control algorithms and also present some new approaches to the design of indirect multivariable adaptive control systems, with special reference to a unified pole-placement approach. The advantage of indirect strategy lies in the possibility of reduction in the number of parameters to be estimated. The reduction in the size of the parameter estimation problem is a crucial issue for application of multivariable adaptive control. There is also another technique given by Djaferis et al. [153] to tackle this problem which is applicable to both direct and indirect strategies and involves restricted use of inputs and outputs. The resulting design reduces significantly the complexity of the scheme. This approach is also promising. Although no global stability analysis has not been carried out, simulation results are encouraging. For implementation, one should know the order of the system and an upper bound on the observability index.

Apart from reduction of the number of parameters to be estimated, Elliott and Wolovich [152] also emphasize other parameterization issues such as types of control structures capable of meeting performance objectives and also type of prior knowledge required regarding the unknown process for implementation. Although only discrete-time systems have been dealt with, there is a straightforward link to continuous-time systems as well. There is also a correspondence with the delay operator formulations given by Goodwin et al. [154,155]. Much work still needs to be done in reducing the excessive size of the estimation problem to exploit fully the practicability of multivariable adaptive control.

Singh and Narendra [143] discuss the design of a dynamic precompensator $W_C(s)$ for an $m \times m$ MIMO plant $W_P(s)$ such that, for almost all cases, $W_P(s)W_C(s)$ has a diagonal Hermitian form, that is, almost all plants can be adaptively decoupled. They also consider a sufficient condition which is weaker than that prescribed by earlier workers for the outputs of the plant to follow asymptotically the outputs of a linear time-invariant reference model. Instead of the interactor matrix of the plant transfer matrix being known, the weaker condition requires only that the relative degree of each element of $W_P(s)$ be known. The prior information required will then consist of this weaker condition as well as the other usual sufficient conditions, that is,

1. The frequency gain matrix K_P satisfies a sign definiteness condition.
2. An upper bound ν on the observability index of $W_P(s)$ is known.
3. The zeros of $W_P(s)$ are located in the open left half-plane.

With the scheme presented in Singh and Narendra [143], almost all MIMO plants can be adaptively controlled, enlarging the class of plants considered by Monopoli and Hsing [156] and Goodwin et al. [83] and using less prior information than that assumed by Elliott and Wolovich [150], Goodwin and Long [154], and Morse [157]. The approach given by Singh and Narendra can also be applied to discrete-time systems.

8.9 DISTRIBUTED PARAMETER SYSTEMS

Adaptive control of distributed parameter systems (DPS) is still an open question and hardly any work is reported in the literature. The problems which arise in such systems are as follows:

1. The controlled plant consists of a set of interconnected nonlinear DPS.
2. The distributed parameter control system is a subsystem of a large-scale system and must adapt to varying control conditions, variations in the environment and process uncertainties.
3. The information necessary for control is obtained on-line by the measurements in the finite number points in the DPS.

Hulkó et al. [158] discuss the preceding problems, which are present in some industrial processes. They introduce the idea of X-representation of DPS on the basis of measurement of input/output quantities at a finite number of points in the DPS. Using this representation, nonlinear DPS adaptive control problems are solved for different amounts of a priori information. The paper considers the transformation of a priori information, expressed as linear partial differential equations, to the X-representation form by using the extensive results of Butkovskij [159]. The controlled plant is considered as an interconnected set of nonlinear DPS. This approach is applied for the hierarchical control of a glass furnace physical model. The mathematical model is given by a set of three nonlinear partial differential equations. The proposed adaptive control scheme has only the usual computational requirements and is easily implementable. There is considerable scope for further research, as the problem of adaptive control of DPS has not been investigated to any significant extent.

8.10 MODEL UPDATING

The performance of a MRAC system can be improved by model updating. For systems whose structure does not match that of the reference model, there is no unique set of parameters of the controller, which can result in a zero error between the output of the plant y_P and that of the model y_M. These oscillations can be reduced by model updating, a concept introduced by van den Bosch and Jongkind [160] for continuous-time MRAC. In this approach a fixed value was used for the update interval. Every T_{up} seconds the state of the system was introduced into the reference model. However, the choice of the value of the fixed update interval is critical.

In a later paper, van den Bosch and Tjahjadi [161] used a different strategy based on the Lyapunov function. Model updating replaces the state of the reference model with the actual state of the plant at some points in time. Finally, there is no difference between y_P and y_M and no need for parameter adjustments. New reference trajectories are calculated, starting in the actual state of the system. Thus, model updating avoids unnecessary control efforts and anticipates disturbances better. Then the adaptive controller can react faster to (1) differences in the parameters, (2) differences in the structure of the plant and of the reference model, and (3) disturbances.

In van den Bosch and Tjahjadi [161], the criteria are considered for determining an appropriate point in time in which to apply model updating. It has also been proved that model updating does not affect the asymptotic stability of the MRAC system. A feasibility study of satellite attitude control (which is also considered in Section 9.14) shows that model updating results in a significant reduction in the energy consumption, and the selection of the adaptive gains becomes less critical.

Another related work is by Tsypkin [162], who considers the problem of synthesis of an optimal adjustable model for linear dynamic plant identification. A noise transfer function is introduced. An optimal structure of the adjustable model is that structure which minimizes some error functional (e.g., the error variance). The properties and specific features of the optimal adjustable model are also discussed. This approach eliminates the ambiguity in choosing identification models for dynamic plants and ensures that different identification algorithms yield unbiased estimation of the basic plant parameters with minimum attainable error variance. This idea is briefly referred to in Section 9.6 while we consider the design of an adaptive lateral autopilot for a missile.

8.11 PERSISTENT EXCITATION

The concept of persistent excitation is of great importance in many contexts in adaptive systems. This idea was first introduced in identification problems and then in control problems. In these cases the convergence of a parameter vector to a desired value is assured by persistent excitation. This has also been referred to in earlier chapters and in Section 8.5. Furthermore, robustness is also dependent on the condition of persistent excitation.

The problem can be stated in simple terms: Suppose we are required to identify a fourth-order SISO system. There are eight real parameters, that is, the numerator and denominator coefficients to be found. If we excite the system by a single sinusoid, there is no algorithm—adaptive or nonadaptive—which would be able to determine all the eight parameters. Then we would say that the input of an adaptive identifier was not persistently exciting. We can regard a signal as persistently exciting, if it is "frequency rich," that is, it contains sufficient frequencies to enable continuously updatable estimation of all parameters rather than a restricted number. That is to say, the input should be (recurrently) "rich" enough to excite all the "modes" of the system. Such an input containing sufficient number of distinct frequencies is represented by

$$u(t) = \sum C_i \sin(\omega_i t + b_i)$$

A broad class of input functions having the property of persistent excitation is characterized by Mareels [163]. The concept of a <u>sufficiently exciting input</u> is precisely defined in two alternative ways:

DEFINITION 1. The function $x \in \mathcal{L}_c^2(\mathcal{R}^+, \mathcal{C}^n)$ is "sufficiently rich" iff there exist positive constants a and S and a positive constant t_1 such that for all $s \geq t_1$,

$$\int_s^{s+S} x(t)x(t)^+ \, dt \geq aI_n > 0$$

where $\mathcal{L}_c^2$ denotes the linear space of functions, defined on $\mathcal{R}^+$, which are square Lebesgue integrable on (0, T) for all finite $T \in \mathcal{R}^+$; and $x(t)^+$ denotes the Hermitian conjugate (transpose and complex conjugate) of x(t).

DEFINITION 2. The distribution $x \in D^*(\mathcal{R}^+, \mathcal{C}^n)$ is said to be "sufficiently rich" iff there exist positive constants a and S, and a non-negative constant t, such that for all $x \geq t$ and all vectors $d \in \mathcal{C}^n$ with $||d|| = 1$, there exists a test function $\Phi \in D((s, s+S), \mathcal{C})$ such that

$$\int_s^{s+S} (d^T x(t))\Phi(t) \, dt \geq a > 0$$

where $D^*(\mathcal{R}^+, \mathcal{C}^n)$ denotes the linear space of distributions defined on $[0, +\infty)$ taking values in $\mathcal{C}^n$; and d^T is the transpose of d.

Definition 2 is a straightforward generalization of Definition 1. If x is sufficiently rich in the sense of Definition 1, then it is also sufficiently rich in the sense of Definition 2. Furthermore, if x is not sufficiently rich according to Definition 1, then it cannot be sufficiently rich according to Definition 2.

After precisely defining a "sufficiently exciting input," Mareels presents necessary and sufficient conditions for an input function of a single-input, linear, stationary system to be sufficiently exciting. In contrast to earlier workers, no assumption about the stability of the linear system to be identified or controlled is required. Both continuous-time and discrete-time cases are dealt with in a deterministic setting.

The properties of persistently exciting inputs are also studied by Anderson [84], Morgan and Narendra [78, 164], and Yuan and Wonham [81]. These ideas are applied in adaptive identification cases by Sondhi and Mitra [165] and Kreisselmeier [80]. They deal with deterministic inputs. Bitmead and Anderson [166] consider random inputs.

Using error models, Narendra and Annaswamy [167] have shown that the persistent excitation of u guarantees the exponential convergence of the parameter error vector $\phi(t)$ to zero in the ideal case, or the boundedness of $\phi(t)$ in the presence of external disturbances. Anderson and Johnstone [168] have used this idea to demonstrate the robustness of identification and control in the presence of mild perturbations.

In adaptive systems we are concerned with an internal signal of the adaptive loop and it is the persistent excitation of this internal signal which is to be assured. This poses a difficulty. As discussed in Section 8.5, this difficulty has been overcome by Anderson and Johnson [82], who have obtained an equivalent condition to demonstrate the persistent excitation of the corresponding signals in the model, and by Boyd and Sastry [85], who show that when the reference input (which is not an internal signal) has as many spectral lines as there are unknown parameters, then it is sufficiently rich, resulting in exponential parameter and error convergence.

Anderson [169], in a recent paper, discusses elaborately what happens in several adaptive algorithms when persistency of excitation conditions are lost. In particular, four adaptive algorithms are considered: equation error identification, output error identification, reference trajectory following, and adaptive pole positioning. In the absence of persistency of excitation, the output-error algorithm will not only fail to identify the parameters but also a bursting phenomenon is encountered in practice. The output can, as a result of noise, nonlinearities, round-off errors, and so on, exhibit "bursts" of unstable, possibly oscillatory behavior between quiescent periods. Signals will still be bounded, but because of bursting, correct tracking of the reference trajectory by the plant output will not occur, except during the quiescent periods between the bursts of instability. Such a conclusion appears to correspond to reports about the field performance of adaptive controllers. In the case of adaptive pole-positioning problem also, bursting phenomenon will be likely, when the external control is not persistently exciting.

These results have been demonstrated theoretically and with simulation data. It has been shown that when inputs (including a reference trajectory) are not frequency rich, there is a possibility that estimates of individual system parameters can drift, because of noise or other unmodeled or unmodelable effects. Bursting phenomena must be expected if the adaptive system has a closed loop, the stability of which is dependent on the parameter estimates. However, there may be long intervals in which no bursting occurs. The time of occurrence, duration, and maximum amplitude of the burst are not predictable. The results confirm that some measure of robustness can be guaranteed if persistency of excitation conditions are fulfilled.

The case when external disturbances are present has to be dealt with by a different approach, since persistent excitation is defined only for bounded signals, and we cannot assume a priori that the plant signals are bounded. Narendra and Annaswamy [170] have pointed out that we have to consider the behavior of a set of nonlinear error differential equations if the concept of persistent excitation is to be used. They present some new results [171] which demonstrate global boundedness of all signals in the adaptive system in the presence of bounded disturbances when the reference input is sufficiently persistently exciting. In particular, it is shown that if the persistent excitation of the model output is larger than the disturbance in some quantitative sense, then the signals of the adaptive system will be bounded. But if the maximum amplitude of the disturbance is greater than that of the model output, then the system can have unbounded signals. This has been shown by analyzing the case of a first-order plant and then studying in detail the corresponding second-order nonlinear differential equation. This approach can be extended to the higher-order adaptive control problem. It is, however, necessary to determine the bounds on the solutions.

Further work on the problem of persistent excitation is needed as this concept is playing an important role in many investigations currently undertaken on different aspects of adaptive systems.

8.12 CONCLUDING REMARKS

We have considered in this chapter the areas of current concern in adaptive control so as to get a picture of the state of the art of the subject. Recent trends in research are given in different aspects of self-tuning, robustness studies, parameter convergence, nonlinear systems, stochastic systems, distributed parameter systems, multivariable systems, model updating, persistent excitation, and so on. There are also many pointers to areas for further work. This and the succeeding chapter will prove informative to researchers and practicing engineers.

REFERENCES

1. Tsay, Y. T., and Shieh, L. S., State-space approach to self-tuning feedback control with pole-assignment, IEE Proc., 128, Pt. D: 93-101 (1981).
2. Warwick, K., Self-tuning regulators—A state space approach, Int. J. Contr., 33: 839-858 (1981).
3. Hesketh, T., State-space pole-placing STR using input-output values, IEE Proc., 129, Pt. D: 123-128 (1982).
4. Warwick, K., Self-tuning property of state-space self-tuners, IEE Proc., 129, Pt. D: 96-100 (1982).
5. Warwick, K., and Westcott, J. H., Arbitrary linear output feedback self-tuners, Proc. IFAC Symp. on Identification and System Parameter Estimation, Washington, D.C., pp. 1169-1174 (1982).
6. Shieh, L. S., Wang, C. T., and Tsay, Y. T., Multivariable state-feedback self-tuning control, Proc. IFAC Symp. on Identification and System Parameter Estimation, Washington, D.C., pp. 1389-1394 (1982).
7. Toivonen, H. T., Variance constrained self-tuning control, Automatica, 19: 415-418 (1983).
8. Modén, P. E., and Söderström, T., Stationary performance of linear stochastic systems using single-step optimal control, IEEE Trans. Auto. Contr., AC-27: 214-216 (1982).
9. Toivonen, H. T., A self-tuning regulator with on-line cost function adaptation, Int. J. Syst. Sci., 15: 1189-1195 (1984).
10. Dexter, A. L., Self-tuning control algorithm for single-chip microcomputer, IEE Proc., 130, Pt. D: 255-260 (1983).
11. Cragon, H. G., The elements of single-chip microcomputer architecture, IEEE Trans. Computers, 13: 27-41 (1980).
12. Latawiec, K., and Chyra, M., On low-frequency and long-run effects in self-tuning control, Automatica, 19: 419-424 (1983).
13. Allidina, A. Y., and Hughes, F. M., Generalized self-tuning controller with pole-assignment, IEE Proc., 127, Pt. D: 13-18 (1980).
14. Yan, S. Y., Adachi, N., and Tokumaru, H., Self-tuning controller with dead-beat convergence, Int. J. Contr., 39: 159-172 (1984).
15. Kučera, V., Discrete Linear Control: The Polynomial Equation Approach, Wiley, New York, 1979.
16. Yan, S. Y., Adachi, N., and Tokumaru, H., On finite time setting adaptive observers, SICE (in Japanese), 19: 124 (1983).

17. Minamide, N., Design of a dead-beat tracking system, Int. J. Contr., 39: 63-82 (1984).
18. Peterka, V., Predictor based self-tuning control, Automatica, 20: 39-58 (1984).
19. Åström, K. J., Self-tuning regulators: Design principles and applications, in Applications of Adaptive Control (K. S. Narendra and R. V. Monopoli, eds.), Academic Press, New York, 1980.
20. Åström, K. J., Design principles for self-tuning regulators, in Methods and Applications in Adaptive Control (H. Unbehauen, ed.), Springer-Verlag, Berlin, 1980.
21. Karny, M., Identification with interruptions as an anti-bursting device, Kybernetika, 18: 320 (1982).
22. Xianya, X., and Evans, R. J., Discrete-time adaptive control for deterministic time-varying systems, Automatica, 20: 309-319 (1984).
23. Kanniah, J., and Malik, O. P., Self-tuning regulators with periodically constant control, Int. J. Contr., 40: 585-602 (1984).
24. Kanniah, J., Malik, O. P., and Hope, G. S., Excitation control of synchronous generators using adaptive regulators, Parts I and II, IEEE Trans. Power App. and Syst., PAS-103: 897-910 (1984).
25. Kanniah, J., Malik, O. P., and Hope, G. S., Self-tuning regulator based on dual-rate sampling, IEEE Trans. Auto. Contr., AC-29: 755-759 (1984).
25a. Kanniah, J., Malik, O. P., and Hope, G. S., Microprocessor-based universal regulator using dual-rate sampling, IEEE Trans. Ind. Electronics, IE-31: 306-312 (1984).
26. Mäkilä, P. M., A self-tuning regulator based on optimal output feedback theory, Automatica, 20: 671-679 (1984).
27. Landau, I. D., M'Saad, M., and Ortega, R., Adaptive controllers for discrete-time systems with arbitrary zeros: A survey, IFAC Workshop on Adaptive Systems in Control and Signal Processing, San Francisco (1983).
28. Mäkilä, P. M., Linear quadratic design of structure-constrained controllers, Proc. 2nd American Control Conf., San Francisco (1983).
29. Anderson, B. D. O., and Moore, J. B., Linear Optimal Control, Prentice-Hall, Englewood Cliffs, N. J., 1971.
30. Menga, G., and Mosca, E., MUSMAR: Multivariable adaptive regulators based on multistep cost functionals, in Advances in Control (D. G. Lainiotis and N. S. Tzannes, eds.), D. Reidel, Dordrecht, pp. 334-341, 1980.
31. Mosca, E., and Zappa, G., MUSMAR: Basic convergence and consistency properties, in Lecture Notes in Control, 28: 189-199, Springer-Verlag, Berlin, 1980.
32. Mosca, E., and Zappa, G., A robust adaptive control algorithm for microcomputer based implementation, Control and Computers, 10: 55-56 (1982).

33. Mosca, E., and Zappa, G., Over-parameterization, positive realness, and multistep minimum variance adaptive regulators, in Nonlinear Stochastic Problems (R. S. Bucy and J. M. F. Moura, eds.), D. Reidel, Dordrecht, 1983.
34. Mosca, E., and Zappa, G., An MV adaptive controller for plants with time-varying i/o transport delay, IFAC Workshop on Adaptive Systems in Control and Signal Processing, San Francisco (1983).
35. Mosca, E., and Zappa, G., Removal of a positive realness condition in minimum-variance adaptive regulators by multistep horizon, IEEE Trans. Auto. Contr., AC-29: 844-846 (1984).
36. Mosca, E., Multivariable adaptive regulators based on multistep cost functionals, in Nonlinear Stochastic Problems (R. S. Bucy and J. M. F. Moura, eds.), D. Reidel, Dordrecht, 1983.
37. Greco, C., Menga, G., Mosca, E., and Zappa, G., Performance improvements of self-tuning controllers by multi-step horizons, the MUSMAR approach, Automatica, 20: 681-699 (1984).
38. McDermott, P. E., and Mellichamp, D. A., An auto-pole-placement self-tuning controller, Int. J. Contr., 40: 1131-1147 (1984).
39. Costin, M. H., and Buchner, M. R., Distributed control using self-tuning regulators, Proc. IFAC Workshop on Adaptive Systems in Control and Signal Processing, San Francisco, pp. 301-307 (1983).
39a. Tuffs, P. S., and Clarke, D. W., Self-tuning control of offset: A unified approach, IEE Proc., 132, Pt. D: 100-110 (1985).
39b. Clarke, D. W., Model-following and pole-placement self-tuners, Opt. Contr. Appl. and Meth., 3: (1982).
39c. Wellstead, P. E., and Sanoff, S. P., Extended self-tuning algorithm, Int. J. Contr., 34: 433-445 (1981).
40. Athans, M., and Valavani, L. S., Some critical questions about deterministic and stochastic adaptive control algorithms, Proc. IFAC Symp. on Identification and System Parameter Estimation, Washington, D.C., pp. 241-246 (1982).
41. Ortega, R., Variable structure systems—An alternative to robust adaptive control?, Coll. Nat. Belle Ile (France) (1982).
42. Anderson, B. D. O., and Johnstone, R. M., Adaptive systems and time-varying plants, Int. J. Contr., 37: 367-377 (1983).
43. Ortega, R., Assessment of stability robustness for adaptive controllers, IEEE Trans. Auto. Contr., AC-28: 1106-1109 (1983).
44. Åström, K. J., Robustness of a design method based on assignment of poles and zeros, IEEE Trans. Auto. Contr., AC-25: 588-591 (1980).
45. Ortega, R., Robustness enhancement of adaptive controllers by incorporation of process a priori knowledge, Syst. Contr. Lett., 4: 135-141 (1984).
46. Ortega, R., Correction to "Robustness enhancement of adaptive controllers by incorporating of process a priori knowledge," Syst. Contr. Lett., 5: 73 (1984).

47. Safonov, M. G., Stability Robustness of Multivariable Feedback Systems, MIT Press, Cambridge, MA (1980).
48. Ortega, R., and Landau, I. D., On the design of robustly performing adaptive controllers for partially modelled systems, Proc. 22nd IEEE Conf. Decision and Control, San Antonio, TX (1983).
49. Ortega, R., and Landau, I. D., On the model mismatch tolerance of various parameter adaptation algorithms in direct control schemes: A sectoricity approach, Proc. IFAC Workshop on Adaptive Control, San Francisco (1983).
50. Kokotovic, P. V., and Ioannou, P. A., Robustness re-design of continuous-time adaptive schemes, Proc. IEEE Conf. Decision and Control, San Diego, pp. 522-527 (1981).
51. Ioannou, P. A., Robustness of model reference adaptive systems with respect to modelling errors, Doctoral Thesis, University of Illinois, Urbana-Champaign (1982).
52. Ioannou, P. A., and Kokotovic, P. V., Singular perturbations and robust redesign of adaptive control, Proc. 21st IEEE Conf. Decision and Control, Orlando, Florida (1982).
53. Ioannou, P. A., and Kokotovic, P. V., Robust redesign of adaptive control, IEEE Trans. Auto. Contr., AC-29: 202-211 (1984).
54. Ioannou, P. A., and Kokotovic, P. V., Instability analysis and improvement of robustness of adaptive control, Automatica, 20: 583-594 (1984).
55. Anderson, B. D. O., and Johnstone, R. N., Adaptive systems and time-varying plants, Report, Department of Electrical and Computer Engineering, University of Newcastle, Newcastle, N.S.W., Australia (1982).
56. Rohrs, C. E., Valavani, L. S., Athans, M., and Stein, G., Analytical verification of undesirable properties of direct MRAC algorithms, Proc. 20th IEEE Conf. Decision and Control, pp. 1272-1284 (1981).
57. Rohrs, C. E., Adaptive control in the presence of unmodelled dynamics, Doctoral Thesis, MIT, Cambridge, Mass. (1982).
58. Rohrs, C. E., A frequency selective adaptive controller, Proc. 22nd IEEE Conf. Decision and Control, San Antonio, TX, pp. 991-996 (1983).
59. Cook, P. A., and Chen, Z. J., Robustness properties of model reference adaptive control systems, IEE Proc., 129, Pt. D: 305-309 (1982).
60. Chen, Z. J., Regarding robustness studies of adaptive control systems, Doctoral Thesis, UMIST, Manchester, U.K. (1983).
61. Rohrs, C. E., Valavani, L. S., Athans, M., and Stein, G., Robustness of adaptive control algorithms in the presence of unmodelled dynamics, Proc. 21st IEEE Conf. Decision and Control, pp. 3-11 (1982).

62. Chen, Z. J., and Cook, P. A., Robustness of MRAC systems with unmodelled dynamics, Int. J. Contr., 39: 201-214 (1984).
63. Rohrs, C. E., Athans, M., Valavani, L. S., and Stein, G., Some design guidelines for discrete-time adaptive controllers, Automatica, 20: 653-660 (1984).
64. Kosut, R. L., and Johnson, C. R., Jr., An input-output view of robustness in adaptive control, Automatica, 20: 569-581 (1984).
65. Silveria, H. M., and Doraiswami, R., New structure for adaptive servomechanism controller, IEE Proc., 131, Pt. D: 64-68 (1984).
66. Clarke, D. W., Self-tuning control of non-minimum phase systems, Automatica, 20: 501-517 (1984).
67. Lim, K. W., Robustness of self-tuning controllers, Report No. OUEL 1422/82, Oxford University (1982).
68. Gawthrop, P. J., and Lim, K. W., Robustness of self-tuning controllers, IEE Proc., 129, Pt. D: 21-29 (1982).
69. Jacobs, O. L. R., and Saratchandran, P., Comparison of adaptive controllers, Automatica, 16: 89-97 (1980).
70. Wittenmark, B., Stochastic adaptive control methods—A survey, Int. J. Contr., 21: 705-730 (1975).
71. Åström, K. J., and Wittenmark, B., On self-tuning regulators, Automatica, 9: 185-199 (1973).
72. de la Sen, M., On-line optimization of the free parameters in discrete adaptive control systems, IEE Proc., 131, Pt. D: 146-157 (1984).
73. de la Sen, M., A method for improving the adaptation transient using adaptive sampling, Int. J. Contr., 40: 639-665 (1984).
73a. de la Sen, M., On the use of optimization and adaptive sampling tools for improving the transients in the adaptive control problem, Proc. Int. Conf. on Computers, Systems and Signal Processing, Bangalore, India, pp. 324-327 (1984).
73b. de la Sen, M., A simple method to improve the transients in adaptive systems, IEEE Proc., 72: 131-134 (1984).
74. Schumann, R., Lachmann, K.-H., and Isermann, R., Towards applicability of parameter-adaptive control algorithms, Proc. 8th IFAC World Congress, Kyoto, pp. 903-910 (1981).
75. Narendra, K. S., and Valavani, L. S., Stable adaptive controller design—Direct control, IEEE Trans. Auto. Contr., AC-23: 570-583 (1978).
76. Narendra, K. S., Lin, Y.-H., and Valavani, L. S., Stable adaptive controller design—Part II: Proof of stability, IEEE Trans. Auto. Contr., AC-25: 440-448 (1980).
77. Morse, A. S., Global stability of parameter adaptive control systems, IEEE Trans. Auto. Contr., AC-25: 433-439 (1980).

78. Morgan, A. P., and Narendra, K. S., On the uniform asymptotic stability of certain linear non-autonomous differential equations, SIAM J. Contr. Optim., 15: 5-24 (1977).
79. Anderson, B. D. O., Exponential stability of linear equations arising in adaptive identification, IEEE Trans. Auto. Contr., AC-22: 83-88 (1977).
80. Kreisselmeier, G., Adaptive observers with exponential rate of convergence, IEEE Trans. Auto. Contr., AC-22: 2-8 (1977).
81. Yuan, J.S.-C., and Wonham, W. M., Probing signals for model-reference identification, IEEE Trans. Auto. Contr., AC-22: 530-538 (1977).
82. Anderson, B. D. O., and Johnson, C. R., Jr., Exponential convergence of adaptive identification and control algorithms, Automatica, 18: 1-13 (1982).
83. Goodwin, G. C., Ramadge, P. J., and Caines, P. E., Discrete time multivariable adaptive control, IEEE Trans. Auto. Contr., AC-25: 449-456 (1980).
84. Anderson, B. D. O., Exponential stability of linear equations arising in adaptive identification, IEEE Trans. Auto. Contr., AC-22: 83-88 (1977).
85. Boyd, S., and Sastry, S., Parameter convergence in adaptive control, Syst. Contr. Lett., 3: 311-319 (1983).
85a. Narendra, K. S., and Lin, Y.-H., Stable discrete adaptive control, IEEE Trans. Auto. Contr., AC-25: 456-461 (1980).
86. Becker, A. H., Kumar, P. R., and Wei, C. Z., Adaptive control with the stochastic approximation algorithm: Geometry and convergence, IEEE Trans. Auto. Contr., AC-30: 330-338 (1985).
87. Goodwin, G. C., Hill, D. L., and Palaniswami, M., A perspective on convergence of adaptive control algorithms, Automatica, 20: 519-531 (1984).
88. Anderson, B. D. O., Dasgupta, S., and Tsoi, A. C., On the convergence of a model reference adaptive algorithm with unknown high frequency gain, Syst. Contr. Lett., 5: 303-307 (1985).
89. Elliott, H., Cristi, R., and Das, M., Global stability of adaptive pole-placement algorithms, IEEE Trans. Auto. Contr., AC-30: 348-356 (1985).
90. Goodwin, G. C., and Teoh, E. K., Persistency of excitation in the presence of possibly unbounded signals, IEEE Trans. Auto. Contr., AC-30: 395-397 (1985).
91. Trulsson, E., Adaptive control based on explicit criterion minimization, Doctoral Thesis, Linkoping University, Sweden (1983).
92. Chen, H. F., Recursive system identification and adaptive control by use of the modified least squares algorithm, Research Report, Dept. of Elec. Eng., McGill University, Montreal, Canada (1983).

93. Chen, H. F., and Caines, P. E., The strong consistency of the stochastic gradient algorithm of adaptive control, IEEE Trans. Auto. Contr., AC-30: 185-189 (1985).
94. Chen, H. F., and Caines, P. E., On the adaptive control of a class of systems with random parameters and disturbances, Proc. IFAC Workshop Adaptive Systems in Control and Signal Processing, San Francisco (1983).
94a. Chen, H. F., and Caines, P. E., Optimal stochastic control for discrete time systems with random parameters, Research Report, Dept. of Elec. Eng., McGill University, Montreal, Canada (1983).
94b. Chen, H. F., and Caines, P. E., Optimal adaptive LQG control for systems with finite state process parameters, IEEE Trans. Auto. Contr., AC-30: 185-189 (1985).
95. Goodwin, G. C., Long, R. S., and McInnis, B., Adaptive control of bilinear systems, Report No. 8017, Dept. of Electrical and Computer Eng., University of Newcastle, Newcastle, N.S.W., Australia (1980).
96. Goodwin, G. C., McInnis, B., and Long, R. S., Adaptive control algorithms for waste water treatment and pH neutralization, Opt. Contr. Appl. Meth., 3: 443-459 (1982).
97. Evans, R. J., Goodwin, G. C., and Betz, R. E., Discrete-time adaptive control for classes of nonlinear systems, in Information and Optimization of Systems (A. Bensonssan and J. L. Lions, eds.), Springer-Verlag, Berlin, 1981.
98. Anbumani, K., Patnaik, L. M., and Sarma, I. G., Self-tuning minimum variance control of nonlinear systems of the Hammerstein model, IEEE Trans. Auto. Contr., AC-26: 959-961 (1981).
99. Anbumani, K., Sarma, I. G., and Patnaik, L. M., Self-tuning cascade control of linear and nonlinear systems, Proc. IFAC Symp. Digital Control, New Delhi, pp. 53-56 (1982).
100. Korpela, K., and Koivo, H. N., On self-tuning control of a class of nonlinear scalar processes, Proc. IFAC Symp. on Identification and System Parameter Estimation, Washington, D.C., pp. 265-270 (1982).
101. Grimble, M. J., Weighted minimum variance control of nonlinear systems, Research Report No. ICU/15, Ind. Contr. Unit, Dept. of Electronic and Electrical Eng., University of Strathclyde, Glasgow, U.K. (1982).
102. Keviczky, L., and Haber, R., Adaptive dual extreme control by Hammerstein models, Proc. IFAC Symp. Stochastic Control, Budapest, pp. 333-341 (1974).
103. Haber, R., and Keviczky, L., Adaptive dual extremem control by finite order Volterra model, Prob. Contr. Inform. Theor., 3: 247-260 (1974).

104. Sternby, J., Analysis of an extremal controller for Hammerstein models, Report No. 7142, Lund Inst. of Tech., Lund, Sweden (1978).
105. Sternby, J., Extremum control systems—An area for adaptive control, Proc. JACC (1980).
106. Kung, M.-C., and Womack, B. F., Stability analysis of a discrete-time adaptive control algorithm having a polynomial input, IEEE Trans. Auto. Contr., AC-28: 1110-1112 (1983).
107. Kung, M.-C., and Womack, B. F., Discrete-time adaptive control of linear systems with preload nonlinearity, Automatica, 20: 477-479 (1984).
108. Urwin, P., and Swanick, B. H., Adaptive control of systems with certain nonlinear structures, Int. J. Contr., 39: 31-55 (1984).
109. Balestrino, A., DeMaria, G., and Zinober, A. S. I., Nonlinear adaptive model following control, Automatica, 20: 559-568 (1984).
110. Ambrosino, G., Calentano, G., and Garofalo, F., Adaptive model following control of plants with nonlinearities of known form, IEE Proc., 132, Pt. D: 11-13 (1985).
111. Young, K.-K. D., Design of variable structure model following control systems, IEEE Trans. Auto. Contr., AC-23: 1079-1085 (1978).
112. Leitmann, G., and Gutman, S., Stabilizing feedback control for dynamic systems with bounded uncertainty, Proc. IEEE Conf. Decision and Control, pp. 94-99 (1976).
113. Gutman, S., Uncertain dynamic systems—A Lyapunov min-max approach, IEEE Trans. Auto. Contr., AC-24: 437-443 (1979).
114. Balestrino, A., DeMaria, G., and Sciavicco, L., Hyperstable adaptive model following control of nonlinear plants, Syst. Contr. Lett., 1: 232-236 (1982).
114a. Ambrosino, G., Celentano, G., and Garofalo, F., Robust model tracking control for a class of nonlinear plants, IEEE Trans. Auto. Contr., AC-30: 275-279 (1985).
114b. Raju, G. V. S., and Riad, K. A., Signal synthesis MRAC systems, Proc. IEEE Int. Conf. on Systems, Man and Cybernetics, New Delhi, India, pp. 934-939 (1983).
115. Monopoli, R. V., and Kessler, F. M., Measurement noise in model reference adaptive systems, Proc. JACC, pp. 1599-1603 (1977).
116. Gutmann, R. L., Monopoli, R. V., and Van Allen, R., Adaptive control of a laser pointing and tracking system: A feasibility study, in Applications of Adaptive Control (K. S. Narendra and R. V. Monopoli, eds.), Academic Press, New York, 1980.
117. Egardt, B., Stability of Adaptive Controllers, Springer-Verlag, Berlin, 1979.

118. Egardt, B., Stability analysis of adaptive control systems with disturbances, Proc. JACC, San Francisco (1980).
119. Peterson, B. B., and Narendra, K. S., Bounded error adaptive control, IEEE Trans. Auto. Contr., AC-27: 1161-1169 (1982).
120. Kreisselmeier, G., and Narendra, K.S., Stable model reference adaptive control in the presence of bounded disturbances, IEEE Trans. Auto. Contr., AC-27: 1169-1176 (1982).
121. Goodwin, G. C., Hill, D. J., and Xianya, X., Stochastic adaptive control for exponentially convergent time-varying systems, Tech. Report No. 8407, Dept. of Electrical and Computer Eng., University of Newcastle, Newcastle, N.S.W., Australia (1984).
122. Goodwin, G. C., and Chan, S. W., Model reference adaptive control of systems having purely deterministic disturbances, IEEE Trans. Auto. Contr., AC-28: 855-858 (1983).
123. Samson, C., Stability analysis of adaptively controlled systems subject to bounded disturbances, Automatica, 19: 81-86 (1983).
124. Samson, C., An adaptive LQ controller for non-minimum phase systems, Int. J. Contr., 35: 1-28 (1982).
125. Goodwin, G. C., Ramadge, P. J., and Caines, P. E., Discrete-time stochastic adaptive control, SIAM J. Contr. Optim., 19: 829-853 (1981).
126. Sin, K. S., and Goodwin, G. C., Stochastic adaptive control using a modified least square algorithm, Automatica, 18: 315-321 (1982).
127. Gawthrop, P. J., On the stability and convergence of self-tuning controller, Int. J. Contr., 31: 973-998 (1980).
128. Caines, P. E., Stochastic adaptive control: Randomly varying parameters and continuously disturbed control, Proc. IFAC World Congress, Kyoto (1981).
129. Fuchs, J. J.-J., Indirect stochastic adaptive control: The general delay—White noise case, IEEE Trans. Auto. Contr., AC-27: 219-223 (1982).
130. Fuchs, J. J.-J., Indirect stochastic adaptive control: The general delay—Colored noise case, IEEE Trans. Auto. Contr., AC-27: 470-472 (1982).
131. Goodwin, G. C., Sin, K. S., and Saluja, K. K., Stochastic adaptive control and prediction—The general delay—Colored noise case, IEEE Trans. Auto. Contr., AC-25: 946-950 (1980).
132. Sin, K. K., Goodwin, G. C., and Bitmead, R. R., An adaptive d-step ahead predictor based on least squares, IEEE Trans. Auto. Contr., AC-25: 1161-1165 (1980).
133. Salyga, Val. I., Baranov, V. V., and Kuzima, O. I., Methods of Markovian decision processes in the problem of identification and adaptive control of stochastic systems, Proc. 6th IFAC Symp. Identification and System Parameter Estimation, Washington, D.C., pp. 1365-1370 (1982).

134. Caines, P. E., and Lafortune, S., Adaptive control with recursive identification for stochastic linear systems, IEEE Trans. Auto. Contr., AC-29: 312-321 (1984).
135. Solo, V., The convergence of AML, IEEE Trans. Auto. Contr., AC-24: 958-962 (1979).
136. Chen, H. F., Recursive system identification and adaptive control by use of the modified least squares algorithms, SIAM J. Contr. Optim., 22: 758-770 (1984).
137. Xianya, X., and Evans, R. J., Adaptive control of discrete-time time-varying systems with unknown deterministic disturbances, IEE Proc., 131, Pt. D: 81-84 (1984).
138. Xianya, X., and Evans, R. J., Discrete-time stochastic adaptive control for time-varying systems, IEEE Trans. Auto. Contr., AC-29: 638-640 (1984).
139. Dugard, L., Goodwin, G. C., and Xianya, X., The role of the interactor matrix in multivariable stochastic adaptive control, Automatica, 20: 701-709 (1984).
140. Ydstie, B. E., Robust adaptive control of chemical processes, Doctoral Thesis, Imperial College of Science and Technology, London (1982).
141. Johannson, R., Parametric models for linear multivariable systems for adaptive control, Proc. IEEE Conf. Decision and Control, Orlando, Fla. (1982).
142. Dugard, L., Goodwin, G. C., and de Souza, C. E., Prior knowledge in MRAC of MIMO systems, Proc. IEEE Conf. Decision and Control, San Antonio, Tex. (1983).
143. Singh, R. P., and Narendra, K. S., Prior information in the design of multivariable adaptive controllers, IEEE Trans. Auto. Contr., AC-29: 1108-1111 (1984).
144. Lainiotis, D. G., Partitioning: A unifying framework for adaptive systems, Pt. I: Estimation, IEEE Proc., 64: 1126-1143 (1976); Pt. II: Control, IEEE Proc., 64: 1182-1198 (1976).
145. Mini-issue on NASA's Advanced Control Law Program, IEEE Trans. Auto. Contr., AC-22: 752-806 (1977).
146. Watanabe, K., A passive type multiple-model adaptive control (MMAC) of linear discrete-time stochastic systems with uncertain observation subsystems, Int. J. Syst. Sci., 15: 647-659 (1984).
147. Lainiotis, D. G., Upadhyay, T. N., and Deshpande, J. G., Optimal adaptive control of linear systems with unknown measurement subsystems, Inf. Sci., 6: 3 (1973).
148. Kidd, P. T., and Munro, N., Design of model reference systems using frequency-domain method, IEE Proc., 130, Pt. D: 273-280 (1983).
149. Rosenbrock, H. H., Computer-Aided Control System Design, Academic Press, New York, 1974.

150. Elliott, H., and Wolovich, W. A., A parameter adaptive control structure for linear multivariable systems, IEEE Trans. Auto. Contr., AC-27: 340-352 (1982).
151. Elliott, H., Wolovich, W. A., and Das, M., Arbitrary adaptive pole-placement for linear multivariable systems, IEEE Trans. Auto. Contr., AC-29: 221-229 (1984).
152. Elliott, H., and Wolovich, W. A., Parameterization issues in multivariable adaptive control, Automatica, 20: 533-545 (1984).
153. Djaferis, T., Das, M., and Elliott, H., Reduced order adaptive pole-placement for multivariable systems, IEEE Trans. Auto. Contr., AC-29: 637-638 (1984).
154. Goodwin, G. C., and Long, R. S., Generalization of results on multivariable adaptive control, IEEE Trans. Auto. Contr., AC-25: 1241-1245 (1980).
155. Goodwin, G. C., and Sin, K. S., Adaptive Filtering, Prediction and Control, Prentice-Hall, Englewood Cliffs, N.J., 1984.
156. Monopoli, R. V., and Hsing, C. C., Parameter adaptive control of multivariable systems, Int. J. Contr., 22: 313-327 (1975).
157. Morse, A. S., Parameterizations for multivariable adaptive control Proc. IEEE Conf. Decision and Control, San Diego, Calif. (1981).
158. Hulkó, G., Rohal-Ilkiv, B., and Sapák, L., On adaptive control of distributed parameter systems, Proc. 8th IFAC World Congress, Kyoto, Japan, pp. 887-896 (1981).
159. Butkovskij, A. G., Characteristics of Distributed Parameter Systems (in Russian), Nauka, Moscow (1979).
160. Van den Bosch, P. P. J., and Jongkind, W., Model reference adaptive satellite attitude control, in Methods and Applications in Adaptive Control (H. Unbehauen, ed.), Springer-Verlag, Berlin, pp. 209-218, 1980.
161. Van den Bosch, P. P. J., and Tjahjadi, P. I., Model updating improves MRAC performance, IEEE Trans. Auto. Contr., AC-29: 1106-1108 (1984).
162. Tsypkin, Ya. Z., Synthesis of the optimal adjustable model in identification problems, Autom. Rem. Contr., 42: 1630-1642 (1981).
163. Mareels, I., Sufficiency of excitation, Syst. Contr. Lett., 5: 159-163 (1984).
164. Morgan, A. P., and Narendra, K. S., On the stability of non-autonomous differential equations $\dot{x} = [A + B(t)]x$, with skew-symmetric matrix, B(t), SIAM J. Contr. Optim., 15: 163-176 (1977).
165. Sondhi, M. M., and Mitra, D., New results on the performance of a well-known class of adaptive filters, IEEE Proc., 64: 1583-1597 (1976).
166. Bitmead, R. R., and Anderson, B. D. O., Lyapunov techniques for the exponential stability of linear difference equations with random coefficients, IEEE Trans. Auto. Contr., AC-25: 782-787 (1980).

167. Narendra, K. S., and Annaswamy, A. M., Recent trends in adaptive control theory, SICE (Japan), 23: 441-448 (1985).
168. Anderson, B. D. O., and Johnstone, R. M., Robust Lyapunov results and adaptive systems, Proc. IEEE Conf. Decision and Control, San Diego, Calif. (1981).
169. Narendra, K. S., and Annaswamy, A. M., Persistent excitation and adaptive systems, Tech. Report No. 8302, Center for Systems Science, Yale University, New Haven, Conn. (1983).
170. Narendra, K. S., and Annaswamy, A. M., Robust adaptive control, Tech. Report, Center for Systems Science, Yale University, New Haven, Conn. (1985).
171. Anderson, B. D. O., Adaptive systems, lack of persistency of excitation, and bursting phenomena, Automatica, 21: 247-258 (1985).

BIBLIOGRAPHY

Anderson, B. D. O., and Johnstone, R. M., When will adaptive systems really adapt? The robustness issue, Proc. 2nd Conf. on Control Eng., I. E. (Australia), pp. 59-66 (1982).

Anderson, B. D. O., and Johnstone, R. M., Global adaptive pole-positioning, IEEE Trans. Auto. Contr., AC-30: 11-22 (1985).

Borkar, V., and Varaiya, P., Adaptive control of Markov chains, Part I: Finite parameter set, IEEE Trans. Auto. Contr., AC-24: 953-957 (1979).

Chan, S. W., and Goodwin, G. C., On the role of interactor matrix in MIMO adaptive control, IEEE Trans. Auto. Contr., AC-27: 713-714 (1982).

Chen, Z. J., Regarding robustness studies of adaptive control systems, Doctoral Thesis, UMIST, Manchester, U.K. (1983).

Corless, M., and Leitmann, G., Adaptive control of systems containing uncertain functions with uncertain bounds, J. Opt. Theor. Appl., 41: 155-168 (1983).

Corless, M., and Leitmann, G., Adaptive control for uncertain systems, in Mathematical Theory of Dynamical Systems and Microphysics (A. Blaquiere and G. Leitmann, eds.), Academic Press, New York, 1984.

Corless, M., and Leitmann, G., Adaptive control for uncertain dynamic systems, Proc. Int. Conf. on Computers, Systems and Signal Processing, Bangalore, India, pp. 313-316 (1984).

Dasgupta, S., Anderson, B. D. O., and Tsoi, A. C., Input conditions for continuous-time adaptive system problems, Proc. IEEE Conf. Decision and Control, San Antonio, Tex. (1983).

de Larminat, P., On the stabilizability condition in indirect adaptive control, Proc. IFAC Symp. Identification and System Parameter Estimation, Washington, D.C., pp. 469-476 (1982).

Dugard, L., and Goodwin, G. C., Global convergence of Landau's "output error with adjustable compensator" adaptive algorithm, IEEE Trans. Auto. Contr., AC-30: 593-595 (1985).

Elliott, H., Das, M., and Ruiz, G., A frequency domain analysis of direct adaptive pole placement algorithms in the presence of unmodeled dynamics, Proc. IFAC Workshop Adaptive Systems in Control and Signal Processing, San Francisco, pp. 43-48 (1983).

Eweda, E., and Macchi, O., Convergence of an adaptive linear estimation algorithm, IEEE Trans. Auto. Contr., AC-29: 119-127 (1984).

Fuchs, J. J.-J., Adaptive control and optimal input design: A realistic approach, Opt. Contr. Appl. Meth., 3 (1982).

Fujii, S., On the trend of the researches in adaptive control, Syst. Contr. Japan, 25: 715-726 (1981).

Goodwin, G. C., and Dugard, L., Stochastic adaptive control with known and unknown interactor matrices, Proc. Workshop Adaptive Systems in Control and Signal Processing, San Francisco, pp. 351-355 (1983).

Goodwin, G. C., Norton, J. P., and Viswanathan, M. N., Persistency of excitation for non-minimal models of systems having purely deterministic disturbances, Tech. Report, Dept. of Elec. and Comp. Eng., University cf Newcastle, N.S.W., Australia (1984).

Goodwin, G. C., Ramadge, P. J., and Caines, P. E., A globally convergent adaptive predictor, Automatica, 17: 135-140 (1981).

Heymann, M., Lewis, J. H., and Meyer, G., Remarks on the adaptive control of linear plants with unknown high frequency gain, Syst. Contr. Lett., 5: 357-362 (1985).

Hodgson, A. J. F., Problems of integrity in applications of adaptive controllers, Doctoral Thesis, Dept. of Engineering Science, University of Oxford (1982).

Ioannou, P. A., and Kokotovic, P. V., Adaptive Systems with Reduced Models, Springer-Verlag, Berlin, 1983.

Ioannou, P. A., Robustness re-design of adaptive control in the presence of disturbances and unmodelled dynamics, Proc. IFAC Workshop Adaptive Systems in Control and Signal Processing, San Francisco, pp. 19-25 (1983).

Irving, E., Dang Van Mien, H., and Redjah, M., Some practical solutions for the robustness problem of multivariable adaptive control, Proc. IFAC Workshop Adaptive Systems in Control and Signal Processing, San Francisco, pp. 269-270 (1983).

Johannson, R., Multivariable adaptive control, Doctoral Thesis, Dept. of Auto. Contr., Lund Inst. of Tech., Lund, Sweden (1983).

Johnson, C. R., Jr., Lawrence, D. A., and Lyons, J. P., A flaw in the reduced order behaviour of a direct adaptive pole placer, Proc. IEEE Conf. Decision and Control, Orlando, Fla., p. 271 (1982).

Johnstone, R. M., and Anderson, B. D. O., Global adaptive pole placement: Detailed analysis of a first order system, IEEE Trans. Auto. Contr., AC-28: 852-855 (1983).

Khalil, H. K., On the robustness of output feedback control methods to modelling errors, IEEE Trans. Auto. Contr., AC-26: 524-526 (1981).

Kokotovic, P. V., Recent trends in feedback design: An overview, Automatica, 21: 225-236 (1985).

Kosut, R. L., Robustness issues in adaptive control, Proc. IFAC Workshop Adaptive Systems in Control and Signal Processing, San Francisco, pp. 13-18 (1983).

Kosut, R. L., and Friedlander, B., Performance robustness properties of adaptive control systems, Proc. IEEE Conf. Decision and Control, Orlando, Fla., pp. 18-23 (1982).

Kosut, R. L., Johnson, C. R., Jr., and Anderson, B. D. O., Conditions for local stability and robustness of adaptive control systems, Proc. 22nd IEEE Conf. Decision and Control, pp. 972-975 (1983).

Kumar, P. R., Simultaneous identification and adaptive control of unknown systems over finite parameter sets, IEEE Trans. Auto. Contr., AC-28: 68-76 (1983).

Kushner, H. J., and Kumar, R., Convergence and rate of convergence of a recursive identification and adaptive control method which uses truncated estimators, IEEE Trans. Auto. Contr., AC-27: 775-782 (1982).

Kuzucu, A., and Roch, A., Suboptimal adaptive feedback control of nonlinear systems, in Methods and Applications in Adaptive Control (H. Unbeheuen, ed.), Springer-Verlag, Berlin, pp. 131-140, 1980.

Lachmann, K.-H., Parameter-adaptive control of a class of nonlinear processes, Proc. IFAC Symp. Identification and System Parameter Estimation, Washington, D.C., pp. 449-455 (1982).

Lyublinskii, B. S., and Fradkov, A. L., Adaptive control of nonlinear static objects with implicitly specified response, Auto. Rem. Contr., 44: 510-518 (1983).

Mannerfelt, C. F., Robust control design with simplified models, Report No. TFRT-1021, Dept. of Auto. Contr., Lund Inst. of Tech., Lund, Sweden (1981).

Minamide, N., Nikiforuk, P. N., and Gupta, M. M., Design of an adaptive observer and its application to an adaptive pole-placement controller, Int. J. Contr., 37: 349-366 (1983).

Minamide, N., Nikiforuk, P. N., and Gupta, M. M., Design of adaptive pole-placement controllers for plants of unknown order, Syst. Contr. Lett., 4: 347-358 (1984).

Moore, J. B., Persistence of excitation in extended least squares, IEEE Trans. Auto. Contr., AC-28: 60-68 (1983).

Morse, A. S., An adaptive control for globally stabilizing systems with unknown high frequency gains, Report No. 8402, Dept. of Electrical Eng., Yale University, New Haven, Conn. (1984).

Mudgett, D. R., and Morse, A. S., Adaptive stabilization of linear systems with unknown high frequency gains, IEEE Trans. Auto. Contr., AC-30: 549-554 (1985).

Nussbaum, R. D., Some remarks on a conjecture in parameter adaptive control, Syst. Contr. Lett., 3: 243-246 (1983).

Ortega, R., and Landau, I. D., On the model-process mismatch tolerance of various parameter adaptation algorithms in direct control schemes: A sectoricity approach, Proc. IFAC Workshop Adaptive Systems in Control and Signal Processing, San Francisco, pp. 35-42 (1983).

Owens, D. H., and Chotai, A., High performance controllers for unknown multivariable systems, Research Report No. 130, Dept. of Control Eng., University of Sheffield, U.K. (1980).

Owens, D. H., and Chotai, A., Robust control of unknown or large-scale systems using transient data only, Research Report No. 134, Dept. of Control Eng., University of Sheffield, U.K. (1980).

Owens, D. H., and Chotai, A., Robust control of unknown discrete multivariable systems, Research Report No. 135, Dept. of Control Eng., University of Sheffield, U.K. (1980).

Praly, L., Robustness of indirect adaptive control based on pole-placement design, Proc. IFAC Workshop Adaptive Systems in Control and Signal Processing, San Francisco, pp. 55-60 (1983).

Praly, L., Towards a globally stable direct adaptive control scheme for not necessarily minimum phase systems, IEEE Trans. Auto. Contr., AC-29: 946-949 (1984).

Pugarchev, V. S., Adaptive and stochastic control: Conditionally optimal estimation in systems with randomly varying structure, Proc. IFAC World Congress, Budapest, Hungary (1984).

Riedle, B., and Kokotovic, P. V., Disturbance cancellation and drift of adaptive gains, Proc. IFAC Workshop Adaptive Systems in Control and Signal Processing, San Francisco, pp. 33-34 (1983).

Riedle, B., and Kokotovic, P. V., A stability-instability boundary for disturbance-free slow adaptation and unmodelled dynamics, Proc. 23rd IEEE Conf. Decision and Control, Las Vegas, Nevada, pp. 998-1001 (1984).

Sagalovsky, B., Adaptive control and parameter estimation in Markov chains: A linear case, IEEE Trans. Auto. Contr., AC-27: 414-419 (1982).

Sastry, S. S., Model reference adaptive control—Stability, convergence, and robustness, Report No. UCB/ERL/M83/59, University of California, Berkeley (1983).

Sobel, K., Kaufman, H., and Yekutiel, O., Design of multivariable adaptive control systems without the need for parameter identification, in Methods and Applications in Adaptive Control (H. Unbehauen, ed.), Springer-Verlag, Berlin, pp. 102-111, 1980.

Söderström, T., On adaptive control of stochastic systems with infrequent output sampling, Report No. UPTEC 8004R, Inst. of Tech., Uppsala University, Sweden (1980).

Watanabe, K., Yoshimura, T., and Soeda, T., Optimal adaptive estimation and stochastic control for distributed-parameter systems, IEEE Trans. Auto. Contr., AC-27: 216-219 (1982).

Yang, D. R., and Lee, W.-K., Effects of model structure, non-zero dc value, and measurable disturbance on adaptive control, Proc. IFAC Workshop Adaptive Systems in Control and Signal Processing, San Francisco, pp. 253-260 (1983).

Additional Readings

Bartolini, G., and Zolezze, T., Variable structure systems nonlinear in control law, IEEE Trans. Auto. Contr., AC-30: 681-684 (1985).

Goodwin, G. C., and Mayne, D. Q., A continuous-time adaptive controller with integral action, IEEE Trans. Auto. Contr., AC-30: 793-795 (1985).

Kosut, R. L., and Friedlander, B., Robust adaptive control: Conditions for global stability, IEEE Trans. Auto. Contr., AC-30: 610-624 (1985).

Lozano-Leal, R., and Goodwin, G. C., A globally stable adaptive pole-placement algorithm without a persistency of excitation requirement, IEEE Trans. Auto. Contr., AC-30: 795-798 (1985).

Mudgett, D. R., and Morse, A. S., Adaptive stabilization of discrete linear system with an unknown high-frequency gain, IEEE Trans. Auto. Contr., AC-30: 798-799 (1985).

9

Recent Trends in Adaptive Control Applications

9.1 INTRODUCTION

There has been a great research effort to develop high-performance control schemes for unknown plants subject to uncertainties and disturbances. Adaptive control strategies have proved to be one of the most promising techniques which can be applied to control a wide variety of processes in industry. Many adaptive control applications were based, until the early 1970s, on analog realization. These were not very successful and so not very convincing. The development of sophisticated and at the same time inexpensive microprocessors led to the reactivation of the adaptive control methodology. The use of digital hardware provides great flexibility in design, improved performance, and reliability compared to classical analog hardware. There are already several adaptive regulators on the market [1]. They have superior performance compared with conventional regulators. Experience in their use is also increasing [2, 3]. (These are discussed in later sections.) Several applications have already shown that adaptive control techniques are very attractive from both the theoretical and practical points of view. However, additional research is required to ensure good control performance under circumstances which may arise in practice.

For adaptive control methodologies to become a successful and effective industrial control technology, many simulation, feasibility, and pilot plant studies, as well as robustness studies, should be carried out to demonstrate the practical operability and reliability of adaptive control algorithms. This will stimulate industrial interest and generate confidence in the adoption of such adaptive control techniques. There have been a significant number of contributions of this type in recent years, and much information is available in the technical literature.

The aim of the present chapter is to discuss certain interesting developments in the past five years to give an overview of the current status of adaptive control applications. The chapter is organized as follows: Section 9.2 deals with a general-purpose adaptive controller which is commercially available. Section 9.3 presents some applications in power systems. Electric generator control is discussed in Section 9.4, while Section 9.5 considers electric drives. Some aspects of control in airborne vehicles, including missiles are described in Section 9.6. Ship control forms the subject matter of Section 9.7. Various process control problems are discussed in Section 9.8. Sections 9.9 through 9.12 consider some control problems in the pulp and paper industry, cement industry, heating and air-conditioning systems, and steel plants, respectively. Section 9.13 discusses the adaptive control of industrial robots. Satellite attitude control is taken up in Section 9.14. Applications in transportation systems are included in Section 9.15. Section 9.16 considers control of power plants. Section 9.17 deals with some valuable medical applications. Biotechnical systems are taken up in Section 9.18. A variety of other applications not covered above are given in Section 9.19. Concluding remarks are included in Section 9.20. The references and bibliography provide sources of more information. It must be pointed out that the "fields" of applications mentioned in this chapter do not constitute a unique division; examples given are not necessarily the most important or the most spectacular ones. This chapter is simply meant to demonstrate the wide variety of applications that have been proposed and used in the 1980s.

9.2 A GENERAL-PURPOSE ADAPTIVE REGULATOR

Adaptive regulators are now commercially available on the market. One of the very first general-purpose adaptive regulators was introduced in 1982 by ASEA, Sweden [1,2] and it was known by the trade name NOVATUNE. This is a multiloop, microcomputer-based system, in which some flexibility has been built in to meet different application requirements. (See Fig. 9.1.)

NOVATUNE contains three different versions of one basic control algorithm. It can handle up to fifteen adaptive control loops. It is implemented with a modularized hardware which admits various input/output configurations. The software consists of a library of well-defined function modules, which performs such functions as logic, arithmetic, switching, and so on. Particular modules realize self-tuning adaptive regulators of different complexity. The system is "configured" for a particular application by "soft-wiring" different function modules together using an interactive dialogue. The procedure includes assignment of execution conditions (e.g., cycle times) to groups of modules. The resulting control scheme, containing up to fifteen control loops is stored in a nonvolatile memory of the EEPROM type. Parameter values, which define the precise function of the different modules, are also stored. All the three different versions of NOVATUNE are built around the same basic algorithm, but they offer different degrees of flexibility to the user.

(a)

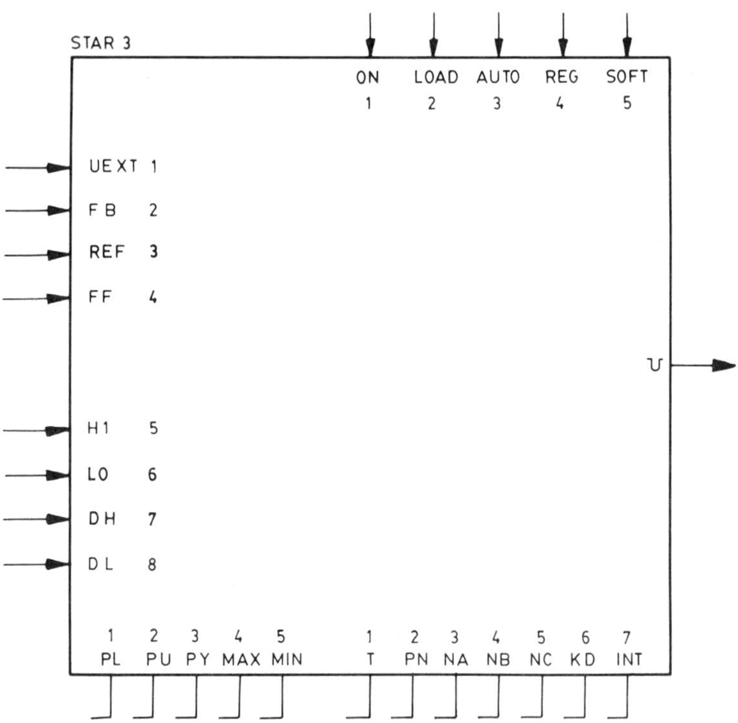

(b)

FIG. 9.1 (a) NOVATUNE self-tuning adaptive regulator (courtesy of ASEA, Sweden). (b) Block diagram of the system (from Ref. 2).

Some of the features of the most complex version of NOVATUNE, that is, STAR 3 (Self-Tuning Adaptive Regulator 3) are illustrated in a block diagram in Fig. 9.1b. The regulator has a number of input signals: UEXT, manual control signal; FB, feedback signal (process variable); REF, reference signal (set point); FF, feedforward signal.

The regulator is capable of handling feedforward information in situations where important disturbances can be measured. The other inputs, HI, LO, DH, and DL, are the different limitations on the control output, U.

A number of on/off signals affect the control algorithm and these are shown on the top of the block diagram in Fig. 9.1. ON and AUTO are self-explanatory. REG is used to switch adaptation on and off, depending on whether the regulator is used in a "one-shot" fashion to tune the parameters of the controller or for "continuous adaptation." The choice between "one-shot" and "continuous adaptation" affects not only the realization of the controller but also the algorithm design since the latter has to handle abnormal situations such as periods of almost no excitation of the process and occasions with large disturbances.

LOAD is tied to a feature which is valuable in batch processes. A change in signal value causes the stored set of parameters to be loaded into the regulator. SOFT is used for soft start-up of the regulator during commissioning.

At the bottom of the regulator block diagram in Fig. 9.1, there are a number of parameters. These are set to constant values and define the details of the control algorithm. The following are the remarks on these parameters:

PN, PL: The adaptive regulator STAR 3 is a minimum-variance self-tuning regulator. The parameters PL and PN come into effect when the algorithm departs from the basic one. PL defines a desired closed-loop pole location. PN imposes a penalty on excessive control signal deviations.

PU, PY: These parameters concern the resolution of actuator and sensor signals, respectively.

MAX, MIN: These define the scaling of the control output.

NA, NB, NC: These parameters define the regulator structure.

INT: This parameter decides whether integral action should be included into the regulator or not.

The choices of the preceding parameters are not crucial. However, a few of the parameters T and KD should be chosen very carefully:

T: The sampling interval is related to the desired closed-loop bandwidth and is therefore important.

KD: This is related to the time delay of the process. The a priori knowledge of the system to be controlled is based on the choice of KD. Experience shows that this is not too difficult to find out in practice. It is also possible to handle time delays which vary with the operating point.

The NOVATUNE is available in different hardware implementations, which are all modularized to fit the specific requirements for each application, thus providing great flexibility in the control system design. It is also possible to include regulator modules with limited flexibility so as to design simple control systems with moderate performance requirements.

The general-purpose character of NOVATUNE has been demonstrated by Bengtsson and Egardt [2], who describe the application of NOVATUNE in chemical reactor control and in elongation control in a skin pass mill. Breitholtz and Egardt [1] report about the application of NOVATUNE for the control of moisture content in a pulp drier and as a standard building block in a large instrumentation system in a chemical plant. These are briefly discussed in Sections 9.8, 9.9, and 9.12. These applications demonstrate the radically different ways of using an adaptive regulator such as NOVATUNE. These NOVATUNE regulators have also been applied to other industrial processes since their release in 1982. Experience with installations using this commercially available system shows that adaptive controllers have indeed reached a mature stage and these can be used as a standard tool by the control engineer. This is especially so if simple design rules are formulated. Such simple design rules are actually employed extensively in the NOVATUNE applications [2].

9.3 POWER SYSTEMS

9.3.1 Load-Frequency Control

Load-frequency control is an important problem in an interconnected power system. A power system consists of different coherent groups of generators interconnected by tie lines. Each coherent group is termed an <u>area</u> and should be able to meet its own load changes. The relationship between the area frequency and the total generation for changes in area load is dependent on the combined effect of all governors on the turbines within the area. The steady-state and dynamic performance of the tie-line controller are determined by this combined governing characteristic.

Whenever changes in frequency occur due to changing load, the area has to be brought back to normal frequency by using supplementary control. The objective of load-frequency control is to regulate the power output of electrical generators within a specified area, whenever changes occur in system frequency, and tie-line loading, so as to maintain the specified frequency and the specified interchange with other areas within fixed limits. Power system parameters are dependent on the operating conditions. It is therefore necessary to track the operating points, update the parameters, and then compute the control, so as to maintain optimum performance of the power system. Adaptive control strategy is best suited for this purpose. The earlier techniques, including the integral controller which was commonly used, are based on a fixed-parameter model of the power system using linearization around a nominal operating point; such controllers are not as effective as adaptive controllers.

Sheirah and Abd-El-Fattah [4] have proposed a modified STR for load-frequency control of interconnected power systems. The control criterion in the proposed algorithm is to track and minimize the variances of the area frequency error and the total tie-line power deviation simultaneously, using time-varying weights on these two variables employed in computing the area control error. The system dispatcher can select which area's output is nearest to its minimum variance. The STR is able to compute very quickly for unpredictable load variations of random magnitude and duration. This regulator scheme is two-way adaptive. First, there is on-line continuous updating of the area model used in the regulator. Second, the weights assigned to the frequency error and tie-line power deviations are changed periodically (Fig. 9.2). Studies on a two-area power system (Fig. 9.3) show that this regulator has better performance than a conventional load-frequency controller. This can be implemented easily with microprocessors, as only locally available measurements are required for each area controller.

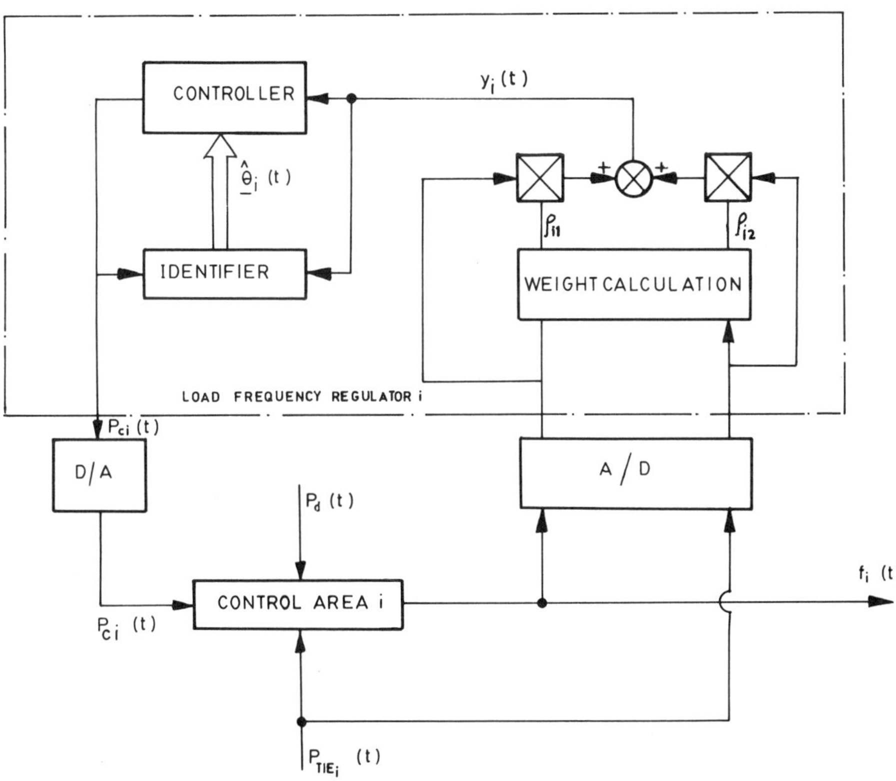

FIG. 9.2 Load frequency controller. (From Ref. 4.)

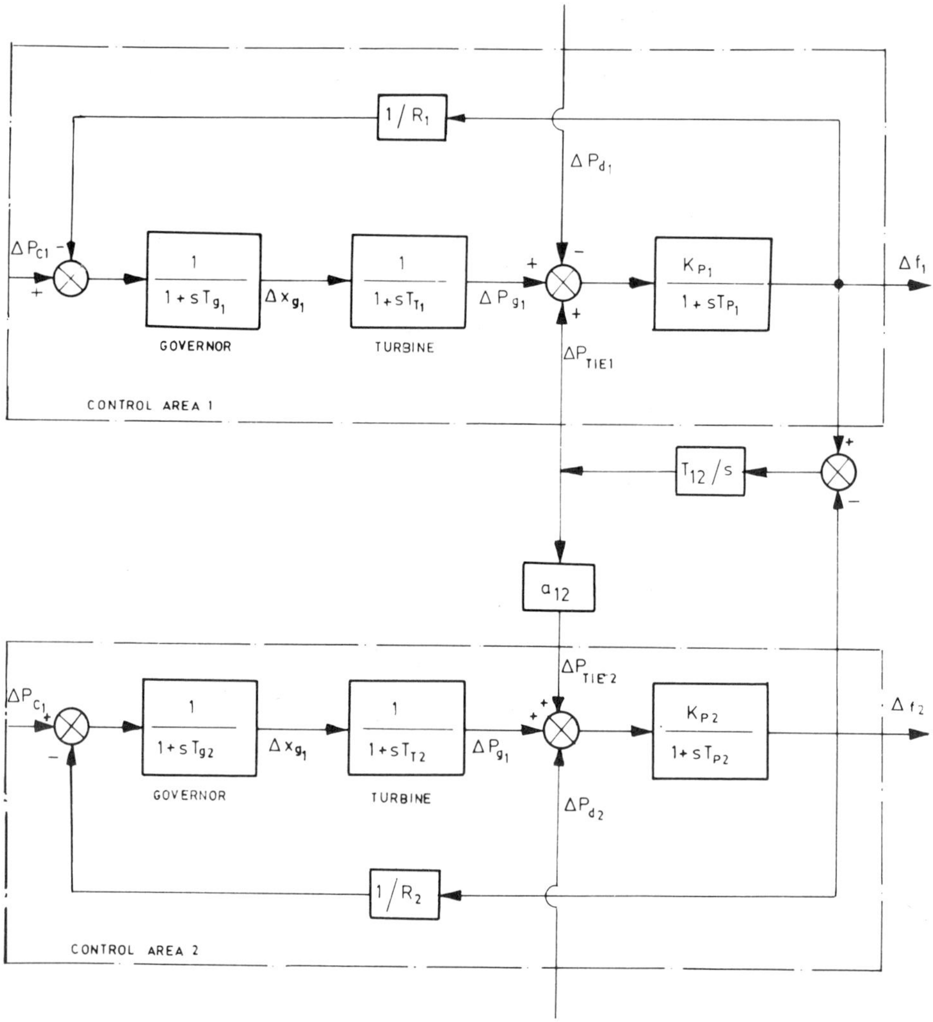

FIG. 9.3 Simplified two-area power system. (From Ref. 4.)

Aly et al. [4a] consider load-frequency control using minimum-variance regulators. They employ a stochastic model of the power system, instead of a deterministic model as used by some earlier workers. They develop a self-tuning regulator (STR) and a detuned minimum-variance regulator (DMVR). In the DMVR, some of the closed-loop poles can be assigned to specified locations. The latter has the advantage that the closed-loop response of the system can be specified by the designer to take into account some practical constraints. In the simulation study a two-area power system is considered, and the responses with both the regulators are compared. It is shown that both the STRs and DMVRs can effectively control the deviations of frequency and tie-line power in an interconnected power system subjected to stochastic load changes. However, it is found that the detuned minimum-variance regulator with extended least-squares estimation technique is more effective than the self-tuning regulator with the least-squares method of estimation.

Kanniah et al. [5] have considered a microprocessor-based adaptive load-frequency control. The control criterion is to minimize area control error (ACE), inadvertent interchange (II) and the time deviation (TD) with minimum area supplementary control (ASC) activity. These can be expressed as follows:

$$\mathrm{ACE} = (\Delta P_{tie} + B\,\Delta f) \longrightarrow 0$$

$$\mathrm{II} = \left(\int P_{tie}\,dt\right) \longrightarrow 0$$

$$\mathrm{TD} = \left(\frac{1}{60}\int \Delta f\,dt\right) \longrightarrow 0$$

while keeping

$$\mathrm{ASC} = f(\mathrm{ACE}, \mathrm{II}, \mathrm{TD}) \longrightarrow \text{minimum}$$

ACE represents generation deficiencies in the area at every instant of time.

The performance criterion can be restated as follows:

$$\mathrm{IACE} = \int \mathrm{ACE}\,dt$$

$$= [\mathrm{II} + 60B(\mathrm{TD})] \longrightarrow 0$$

This cannot be accomplished, in practice, because of metering errors and other inaccuracies in both magnitude and phase, as well as inaccuracies in the frequency bias B. Further, ASC action should change generation so as to keep ACE at a minimum. From a practical point of view, this is also not feasible because the maximum permissible rate of change of generation is limited and the mechanical parts usually have dead zones which introduce limit cycles. It may also not be desirable to control the generation to follow rapidly varying loads because this would result in unnecessary wear and tear of the valve control equipment. Furthermore, other constraints and uncertainties result in significant ACE, additional generation changes, and consequent system oscillations.

A two-area power system is considered for application of a minimum-variance STR strategy. It is assumed that one valve position is to be controlled. However, this can be easily extended to situations where several valves are to be controlled, in either thermal or hydro sets. Control strategy includes various unknown disturbances which are assumed to be uncorrelated. The selection of an appropriate sampling interval is important for STR implementation.

In the experimental setup (Fig. 9.4) the area characteristics were represented by an equivalent turboalternator modeled by a micromachine set, supplying a local resistive load connected to its terminals. Steam-turbine dynamics were realized using standard analog components. Boiler dynamics were not considered, since the transients lasted for a short enough period during the tests. The microalternator was connected to a large system (corresponding to the second area) through a tie line realized by interconnecting several inductors. The microalternator set was controlled by the STR so as to control the ACE of the local system against disturbances in local load and other disturbances. The control input signal regulated the mechanical input to the microalternator.

Comparison is made of the system response obtained from the proposed STR-based system and the conventional integral controller. The test results show that the overshoot and settling time are significantly less with the STR, resulting in superior performance in both stability and transient response. Further, the proposed regulator is self-optimizing; that is, no detailed studies have to be made to design the regulator. Such a controller is simple and relatively inexpensive and is well suited for computerized control of power systems, since it can be easily incorporated in any acceptable control hierarchy. However, the development of software requires considerable programming effort for data acquisition, identification, and control.

Agathoklis and Hamza [6] present a comparison of three algorithms for load-frequency control. The three different controllers are: (1) an integral controller, (2) a self-tuning controller, and (3) a combination of an integral controller and a self-tuning regulator with a supervisor. The supervisor selected the weighting factors between the integral and self-tuning regulator as a function of the input signal characteristics. To satisfy practical constraints, the magnitude of the control signal was penalized in the objective function and its rate of change was limited. A combined STR and integral controller has been adopted to retain the advantages of both of them. If the STR only is used in the system, the integral of the frequency error and the inadvertent energy interchange increase more rapidly with time than when the integral controller is used. However, the STR, being a stochastic controller, can operate satisfactorily in the presence of stochastic disturbances. Hence, by combining the integral and the self-tuning regulators,

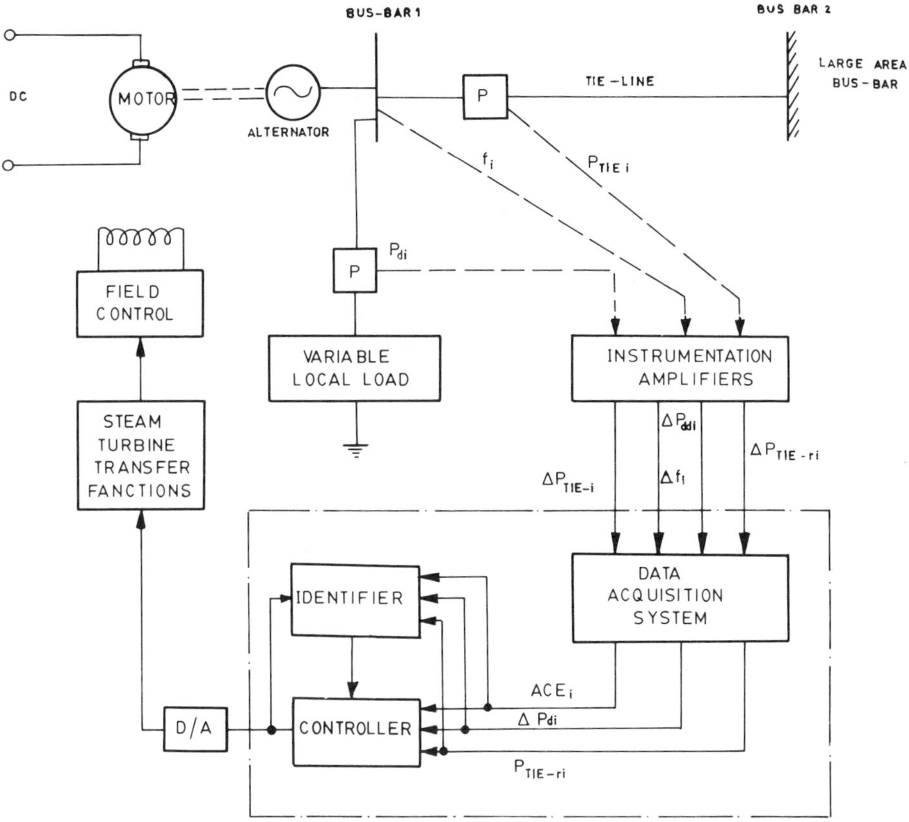

FIG. 9.4 Experimental setup representing a two-area system. (From Ref. 5.)

a better controller than both can be obtained. Figure 9.5 shows such a scheme.

The choice of the weighting factors γ_{1i} and γ_{2i} is made from simulation studies and depends on the characteristics of the input signal, particularly the ratio between deterministic and stochastic load changes. The "supervisor" is used to analyze the input signals and select the appropriate weighting factors. A delimiter is used in the simulation, as the governor turbine combination cannot follow rapid changes in the control signal.

Using simulation studies, the performance of this combined controller is compared with that of the integral controller only and that of a self-tuning regulator only. It is found that the combined controller has improved performance for the integral of the frequency error and the inadvertent energy interchange.

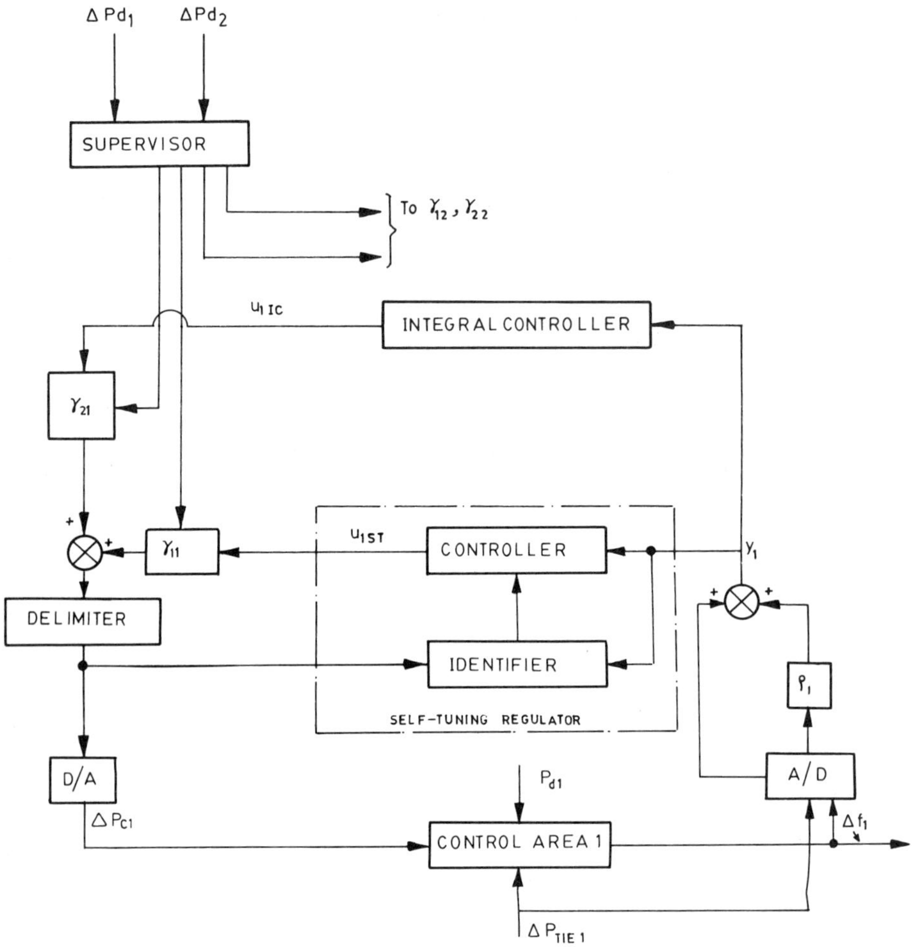

FIG. 9.5 Area 1 regulator. (From Ref. 6.)

Vajk et al. [7] present the load-frequency control of the Hungarian electric power system where all a priori information about the system is used in the regulator design. A combined regulator was developed based on the self-tuning predictor of the area requirement and the RAFT- (required average for finite time) type regulator [8] for the area control error. This adaptive regulator is implemented in the central computer control system of the Hungarian power system and it is a standard facility now. The following are the objectives of the controller:

1. The area regulates its own load fluctuations.
2. It contributes to the control of system frequency.
3. During the accounting intervals the exported or imported energy is of scheduled value.
4. The regulator satisfies the requirement with minimum cost.
5. The regulator reduces the commands sent to the power station without compromising other control objectives.

The power control center allocates the total required generation among the power stations in such a way that operation will be quick and generation cost will be minimized. The working points of the power stations are chosen on the basis of economic load dispatch (ELD). The load-frequency controller moves the plants around these points. The cost is minimal if all the plants are operated at the same incremental cost value. Quick operation is possible if the commands are executed by as many power units as possible. The LFC and ELD are coordinated to minimize the control action.

The preceding control strategy is realized in the process control computer. The measured signals are the tie-line measurements, the system frequency, the generation of individual power plants, and the number of controllable blocks within a plant. Other basic data for the controller are the daily scheduled tie-line power and frequency, the points of the power plant incremental cost curves, and so on. Real-time experiments show that the adaptive regulator performs well. The adaptive strategy leads to an improved dynamic performance, more efficient use of control action under a wide range of conditions. Further the scheduled value of the exchange energy is guaranteed, as is the area requirement, with minimum cost. It is possible to include a second adaptive predictor for the short-time load forecast necessary to the ELD of the power system.

9.3.2 Power System Stabilizer for Large Disturbances

During large disturbances, a power system stabilizer is required to supplement the excitation control of a synchronous machine. Supplementary excitation control, commonly referred to as a power system stabilizer (PSS), is an important method of improving the damping of a power system and therefore extending the stability limits. In recent years many investigations have been done on the design of excitation control for the improved dynamic performance of power systems. The design of power system stabilizers and the tuning of their settings have attracted special attention.

An adaptive stabilizer having a self-tuning property has been proposed by Ghosh et al. [9]. When applied to a power system, it produced very good damping of large system swings.

In the case of a linear controller, its parameters must change as the operating point of the machine changes. Even for a single machine, it is not easy to characterize the system state. The parameters of the stabilizer can be chosen from a set of precomputed values. However, in the case of a multimachine system, it is very difficult to characterize the system state and it is almost impossible to have a set of precomputed stabilizer parameter values.

An estimate of the system dynamics can be obtained continually, using explicit on-line identification. On the basis of this up-to-date model of the system, it is possible to find the stabilizer parameters and the linear controller can be tuned optimally to damp any disturbances which may arise.

In the scheme proposed by Ghosh et al., a fixed-parameter conventional AVR is combined with an adaptive stabilizer. Thus, the advantage of an AVR are available in steady-state performance and in limiting the first swing following a fault. The adaptation of the stabilizer parameters affects only the subsequent damping. The proposed excitation control scheme is shown in Fig. 9.6. The transfer function between the system input u and the output y_p is identified using the recursive least-squares method. The stabilizer parameters are computed each sampling period, based on the updated estimates of the system parameters. The control u is then computed by using a pole-shifting algorithm which radially shifts the poles toward the origin. This technique has the advantage of having a single parameter, instead of an entire set of parameters in a polynomial, available to the user to influence the value of the controller gain. The gain is adjusted so that the controller is limited only for the first few swings following a large disturbance. The paper considers certain problems which arise with the direct application of explicit identification. One of the problems is the action of the identifier during a fault, as large errors occur in the estimate immediately following a fault. The very time at which the adaptive controller should operate effectively to produce good damping is the time that the control parameters are in error. This situation is averted by turning off the identification for a short time following the fault. The onset of the fault is detected by a sudden rise of the prediction error. The authors suggest that identification recommence after a fixed interval of about 0.5 s. This choice allows successful adaptive control even during severe disturbances.

Simulation studies were done on a synchronous generator connected to an infinite bus through a double-circuit long transmission line. Effects of (1) 35% step change in load, (2) 10% step change in reference voltage, (3) a short-circuit at the midpoint of one circuit of the transmission line, with a successful reclosure, and (4) a short-circuit, with unsuccessful reclosure, that is, with a line loss, are studied to confirm that the stabilizer is able to maintain stability in all cases.

Experimental tests were performed on a micro-alternator connected to a constant-voltage utility system through a pair of transmission lines modeled to be equivalent to 250 km of a 760 kV line. Elaborate results show that the proposed stabilizer gives consistently improved damping even during large disturbances like a line loss, the controller adapting through

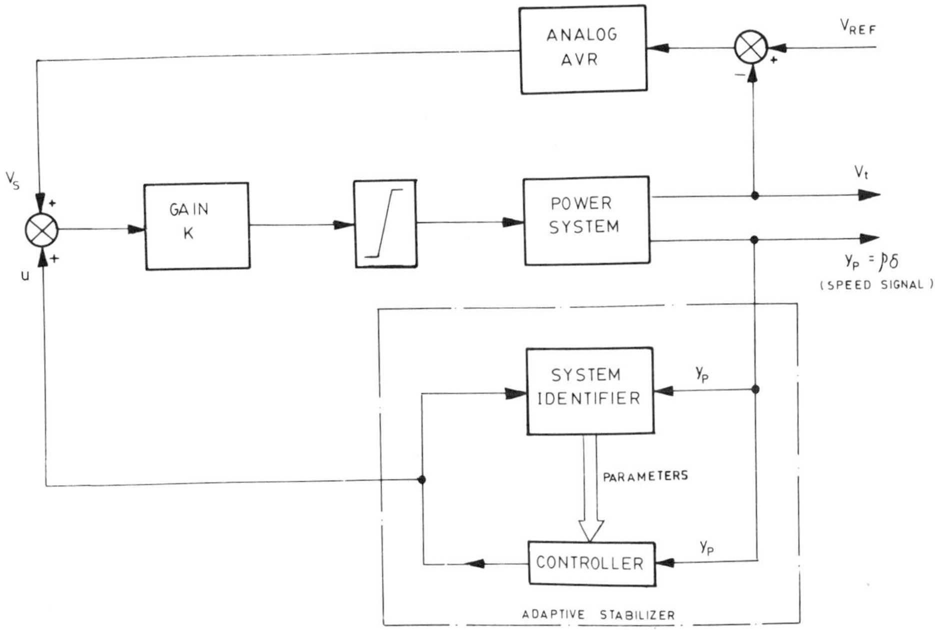

FIG. 9.6 Excitation control scheme. (From Ref. 9.)

the fault. This adaptive stabilizer produced performance which could not be equaled by any fixed-parameter controller.

The work of this research group is also reported in Ghosh et al. [9a] and Malik et al. [9b]. In the first paper comparative results of studies are given with an adaptive stabilizer based on different adaptive control algorithms and a conventional fixed-parameter stabilizer. It is concluded that explicitly identified controllers are preferable for power system applications. Minimum-variance controllers can only be used for small disturbance cases, provided the system is minimum phase. Pole-assigned controllers can be used only in cases where the system is known well, although it can damp out large disturbances efficiently. Linear quadratic and pole-shifting controllers can be most useful when the system is not known well, but has stable open-loop poles. It is also pointed out that adaptive controllers are designed primarily as transient controllers and so special attention must be paid to good steady-state control. This can be done by including an analog voltage regulator which maintains the steady-state performance, while the adaptive controller works as a transient gain stabilizer.

In Malik et al. [9b] details are given of the implementation of adaptive stabilizers using microprocessors. Two different adaptive control techniques—one using the linear quadratic optimization criterion and the other based on shifting the open-loop poles radially to form the closed-loop poles of the system—are investigated. Experimental setup and software structure are discussed. Experimental results on a power system model show good damping of oscillations resulting from large disturbances. With the advent of 16 bit microprocessors, and with the cost of hardware coming down, adaptive control algorithms are now feasible for implementation of such power system stabilizers.

The use of high-gain and low-time-constant voltage regulators in synchronous generators employing thyristor excitation results in the injection of negative damping into the system, leading to dynamic instability. Sengupta et al. [9c] present a new criterion for designing a power system stabilizer, based on the cancellation of the negative damping torque inherent in a synchronous generator and automatic voltage regulator. This technique guarantees positive damping for all conditions. Significant improvement is achieved by overcompensation to the extent of about 50%, after canceling the AVR negative damping. The computation of the gains for different operating points and systems is quite fast and so this technique can be implemented as an adaptive controller. A microprocessor or even an analog circuit can be applied to update the stabilizer gains, using suitable limiters in case the terminal voltage is drastically changed.

Extensive computations and simulations have been performed, and experimental work has been conducted on a computer-controlled micromachine system with different operating conditions, including a three-phase short-circuit lasting 120 ms. These results illustrate the effectiveness of the adaptive stabilizer. The postfault recovery of the terminal voltage and the damping of the rotor are the signals that have been presented for comparison, under different operating conditions.

A combination of speed and acceleration signals has been used for performance calculations. Computational results are given in terms of damping and synchronizing torques and also as eigenvalues. It is necessary to calculate the eigenvalues to consider the effects of the stabilizer gains on other vulnerable modes, especially the exciter mode. Synchronizing and damping torques provide information about the rotor mode. The transient response, following a small perturbation, is also given.

Although the preceding study considers speed and acceleration feedback, we can also employ other signals such as changes in output power, frequency, and armature current for the effective cancellation of negative damping contributed by the voltage regulator. This method can also be extended to a multimachine problem.

9.4 ELECTRIC GENERATOR CONTROL

9.4.1 Application of MRAC

Model reference adaptive control of generators improves steady-state stability of the power transmission system and also reduces unit stress in generators operating with delicate stability conditions. Irving et al. [10] use the hyperstability approach to design the adaptive control of generator excitation, using voltage, speed, and active power as input signals. These additional signals lead to an increase in the "steady-state stability" domain even with a regulator having constant parameters. The control system parameters, including the additional signals, are chosen to cover all possible operating conditions and network configurations. There is a reduction of the oscillations of the H.P. turbine valves when the generator is affected by network perturbations. Wear and tear on the large units, especially in nuclear power plants, will thus be reduced. Experimental results confirm the superiority of adaptive control. The choice of MRAC without identification and with adequate additional devices has been proved to be particularly suitable for improving unit voltage regulation performance. The "parallel" MRAC structure eliminates the unmeasurable disturbances in a sudden manner which is harmful to unit reliability. Hence the "series-parallel" structure of MRAC is used and it has the following additional advantages: (1) this leads to a very simple adaptation algorithm, (2) the progressive and steady way in which network disturbances are eliminated considerably reduces wear on the unit. The use of state variables of the generator simplified model allows the design of a much more effective regulation structure than Monopoli's design with high-order derivatives. As the complete reference model cannot be realized using the adjustable closed-loop control system, the concept of following a "partial" reference model has been proposed. However, the necessity of using a partial reference model does not involve too large a degradation of the closed-loop system dynamic performance owing to the appropriate choice of this partial reference model. Validity of the partial reference model method has also been demonstrated.

The requirements of the control system are as follows:

1. In spite of varying parameters, the behavior of the stator voltage, electric power, and unit speed must be sufficiently damped (static stability).
2. We have to find a design such that the static gain between the voltage and its set-point variable equals unity.
3. In spite of sudden variations in the mechanical power (the closing down of the valve after a network fault), the unit speed must present the least possible variation (transient stability).
4. The design must work with the same efficiency for different static values of the set-point variable and mechanical power, which implies a controller working on variations of the different signals and rejection of continuous components.
5. It is not desirable to maintain stability by any action on the mechanical power (i.e., the valve of the unit) because, first, the reliability of the unit may be affected and second, this is not an efficient way to tackle the problem because of reasons relating to the thermal time constant.

For these reasons it is proposed to act on the field voltage to fulfill the preceding requirements. Experiments have been performed on a micronetwork, which is very convenient for reproducing different working conditions of the generator. The results have been compared with those with a conventional regulator to demonstrate the improvements brought about in the general operating conditions of the power network on account of the adaptive controller.

This method has great potential on account of the need for controlling large-sized generators of the order of 1500 MW or more in future power networks, and there is need for even more thorough investigations.

9.4.2 Application of STR

Kanniah et al. [11] have employed self-tuning regulators for the excitation control of generators. In such generator control, there arises a conflict between the small sampling interval necessary to track the machine dynamics and large computation time required by the process computer. To overcome this problem, a dual-rate sampling STR (DSSTR) has been applied. This novel idea has been briefly discussed in Section 8.2.8. Such a device will have a smaller sampling interval and a larger control interval, as shown in Figs. 9.7 and 9.8. This is very appropriate for use with a synchronous generator, which is a fast-responding plant governed by relatively low time constants. Figure 9.9 shows how DSSTR is applied for synchronous machine excitation control. It has been shown that the number of parameters to be identified in a dual-rate sampling STR is, in general, fewer than that for a conventional STR, thus resulting in less computation time. Further, the prediction depends on more recent values, which is not the case with a conventional STR.

Simulation studies are performed for a test system (Fig. 9.10) with a conventional STR, a dual-rate sampling STR, and an analog regulator, and these clearly demonstrate the superior performance of the excitation control scheme based on a dual-rate sampling STR for different operating conditions of the power system network, and even in the presence of disturbances. In the second part of this paper, the preceding scheme has been implemented using M6800 microprocessor. Details of hardware and software are given. The regulator is used to control a microalternator set connected to a large system through a tie line. This setup was subjected to various disturbances:

1. Step input disturbance, such as changes in input to the generator
2. Tie-line disturbances, such as tie-line reactance changes
3. Terminal voltage reference disturbances
4. Three-phase to ground fault

All the test results show that the responses of the dual-rate sampling STR-based system are consistently superior ro the conventional analog regulator

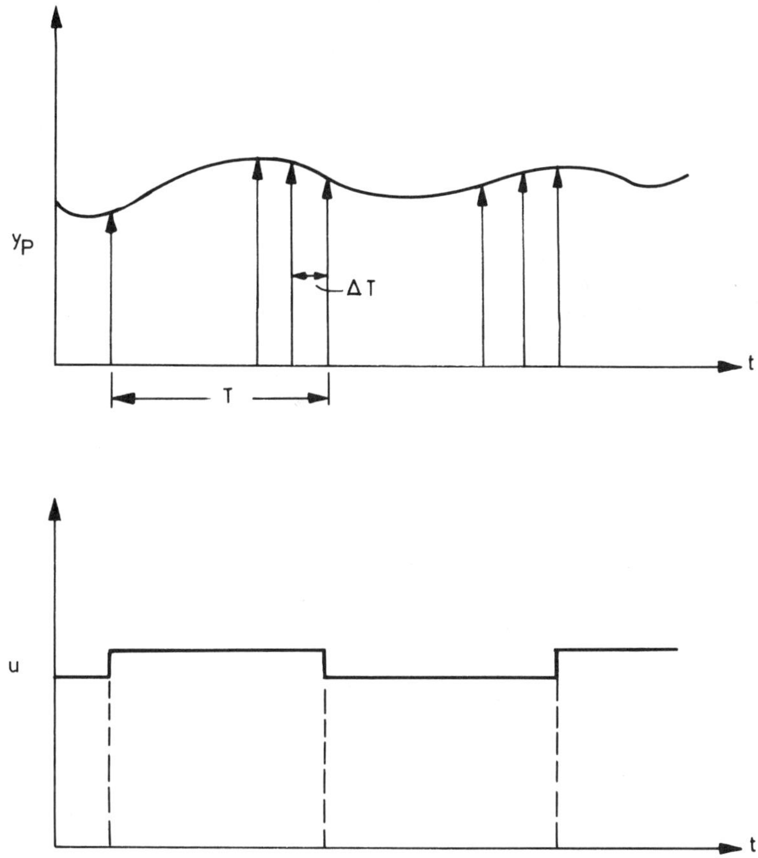

FIG. 9.7 Dual-rate sampling. (From Ref. 11. Copyright © 1984 IEEE.)

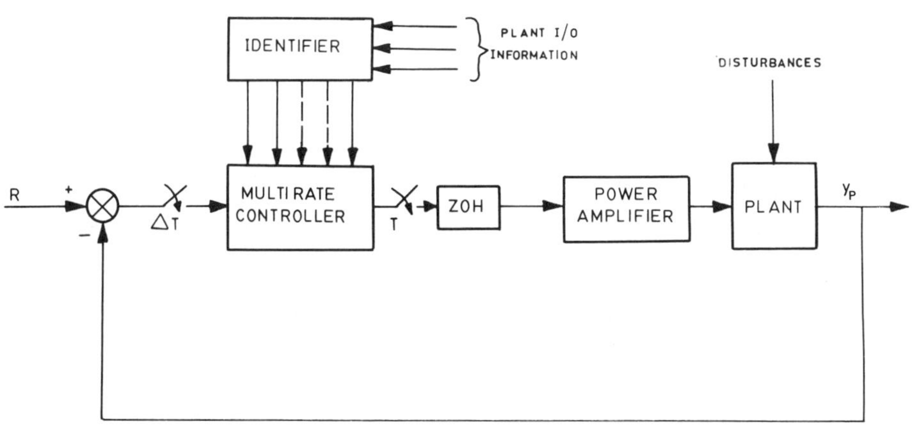

FIG. 9.8 Dual-rate sampling STR (DSSTR). (From Ref. 11. Copyright © 1984 IEEE.)

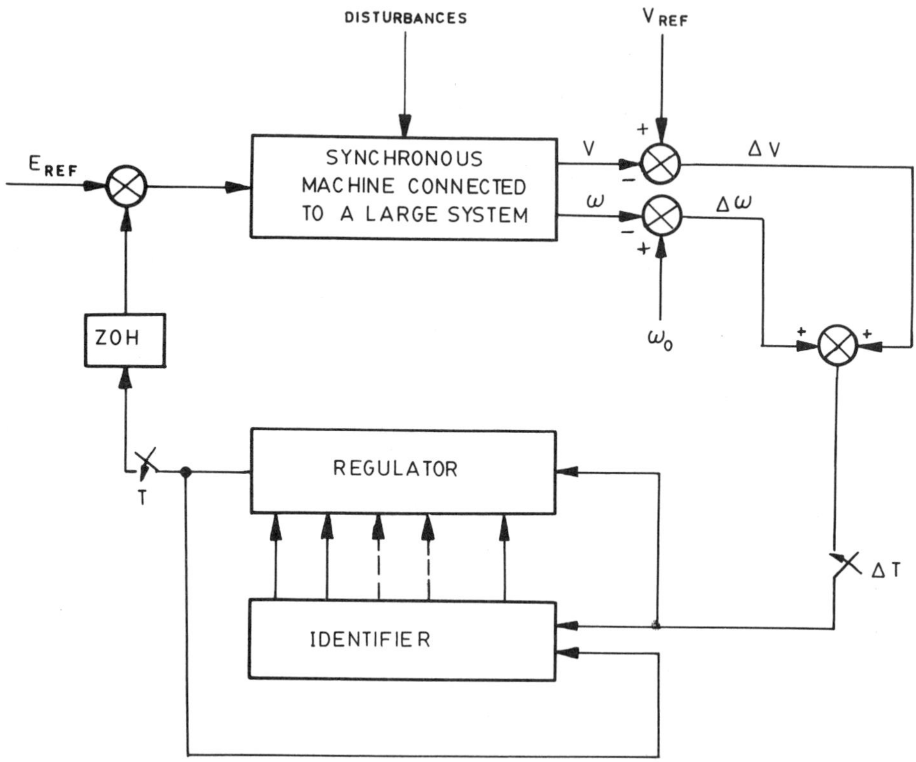

FIG. 9.9 DSSTR applied for synchronous machine excitation control. (From Ref. 11. Copyright © 1984 IEEE.)

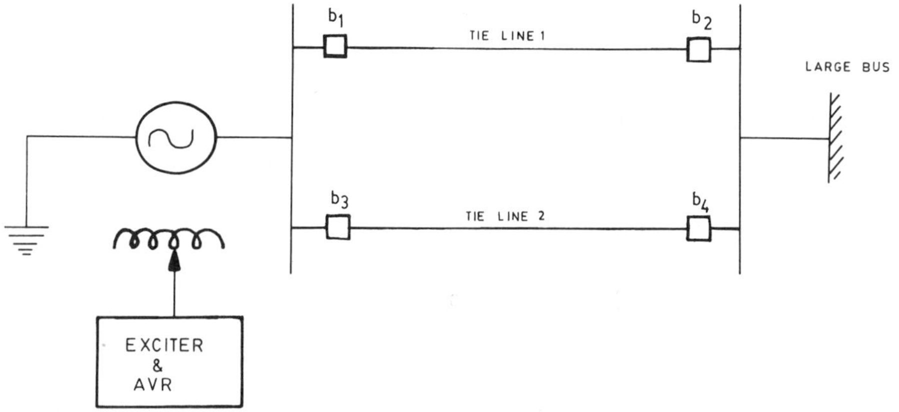

FIG. 9.10 Test system. (From Ref. 11. Copyright © 1984 IEEE.)

responses in terms of quickness of damping of load angle and the magnitude of angle overshoot.

An important characteristic of the dual-rate sampling STR is that it is a self-optimizing regulator. Such a regulator will easily fit into any part of a computer control hierarchy and there is scope for further study of other applications in adaptive control of power apparatus and systems.

Hanus et al. [12] show that the dynamic behavior of a synchronous machine is characterized by a third-order, nonlinear MIMO, non-minimum-phase system which varies greatly with load. They consider an adaptive controller to deal with the dynamics corresponding to any operating point. Simulation results show satisfactory control for the case of a minimum-phase system.

Dash and Khincha [12a] present the design of an adaptive regulator for the excitation control of a synchronous generator connected to an infinite bus through a double-circuit transmission line. A simplified version of a minimum-variance self-tuning regulator is applied. A stochastic approximation variant of the recursive least-squares estimator is used with a variable forgetting factor. The method is based on only three-parameter identification. This scheme is simply implementable with a single-chip microcomputer, with restricted processing power and limited on-chip data storage capability. The use of integer arithmetic allows a considerable reduction in the amount of data and program memory. Simulation results are given for a power system subjected to stochastic load variations and large disturbances such as a three-phase short-circuit, switching off of one transmission line, and change in the generator terminal voltages. The responses in all these cases indicate that the controller has acceptable performance and can adapt rapidly for any sudden changes in the system operating conditions.

9.4.3 Turbogenerator Power Plant

Hahn et al. [13] have considered a multivariable STC for a pilot plant consisting of a turbogenerator set which has a synchronous generator coupled to an air pressure turbo-engine (Fig. 9.11). The objective is to control two output signals, the frequency and the phase voltage, by means of two input signals, namely, the field current of the synchronous generator and the pressure of the turbine. The symmetric load can be switched between two values. The dynamic behavior of the plant is highly nonlinear on account of the following features:

1. The actuating valve has a nonlinear characteristic.
2. The output voltage is proportional to the product of the magnetic field and the frequency.
3. The magnetic field itself is a nonlinear function of the field current.

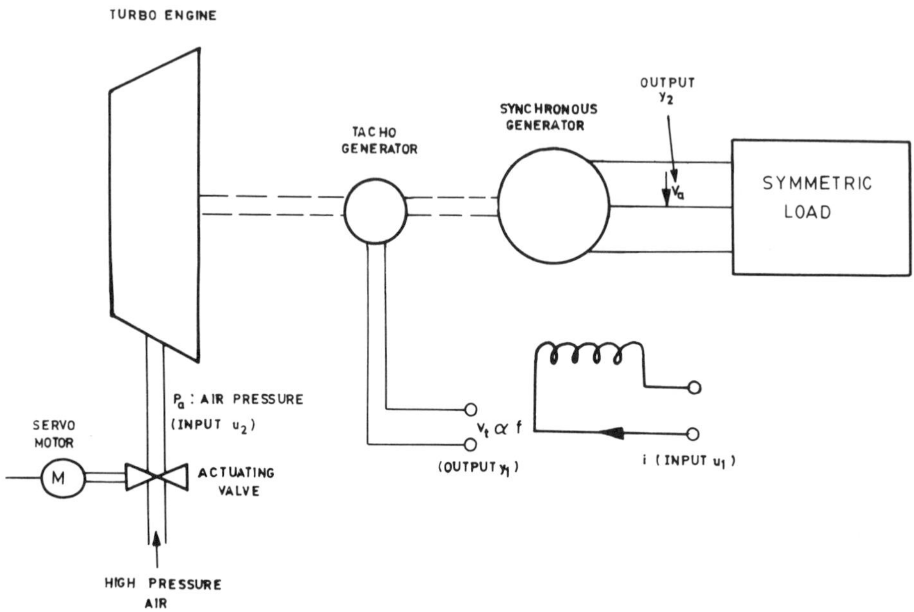

FIG. 9.11 Turbo-generator pilot plant. (From Ref. 13.)

Adaptive control is therefore an appropriate method to control the plant in all situations involving different set points, different load conditions, and parameter variations. Further, since the plant is found to be non-minimum phase, and instability has actually been observed during experiments, Koivo's scheme of generalized minimum-variance multivariable STC is applied. With a slight modification, this scheme is shown to be applicable to cases of varying dead times also, in contrast to Koivo's original scheme. The algorithm proposed by Ljung et al. [14] is used to reduce storage. This computes the Kalman gain factor recursively. A variable forgetting factor has been introduced.

The preceding adaptive scheme is able to control the plant in all situations in an acceptable manner. However, it is found that the set point changes and deterministic disturbance rejection can be accomplished in a faster way if the sampling rate is reduced. This does not seem possible with the full least-squares estimator and the process computer used. It is felt that an estimation procedure using stochastic approximation algorithm is preferable, as this is much simpler, especially in the multivariable case. This experimental study confirms the feasibility of self-tuning schemes for control of electromechanical systems and can be applied to control turbo-generators in a real power plant.

9.5 ELECTRIC DRIVES

High performance is usually demanded of dc and ac drives. The plants usually exhibit nonlinear and time-varying behavior. It is difficult to determine, in some cases, the values of the dynamic parameters for processes with known structures, for example, the squirrel-cage induction motor supplied with variable frequency and voltage. Furthermore, large variations of plant parameters can also occur, such as change of moment of inertia of the load, change of the field current, change of the gain of the thyristor rectifiers, and so on. In these situations, adaptive control schemes are most suited to maintain high performance of the electric drives.

9.5.1 DC Drives

Courtiol and Landau [15] present a general approach for the design of adaptive control systems for electric drives subject to rapid, wide, and unmeasurable variations in plant parameters. They apply model reference technique. The implementation is realized as for a signal synthesis MRAC system. The adaptation mechanism operates by the introduction of a supplementary signal which modifies the output signal of a linear model-following regulator. A high-speed adaptive control system for dc drives is given in Fig. 9.12. This scheme can also be simplified. The same methodology can be extended for the speed control of ac drives as well as for the adaptive control of other variables such as position, acceleration, armature and field current, voltage, and so on. The adaptive controller maintains constant dynamic performance in the presence of unpredictable and unmeasurable variations of the moment of inertia and field current, and in the presence of nonlinearities occurring at low speed.

Experimental work is done by applying the adaptive speed controller to a 5.3 kW dc electric drive. The speed responses to step command inputs obtained with the conventional PI controllers are compared with those obtained with an adaptive speed controller. The results confirm the superior performance of the adaptive controller. The adaptive regulators can be used in electrical drives such as machine tools, rolling machines, winding machines, and so on, which cannot be controlled satisfactorily by conventional PI regulators.

9.5.2 Induction Motor Drive

A new approach to the problem of induction motor speed control is made possible by applying adaptive strategies. Evans et al. [16] present a nonlinear adaptive control scheme capable of working effectively when either motor and/or load parameters are not known. This control system is applied to an inverter-fed induction motor. This approach is fundamentally different from previous work on induction motor control in that the objective is to control the shaft speed directly. Further, inverter control variables,

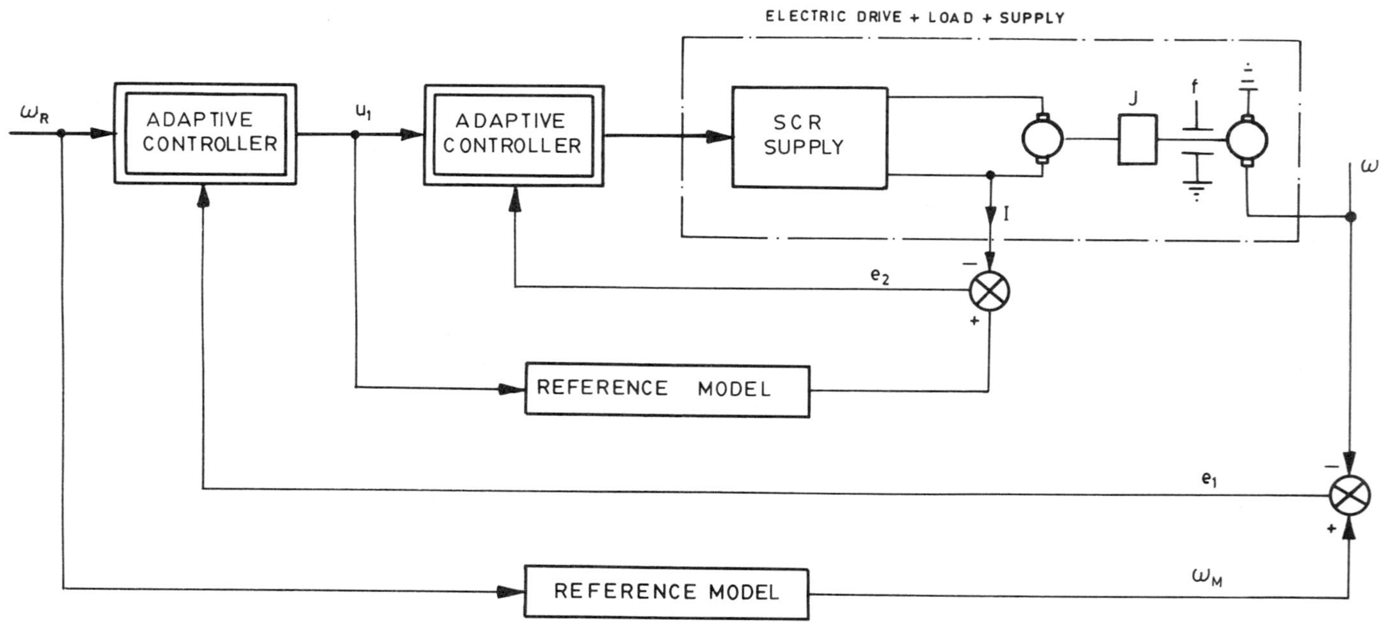

FIG. 9.12 Adaptive control system configuration with two adaptive loops. (From Ref. 15. Reprinted with permission from Pergamon Press, copyright 1975.)

frequency, and voltage are related to the shaft speed and motor current by means of the steady-state model. If we make the controller adaptive, we do not need to know the load or machine parameters. However, machine parameters are available from the manufacturer and so we can make full use of them.

The following difficulties arise in designing the inverter and controller for such systems:

1. The second order effects caused by the nonsinusoidal nature of the source
2. The highly nonlinear relationships between the control inputs to the motor (voltage and frequencies) and the controlled outputs (torque, speed)

The inverter has two modes of operation: (1) as a variable-frequency-controlled voltage source (CVS) or (2) as a variable-frequency-controlled current source (CCS). In the latter case a large inductor is included in the link and it acts as a controlled current source. Figure 9.13 shows a typical inverter-fed induction motor configuration (voltage controlled). The controller measures the motor shaft speed, compares it with the desired one, and uses the error signal to (1) update the estimates of the load and motor parameters, and (2) adjust the inverter voltage and frequency, such

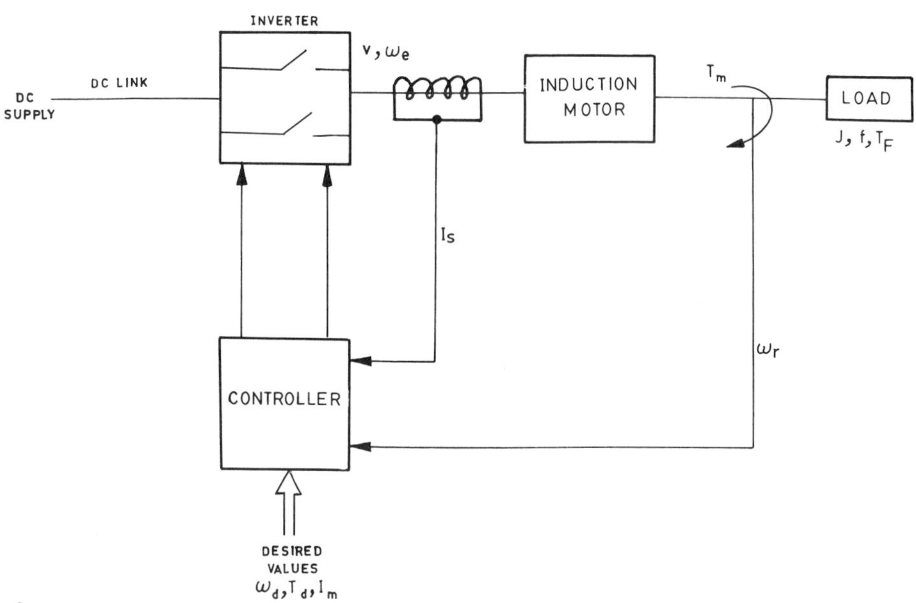

FIG. 9.13 Typical inverter-fed (voltage-controlled) induction motor configuration. (From Ref. 16. Copyright © 1983 IEEE.)

that the motor operates at the desired speed, while satisfying the constraints: (a) the stator current never exceeds I_m, and (b) the machine flux is near maximum at all times, so as to make optimum use of the machine iron.

Adaptive control algorithms are derived for the case when motor and load parameters are unknown. As the motor parameters are usually known, the algorithms are easier to implement. The case of a linear load is considered. The nonlinear load case can also be treated.

A digital simulation has been carried out to test the performance of the adaptive algorithm for the linear, unknown load. Two cases are considered: one allowing regeneration and another with no regeneration. It is found that the speed changes are achieved in minimum time with the motor running at maximum slip (corresponding to maximum current) and hence maximum torque. The speed control property of the adaptive control algorithm is quite good in both the cases. The regenerative case has a significantly faster response. In both the cases, J and f parameters converge to the actual values, through T_F does not. Despite this, the speed control property is very encouraging.

9.5.3 Reluctance Motor

Adaptive microprocessor control of a linear helical reluctance motor has been considered by Paul et al. [17]. They have compared the performance of open-loop control, nonlinear sampled-data control, two-mode control, and adaptive control and show how an adaptive controller is capable of accommodating a wide variation of load inertia and load torque. Whereas the other controllers may give satisfactory dynamic and static performance provided the output inertia and load torque do not vary, the adaptive controller can cope with a load mass about three times the rotor mass.

It is found that system stability deteriorates with increasing load inertia since the rotor mass determines the damping ratio and natural frequency of the motor. We may reduce the controller gain to suit the worst load inertia condition; but this leads to a very slow response for small loads. It is therefore necessary to use an adaptive controller so that the controller gain varies with the load mass.

The adaptive controller is implemented on the basis of the ratio $(\Delta\theta_0/\Delta\theta_P)$, that is, the variation of rotor position with respect to the movement of the microprocessor pointer through a look-up table, which is used to detect inertia variation. This ratio should be maintained as close as possible to unity, to ensure that the load angle does not exceed an acceptable limit. This ratio is used to adapt the output of the controller. To start with, the controller output is set to its no-load value. Then if $|\Delta\theta_0/\Delta\theta_P| - 1$ exceeds a certain threshold, indicating a different load condition, then the output is reduced over each sample interval until $\Delta\theta_0/\Delta\theta_P$ is close enough to unity. The adaptive controller configuration is shown in Fig. 9.14.

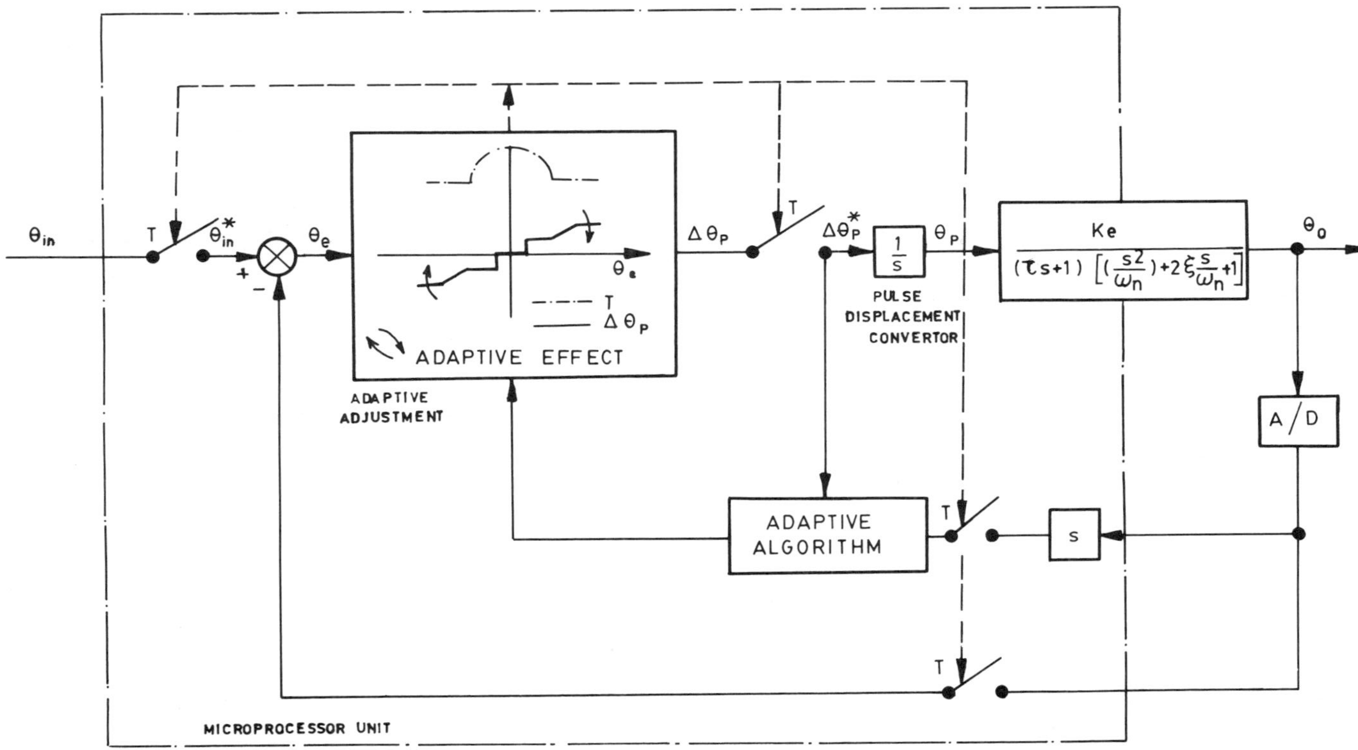

FIG. 9.14 Adaptive controller configuration. (From Ref. 17.)

The adaptive controller is shown to achieve stabilized operation and the motor does not lose synchronism even when the load mass varies from 2.9 kg to 7.9 kg, whereas in the nonadaptive case the system is unstable.

9.6 AIRBORNE VEHICLES

9.6.1 Flight Control

Adaptive control of airborne vehicles received significant attention during the 1960s, with examples ranging from very simple techniques, based largely on heuristic or intuitive reasoning [18], to quite sophisticated and very complicated procedures based on optimal and learning control theory [19]. A special issue of the IEEE Transactions on Automatic Control in 1977 deals with the design and performance of adaptive flight control systems [20]. Some of these designs have received justifiable criticism [21,22], as Gressang [21] states that any control system which ignores available information about the aerodynamic forces and moments (which are important in flight control) is not very practical and is unnecessarily complex.

Young [23] deals with an adaptive autostabilization system for an airborne vehicle whose dynamic behavior may vary rapidly and over a wide range during normal operation, as in the case of a high-performance missile during an eighty-second mission. This mission includes a rocket boost phase, with accompanying large variations in the model parameters caused by changes in dynamic pressure, Mach number, altitude, and mass. The overall adaptive scheme is hybrid (analog-digital) in character. This approach is extendable to multivariable lateral autostabilization systems also.

Kanai et al. [24] consider the design of a model reference adaptive flight controller with an observer for a linear dynamic flight system using only the input and output data. In designing a flight controller for high-performance aircraft, it is necessary to know the aircraft dynamics exactly. However, this is not possible on account of varying flight conditions and uncertain environment. In such cases of unknown parameters and inaccessible state variables, we should have a scheme for on-line estimation of these unknown values from only measured input and output data. In this paper an adaptive flight controller is synthesized such that the tracking error between the output of the short-period motion, which combines a normal acceleration signal and a pitch rate signal, and that of a model, is regulated to zero and the adaptive observer which is constructed by introducing unknown initial states into the recursive algirhtm is employed to estimate parameters and state variables simultaneously. The computational results for two cases of a typical high-performance supersonic aircraft show that the convergence of the adaptive observer and the controller is quite rapid and the structure of the controller is simple. Simulation results demonstrate the effectiveness of the proposed scheme.

9.6.2 Wing Flutter Control

Johnson et al. [25] apply the self-tuning approach to adaptively control the aircraft wing/store flutter. An STR has been designed and tested for a very simple reduced-order model of single-mode flutter on an advanced aircraft. Later the design is validated using the full-scale simulation of the aircraft dynamics which included rigid body and flexure modes. With the reduced-order model, the self-tuning design successfully detected and controlled oscillatory instabilities in the following three cases: (1) constant plant parameters, representing level flight at and above flutter speed; (2) slowly varying parameters, representing maneuvers which cause the aircraft to exceed its flutter speed gradually; and (3) abruptly changing parameters, representing the sudden onset of flutter on account of release of stores, which result in the unloaded wing being above the critical flutter speed.

On the reduced-order model, the STR design ensures stability for parameter variations of up to 30 percent in these cases. The scheme is both robust and adaptive. These initial results are encouraging. The main advantage of the STR is its simplicity; however, the main disadvantage is the inadequacy of prior performance guarantees. A comparison is made of the performance of the STR and the maximum-likelihood estimator. The STR is easier to implement and requires much less storage and computation time. The STR approach needs further investigation before it can be successfully applied to the design of aircraft control systems.

9.6.3 Autopilots for Surface-to-Air Missiles

An autopilot stabilizes and guides the missile by adjusting fin deflections, which cause the missile body to rotate and translate in the desired direction. That is, the autopilot controls the orientation of the missile from inputs that depend on the specific phase of flight. In this type of application, we note that the dynamic pressure varies considerably over the range of possible mission flight profiles, so that the missile parameters will change as the missile goes through a changing environment. In older interceptor-missile autopilot designs, band switching of the autopilot gain has been used between two or more values to compensate for the air-frame response with Mach number and altitude. In these, the gain of the autopilot is designed to depend on the Mach number and/or dynamic pressure, requiring considerable analysis to find the gains for various flight conditions. In the case of an adaptive autopilot, the adjustments are done on-line automatically. There are three separate autopilots in interceptor missiles to control roll, pitch, and yaw. For instance, the roll autopilot can adaptively keep constant the product of its gain and the roll moment effectiveness [26].

Pramanik and Seshadri [26a] consider the design of a stable signal synthesis MRAC system for adaptive lateral autopilot for missiles under atmospheric disturbances. A new method based on error sampling technique has been applied to construct the model of the system to avoid the non-minimum-phase transfer function in the MRAC. The method is based on the noise transfer function due to Tsypkin [26b] in which the model itself is modified. Such an optimal model is synthesized through a relatively simple feedback from the plant output and the model output. The impulse invariant sampling technique adopted by them is shown to have the property of preserving the controllability of the original system. A practical decentralized model-updating procedure has been given for the realization of the adaptive algorithm. This approach leads to a reduction of the model order and implementation becomes easy and fast. This method is applied to missile lateral autopilot design. Simulation results show the effectiveness of the adaptive algorithm.

9.6.4 Gimbal Bearing Friction Compensation

Airborne pointing and tracking systems suffer from line-of-sight stabilization errors caused by gimbal bearing friction. Similarly, friction can limit the smoothness of robot motion trajectories. Walrath [27] describes a novel technique to improve stabilization performance. A model in system software predicts real-time friction torque values. This dynamic friction model is adaptively adjusted into agreement with actual friction behavior by processing inputs from conventional sensors. Measurements from these sensors cause on-line adjustment of model parameters, resulting in adaptive compensator action. The output of the model is used to generate conventional stabilization subsystem commands and also an additional gimbal motor torque equal and opposite to the actual friction disturbance. Thus the model-referenced compensator operates in a predictive, adaptive, feedforward manner to precondition the stabilization subsystem. The stabilization errors are reduced well below the levels which are achievable through conventional uncompensated methods, as illustrated by extensive test results.

This adaptive, real-time friction compensator can be integrated into the software of an airborne pointing and tracking system. Additional hardware is not required for the compensator because of its use of existing system components and computing facilities.

9.7 SHIP CONTROL

Conventional autopilots at sea do not perform satisfactorily, as they do not take into account the variations in the ship's dynamic characteristics such as speed, draft, course, loading condition, and the environment in which it operates, such as waves, wind, and currents. Adaptive autopilots can cope with these variations.

9.7.1 Application of MRAC

Van Amerongen [28] discusses the model reference approach to adaptive steering of ships. The main advantage is the simplified controller adjustment, which results in safer operation and reduced fuel cost. A mathematical model is formulated for the steering dynamics of the ship, including disturbances such as wind, waves, and current. Criteria for optimal steering are given with particular reference to course-changing and course-keeping maneuvers. Algorithms are given for direct adaptation of the controller gains applicable after setpoint changes as in course changing. Indirect parameter adjustment is possible by means of on-line identification and controller optimization, as well as adaptive state estimation, when the input is constant as in course keeping. This approach can also be applied to a certain class of nonlinear systems. Full-scale trials at the sea and tests with a scale model in a towing tank confirm that adaptive steering of ships results in improved maneuvering, easier adjustment, and better fuel economy. The advanced filter algorithm, combining MRAC with Kalman filtering, provides the major contribution to reduction in fuel costs. Fuel savings of up to 5% are possible compared to conventional PID controllers. Adaptive autopilots are easily realizable because of the availability of inexpensive digital hardware, and it is expected that such MRAC systems will be increasingly applied for steering of ships.

Earlier work of van Amerongen et al. is reported in several papers [29-35]. Herther et al. [36] discuss a self-adjusting digital autopilot for better maneuvering and improved course and track keeping.

9.7.2 Application of STR

Self-tuning regulators have been applied with encouraging results for autopilots by Källström et al. [37] and Brink et al. [38]. Källström et al. [37] have designed two adaptive autopilots for tankers, incorporating velocity scheduling, a self-tuning regulator for steady-state course keeping, a high-gain tuning regulator with variable structure, and a Kalman filter. Simulation results as well as full-scale experiments with three different tankers demonstrate that the adaptive autopilots work efficiently under different load, speed, and weather conditions and also guarantee smooth and precise turns, which implies a much safer handling of the ship.

Self-tuning controllers have been used by Mort and Linkens [39] for course and track keeping. Mort and Linkens [40] describe the application of the algorithm of Clarke and Gawthrop to the control of surface ship course and track keeping and maneuvering, considering the disturbances due to sea wave conditions.

It should be noted that a surface ship is a time-varying, nonlinear, multivariable stochastic system. In some ships there are significant cross-coupling effects between rudder and fin stabilizers so that the control of yaw and roll motion cannot be considered separately. There is considerable scope for further research to develop suitable controllers to tackel these problems.

Fung and Grimble [41] deal with dynamic ship positioning control using self-tuners to maintain the position and heading of a ship, or a floating platform, above a preselected fixed position on the seabed by using the vessel's thrusters. Such problems occur in drilling in deep water. This type of control can also be used for tracking pipelines or cables. The control scheme consists of a self-tuning filter, a constant-gain Kalman filter, and a state estimate feedback control law.

Saelid and Jenssen [42] describe the design and analysis of an adaptive autopilot based on a model obtained from physical laws and physically meaningful parameters. The model has six states and four parameters. The states are estimated by a scheduled-gain state estimator. This reduces the computing requirements. The parameters are estimated by a prediction error method. The model of the vessel has two parts: A low-frequency part accounts for the steering dynamics when oscillatory water waves are not present. A high-frequency part models the oscillatory part of the yaw motion due to the water waves. The models are used in a Kalman filter and the rudder control is computed from linear quadratic theory based on the low-frequency part of the state vector. This yields a very effective filtering of the wave component of the yaw motion. Knowledge of the vessel model parameters and the dominating wave frequency is required for proper operation of the filter-controller structure. The convergence properties of the autopilot are studied by Ljung's method. Simulation studies confirm the effective operation of this method. The simulator includes realistic wave excitations in yaw, unknown steady moments from wind and current, and white process noise excitations of the sway- and yaw-rate equations. This adaptive autopilot design is found to be robust.

Lim and Forsythe [43] have presented a self-tuning ship-autopilot design which reconciles the conflicting demands of course-keeping and of maneuvering by using two modes of operation. Course changing demands a fast and accurate response, especially when a ship is being maneuvered in a crowded and confined waterway. Course keeping requires a control that maintains a reasonable course with minimum rudder activity. During the course-keeping mode the performance criterion is formulated to minimize heading error, propulsion losses, and rudder activity. In the maneuvering phase we are concerned only with minimizing the heading error. The autopilot must adapt to changes in the ship's state or environment (speed, loading, depth of water, sea state, etc.). It should also be capable of dealing with nonzero mean disturbances.

The design is based on the STC strategy of Clarke and Gawthrop, which includes explicitly the controller output in the performance criterion. This is relevant to course keeping for which we have to minimize propulsion losses caused by hull and rudder drags. With some modification, we can get a fast system response as required by maneuvering operations. A mathematical description of the ship simulation model including its steering gear is given. The nature of the external stochastic disturbances caused by ocean waves and the mathematical model for generating nonlinear disturbing motions are considered.

The design of the self-tuning system is a two-stage process:

1. Design a controller to meet a suitable performance criterion, assuming that the plant parameters have the nominal values.
2. Tune the parameters of the control algorithm automatically to ensure the optimum performance as specified by the criterion, despite the absence of detailed knowledge of external disturbances, plant parameters, and their variations with time. Some on-line identification technique, that is, RLS algorithm, should be incorporated.

The design of the self-tuning control should also eliminate steady-state output errors by appropriately modifying the algorithm.

Simulation studies corresponding a Mariner-class cargo vessel demonstrate the high degree of adaptability of the autopilot. The choice of control variable weighting factor has a significant effect on both course-keeping and maneuvering operations. Further work is necessary regarding the choice of this weighting factor not only to ensure adequate closed loop stability when maneuvering, but also to minimize propulsion losses due to steering during course keeping. Sea trials are necessary for a full evaluation of the self-tuning autopilot.

9.8 PROCESS CONTROL

Chemical process control systems are more complex than other automatic control systems, because they have to deal with systems which (1) are sometimes non-minimum phase, (2) have variable time constants, (3) have large time delays that may also vary, (4) can be strongly nonlinear and interactive, and (5) are subject to stochastic disturbances. Such systems should also have a fast stable response to set-point changes and sudden large load disturbances. Nonlinearities occur in valves, in relationships between reaction rates and temperatures. Many chemical processes exhibit variable distance velocity lags, catalyst decay, and other gradual drifts, all of which violate the assumption of a linear, time-invariant system [44]. Seaborg et al. [45] give a survey of adaptive strategies for process control.

Robust stable adaptive controllers are very appropriate to deal with the problems in process industries.

9.8.1 Distillation Column

One such complex process is the distillation column product composition control, which involves severe nonlinear behavior as well as a large number of state variables to adequately describe its dynamic behavior. If the operating regions are changed or the process characteristics change over periods of time, and because of initial uncertainty in the plant dynamics, adaptive control would become necessary.

Morris et al. [46] apply self-tuning for the separate and simultaneous regulation of distillation column terminal compositions. They present results of both comprehensive simulation and pilot plant tests. The main objective is to control the top and bottom product compositions at their desired values, despite feed-flow rate disturbances.

Morris et al. [47] apply multivariable self-tuning strategies to the distillation column, the performance of which depends on the choice of a specific compensating function. Further work on a multirate, multivariable self-tuning scheme is possible. Gerry et al. [48] reported about a successful application of a single-loop adaptive control of a distillation column. The extension to a multivariable algorithm is at present restricted by the system interactions, because of the presence of time delays, which are not dynamically decoupled.

Coppus et al. [49] discuss Davison's technique [49a] for the synthesis of robust multivariable feedback-feedforward controllers for an unknown plant whose state-space description is difficult to obtain. No assumptions are required regarding the plant model order or minimum-phase behavior. This technique is applied to the control of a pilot-scale binary distillation column. The controller performance is evaluated experimentally. Robust set-point tracking and regulatory control is obtained despite large upsets in the process operating conditions, consequent parameter variations, little information on the distillation column model, significant measurement time delay, process time lags, and nonlinear characteristics.

Dahlquist [50] investigates the application of SISO and MIMO self-tuning regulators to the control of top and bottom compositions of a distillation column. The results showed robustness of this adaptive scheme. Weimer et al. [51] apply a multivariable MRAC strategy to a pilot distillation column. In such plants only a little a priori knowledge about the process dynamics is available. By using adaptive schemes the controller design is simplified. Further the controller can adapt itself to the slowly time-varying behavior of the plant and to set-point changes. On account of the nonlinearity and initial uncertainty of the plant dynamics, adaptive control is appropriate. This control scheme is based on a discrete version of the algorithm suggested by Monopoli and Hsing [52]. The control objective is to maintain the top and bottom product compositions constant under varying flow rate, composition, and temperature at the inlet. Smooth control action is achieved by augmenting the plant output by filtered plant input signals. The control scheme yields satisfactory disturbance rejection. The performance of this MRAC scheme is compared with that of the STR strategy [53] for the same column. It is found that although the results are similar, a large storage and five times more computation time are required compared to MRAC, in view of the full least-squares estimation combined with self-tuning control.

Martin-Sanchez and Shah [54] have reported about the application of adaptive predictive control systems (APCS) to a binary distillation column in the presence of step and stochastic feed-flow disturbances, set-point changes and presence of measurement noise. Experimental results for both SISO control and MIMO control show the simplicity of the adaptive algorithm and confirm the excellent performance in the control of a nonlinear multivariable plant with time delays in a typical industrial environment. It is shown that the APCS outperforms well-tuned conventional PID controllers, thus demonstrating that adaptive predictive control is now beyond the theoretical stage and is worth exploiting further.

9.8.2 pH Neutralization

Buchholt and Kümmel [55] use a self-tuning regulator for the control of a pH neutralization process. The process is difficult to control due to great variations in process sensitivity with pH and due to process nonlinearities. Experiments on a pilot plant demonstrate that a self-tuning pH regulator incorporating an exponential forgetting factor and quadratic optimal control has satisfactory performance even though the process is nonlinear. It is also shown that the self-tuning algorithm can adapt to changes in process conditions. The STR does not need exact knowledge of composition of the acid stream. Nor does it require an a priori mathematical model because the regulator parameters are adjusted in real time on the basis of input/output measurements.

Proudfoot et al. [56] describe a self-tuning PI controller applied to control a full-scale pH neutralization process in an industrial effluent treatment plant. Extended trials show that the performance of this is superior to that previously achieved using standard PI controllers, since the self-tuning algorithm can easily find suitable fixed parameter values for the controller, whereas human operators found the tuning difficult, if not impossible. The details of the Waste Brine Treatment plant are as follows:

The incoming effluent at approximately pH 2.5 is neutralized by the addition of sodium hydroxide solution (NaOH) to a set point between pH 8 and 9. The output of the pH controller u(t) acts as a set point to a local flow controller which controls the addition of the reagent (NaOH) to the neutralization vessel, where the mixing process can be approximated by a first-order lag of time constant T and a pure time delay of approximately 0.05 T. There is also a distance-velocity lag associated with the reagent delivery, and this becomes appreciable when the reagent flow is low. The main nonlinearity of the system is the steady-state relationship between the amount of reagent added and the final pH. Flow into the vessel normally varies between 200 m^3/h and 400 m^3/h but can rise to as high as 700 m^3/h for periods of several hours. There are other disturbances which can vary the flow up to 100 m^3/h over a relatively short time (1/2 hour). The self-tuning PI controller has been used on-line for extended trials over several months and it has been found that the control performance is good for long periods of time, without the need for retuning. This algorithm can be of use in controlling many other similar processes.

Another relevant paper in this field is by Goodwin et al. [57].

9.8.3 Double Effect Evaporator

Buchholt et al. [58] have applied a multivariable self-tuning regulator to control a pilot plant double-effect evaporator. Although classical multivariable control strategies can be applied to this problem, it is difficult and very time-consuming to develop an appropriate process model based on physical properties. A self-tuning algorithm is preferable and works satisfactorily even in the presence of stochastic disturbances in the inlet flow to the evaporator. The STR consists of a recursive least-squares algorithm combined with a single-step optimal control strategy which minimizes a quadratic criterion. The algorithm is very fast and can be adjusted by an external operator to obtain satisfactory regulator performance. This approach is applicable to cases where the development of a process model is cumbersome or impossible.

9.8.4 CO_2 Absorption-Desorption Plant

The STR with a variable forgetting factor has been successfully used by Fortescue et al. [44, 59] to control several loops on a CO_2 absorption-desorption plant. The loop that has been studied in greatest detail is control of the pressure in the desorption column. This was chosen since it exhibits strong nonlinearity, varying operating conditions, and drifting parameters, as the load changes. The noise level is low so that the process is nearly deterministic in character. Very little dynamic information about the process is available during steady-state operation. If a constant forgetting factor were used, there will be sudden periodic bursts in the parameters. The variable forgetting factor gives improved performance and the system is able to follow sudden changes in the process dynamics which may arise during a set-point change on a nonlinear plant or a sudden change in load. The forgetting factor decreases to a lower value when there is a discrepancy between the model prediction and plant measurement. When the model parameters have converged to their new values and the prediction improves, the forgetting factor attains the former value, close to 1.

This modified version of the STR can be successfully used in chemical processes of a similar nature. Conventional PI controllers may not be able to give acceptable performance, since many plants have one or two loops with nonlinearities and/or frequent changes in load and set points. Satisfactory control is possible only if the regulator parameters are retuned by the operator periodically. However, for such cases adaptive control may be a viable and stable alternative to periodical manual tuning.

It is interesting to note that three pilot plants, a CO_2 absorber-desorber unit, a fractional crystallization unit, and a fixed-bed catalytic reactor are linked to a computer [59] and surveillance and control of all plants can be carried out simultaneously.

9.8.5 Batch Polymerization Reactor

The batch suspension polymerization reactor has highly nonlinear, unstable, and time-varying characteristics. In general, such process control problems require tracking and regulation of complex, often poorly understood processes with significant time delays and external disturbances. State-space description of such nonlinear, time-varying plants with time delays is both difficult and inadequate, as a dynamic model of the process is not available. A conventional PID controller cannot cope with this problem. Hence adaptive control of such unknown processes offers a very promising solution.

Kiparissides and Shah [60] have evaluated two adaptive control techniques by application to a realistic mathematical model of a suspension polyvinyl chloride (PVC) reactor. Both techniques, the self-tuning regulator of Åström and Wittenmark [61] and a globally stable adaptive controller algorithm suggested by Goodwin et al. [62], prove to be very robust and give excellent control of the temperature or the rate of conversion in the PVC reactor by manipulating the heat removal rate from the reactor jacket. Both adaptive methods require only input and output measurements for control. The overall control problem can be stated as follows:

1. Maintain the polymerization temperature as close as possible to its desired profile in spite of changes in the heat transfer coefficient or coolant temperature, or the presence of large time delays.
2. Minimize the duration of the initial heating period and the overshoot to the subsequent constant temperature phase.
3. Determine the optimum temperature profile, which produces a polymer with "desired" properties in minimum time.

The two adaptive techniques have given satisfactory performance both for regulation and set-point changes in the temperature and the conversion rate, even in the presence of measurement noise and highly nonlinear reactor dynamics. In comparison, the performance of a fixed-gain PID controller under such conditions is not satisfactory. In view of the superior performance of the two adaptive control strategies, either of these two could be implemented on an industrial-scale PVC suspension reactor.

In his recent comments on this work, Kidd [63] questions whether the use of adaptive control is justified and also raises some psychological issues such as the withering of the innovative abilities of an engineer arising out of such applications. The authors have rightly answered the first point by stating that adaptive control of such unknown, nonlinear, and time-varying processes has been demonstrated to be definitely superior to conventional PID controllers. They do agree that some other intermediate engineering solutions or ad hoc methods such as gain scheduling algorithms may also offer improved results. Regarding the second point, they have justly refuted the criticism by asserting that considerable insight, skill, creative probing, and understanding is inherent in developing and applying adaptive controllers since these are surely not "plug-in smart controllers."

9.8.6 Fixed-Bed Autothermal Reactor

McDermott and Mellichamp [64] discuss the application of their novel auto-pole-placement self-tuning controller (see also Section 8.2.11) to a fixed-bed autothermal reactor, which is a highly nonlinear distributed process. This is a very challenging problem since it exhibits non-minimum-phase characteristics when operating at an unstable state. In this tubular autothermal reactor, there is internal countercurrent heat exchange. Feed gases enter at the bottom of the reactor, travel up through an annular section, turn around at the top of the reactor, and pass down through the catalyst bed. Heat is generated in the bed by an exothermic reaction and is transferred radially, preheating the feed gases in the annulus. This thermal coupling results in one unstable and two stable steady states. The manipulated variable for the SISO controller is a small heater located at the turnaround point at the top of the reactor. The controlled variable is a measured temperature at a specific point in the catalyst bed.

The mathematical model of this reactor consists of five partial differential equations in three dimensions. By eliminating two spatial coordinates, a thirty-sixth-order nonlinear ordinary differential equation is obtained. A random signal is added to the flow rate of one of the feed components to simulate process noise. Several simulation studies are made to consider the effects of pole locations, unknown load disturbances, measurement locations, parameter variations, and also on-line tuning of the optimal closed-loop pole. The operation involves the transition of a highly nonlinear non-minimum-phase process from a stable to an unstable operating point. During this transition, there is not only a change in the process dominant (open-loop) time constant from the right to the left half-plane and an increase in the process dead time, but also a gain reversal. This auto-pole-placement STC is shown to be able to deal with this combination of process characteristic changes over such a relatively short time span, which no other self-tuning strategy can cope with. There is therefore great scope for further work on this new approach, to find new areas of application.

9.8.7 Temperature Control in a Chemical Reactor

The general-purpose adaptive controller NOVATUNE, which was discussed in Section 9.2, has been successfully applied for temperature control in a chemical reactor producing ethylene oxide [2]. The process is highly exothermic and is very much time dependent. It is essential for the temperature to be accurately controlled under different process conditions and production levels for stable operation and an improved yield. This results in a prolonged life cycle for the catalyst and also optimal performance.

Further, a stable temperature value improves the process safety, since there is a possibility of an explosion if temperature goes out of control. The conventional PID temperature and flow controllers were not having satisfactory performance, and hence there was a need for an improved control scheme based on NOVATUNE.

The NOVATUNE system is used to control two chemical reactors of the same type, but with different production capacities. The objectives of control is to maintain the oil temperature in the cooling system at a constant set-point value. Oil flow should also be kept constant, at the same time. The temperature and flow are regulated by manipulating three valves. The process is highly interactive since any change in any one of these valves will affect both the temperature and the flow on account of the interacting pressures. There is also a considerable time delay between the valves and their effect on the temperature. The values of these delays are not known and had to be estimated roughly. Further, the dynamics of the process was also basically unknown.

The disturbances in the process were due to the variations in the incoming gas and load changes. Large disturbances are caused by changes in production level or by shutdowns due to some failure in the process equipment. During such shutdowns, the process temperature must be maintained at its appropriate value so that production can restart once the defect in equipment is rectified. The stringent requirements to be met are a deviation less than ±0.1°C during normal operation and less than ±0.5°C during large disturbances or production changes.

The control structure consists of two single-loop self-tuners, one controlling the flow and the other controlling the temperature. In this scheme interaction was not expected to cause any difficulty, as the flow control loop was estimated to be about three times faster than the temperature control loop.

The self-tuning regulator is provided with internal protection against distorted information in the measured variables. The adaptation is turned off when the control error exceeds a predetermined value. The model is refreshed using the LOAD logical input if the control error hits one of the alarm limits. See Bengtsson and Egardt [2] for detailed information.

This control system has been running satisfactorily since 1983 in Berol Kemi in Sweden, and the system can be kept in automatic mode almost all the time, even during production changes.

9.8.8 Instrumentation in a Chemical Plant

The NOVATUNE adaptive controller has also been applied in an instrumentation system in a chemical plant of Berol Kemi in Sweden [1]. The NOVATUNE units are used as process stations and are connected to a central computer with man-machine functions.

The NOVATUNEs are connected to the process via remote I/O units, termed the process terminals. These terminals contain analog as well as digital I/O and are connected to the NOVATUNEs via serial communication links. Apart from basic data acquisition, the process terminals also take care of thermocouple linearization.

A number of centralized functions are handled by two computers in a dual configuration with one computer as a standby. The computer has features such as alarm and event checking, data logging, and so on.

The control system consists of forty control loops, out of which thirty are directly associated with the process flow through reactors and distillation columns. The remaining ten loops are for the support systems for steam, and so on. A majority of the loops handle temperatures, flows, levels, and pressures. Most of these loops are not crucial and are equipped with PID regulators. The remaining loops are important for the overall performance of the plant and are controlled by adaptive regulators. In all, six temperature control loops are equipped with adaptive controllers. These controllers affect heating (steam) as well as cooling (water). The disturbances from surrounding processes must be handled properly. In one control loop the pressure in a condenser is controlled using an adaptive controller in cascade with a conventional flow regulator. This control loop should be fast enough to handle the situation arising when a reactor is started or stopped, when the process is severely affected. It is found that the adaptive control can work satisfactorily in this loop.

Adaptive regulators have great potential as a standard building block in large instrumentation systems in various plants such as chemical, metallurgical, paper, water, waste water, automation, and so on.

9.9 PULP AND PAPER INDUSTRY

9.9.1 Plants with Varying Dead Time

Some industrial processes are subject not only to varying gain and time constants but also to varying dead time. In the pulp and paper industry, the time delay between the percent solids of the strong liquor and the feed liquor flow rate in a black liquor evaporate may change owing to unexplained causes. In pulp manufacturing, control in batch digesters is difficult because of changing mixing conditions in the blow tank. Bleach plant towers are also processes where dead time can change, mainly owing to the presence of channeling in the tower. In addition to changing the dead time, this channeling has a detrimental effect on the quality of the pulp. In general, a time-varying dead time will appear every time the quality of the outflow of a vessel subject to changing mixing conditions has to be controlled. It is difficult to estimate the dead time in such processes. The conventional STR cannot be applied to such processes, since it requires knowledge of the process time delay. Dumont [65] proposes an adaptive dead-time compensator (ADTC) closely related to that developed by Vogel and Edgar [66]. It is based on a recursive least-squares estimator and a pole-placement algorithm. A variable forgetting factor is used for tracking sudden changes in process dynamics. Simulations show that for the servo problem, parameter estimates are consistent and can yield the desired closed-loop behavior. For the regulator problem and a given class of systems, this method seems to be able to yield near minimum variance control, without requiring knowledge of the dead time.

9.9.2 Headbox Control

The headbox in a paper machine is a crucial part, largely influencing the paper quality. Its purpose is to distribute the water-fiber suspension on the wire as evenly and steadily as possible. Self-tuning regulators are well suited for headbox control, as they can deal with slowly changing process characteristics. D'Hulster et al. [67] apply a self-tuning multivariable deadbeat controller containing an integrator for control of a paper headbox. Comparison is made with conventional PI control and it is shown that adaptive control gives much better results.

Halme and Selkainaho [68] apply a nonlinear filter coupled with a state-space controller to a pilot plant headbox. The nonlinear filter estimates both parameters and states. A multivariable adaptive regulator is then tuned to minimize a quadratic criterion. More details of the nonlinear filter are given by Selkainaho et al. [69]. All adaptive controllers outperform conventional PID controllers. However, caution must be taken to avoid the oscillatory mode, which can be serious in an industrial headbox.

Angelov et al. [70] propose a new method for the synthesis of a quasi-optimal adaptive control applying two different criteria for some technological parameters in paper machine operation. The method is applied to the control of the headbox and drying section of the paper machine which produces printing and writing paper. The most important criterion for the headbox is fast response and that for the drying section is minimum fresh steam consumption with fixed time of the transient process. The phase plane method is used in the synthesis. A modified switching curve is used, which is dependent only on the initial conditions of the control parameters in the headbox. This switching curve is approximated by three segments by the equivalent signum-function method. The quasi-optimal control algorithm is applied with the help of a standard microprocessor multiloop controller. Results indicate that significant improvement of the process is possible; however, it is pointed out that the models used for the headbox and the drying section are not accurate. In spite of this, it is found that moisture is stabilized, transient process time is reduced, and overshoot in the system is acceptable.

Dumont [71] reviews the application of advanced control methods including adaptive control strategies in the pulp and paper industry. He thinks that adaptive control methods, together with new sensors, will form the basis for the new generation of control systems for paper and pulp mills. Some of the difficulties arising in the adoption of these techniques and future trends are also discussed.

9.9.3 Chip Refiner Motor Load

Dumont [72] discusses a motor load control scheme for a chip refiner in the wood pulp industry. The refiner motor load loop is a process whose gain is subject to a slow drift due to plate wear and also sudden, unpredictable changes in sign due to the collapse of the pulp pad. The gap at which the pad collapses depends on the wood species, chip size and density, refining zone consistency, plate wear, and so on, and is not predictable. The main objective of control is to allow set-point changes and compensate for pulselike disturbances. A self-tuning regulator has been designed to control motor load by manipulating the plate gap. It uses a recursive least-squares estimator with a variable forgetting factor, in conjunction with a Dahlin algorithm [73]. It is found that such a regulator can successfully compensate for large gain variations and take appropriate corrective action when gain sign changes occur. Results are based on plant trials performed on an industrial thermomechanical pulp production unit. This approach does not require complex logic or expensive dedicated sensors. The novelty of this scheme lies in the use of a self-tuning regulator with a Dahlin algorithm. This is probably the first application of STR to a process whose gain changes sign and also the first industrial use of the Fortescue's variable forgetting factor idea. The use of the variable forgetting factor is shown to eliminate the appearance of bursts in the estimates after a long period of steady operation.

9.9.4 Pulp Drier Control

The general-purpose adaptive controller, NOVATUNE, which was discussed in Section 9.2, has been applied to control the moisture content in a pulp drier [1]. The purpose of the plant is to dry a wet pulp sheet passing a drying section and a cooling section. The speed of the sheet is kept constant by a separate control scheme. The moisture content of the sheet is about 55% to begin with and after the drying process, it is about 10-12%. The drier works as a heat exchange system. Cold air is fed to the drier by flow-controlled fans. This air communicates thermally (without mixing) with the somewhat superheated steam entering the drier through a control valve. The temperature of the pulp sheet increases, as the sheet is heated by this drying air. Sometimes the evaporated water will recondense on the surface of the sheet, thus causing a serious process disturbance, leading at times to production breakdown. Another important disturbance is the variation in the pH value in the preceding process sections. Further, changes in production rate affect the moisture content of the pulp sheet leaving the drier. Whereas the drying section takes about 9 min to pass the pulp sheet, the cooling section takes 0.6 min. The cooling section represents dynamically a pure delay. Moisture content of the dried pulp is measured by a traversing sensor moving across the pulp sheet back and forth in a diagonal fashion.

The objective of control is to accurately maintain the moisture content at a specified value (say 11%) in the presence of the disturbances mentioned above and other disturbances. The customer requires the variations in the moisture content to be less than ±0.2% independent of the production rate. Conventional control is not effective to achieve this.

A NOVATUNE control system has been designed [1] for this purpose. The moisture content measured by the traversing sensor is software low-pass filtered and connected to the FB input of the STAR. The desired moisture content is chosen by the operator outside the NOVATUNE and software connected to the REF input. The pH value measured in an earlier process section is used as a feedforward signal. This is software connected to the FF input. The control signal defines the desired pressure in the steam battery to achieve the set-point value. This control signal acts on the control valve, changing the steam flow continuously. For more details, see Breitholtz and Egardt [1].

9.10 CEMENT INDUSTRY

9.10.1 Cement Kiln Control

A cement plant constitutes (1) the raw material mixing system, (2) the cement kiln process, and (3) the clinker grinding process. In recent years, to save energy the wet process has been replaced by a dry method. Westerlund [74] considers the control strategy for a dry process rotary cement kiln, having a capacity of 1000 tons of clinker per day. The digital control system consists of a multivariable, minimum-variance controller and a target adaptive controller for automatic adjustment of the free lime content set point. The disturbances which occur in cement kilns are mainly of a stochastic nature.

The maximum-likelihood method is applied to identify, from real process data, the cement kiln by a multivariate autoregressive average (ARMA) model. This model is used to compute a controller for quality control of clinker production. Improvements both in production time and fuel economy have resulted, without any deterioration in product quality, compared with manual control by two operators.

The controller should be able to minimize variations in clinker quality to keep the free lime content within the specified limits. The control problem has been solved using a two-step procedure. First, a controller minimizing the variations in output is designed and next a procedure is developed for automatic adjustment of the free lime content set point. As the variations in the free lime content do not follow the normal distribution but a logarithmic normal distribution, an appropriate cost function is used. A variance-restricted minimum-variance controller is computed for the process model. In many cases a controller with a simpler structure than the optimal controller structure indicated by the process model can perform satisfactorily, giving the same control quality.

A microcomputer control system has been implemented on the cement kiln process. The "minimum-variance" controller described above is the main control algorithm. It is also necessary to estimate the (logarithmic) variance (round the set point) of the free lime content in the clinker analyzed and read by the operator into the microcomputer every second hour. The new free lime content set point is then specified. Decreasing variance results in increased average product quality and reduced energy consumption. Experimental results validate this. Although this control system is specifically for a cement kiln, this approach is applicable to other plants also.

Another recent application of efficient adaptive control to an extremely challenging task is the implementation of an "extended" self-tuning control (ESTC) by Latawiec et al. [75] to control the coal-fired clinker rotary kilns in a dry-process cement plant (briefly mentioned in Section 8.2.4). The dynamics of the system show non-minimum-phase behavior. The new concept of "extended" STC allows drift rejection capabilities, as well as blow-up countermeasures. Two coupled ESTC based on the "mean deviational" model and fitted with special blow-up countermeasures, drift compensating/disregarding means, nonlinear effects and emergency measures has been operating on-line on a long-term basis in a full-size cement plant. The ESTC appears to have a lot of promise in applications to complex industrial processes.

9.10.2 Ball Mill Control

The transfer behavior of a ball mill in a cement plant is very slow and is time variant and nonlinear. Schulz [76] reports on an adaptive control capable of acting without external signals. For on-line process identification, a self-tuning reference model is used. The parameters of the real-time grinding power cascade controller are optimized in a fictitious control loop using the identified reference model. The control speed in the fictitious control loop is a multiple of the control speed in the real process. The adaptation of the reference model to the process behavior does not introduce an additional disturbance into the process. The adaptive control algorithm is based on the optimization of controller and process parameters by the gradient method. A microcomputer has been used as a digital controller. Optimal controller parameters are identified within the control time of real-time control processes by means of time transformation in the fictitious control loop. The results show that high operation security and good degree of fill of a ball mill are possible with the proposed method.

9.10.3 Blending Control

Appropriate control of the ground mix of the raw materials is very important before they are fed to the kiln. If the components of raw materials and their physical characteristics such as grindability change very much, then computer control is necessary to readjust the regulator parameters.

Hetthessy et al. [77] describe a general multivariable adaptive control system for the control of cement raw material blending. The objective of blending control is to decrease the variance of the chemical composition of the raw materials and to reject the disturbances caused by changes in the composition of raw materials. Further, every full silo content is homogenized to reduce the composition variations around the average value applying to batch-operation silos. These two objectives are suitably formulated in an appropriate loss function. By means of linearization and decoupling, single variable adaptive control loops are introduced. The inner loop has a SISO minimum-variance STR and the outer loop consists of a PI regulator with time-varying parameters, based on a simple pole cancellation.

The adaptive composition control system has been implemented on a microprocessor-based system. The parameter and covariance matrix updates are performed by a UD filter and continuous "burst" supervision is also applied. Results are given for the filling of two silos, which demonstrate the effectiveness of the proposed control scheme.

Other relevant papers in this field are by Keviczky et al. [8] and Westerlund et al. [78].

9.11 HEATING AND AIR-CONDITIONING SYSTEMS

9.11.1 Heating Systems

Self-tuning start control schemes are suitable for controlling the start-up of the heating plant in industrial and domestic buildings. Dexter [79] has developed a self-tuning controller based on on-line adaptive predictor. The controller assumes no prior knowledge of the thermal characteristics of the heating plant and building. Results of simulation of a heating plant, building, and environment show that the self-tuning controller has superior performance characteristics compared to conventional start controllers. Low-cost microprocessor implementation is possible for the control of domestic heating equipment.

Dexter and Hayes [80] examine the problem of controlling domestic electric stored-energy heating systems and show that a self-tuning control scheme can adjust the off-peak charging of the store to minimize electrical energy consumption. The regulator uses a statistical estimator, based on a lumped parameter model of the thermal behavior of the building and heating system, to predict the next day's heating energy demand before the start of the off-peak charging period. The long-term performance of the self-tuning charging controller is assessed by digital simulation of a building heated by two widely different commercial storage heating schemes and shows that the control system behaves well and achieves substantial energy savings without the need for manual commissioning. The controller is robust and able to adapt to provide acceptable operation with 10 or 20 days of start-up.

Kurz et al. [81] apply a parameter-adaptive control algorithm to control an air heater element used in climate chambers. The air temperature is controlled by the input voltage of the actuator. The outlet temperature is manipulated by changing the three-way valve which affects the water flow ratio through the cross-flow air heater. The system exhibits nonlinear static behavior. The algorithm is a combination of a recursive least-squares parameter estimator and a parameter-adaptive deadbeat controller and is implemented on an INTEL-SBC 80/10 microcomputer. The control behavior for step changes of the reference signal is satisfactory. After closing the loop, the controlled variable converges to the reference signal. Other experiments show that the algorithms can also adapt to changes in the airflow ratio.

Lozano et al. [82] apply an adaptive pole-placement algorithm for non-minimum-phase systems to a thermal process in which a heating element feeds heat into the airstream circulated by a centrifugal fan along a tube. The control input is the power in the heater and the output is the air temperature at the end of the tube. Three different airflow rates are considered and the parameters are identified for these. Experimental results show satisfactory control in that the tracking error converges to zero when parameters are switched from one set of values to another.

9.11.2 Air-Conditioning Plants

An air-conditioning plant consists of an air heater and air humidifier. The output variables are the temperature and the relative humidity of the conditioned air. This is an interesting system to be controlled, since it consists of many problems such as nonlinearities in static and dynamic behavior, drifts of the output signals due to external disturbance, and so on. Such problems are usually faced in the design of a control system in any industrial environment.

Schumann [83] applies two multivariable self-tuning algorithms for the control of an air-conditioning pilot plant. The algorithms are designed on the basis of linear input/output system models and are applied to a real process with nonlinear static and dynamic behavior. These algorithms combine recursive parameter estimation and control algorithms: a parameter-adaptive deadbeat controller and a parameter-adaptive optimal state controller. These controllers are applied to a two-input/two-output air-conditioning plant. The objective of control is to maintain the temperature and the relative humidity within a certain tolerance band, in the presence of external disturbances, such as changes in the airflow rate or in the state of fresh air. Further, the temperature and the relative humidity must be changeable independently according to the actual set points. Both these control tasks are to be managed by a single controller based on a linearized system model and it must be robust. Under these conditions, it is difficult to tune a conventional PI controller even in the single variable case and it is practically impossible in the multivariable case. A parameter-adaptive multivariable controller and can be applied directly to the plant such that it can adapt itself to parameter changes in the linearized system model which are due to inherent nonlinearities and external disturbances.

For the real-time experiments, a process computer was connected to the air-conditioning plant, thus enabling direct digital parameter-adaptive control. Experimental results demonstrate that the parameter-adaptive controllers can be applied to stabilize the controlled system after a short adaptation phase and to achieve satisfactory control performance for time-varying airflow rates and for time-varying set points of temperature and relative humidity, as long as the operating points are not changing too fast.

9.12 STEEL PLANTS

9.12.1 Control of a Blast Furnace

A self-tuning multistep predictor for the control of a blast furnace has been proposed by De Keyser and van Cauwenberghe [84]. The k-step-ahead predictor not only forecasts the process output at one sampling instant k-step-ahead, but also over a longer interval (including several sampling points) in the future. For different control policies proposed by the operator, the method predicts the process output over a wide range in the future. This enables the process operator to choose the best policy. The predictor can be used in open loop as an operator guide. The method does not require a reliable and extensive knowledge of the physicochemical model of the blast furnace. On the other hand, we can describe the blast furnace by a very simple input/output dynamic model. However, this model is updated continuously in real time and without disturbing the process. It has been shown that for decision making, this approach leads to a more desirable operation of the blast furnace.

One of the main objectives of the scheme is to hold the Si content of the pig iron as close as possible to a prescribed value. This Si content is a good measure of the heat content and the course of the blast furnace process. The control strategy is as follows: minimize the variance of the Si content such that its mean value can be lowered with the same probability for undesired situations. The Si content of the pig iron is fluctuating stochastically. This makes manual control difficult and imprecise, causing the blast furnace to operate far from its economic optimum. The purpose of the self-tuning predictor is to advise the operator whether a planned action will be favorable or not. When an intervention appears necessary, the operator can try several control strategies on the predictor and then apply the best control policy to the blast furnace. Many factors affect the process, such as the composition of the load, wind properties, distribution of the load over the surface of the furnace, and distribution of the heat in the furnace. With the aid of the predictor, the blast furnace can be operated in a more economic way. Even small improvements can result in significant savings, in view of the large quantities produced.

This strategy can be applied for better control of very complex industrial processes, which can hardly be described by a reliable and complete model. The combination of "operator plus predictor" is very important when incorporating new techniques in process control, since the operator's vast experience should not be ignored.

9.12.2 Control of Casting Speed in Ingot Casting

Shiraiwa et al. [85] incorporate a microwave casting speed meter to develop an automatic casting speed control system for the bottom pouring process of steel ingots. The control system is a pulse-width-modulated on-off feedback system and it can be regarded as a kind of sampled-data system. A model reference adaptive control has been applied to compensate the time variation of the process characteristics during the casting operation. The microwave instrument is based on the Doppler radar technique and a circuit of period measurement, and reciprocal calculation has been developed to make the response time faster. It can also measure the cast height. The casting pattern is chosen from among the patterns which are made beforehand, depending on the size of the mold, composition of the steel, and temperature of the molten steel to be cast. This automatic control results in markedly improved surface quality of the ingot and almost null rejection ratio. Such automatic control is in operation in an actual metal industrial application in Japan.

9.12.3 Control of Oxygen-Converter Steel Melting

Bunich [86] proposes adaptive algorithms for calculation of integral burden parameters (consumption of oxygen, lime, ore) which take into account uncontrolled variations of process parameters depending on the physicochemical properties of the burden, heating mode of the facility, and so on. It is noted that the process control quality is determined not only by integral values of the burden consumption but also by their distribution over the melting period. Hence the algorithm must be chosen properly for the processing of current data, apart from making measurements quite accurate. The material and heat balance conditions determine the controller structure, and its parameters (gain coefficients) are updated by the results of the melts in the course of the procedure.

An adaptation algorithm with a static characteristic of the "dead zone" type has been used to adjust the controller. Experimental data of 100 medium-carbon steel melts are used for simulation of the algorithm. The melts were carried out in a 370-ton converter. The following parameters are involved: the scrap weight, the weight of pig iron, the content of silicon and manganese in pig iron, its temperature, as well as the converter idle time between melts. The gain values are determined using the specified content of carbon and phosphorus for the type of steel melted. Simulation results are given with particular reference to the consumption of lime.

9.12.4 Strip Thickness Control

Hrusak et al. [87] consider the design of a thickness control subsystem for cold rolling mills. An adaptive controller minimizing variance of the defined "input error" is proposed. This approach avoids "turn-off" and/or "escape" effects that appear when regulators minimizing variance of "output error" are used. The strip thickness control system suggested is a two-level structure. The roll position control is performed by an analog regulator on the first level, while the digital adaptive controller forms the second control level. The control objective is to compensate for disturbances influencing strip thickness fluctuations, by generating control variables u which would make it possible to achieve a minimum of output thickness error. This is done by formulating and solving an "input error minimum variance" (IEMV) problem. The IEMV controller does not require a separate roll eccentricity elimination control.

The thickness control algorithm has been implemented on a minicomputer. However, it is possible to apply it on a microcomputer. This algorithm has been tested both on 20-roll mills and four-high rolling mills. From experimental results, it is concluded that the proposed algorithm in its general form is preferable for four-high rolling mills, in which the roll eccentricity is very significant. In the case of 20-roll mills, roll eccentricity does not play a significant role.

9.12.5 Elongation Control

The general-purpose adaptive controller NOVATUNE, which was discussed in Section 9.2, has been applied for elongation control in a skin pass mill [2]. The process is mostly batch oriented, with a specified sequence of operations to be performed for each batch. The skin pass mill is at the end of the production line. The temper rolling is done on annealed strip, so as to affect the hardness, yield point, and so on.

The objective of control is to maintain the elongation of the strip constant at the given set point. The elongation is measured as the ratio $(v_2 - v_1)/v_1$, where v_1 is the input speed and v_2 is the output speed of the strip. Because of the reduction in thickness, output speed is slightly higher than input speed. Another objective of control is to transfer the system from one control mode to another, in a very precise manner. The control modes are the sequence of operations performed by the mill, that is,

1. Slow rolling at low speed start-up
2. Acceleration to fast rolling
3. Fast rolling at production speed
4. Intermediate decelerations to slow rolling or even to standstill
5. Deceleration to slow rolling at the end of the strip
6. System at rest waiting for the next strip

Large disturbances in elongation occur during a change in operation mode. Further, a change of mode results in a change in the dynamic behavior, since the process dynamics is dependent on the operation speed, the strip dimensions, and the strip hardness. Each time a new strip enters the mill, there is a step change in the process dynamics. The rapid speed changes during acceleration, and deceleration lead to a rapidly changing process gain. These features of the process make the control problem challenging and only adaptive control can cope with them.

Further to the control problem, we also have measurement and synchronization problems. Since the speed difference (v_2 - v_1), which occur in the expression for elongation, is very small, we should measure the speeds very accurately and also synchronously to avoid errors in elongation during acceleration and deceleration.

A NOVATUNE control system has been successfully applied to solve the preceding problems. The synchronization is possible by using an extended clocking facility. In addition to the time clock, the control functions in NOVATUNE can be run with external clocks defined by external pulse generators and/or an internal clock formed by using standard function modules. The desired accuracy in speed measurement is made possible by accurate time measurement of a specified number of pulses and by extending the measurement period adequately. For details, see Bengtsson and Egardt [2]. This control strategy has been successfully applied to a skin pass mill at Krupp Stahl AG in Bochum, West Germany, in 1983 and has been running satisfactorily since then.

9.13 INDUSTRIAL ROBOTS

The purpose of a robotic manipulator control is to maintain a prescribed motion for the manipulator along a desired time-based trajectory by applying corrective compensating torques to actuators to adjust for any deviations of the manipulator from the trajectory. Some of the control algorithms suggest nonlinear compensation of the interaction forces among the various joints and control of the manipulator at the joint level or the hand level. These control techniques are inadequate since they neglect the changes of the load in a task cycle. These changes in the payload of the controlled system are significant, and so conventional control methods are not satisfactory and result in reduced response speed and damping of the manipulator. Precision and speed of the end effector are affected. Hence, we have to adopt adaptive control techniques to obtain significant improvement in the performance of robotic manipulators.

Control of industrial robots with multifunctional manipulators with high-speed continuous movements is a challenging task, as they are highly interconnected and also nonlinear. Adaptive control makes the robotic manipulator more versatile and even trainable. It will be able to perform satisfactorily under diverse circumstances, particularly in repetitive jobs.

Dubowsky and Des Forges [88] apply model reference adaptive control to robotic manipulators. They use a linear second-order time-invariant differential equation as the reference model for each degree of freedom of the robot arm. The manipulator is controlled by adjusting the position and velocity feedback gains to follow the model. A steepest-descent method is used for updating the feedback gains.

In another paper, Dubowsky and Des Forges [89] present an adaptive control algorithm which minimizes the quadratic function of the error defined as the difference between the desired state vector and the robotic process state vector. The model considered is linear and decoupled and the control action has to be preceded by a "learning period."

Cao [90] gives a simple adaptive concept incorporating an adaptive Kalman filter for the control of the arm of an industrial robot. The scheme is simple and flexible and the global results are satisfactory.

Arimoto and Takegaki [91] obtain adaptive control algorithms based on local parameter optimization for a robot described by a linear time-varying model derived from linearization around the desired trajectory. This method assures the stability of the error system. However, simulation results, referred to a nonlinear model, give errors with a maximum of 1 cm when the trajectory is a straight line with a maximum speed of 0.1 m/s.

Koivo et al. [92] use the Kalman filter in conjunction with an adaptive controller to control the positions and velocities of the joints in an industrial manipulator. The dynamics of a joint of a robotic manipulator are modeled by using difference equations. A Kalman filter is applied for the on-line estimation of the parameters of the model for the motion of the joint. An adaptive controller is designed for each joint to make the trajectory of each joint follow a desired path. This can be accomplished by minimizing the squared error between the actual and the desired paths of the manipulator. A simulation study is made of a gripper of the manipulator which has to follow a circular path. Simulation results demonstrate the feasibility of this technique. The adaptive controller has been designed for independent joint controls so that the control scheme is more accurate in repetitive motions. It is possible to make the manipulator learn by training. We can also change the desired path while the manipulator is in motion. Such a situation arises when an obstacle is recognized along the desired path by the vision system of the manipulator with the consequent need to modify the desired path. The implementation of this control scheme is relatively simple, requiring only a small memory. Adaptive control is best suited for repetitive tasks of the manipulator, such as when the objects to be moved have unknown and variable masses.

Leininger and Wang [93] apply a pole-placement self-tuning method to a six degrees of freedom robotic manipulator. This method does not require that the system be minimum phase and the open-loop mechanism stable, unlike minimum-variance STC. Further, this approach is more robust than the minimum-variance method and so more appropriate for robotic manipulators. The manipulator selected is the JPL-Stanford arm with five revolute joints and one prismatic joint. It has seven links and a gripper (end effector). Potentiometers or optical encoders are used to measure joint positions, and tachometers measure joint velocities. In one example, the manipulator carries a 2-kg load in the gripper and is required to follow a circular path and each complete revolution is to take place in 0.6 s. All six joints should be accurately controlled; otherwise the desired circular path cannot be followed. Each joint is individually controlled with a pole-placement STC. In the second example, which makes more demands on the control system, the load in the gripper has to be released part way through its trajectory and returned to its original position. These examples show the high degree of success that can be obtained by using pole-placement STC under "dynamically active" conditions. Similar applications can be made for (1) joint force/torque control, (2) gripper position/velocity control, (3) gripper torque/force control, and (4) gripper hybrid control (simultaneous force/position control).

Tomizuka et al. [94] discuss MRAC technique for mechanical manipulators described by simplified dynamical equations. Step responses are given as examples.

Koivo and Ten-Huei Guo [95] consider an adaptive self-tuning controller for each joint of a robotic manipulator. They use an autoregressive model to fit the input/output data from the manipulator. They assume that the interaction forces among the joints are negligible. The controller makes the output of each manipulator joint follow a desired trajectory specified at discrete points. This scheme does not require a priori knowledge of the physical parameters of the manipulator or its configuration. The adaptive controller algorithms are designed by the explicit method. The estimates of the model parameters and the controller gains are computed recursively on-line. Simulation studies demonstrate the applicability of the adaptive control scheme to robotic manipulators. Some aspects of the implementation are discussed.

Balestrino et al. [96] give a nonlinear model and a control law based on the unit vector adaptation with a discontinuous process-forcing signal similar to a pulse amplitude modulator. Simulation results for continuous-time control exhibit good quality with a maximum error of 0.3 mm on a straight-line trajectory with a speed of 0.75 m/s.

Nicoló and Katende [97] apply the method due to Miroshnik et al. [98] for an adaptive control scheme with a partially linear control. A nonlinear model is used for the robot. Simulation results show a maximum error of 0.35 mm when the control is continuous and 3.5 mm when the control is discrete time with a sampling time of 10 ms on a straight-line trajectory with a maximum speed of 0.75 m/s.

Dunskaya and Pyatnitskii [99] propose iterative procedures for constructing piecewise-continuous control (learning procedures), which moves the manipulator grip from an initial to a final position. The algorithms permit implementation directly in the manipulator. There is no need for preliminary determination of system parameters (lengths, masses, moment of inertia, etc.). The convergence of the algorithm is proved and the rate of convergence is found. A method of improving the response of the control is also presented. The algorithms were implemented on a computer to find the number of trials required for learning.

Tomizuka and Horowitz [100] apply MRAC systems for the dynamic control of mechanical manipulators which are highly nonlinear and complex. Further, the inertia characteristics of the manipulator depend on the payload, which is often unknown. A continuous-time MRAC scheme is used for nonlinearity compensation and decoupling joint interaction. Performance can be drastically improved by the MRAC scheme. A discrete-time MRAC system based on the independent tracking and regulation algorithm of Landau and Lozano [101] has also been applied, since the use of microcomputers is an economical and reliable method to obtain high-performance manipulator systems. Simulation studies for a three degrees of freedom manipulator indicate the definite advantages of MRAC approaches. Encouraging results have been obtained from preliminary experimental evaluation of the discrete-time MRAC scheme. The adaptive controller can make the response of the manipulator insensitive to payload and configuration changes.

Karnik and Sinha [101a] discuss the application of robots for arc welding processes and consider the problems encountered and different approaches for solving the same. The quality of the weld joint depends on (1) shape and local variation of the seam to be welded, (2) relative position of the welding arc and the joint, (3) arc length, (4) shape and size of the molten weld pool, and (5) position of the arc center relative to the pool. The primary objective of the positional control of the robot arm is to ensure that the path traced by the robot arm is exactly on the seam of the two workpieces. The position of the seam can be sensed by three principal methods, namely, (1) arc sensing, (2) tactile or proximity sensors, and (3) vision or scanning sensors. The last method is the most popular since it is noncontacting and does not require weaving. From the coordinates of the end point obtained from the sensors, signals required for the motors are evaluated and applied. However, the end point of the robot arm does not usually coincide with the desired location because of several factors. For acceptable welding, this positional error should be minimized so that welding is done properly on the seam. Several approaches to compensate for positional error along with their merits and demerits are discussed. Among these, it is shown that the self-tuning regulator, signal-synthesis MRAC, and vernier robot approach are most suitable for the control of the robot arm. The time between the updates depends on the speed of welding, which in turn depends on the type of material, type of joint, and so on. A discussion on the choice of the appropriate microprocessors for implementing the adaptive controller is presented. Details are given of the modified UNIMATE 2000 robot, which has five degrees of freedom. The movement of the robot is achieved by applying the required voltages to the motors corresponding to each axis. The desired position of the gun is the input to the master controller which evaluates the voltages required by the individual motors to attain the position. Each of the five slave controllers receives data from the encoder, closes the positional loop, and controls the acceleration, velocity, and position of the servo for the corresponding axis. Also see the papers by Cook et al. [101b], who consider microcomputer control of an adaptive positioning system for robotic arc welding and Begin and Boillot [101c], who discuss infrared vision applied to adaptive welding.

Nicosia et al. [102,103] use MRAC for industrial robots, which are highly nonlinear. In a later paper [104], they report about a generalized MRAC algorithm using a unit vector approach for the control of robots, whose models are highly interconnected and nonlinear. Case studies of a three degrees of freedom robot are given. A straight-line trajectory and a maximum velocity of 0.75 m/s and maximum acceleration of 0.75 m/s^2 are considered. Maximum errors and torques in the case of discrete-time control are given for sampling times of 10, 15, and 20 ms. The algorithm proposed [104] is simpler than those suggested by others and gives results slightly less precise than those obtained by the decoupling algorithm, which is more complex. Further work on the application of MRAC to robots with elasticity at joints is relevant.

Greco et al. [105] discuss a case study of adaptive control of a robot arm. A hydraulic industrial manipulator is controlled by applying MUSMAR self-tuners. Each robot axis is driven by a hydraulic motor. The model considered describes the joint motion along the x-axis. A 12-bit D/A device converts the signal coming from the digital controller into servovalve excitation current of the hydraulic motor. A 14-bit resolver supplies the angular position of the robot arm to the digital controller. The performance of the MUSMAR self-tuner is compared with that of other conventional self-tuners. In respect of stability, robustness, quantization noise, parameter variations, the MUSMAR self-tuner exhibits much better performance in the control of the robot arm.

Vukobratovic et al. [106] report on control synthesis for robotic systems with a variable payload, first in a nonadaptive manner. If such nonadaptive control cannot stabilize robotic manipulator trajectories, they suggest an adaptive control scheme, which includes an algorithm for on-line identification of variable payload and adaptation of local gains. The efficiency of the dynamic estimator and of the complete adaptive algorithm is illustrated by an example of a hydraulic six degree of freedom manipulation system. An earlier paper by the same authors is also of interest [107].

Lee and Chung [108] consider an adaptive control strategy for computer-based manipulators. In later papers Lee and Chung [109,110] suggest an adaptive perturbation control which can track a time-based joint trajectory as closely as possible for all times over a wide range of manipulator motion and payloads. This strategy differs from other adaptive controls by taking all the interactions among the various joints into consideration. The adaptive control is based on the linearized perturbation equations in the vicinity of a nominal trajectory. The highly coupled nonlinear dynamic equations of a manipulator are linearized about the planned manipulator trajectory to obtain the linearized perturbation system. The desired trajectory is specified by an interpolated and smoothed trajectory whose angular position, angular velocity, and angular acceleration are known at every sampling instant. The controlled system consists of feedforward and feedback components which can be computed separately and simultaneously. Newton-Euler equations of motion are used for the inverse dynamics of the manipulator, and the feedforward component computes the nominal torques, which compensate for all the interaction forces among the various joints along the nominal trajectory. The feedback component computes the perturbation torques, which reduce the position and velocity errors of the manipulator to zero along the nominal trajectory. The feedback component consists of a recursive least square identification scheme to identify the system parameters in the perturbation equations. Then an optimal adaptive self-tuning control is designed to control the linearized perturbation system about the nominal trajectory. The parameters and the feedback gains of the linearized system are updated and adjusted at each sampling period to get the necessary control effort. The total torque applied to the joint actuators will therefore consist of the nominal torques computed from the Newton-Euler equations of motion and the perturbation torques from the optimal adaptive self-tuning control of the linearized system. Because of the parallel structure, computations of the adaptive control can be implemented by low-cost microprocessors. This adaptive control strategy reduces the manipulator control problem from nonlinear control to controlling a linear system about a desired trajectory. A computer simulation study was conducted to evaluate the proposed adaptive control strategy for various loading conditions along a given trajectory for a three-joint PUMA manipulator. In all the cases the adaptive controller shows better performance than the PID controller. The performance evaluation is based on position tracking along the desired joint trajectory and the final position error of the manipulator. This simulation is, of course, a study of an ideal system. Practical implementation requires further investigation of the effects of gear friction, backlash, control device dynamics, and flexible link structure.

Andrew [111] makes some very important comments on adaptive robotics. He points out that existing robotics systems depend on precise mechanisms and are inefficient from the point of view of energy. He opines that adaptive control should permit less expensive hardware and reduced energy consumption, at the cost of increased complexity of software. He also considers some important trends with special reference to some aspects of artificial intelligence applied to robotics.

9.14 SATELLITE ATTITUDE CONTROL

One of the most sophisticated and complex applications of adaptive control is high-precision satellite attitude control. Dodds [112] deals with microprocessor implementation of a three-axis, gas jet, high-efficiency, high-precision satellite attitude control having application in any spacecraft where precise pointing is required within stringent mass and power limitations. Some novel features are incorporated in this scheme. Parabolic switching boundaries are employed with parameters which adapt to a disturbing acceleration estimate. Both fuel consumption and thruster operation frequency are minimized. The disturbing acceleration estimate is obtained from a third-order state estimator, together with angle error and rate estimates. Time optimal recovery from large initial angle errors and rapid response to step changes in disturbing acceleration are achieved. In addition, stable control is achieved with disturbing acceleration approaching the control jet acceleration. A slew control algorithm is also included which enables the same control law to be utilized for fuel optimal slewing through unlimited angles, one axis at a time. Such a control system is incorporated in Europe's x-ray astronomy satellite, Exosat [113], which carries scientific instruments with ultimate angular accuracies of the order of 5 arc sec. High pointing accuracy is maintained without unnecessary fuel usage in the presence of disturbances such as those caused by orbit change jet firings or couplings of control torques from the other axes due to misalignment between the control axes and principal axes. The control system is also able to accept mismatches in jet torque such as produced by shifting of the vehicle center of mass during fuel usage. This also implies the ability to accept a considerable degree of nonlinearity of thruster. Though numerical examples and simulation results are given for Exosat, this approach is applicable to any other satellite mission by appropriate choice of the parameters in the algorithms. When microprocessors of high capacity are space qualified, extremely rapid response to high-level disturbing torque steps will be possible.

Astronomical phenomena occur at irregular intervals and for a very short duration. Satellites are used to observe such phenomena. It is necessary to reorient the satellite to the celestial phenomenon very quickly to make the observation time as long as possible. An attitude control system makes this reorientation possible. Some control systems control each of the three principal axes of the satellite separately, and these require more time for reorientation or slew. A method for a simultaneous slew around the three principal axes was proposed by van den Bosch and Jongkind [114] enabling rapid reorientation in space, by means of one single slew around the so-called Euler axis. An MRAC system with model updating accomplishes the desired control objectives. Brief consideration of this method is given in Section 8.10. Satisfactory control behavior is obtained, even in the presence of disturbances and large parameter variations in the satellite or in the controller. A satellite is a highly nonlinear multivariable system with much interaction among the three axes. The reference model neglects some nonlinearities and the interaction due to gyroscopic coupling. This results in a significant difference between the satellite and its reference model. All states can be measured accurately.

van den Bosch and Tjahjadi [115] consider a feasibility study concerning the Dutch-American-British IRAS satellite. This satellite is provided with reaction wheels to realize slews. They apply the principle of model updating with the MRAC design, to obtain one single large-angle, three-axis slew. Results are given for a conventional PD controller, an MRAC-based design, and an MRAC-based design with model updating. Comparison is made for all these designs in respect to energy consumption and response times, considering wheel-speed saturation as well as neglecting the same. The results clearly establish that the MRAC design with model updating has improved performance with reduction in energy consumption and response times. The selection of adaptive gains is also less critical when model updating is used. In fact, large adaptive gains yield an improved performance. Further, model updating does not affect the asymptotic stability properties of the MRAC design.

9.15 TRANSPORTATION SYSTEMS

9.15.1 Multilocomotive Powered Trains

A decentralized self-tuning control scheme for handling very long multilocomotive powered trains including braking operations has been proposed by Wong and Bayoumi [116]. The objective of the controller is to maintain a specified velocity profile and to minimize the effect of noises due to grade changes while keeping the absolute value of all the coupler forces below a given limit. The system parameters are unknown, since they may change with loading and unloading at stations. The controller is superimposed to an off-line throttling and braking schedule. This is an open-loop strategy to determine the approximate throttling and braking inputs needed for accelerating or decelerating the train and to linearize the nonlinear rolling resistance due to grade and velocity changes.

The control algorithm is tested by simulating a three-locomotive 60-car train operating over various grades and velocity profiles. Simulation studies show that the velocity offset and the oscillating behavior of the coupler forces in the steady state were eliminated with the use of the localized self-tuning controllers. These controllers have a simple structure and have good performance to meet the control objectives for handling train operations. The adaptive controllers have the following advantages:

1. Little a priori information about the system parameters is required for the controller.
2. Only local measurement of the locomotive velocities is required for control of a locomotive. No observer is needed for the estimation of other states.

3. The adaptive controller is relatively insensitive to parameter variations and small nonlinearities.
4. Stochastic disturbances and measurement noise are tolerated.
5. Minimum variance for the velocity derivation can be achieved.
6. The control scheme is easy to implement using microprocessors.

The localized STCs are promising solutions for future train operations.

Murthy and Fenton [117] have proposed a velocity-adaptive lateral controller. The lateral controller should meet the following requirements:

1. Accurate following (tracking errors ≤ 0.1 m) of the desired path in both straight and curved sections
2. A response nominally independent of velocity
3. A response relatively insensitive to changes in the tire-rail interface
4. A comfortable ride during all maneuvers, except possibly in cases of very large disturbances and/or emergency operations

This controller has been implemented using a 16-bit microprocessor and installed in a dual-mode test vehicle.

The controller-vehicle combination was evaluated under field conditions. A wire-follower configuration was employed to provide both a lateral reference and a lateral position error. The vehicle was automatically steered on straight sections at speeds up to 35.8 m/s and small-radius (100-m) curves were traversed at speeds up to 17.9 m/s. Results show excellent lateral control, close tracking, a comfortable ride, and good insensitivity to disturbance force. During full-scale testing over a six-week period, cross- and side-winds up to 10 m/s were frequently present. These winds had an almost negligible effect on the performance of the system. The selected control algorithm was realized using some 10% of the available computation time. Thus the microcomputer can also be utilized for monitoring of system functions such as oil pressure, alternator output, and engine temperature, and it can serve as a source of vehicle-based intelligence to enhance system reliability and reduce the response time to various emergencies.

9.15.2 Electromagnetic Suspension System

Magnetic suspension is one of the new ideas being actively explored for application in transportation systems. Sinha and Hulme [118] consider the model reference adaptive control of a single degree of freedom suspension system with nonlinear characteristics. The design objective is to generate the feedback gains so that the nonlinear system follows the reference model. The control algorithm is implemented through a floating-point arithmetic subroutine. The effectiveness of this adaptive algorithm is simulated using ACSL (Advanced Continuous Simulation Language). The responses obtained in the simulation study demonstrate the feasibility of the proposed adaptive control strategy. Despite the fast response of the suspension system, it is shown that the adaptive technique is amenable to microprocessor control, although care is required in studying the sampling frequency and the word length of ADC and DAC. Further studies are necessary to consider the effect of microprocessor computing time on system stability, and the effects of certain key parameters in real-time control. The work on the single degree of freedom system has to be extended to multidegree of freedom systems. It is expected that adaptive control will have significant influence on the control of multimagnet suspension systems.

9.15.3 Car Steering

The car steering system presents some interesting problems. The dynamics is highly dependent on the speed of the car. The zeros of a sampled model move from inside the unit circle for low speeds to well outside it for higher speeds. The system behavior is therefore changing from minimum-phase to non-minimum phase. Andersson and Ljung [119] have considered adaptive control of such a car steering system as a test case. They use the model developed by Fenton et al. [120], which is of the fifth order. When disturbances like side-wind effects are also considered, the model contains 15 parameters to be tracked. This may lead to slow adaptation. It may therefore be necessary to use lower-order models in the adaptive control mechanism, which raises questions of accuracy. The car steering system can be a good test case for evaluating different adaptive schemes which can cope with these problems.

The adaptive regulators tried by Andersson and Ljung are the minimum-variance self-tuning regulators. The extended least squares method of parameter estimation is used. A forgetting factor ranging between 0.95 and 0.99 has been incorporated. The input is the angle of the steering wheel and the output is the lateral position of the automobile. The objective of the work reported [119] is not to suggest what the "best" adaptive control policy should be, but to stress the sort of difficulties encountered in applying adaptive control to more complex systems.

This study has identified certain important issues which arise in realistic applications of adaptive control:

1. There is a trade-off between the complexity of the model and tracking capability.
2. It is necessary to use lower-order models that are simpler than the actual system.
3. When we use such simplified models to find the regulator, the question of the sensitivity of the chosen regulator arises with respect to the accuracy of the model. This therefore raises a dilemma: To achieve adequately fast adaptation, simplified models must be used. But these simpler models are not good enough for the demanding goals of adaptive control.

The preceding issues indicate that it would be interesting and worthwhile to study different approaches for the application of adaptive control.

9.16 POWER PLANTS

9.16.1 Nuclear Power Plants

Multivariable model reference adaptive control of thermal and nuclear power plants brings about striking improvements in performance. Implementation of real-time multivariable adaptive control algorithms is possible at a reasonably low cost using microprocessors.

Irving and Dang van Mien [121] present a multivariable MRAC algorithm for the control of the steam generator of the second fast breeder French nuclear reactor called Super-Phenix. The plant consists of the reactor heating the sodium of a primary loop, which heats a secondary sodium loop, which, in turn, heats a steam-water loop by the intermediary of the steam generator. The objective of the steam generator control is to supply to the turbine, in the quickest way possible, a specified thermal power equivalent to the specified steam flow rate at specified pressure and temperature. A specified steam temperature at the input of the turbine is equivalent to a given sodium temperature at the sodium output of the steam generator.

The mathematical model based on thermo-hydraulic fluid mechanics is a highly sophisticated nonlinear model with several hundred state variables. However, a simplified linearized model of the vector difference type [122] has been developed for the controlled system. The adaptive control method is of the series-parallel direct model reference type. The adjustment algorithm is the standard least squares estimation method, with hyperstability conditions, and using a forgetting factor and threshold operation. The algorithm is simple and robust. Numerical results are presented to illustrate the quickness of the different responses, showing significant improvements compared with conventional controllers.

9.16.2 Solar Plants

The control problem of processes in the solar plant requires the accurate study of its dynamic performance, because of the variability of the possible solar circuit configurations and of slowly time-varying operating conditions of the controlled system. Adaptive control is an appropriate technique to control such solar plants in the presence of disturbances in the input due to solar radiation and external temperature.

Fortuna et al. [123] deal with the control of an experimental pilot solar plant. They apply least squares and generalized least squares (GLS) methods for on-line identification of the multivariable process model. These algorithms are implemented on a microcomputer. Simulation studies confirm the efficacy of the proposed identification algorithms. They suggest a two-level adaptive control structure, where the high level has the function to identify the process by a fixed program by a microprocessor and to determine the control law, while the lower level has the function to control the process on the basis of this control law. This sort of control strategy appears to be best suited for such a solar plant. Further studies on actual implementation are necessary.

9.16.3 Steam Turbines

Oliva et al. [124] develop a modular adaptive controller for a small steam turbine based on least squares identification, pole assignment, and state variable feedback. The objectives of design are to develop a controller for a nonlinear system whose parameters are known approximately; to devise a scheme which could be split into separate modules for easy implementation on one or more microprocessors; and to use relatively simple building blocks which can be modified as required.

In the controller designed, three completely independent sections are implemented as separate subroutines in the control computer, but they could just as easily be run on separate but interconnected processors.

The plant to be controlled is a single-stage impulse-type de Laval steam turbine. The load was provided by an alternator driven through a reduction gear and delivering power to a resistance bank.

Experimental results are given comparing the results of the proposed adaptive controller with those of other types of controllers. The adaptive algorithm provided good control under all conditions, the only difficulty being the size of the perturbations caused by the PRBS (pseudorandom binary sequence) signal, which is added to the control input when identification and adaptive control routines are being used. The adaptive controller compares favorably with all the other controllers tested.

Other relevant work is given by Oliva [125].

9.17 MEDICAL APPLICATIONS

9.17.1 Adaptive Control of Blood Pressure

The self-tuning approach for adaptive control of blood pressure is suggested by Walker et al. [126]. Three-parameter adaptive algorithms based on RLS and RML procedures have been applied to the control of blood pressure by infusion rate of a vasco-active drug on an anesthetized dog. Closed-loop control is provided by a minimum-variance algorithm. Simulation results and experiments on animals show excellent performance and conform to the behavior predicted by analytical methods. Further work is necessary to extend this technique to more general models and to the multivariable case to accommodate the use of two or more drugs and to the automatic identification of the time delay.

Slate and Sheppard [127] propose an adaptive, multiple-mode, multirate, discrete-data controller for the control of drug infusion rate in hospitalized patients. It is implemented on a microcomputer system using model-based techniques. Simulation results, animal experiments, and clinical evaluation with postsurgical patients show superior performance compared with earlier automated infusion systems.

The MRAC approach is adopted by Kaufman et al. [128,129] for a computer-based feedback system regulating the infusion rate of a drug (nitroprusside) to maintain the desired blood pressure. Since the transfer function parameters are also variable with time, an adaptive algorithm is suitable for maintaining both steady-state and transient specifications. Simulation studies have shown that the MRAC procedures can be successfully applied to the control of blood pressure, in spite of uncertainties in time delays, gains, and time constants. Further work is possible to consider additional outputs such as arterial blood flow rates and to include two or more drugs.

Research in this area is very valuable for constant monitoring and adjustment of physiological variables for critical-care patients who live within very narrow limits of physiological balance. It is also necessary to develop a compact system for use in the operating room during and immediately after open-heart surgery.

9.17.2 Adaptive Control in Cardiology

Adaptive Control of the Total Artificial Heart

McInnis et al. [130] developed a microcomputer-based adaptive control system for the total artificial heart. Replacement of the natural heart with a total heart necessitates the solution of problems in the control of cardiac output and blood pressures, since the surgery opens several essential natural feedback loops. One of the primary natural control mechanisms for cardiac output is the heart rate. The signal which normally controls heart rate is available in the form of the P-wave on the atrium. Because of the great correlation between the P-wave rate and the body's demand for cardiac output, P-wave synchronized control is an important area of research.

In the system developed by McInnis et al. many functions of the control system of the natural heart are duplicated by detecting the P-wave on the atrium, by triggering the beginning of systole after a time delay, and by adjusting the duration and intensity of the systole of the artificial pump. The performance measure of the system is chosen to maximize stroke volume of the artificial pump and to hold blood pressures at desired levels.

The development of an adaptive control system for the pneumatic drive is very desirable, because of the complexity of the system to be controlled and the variability of physiological conditions. When major changes in the operating conditions occur, the parameters of the control algorithm should be adjusted to reflect these changes.

An adaptive control algorithm has been developed which produces maximum pump efficiency while maintaining the hemodynamic conditions within normal physiological limits under varying physiological conditions.

Experimental results both in vivo and in vitro demonstrate the detection of the P-wave and the synchronization of the artificial heart. Computer simulation results are also given which show that the adaptive control algorithm performs well for the systemic circulatory system if the relevant parameters are carefully chosen. This algorithm can be extended to the whole circulatory system (systemic plus pulmonary) by taking the coupling effect between these two systems into account. Some simplification of the adaptive control algorithm is possible by ignoring some unimportant parameters to reduce computing time and memory. Physically meaningful parameter bounds can also be introduced to improve the transient condition of the parameter estimation and controlled outputs.

As increased understanding of the regulation of the heart is achieved, it is possible to develop a refined microcomputer-based artificial device which performs precisely the function of the natural heart it replaces.

Adaptive Control of Assist Devices

A variety of cardio-circulatory assist devices are used by cardiologists. Control engineers are also involved in the problem of controlling these assist devices or artificial hearts in order to aid recovery of the failing heart or to supply blood in proper quantities. A device is therefore necessary for a precise control of hemodynamics.

Kitamura and Akashi [131] propose a design method for a computer-controlled system for mechanical cardio-circulatory assist. In this system a blood pump is driven pneumatically by a hydraulic stepping cylinder, and a minicomputer is used as a real-time controller. The algorithm designed for left ventricular assist has an adaptive control scheme: (1) state estimation of hemodynamics, (2) identification of mechanical parameters in a left ventricle and systemic circulation, and (3) optimization of settings of the pump driver.

The hardware structure of the designed system consists of a minicomputer as a real-time controller, a driving system, a blood pump, artificial valves, and cannulas. The driving system is composed of a hydraulic stepping cylinder and a pulse generator. The position of the cylinder with bellows of steel runs at 300 μm/pulse. The output signals of the pulse generator drive the piston in a trapezoidal motion. The pulse generator has a digital delay circuit, by which the blood pump can be driven synchronously to the electrocardiograph with a given delay time T_d of the pump systole to that of the heart.

Experiments are conducted in a mock circulation system for parameter identification and for adaptive control for left ventricular assist. The chosen performance criterion was constant value control of the aortic pressure. Its reference value was 100 mm Hg. The control error was within 5%. It takes about 11 s in CPU time for a cycle of adaptive control from the data acquisition to the determination of the optimal control law. This work establishes the fact that adaptive control strategies are possible for cardio-circulatory assist.

McInnis et al. [132] present a computer-based adaptive control system for left ventricular bypass assist devices consisting of air-driven diaphragm pumps. The control systems of ventricular assist devices, as well as total artificial hearts, currently in use are still manual or semiautomatic with regard to regulation of the blood pressures of the recipient. The application of recent developments in adaptive control algorithms to stabilize the blood pressures of the circulatory system at desired levels by means of beat-by-beat adjustment of the pump stroke volume is the main contribution of McInnis et al. This is realized by an improved drive system utilizing a hydraulic servoloop to produce the pneumatic pressure that drives the assist pump.

The system provides for (1) synchronization of pumping with ECG signals; and (2) control of atrial pressure at desired levels. The system design is based on a self-tuning PID controller with pole placement. For the system under study, the changes of operating conditions are mainly due to (1) heart rate, (2) strength of the natural ventricles, and (3) peripheral resistance. The system is time varying. A recursive least squares algorithm with exponential forgetting factor is used. To avoid the severe problem of estimator wind-up and bursts, when the process is not well excited, a variable forgetting factor has been suggested. Since a reduced- (second-) order model is used to represent the process, there is always a modeling error in addition to measurement errors. In view of this, a least squares algorithm with dead zone and covariance resetting [133] was used for parameter estimation.

The performance of the system has been demonstrated by using a mock circulatory system to evaluate the controller in the presence of changes in heart rate, in systemic peripheral resistance, and in contractility of the simulated natural left ventricle. When there is an increase in atrial pressure, the system is found to respond with an increase in the stroke volume. Following major changes in the circulatory system, the control algorithm retunes itself and restores the system to the desired state.

Reference may be made also to the earlier work of McInniss et al. [134a, 134b].

From an engineering point of view, we can say from the work reported so far that adaptive control is attractive for medical applications.

9.18 BIOTECHNICAL SYSTEMS

9.18.1 Fermentation Process

The fermentation process is nonlinear and time varying. The ability to control such fermentation processes accurately and automatically in the production of microorganisms is of great importance to several fermentation industries in maintaining the quality, uniformity, and reproducibility of the product, and in increasing yield while reducing production costs. Williams et al. [135] have proposed direct digital adaptive control of a real small-scale plant. The fermentation process is simulated by a deterministic model to determine if the adaptive control technique is feasible. The main objective of the model was to verify that the adaptive control technique would control a chosen state variable through a specified trajectory, so that growth was maximized. The technique was validated on a real plant, and investigations were made under various conditions and control references. The complexity of chemical reactions which take place during the fermentation process and the poor quality of modeling because of its highly complex and poorly understood dynamics provide a good test of the feasibility of adaptive control. In the present scheme, an on-line computer together with a multivariable adaptive control package is interfaced with the fermentation plant. Results are presented to show the successful performance of the adaptive technique achieved by controlling various process variables, in both SISO and MIMO configurations. These studies indicate that the fermentation process can be accurately controlled, but true optimization is dependent on careful choice of control parameters and performance criterion. The choice of estimator and controller sampling times as well as the choice of the weighting matrix for the input variables is important as they affect the quality of control. The MIMO implementation has successfully achieved its control objectives, showing that the adaptive control can function on a reduced-order description of the complex fermentation process. This application has highlighted the fact that adaptive control strategy can be successfully utilized even with limited understanding of complex fermentation and other such microbiological processes. There is scope for further work on other biochemical engineering applications. Reference may be made to the work of Bastin et al. [136,137].

9.18.2 Bacterial Growth Process

Parameter estimation and nonlinear control of microbial growth systems have attracted the attention of researchers in recent years, and a large number of papers and references are available in the proceedings of the first IFAC Workshop on Modelling and Control of Biotechnical Processes [138].

Dochain and Bastin [139] consider adaptive identification and control algorithms for nonlinear bacterial growth systems. A similar idea has also been used for dissolved oxygen adaptive control in waste-water treatment [140], which is discussed in the next section.

The process is described by a nonlinear state space representation and is then approximated by a discrete-time time-varying model which is linear in the parameters and in the control input, though globally nonlinear. The time-varying parameters in this model, that is, the growth rate and a yield coefficient, have a physical meaning and are identified in real time using a standard RLS algorithm. The parameter estimation algorithm is applied in conjunction with minimum-variance and Clark-Gawthrop controllers to obtain adaptive controllers for (1) substrate concentration and (2) production rate. For each of these two cases, simulation experiments confirm the effectiveness of the parameter estimation algorithm and the adaptive control algorithms. A theoretical proof of convergence of the substrate control is given. The case of convergence in production rate control is much more difficult since the algorithm involves two cascaded steps together with the estimation of a truly time-varying parameter.

This approach is well suited to such industrial applications as waste treatment in sugar industries. There is need for further investigations to extend this approach to multistage, high-order, highly nonlinear systems as in the complex five-stage, 12 parameter reaction reported by Bastin et al. [137].

9.19 OTHER APPLICATIONS

9.19.1 Flexible Manufacturing Systems

Flexible manufacturing systems are designed to produce workpieces with very few operators working. Generally, the feed rate of work material is kept low to protect the tool from breakage caused by unpredictable load. It is possible to employ adaptive control systems to prevent the tool from breakage by adjusting cutting conditions automatically. One method is the adaptive control constraint method, which maximizes the feed rate under the condition that several measured or estimated variables of the cutting process are kept below or at the respective constraint values. In a milling machine the cutting force of the milling process varies periodically, and the values of the parameters of the milling process also vary widely. Further, we have to control many variables which generate the tool path, especially since workpieces of complex shape have to be cut. Hence the problems of adaptive control are not simple.

Watanabe and Iwai [141] consider four types of algorithms to find the best control algorithm for an adaptive control constraint system. The system consists of a milling machine tool and a minicomputer. Adaptive control, numerical control, and servo control are simultaneously performed by the computer. The adaptive control program generates the desired feed rate according to the change of the cutting process. The numerical control program generates the increments of the desired x, y coordinates of the table. The servo program calculates the errors of the x, y coordinates of the table from the desired x, y coordinates. The velocities of the table in x, y axes are controlled by the hardware circuits so as to be in keeping with the errors in order to obtain the accurate tool path.

In the adaptive control constraint system, the peak value of the cutting force is kept constant by self-adjustment of the feed rate. The system is a high-order discrete-time system. Its stability has been studied, and the transient responses of the system to the variation of process parameters are simulated on a computer. Study is made of four algorithms: (1) an ordinary continuous integral algorithm, (2) a discrete integral algorithm, (3) a nonlinear algorithm in which the optimum feed rate is calculated by using a division on the assumption that the force is proportional to the feed rate, and (4) an improved division algorithm, which measures the actual feed rate in order to calculate the optimum feed rate more accurately. Experimental results show that the last algorithm is the most stable one and has the best response when process parameters vary. This algorithm is also best suited, as the overload generated by the change of the cutting process is the lowest.

9.19.2 Glass Furnace Control

Haber et al. [142] have presented the modeling, identification, and adaptive self-tuning control of a glass furnace. The special feature is the application of a mobile process computer laboratory. The computer is provided with an interactive process control program package. The objectives are to improve glass production and to provide some innovations in the design of glass furnaces. Identification was performed without interrupting normal production. The proper structure is obtained by repeated parameter estimations. During self-tuning control, the operation becomes more uniform, as confirmed by the local operators. Identification and control developed in these experiments can be widely used by applying microprocessors.

From the point of view of glass technology, the following controls are necessary: (1) control of glass level, (2) temperature control, (3) control of furnace pressure, and (4) control of the proportion of the fuel gas to the combustion air. The most important requirement to be met is to keep the glass level at a constant value, since the quality of the final product depends on the level of the melted glass as a result of gravimetrical feeding. The control of the level is of special importance, even in the presence of the disturbances acting on the process. The disturbing factors appear in the form of slow and rapid stochastic noise effects. The choice of STRs is appropriate to compensate for stochastic disturbances. Using the identification results, the type and structure of the STC are chosen. As a result of adaptive control, the fluctuation in the glass level becomes remarkably less, and the following useful effects are also noted:

1. Distribution of the mixture on the surface of the melt becomes more uniform.
2. The glass can melt faster than earlier.
3. Periodical fluctuation of the vault temperature decreases.
4. The average temperature increases for the same quality of fuel gas.

This technique can also be extended for the analysis and control of other technological processes.

9.19.3 Waste-Water Treatment

Waste-water treatment has become important to reduce energy consumption and to maintain high effluent quality. Large flow-rate variations, changes in the concentration of pollutants, and inability to measure key process parameters lead to difficulties in process control. Ko et al. [140] apply nonlinear adaptive controllers for the control of dissolved oxygen (DO) concentration levels in the activated sludge process, which is one of the important processes in a waste-water treatment plant. This is made possible by the availability of on-line computers and reliable DO concentration sensors. The control of DO concentration above a critical level will supply enough oxygen to maintain the activity of the microorganisms.

Adaptive control makes possible the implementation of control systems, even though the parameters of the process are unknown, and provides for automatic adjustment to step changes in operating conditions or slow continuous variations. The DO dynamics are represented by a bilinear model to which we look for both parameter estimation and control. A least squares parameter estimator is combined with a minimum-variance control algorithm to obtain an adaptive controller. This approach allows us to identify the mass transfer coefficient and the oxygen uptake rate, which provide useful information concerning the state of the activated sludge process. Robustness in the presence of bounded error terms and global convergence are the features of this class of adaptive control algorithm [143].

Simulation experiments demonstrate the effectiveness of the adaptive controller even in the presence of noise for a square wave set point. Another relevant paper in this field is by Goodwin et al. [57].

9.19.4 Dry Etch Process in VLSI Manufacture

Dry etch processes are relatively new in the manufacture of VLSI components. The control parameters cannot be found analytically since the etch process is not yet well understood. Boehm [144] uses adaptive control techniques to solve the problem of working-point variations. The adaptive algorithms are included in a microcomputer as part of a data acquisition and monitoring system. The dry etch equipment is interfaced with a microcomputer containing a real-time control system with adaptive control methods. Four nonlinear process variables are adaptively controlled with a recursive least squares estimation method in conjunction with a deadbeat controller. This system also handles direct digital control and sequencing control program modules. The user can select adequate modeling/controller combinations between different parameter estimation methods and different types of controllers. The user can either store the process data or print them on a printer. The RLS/DB combination mentioned above needs no specific knowledge of the control loop, except of the model order and the scan interval. The object of the adaptive control is to control simultaneously as many variables as possible. The product variable of most interest is the etch rate obtained by a particular gas or gas mixture. This depends on HF power, flow rate, pressure, and temperature, which are interdependent. Since there was no means for direct control, the temperature was regarded as an uncontrolled interference. At the start of the etch process, the wafers had about the same temperature. The other variables are controlled simultaneously by adaptive control. The starting fluctuation of the model parameters and their behavior after set-point changes are shown. It is found that the control parameters adapted reliably even in slightly nonlinear plants. The control quality of adaptively controlled variables was better than the previously used PI control. With this microcomputer-supported adaptive control system, a process engineer saves time in the process development phase, because of

1. No search for controller parameters at different working points in nonlinear systems.
2. Reduction of runs as set points can be changed in one run.
3. Adaptation of controller parameters automatically to new operating levels.
4. Improved reproducibility, and
5. Easier, faster, and more reliable data evaluation.

9.20 CONCLUDING REMARKS

The field of adaptive control is rapidly developing on account of (1) definite needs in certain industries, (2) potential and demonstrated capabilities, and (3) the decreasing cost of microelectronic devices.

Many applications have been reported in the literature in the past two decades, though there has been a spurt in the number since the late 1970s due to the advent of microprocessors. Since a detailed review of the entire range of applications is beyond the scope of this book, the present chapter surveys some recent developments. Some have been successfully applied in real processes, some are at the stage of feasibility and simulation studies, and some are pilot plant studies. A broad view of the present position and future possibilities is given. Many complex processes which were considered difficult, if not impossible, can now be adaptively controlled, because of the availability of versatile microelectronic devices. This trend, which is a veritable revolution, will further expand in the future.

At the present juncture, neither the whole spectrum of possible applications nor the ultimate constraints can be foreseen. There is, however, need for further theoretical foundations and more practical experience.

REFERENCES

1. Breitholtz, C., and Egardt, B., Adaptive control in an instrumentation environment, Tech. Report No. 1028V/0040V, ASEA AB, Västerås, Sweden (1985).
2. Bengtsson, G., and Egardt, B., Experiences with self-tuners in the process industry, Proc. IFAC World Congress, Budapest (1984).
3. Bengtsson, G., Industrial experience with a microprocessor-based self-tuning regulator, Proc. IEEE-CDC, Ft. Lauderdale, Fla. (1979).
4. Sheirah, M. A., and Abd-El-Fattah, M. M., Improved load-frequency self-tuning regulator, Int. J. Contr., 39: 143-158 (1984).

4a. Aly, G., Abdel-Magid, Y. L., and Wali, M. A., Load-frequency control of interconnected power systems via minimum variance regulators, Elec. Power Syst. Res., 7: 1-11 (1984).

5. Kanniah, J., Tripathy, S. C., Malik, O. P., and Hope, G. S., Microprocessor-based adaptive load-frequency control, IEE Proc., 131, Pt. C: 121-128 (1984).
6. Agathoklis, P., and Hamza, M. M., Comparison of three algorithms for load frequency control, Elec. Power Syst. Res., 7: 165-172 (1984).
7. Vajk, I., Vajta, M., Keviczky, L., Haber, R., Hetthessy, J., and Kovacs, K., Self-tuning load-frequency control of the Hungarian power system, Automatica, 21: 129-137 (1985).
8. Keviczky, L., Hetthessy, J., Hilger, N., and Kolostori, J., Self-tuning adaptive control of cement raw material blending, Automatica, 14: 525-532 (1978).
9. Ghosh, A., Ledwich, G., Hope, G. S., and Malik, O. P., Power system stabilizer for large disturbances, IEE Proc., 132, Pt. C: 14-19 (1985).

9a. Ghosh, A., Ledwich, G., Malik, O. P., and Hope, G. S., Power system stabilizer based on adaptive control techniques, Preprints IEEE-PES Winter meeting, Dallas, Tex. (1984).

9b. Malik, O. P., Ghosh, A., and Hope, G. S., Implementation and test results of adaptive power system stabilizers, Proc. IEEE Int. Symp. Circuits and Systems (ISCAS), pp. 140-143 (1984).

9c. Sengupta, D. P., Narahari, N. G., Boyd, I., and Hogg, B. W., An adaptive power system stabilizer which cancels the negative damping torque of a synchronous generator, IEE Proc., 132, Pt. C: 109-117 (1985).

10. Irving, E., Barrat, J. P., Charcossey, C., and Monville, J. P., Improving power network stability and unit stress with adaptive generator control, Automatica, 15: 31-46 (1979).
11. Kanniah, J., Malik, O. P., and Hope, G. S., Excitation control of synchronous generators using adaptive regulators, Parts I, II, IEEE Trans. Power App. and Syst., PAS-103: 897-910 (1984).
12. Hanus, R., Maun, J.-C., and Kinnaert, M., Adaptive control of a synchronous generator, Proc. IFAC Workshop on Adaptive Systems in Control and Signal Processing, San Francisco, pp. 333-340 (1983).

12a. Dash, P. K., and Khincha, H. P., A self-tuning regulator for power system control using a single-chip microcomputer, Proceedings of Int. Conf. on Computers, Systems and Signal Processing, Bangalore, India, pp. 694-698 (1984).
13. Hahn, V., Unbehauen, H., and Stoffer, R., Self-tuning control of a multivariable non-minimum phase pilot plant, Proc. Workshop on Adaptive Control, Florence, Italy (1982).
14. Ljung, L., Morf, F., and Falconer, D., Fast calculation of gain matrices for recursive estimation schemes, Int. J. Contr., 27: 1-19 (1978).
15. Courtiol, B., and Landau, I. D., High-speed adaptation system for controlled electric drives, Automatica, 11: 119-127 (1975).
16. Evans, R. J., Cook, B. J., and Betz, R. E., Nonlinear adaptive control of an inverter-fed induction motor: Linear load case, IEEE Trans. Ind. App., IA-19: 74-83 (1983).
17. Paul, R. J. A., Fleming, P. J., and Elsamahy, A. M., Adaptive microprocessor control of a linear helical reluctance motor, IEE Proc., 132, Pt. B: 106-111 (1985).
18. Bernadyn, J., Off-line adaptive control systems, Tech. Report No. AFFDL-TR-65-182, Wright Patterson Air Force Base, Ohio (1965).
19. Hoppe, S. G., et al., A feasibility study of self-learning adaptive flight control for high performance aircraft, Tech. Report No. AFFDL-TR-67-18, Wright Patterson Air Force Base, Ohio (1967).
20. Mini-Issue on NASA's Advanced Control Law Program, IEEE Trans. Auto. Contr., AC-22: 752-806 (1977).
21. Gressang, R. V., Discussion item, IEEE Trans. Auto. Contr., AC-22: 805 (1977).
22. Rynaski, E. G., Discussion item, IEEE Trans. Auto. Contr., AC-22: 805 (1977).
23. Young, P., A second generation adaptive autostabilization system for air-borne vehicles, Automatica, 17: 459-470 (1981).
24. Kanai, K., Degawa, T., Nikiforuk, P. N., and Gupta, M. M., Synthesis of C^*-Model reference adaptive flight controller, Proc. 8th IFAC World Congress, Kyoto, pp. 1027-1033 (1981).
25. Johnson, T. L., Harvey, C. A., and Stein, G., Self-tuning regulator design for adaptive control of aircraft wing/store flutter, IEEE Trans. Auto. Contr., AC-27: 1014-1024 (1982).
26. Siouris, G. M., Review of "Adaptive Control: The Model Reference Approach," by I. D. Landau, IEEE Trans. Syst., Man and Cybernetics, SMC-14: 169-170 (1984).
26a. Pramanik, S. K., and Seshadri, V., Design of a stable signal synthesis adaptive lateral autopilot for missiles under atmospheric disturbances, Proc. Int. Conf. on Computers, Systems and Signal Processing, Bangalore, India, pp. 1064-1068 (1984).
26b. Tsypkin, Ya. Z., Synthesis of the optimal adjustable model in the identification problems, Auto. Rem. Contr., 42: 1630-1642 (1981).

27. Walrath, C. D., Adaptive bearing friction compensation based on recent knowledge of dynamic friction, Automatica, 20: 717-727 (1984).
28. van Amerongen, J., Adaptive steering of ships—A model reference approach, Automatica, 20: 3-14 (1984).
29. van Amerongen, J., and Udink ten Cate, A. J., Model reference adaptive autopilots for ships, Automatica, 11: 441-449 (1975).
30. van Amerongen, J., Nieuwenhuis, H. C., and Udink ten Cate, A. J., Gradient based model reference adaptive autopilots for ships, Proc. 6th IFAC Congress, Boston (1975).
31. van Amerongen, J., and van Nauta Lemke, H. R., Optimum steering of ships with an adaptive autopilot, Proc. 5th Ship Control Systems Symp., Annapolis, Md. (1978).
32. van Amerongen, J., and van Nauta Lemke, H. R., Experience with a digital model reference adaptive autopilot, Proc. Int. Symp. on Ship Operation Automation, Tokyo (1979).
33. van Amerongen, J., Model reference adaptive control applied to steering of ships, in Methods and Applications in Adaptive Control (H. Unbehauen, ed.), Springer-Verlag, Berlin, pp. 199-208, 1980.
34. van Amerongen, J., A model reference adaptive autopilot for ships—Practical results, Proc. IFAC World Congress, Tokyo (1981).
35. van Amerongen, J., Adaptive steering of ships—A model reference approach to improved maneuvering and economical course keeping, Doctoral Thesis, Delft University of Technology, The Netherlands (1982).
36. Herther, J. C., Warnock, F. E., Howard, K. W., and van Velde, W., Digipilot—A self-adjusting digital autopilot for better maneuvering and improved course and track keeping, Proc. Int. Symp. on Ship Steering Automatic Control, Genoa, Italy (1980).
37. Källström, C. G., Åström, K. J., Thorell, N. E., Erikson, J., and Sten, L., Adaptive autopilots for tankers, Automatica, 15: 241-254 (1979).
38. Brink, A. W., Bass, G. E., Tiano, A., and Volta, E., Adaptive automatic course-keeping control of a super-tanker and a container ship—A simulation study, Proc. 5th Ship Control Systems Symp., Annapolis, Md. (1978).
39. Mort, N., and Linkens, D. A., Self-tuning controllers for surface ship course- and track-keeping, Proc. Int. Symp. on Ship Steering Automatic Control, Genoa, Italy (1980).
40. Mort, N., and Linkens, D. A., Self-tuning controllers for surface ship course-keeping and maneuvering, in Self-Tuning and Adaptive Control (C. J. Harris and S. A. Billings, eds.), Peter Peregrinus, London, 1981.
41. Fung, P. T. K., and Grimble, M. J., Self-tuning control of ship positioning systems, in Self-Tuning and Adaptive Control (C. J. Harris and S. A. Billings, eds.), Peter Peregrinus, London, 1981.

42. Saelid, S., and Henssen, N. A., Adaptive autopilot with wave filter, Proc. 6th IFAC Symp. on Identification and System Parameter Estimation, Washington, D.C., pp. 1365-1370 (1982).
43. Lim, C. C., and Forsythe, W., Autopilot for ship control, Part 1: Theoretical design; Part 2: Simulation studies, IEE Proc., 130, Pt. D: 281-294 (1983).
44. Fortescu, T. R., Kirshenbaum, L. S., and Ydstie, B. E., Implementation of self-tuning regulators with variable forgetting factors, Automatica, 17: 831-835 (1981).
45. Seaborg, D. E., Shah, S. L., and Edgar, T. F., Adaptive strategies for process control—A survey, Proc. Annual A.I.Ch.E. Meeting, Washington, D.C. (1983).
46. Morris, A., Nazer, Y., and Wood, R. K., Single and multivariable applications of self-tuning controllers, in Self-Tuning and Adaptive Control (C. J. Harris and S. A. Billings, eds.), Peter Peregrinus, London, 1981.
47. Morris, A. J., Nazer, Y., and Wood, R. K., Multivariable self-tuning process control, Opt. Contr. Appl. Meth., 3: 363-387 (1982).
48. Gerry, J. P., Vogel, E. F., and Edgar, T. F., Adaptive control of a pilot-scale distillation column, Proc. A.I.Ch.E. National Meeting, Houston, Tex. (1983).
49. Coppus, G. W. M., Shah, S. L., and Wood, R. K., Robust multivariable control of a binary distillation column, IEE Proc., 130, Pt. D: 201-208 (1983).
49a. Davison, E. J., Multivariable tuning regulators: The feedforward and robust control of a general servomechanism problem, IEEE Trans. Auto. Contr., AC-21: 35-47 (1976).
50. Dahlquist, S. A., Application of self-tuning regulators to the control of distillation columns, Proc. 6th IFAC/IFIP Conf. on Digital Computer Applications to Process Control, Dusseldorf (1980).
51. Wiemer, P., Hahn, V., Schmid, Ch., and Unbehauen, H., Application of multivariable MRAC to a binary distillation column, Proc. IFAC Workshop on Adaptive Systems in Control and Signal Processing, San Francisco, pp. 309-314 (1983).
52. Monopoli, R. V., and Hsing, C. C., Parameter adaptive control of multivariable systems, Int. J. Contr., 22: 313-327 (1975).
53. Koivo, A. J., A multivariable self-tuning controller, Automatica, 16: 351-366 (1980).
54. Martin-Sanchez, J. M., and Shah, S. L., Multivariable adaptive predictive control of a binary distillation column, Automatica, 20: 607-620 (1984).
55. Buchholt, F., and Kümmel, M., Self-tuning control of a pH-neutralization process, Automatica, 15: 665-671 (1979).
56. Proudfoot, C. G., Gawthrop, P. J., and Jacobs, O. L. R., Self-tuning PI control of a pH-neutralization process, IEE Proc., 130, Pt. D: 267-272 (1983).

57. Goodwin, G. C., McInnis, B., and Long, R. S., Adaptive control algorithms for waste-water treatment and pH neutralization, Opt. Contr. Appl. Meth., 3: 443-459 (1982).
58. Buchholt, F., and Kümmel, M., A multivariable STR to control a double-effect evaporator, Automatica, 17: 737-743 (1981).
59. Kirshenbaum, L. S., and Fortescue, T. R., Implementation of on-line control in chemical process plants, Automatica, 17: 778-788 (1981).
60. Kiparissides, C., and Shah, S. L., Self-tuning and stable adaptive control of a batch polymerization reactor, Automatica, 19: 225-235 (1983).
61. Åström, K. J., and Wittenmark, B., On self-tuning regulators, Automatica, 9: 185-199 (1973).
62. Goodwin, G. C., Ramadge, P. J., and Caines, P. E., Discrete-time multivariable adaptive control, IEEE Trans. Auto. Contr., AC-25: 956-960 (1980).
63. Kidd, P. T., Comments on "Self-tuning and stable adaptive control of a batch polymerization reactor," Automatica, 20: 481-482 (1984).
64. McDermott, P. E., and Mellichamp, D. A., An auto-pole-placement self-tuning controller, Int. J. Contr., 40: 1131-1147 (1984).
65. Dumont, G. A., Adaptive dead-time compensation, Proc. IFAC Symp. Identification and System Parameter Estimation, Washington, D.C., pp. 475-480 (1982).
66. Vogel, E. F., and Edgar, T. F., An adaptive dead-time compensator for process control, Proc. JACC, Paper No. TP5-E (1980).
67. D'Hulster, F. M., De Keyser, R. M. C., and van Cauwenberghe, A. R., Simulations of adaptive control for a paper machine head-box, Automatica, 19: 407-414 (1983).
68. Halme, A., and Selkainaho, J., Application of a nonlinear filter to multivariable parameter adaptive control in a distributed microcomputer system, Proc. 6th IFAC Symp. on Identification and System Parameter Estimation, Arlington, Va. (1982).
69. Selkainaho, J., Halme, A., and Behbehani, F., An efficient nonlinear filter with application experiences on multivariable adaptive control and fault detection, IFAC Workshop on Adaptive Systems in Control and Signal Processing, San Francisco (1983).
70. Angelov, V. B., Valtchev, V. S., and Guntchev, L. At., Optimal paper machine control with adaptive properties, Proc. of 5th Int. IFAC/IMEKO Conf. on Instrumentation and Automation in Paper, Rubber, Plastics and Polymerization Industries, Antwerp, Belgium, pp. 341-346 (1983).
71. Dumont, G., Application of advanced control methods in the pulp and paper industry, Proc. 5th Int. IFAC/IMEKO Conf. on Instrumentation and Automation in Paper, Rubber, Plastics and Polymerization Industries, Antwerp, Belgium, pp. 329-340 (1983).

72. Dumont, G. A., Self-tuning control of a chip refiner motor load, Automatica, 18: 307-314 (1982).
73. Dahlin, E. B., Designing and tuning digital controllers, Instr. Contr. Syst., 41: 77 (1968).
74. Westerlund, T., A digital quality control system for an industrial dry process rotary cement kiln, IEEE Trans. Auto. Contr., AC-26: 885-890 (1981).
75. Latawiec, K., and Chyra, M., On low frequency and long-run effects in self-tuning control, Automatica, 19: 419-424 (1983).
76. Schulz, R., Adaptive control of a ball mill with self-tuning reference model, Proc. IFAC/IFIP Symp. on Real Time Digital Control Applications, Guadalajara, Mexico, pp. 203-208 (1983).
77. Hetthessy, J., Vajk, I., Haber, R., Hilger, N., and Keviczky, L., A microprocessor-based adaptive composition control system, Proc. IFAC/IFIP Symp. on Real Time Digital Control Applications, Guadalajara, Mexico, pp. 209-214 (1983).
78. Westerlund, T., Toivonen, H., and Nyman, K.-E., Stochastic modelling and self-tuning control of a continuous cement raw material mixing system, Modeling, Identification and Control, 1: 17 (1980).
79. Dexter, A. L., Self-tuning optimum start control of heating plant, Automatica, 17: 483-492 (1981).
80. Dexter, A. L., and Hayes, R. G., Self-tuning charge control scheme for domestic stored-energy heating systems, IEE Proc., 128, Pt. D: 292-300 (1981).
81. Kurz, H., Isermann, R., and Schumann, R., Experimental comparison and application of various parameter-adaptive control algorithms, Automatica, 16: 117-133 (1980).
82. Lozano, R., and Bonilla, M., Global adaptive pole-placement for non-minimum phase plants—Application to a thermal process, Proc. IFAC Workshop on Adaptive Systems in Control and Signal Processing, San Francisco, pp. 331-392 (1983).
83. Schumann, R., Digital parameter adaptive control of an air-conditioning plant, Automatica, 18: 569-575 (1982).
84. De Keyser, R. M. C., and van Cauwenberghe, A. R., A self-tuning multi-step predictor application, Automatica, 17: 167-174 (1981).
85. Shiraiwa, T., Sakmoto, Y., Kobayashi, S., Anezaki, S., Kato, H., and Kuwabaro, A., Automatic control of casting speed in ingot casting, Automatica, 17: 613-618 (1981).
86. Bunich, A. L., Adaptive control of oxygen-converter steel melting, Proc. 4th IFAC Symp. on Automation in Mining, Mineral and Metal Processing, Helsinki, pp. 287-290 (1983).
87. Hrusak, J., Mošna, J., Janeček, E., and Simandl, M., A new adaptive controller for cold rolling mills, Proc. IFAC/IFIP Symp. on Real Time Digital Control Applications, Guadalajara, Mexico, pp. 69-74 (1983).

88. Dubowsky, S., and Des Forges, D. T., The application of MRAC to robotic manipulators, ASME Trans., J. Dyn. Syst. Meas. Contr., 101: 193-200 (1979).
89. Dubowsky, S., and Des Forges, D. T., Robotic manipulator control systems with invariant dynamic characteristics, Proc. 5th World Congress on Machines and Mechanisms (IFIOMM), Montreal (1979).
90. Cao, C. T., A simple adaptive concept for the control of an industrial robot, in Methods and Applications in Adaptive Control (H. Unbehauen, ed.), Springer-Verlag, Berlin, pp. 270-279, 1980.
91. Arimoto, S., and Takegaki, M., An adaptive method for trajectory control of manipulators, Proc. 8th IFAC World Congress, Kyoto, pp. 1921-1926 (1981).
92. Koivo, A. J., Zhang, Q., and Guo, T.-H., Industrial manipulator control using Kalman filter and adaptive controller, Proc. IFAC Symp. Identification and System Parameter Estimation, Washington, D.C., pp. 265-270 (1982).
93. Leininger, G. G., and Wang, S. P., Pole-placement self-tuning control of manipulators, Proc. IFAC Symp. on CAD of Multivariable Systems, West Lafayette, Ind., pp. 27-30 (1982).
94. Tomizuka, M., Horowitz, R., and Landau, I. D., On the use of MRAC techniques for mechanical manipulators, Proc. 2nd IASTED Symp. on Identification, Control, and Robotics, Davos (1982).
95. Koivo, A. J., and Guo, T.-H., Adaptive linear controller for robotic manipulators, IEEE Trans. Auto. Contr., AC-28: 162-171 (1983).
96. Balestrino, A., De Maria, G., and Sciavicco, L., An adaptive model following system for robotic manipulators, ASME Trans., J. Dyn. Syst. Meas. Contr., 105: 143-151 (1983).
97. Nicolò, F., and Katende, J., A robust MRAC for industrial robots, Proc. 2nd IASTED Int. Symp. on Robotics and Automation, Lugano (1983).
98. Miroshnik, I., Nicolò, F., and Nicosia, S., Robust model reference adaptive control, Proc. Workshop on Adaptive Control, Florence, Italy (1982).
99. Dunskaya, N. V., and Pyatnitskii, E. S., Adaptive manipulator control, Auto. Rem. Contr., 44: 240-249 (1983).
100. Tomuzuka, M., and Horowitz, R., Model reference adaptive control of mechanical manipulators, Proc. IFAC Workshop on Adaptive Systems in Control and Signal Processing, San Francisco, pp. 27-32 (1983).
101. Landau, I. D., and Lozano, R., Unification of discrete-time explicit MRAC designs, Automatica, 17: 593-611 (1981).
101a. Karnik, A. M., and Sinha, N. K., The adaptive control of a robot arm for arc welding process, Proc. IEEE Int. Conf. on Systems, Man and Cybernetics, Bombay, India, pp. 614-618 (1983).

101b. Cook, G. E., et al., Microcomputer control of an adaptive positioning system for robotic arc welding, Proc. IEEE Conf. on Mini- and Microcomputers, San Francisco, pp. 324-329 (1981).

101c. Begin, G., and Boillot, J., IR vision applied to adaptive welding, presented at Robotics International, Central Ontario Section, ASEA, Malton (1983).

102. Nicosia, S., Nicolo, F., and Lentini, D., Dynamical control of industrial robots with elastic and dissipative joints, Proc. 8th IFAC World Congress, Kyoto (1981).

103. Nicosia, S., and Tomei, P., MRAC algorithms applied to industrial robots, Report No. 07.83, Department Informatica e Sistemistica, University of Rome (1983).

104. Nicosia, S., and Tomei, P., Model reference adaptive control algorithms for industrial robots, Automatica, 20: 635-644 (1984).

105. Greco, C., Menga, G., Mosca, E., and Zappa, G., Performance improvements of self-tuning controllers by multi-step horizons: The MUSMAR approach, Automatica, 20: 681-699 (1984).

106. Vukobratović, M., Stokić, D., and Kirćanski, N., Towards non-adaptive and adaptive control of manipulation robots, IEEE Trans. Auto. Contr., AC-29: 841-844 (1984).

107. Vukobratović, M., and Kirćanski, N., An engineering approach to adaptive control synthesis for multi-degree-of-freedom manipulation robots, Proc. 3rd IFAC/IFORS Symp. on Large Scale Systems, Warsaw (1983).

108. Lee, C. S. G., and Chung, M. J., An adaptive control strategy for computer-based manipulators, Proc. IEEE Conf. Decision and Control, Orlando, Fla., pp. 95-100 (1982).

109. Lee, C. S. G., and Chung, M. J., An adaptive control strategy for mechanical manipulators, IEEE Trans. Auto. Contr., AC-29: 837-840 (1984).

110. Lee, C. S. G., and Chung, M. J., Adaptive perturbation control with feedforward compensation for robot manipulators, Simulation, 44: 127-136 (1985).

111. Andrew, A. M., Some comments on adaptive robotics, Proc. IFAC Symp. on Artificial Intelligence, Leningrad, pp. 239-243 (1983).

112. Dodds, S. J., Adaptive, high precision, satellite attitude control for microprocessor implementation, Automatica, 17: 563-573 (1981).

113. Slachmylders, E., The Exosat attitude and orbit control system, Proc. 7th IFAC World Congress, Helsinki (1978).

114. van den Bosch, P. P. J., and Jongkind, W., Model reference adaptive satellite attitude control, in Methods and Applications in Adaptive Control, Springer-Verlag, Berlin, pp. 209-218, 1980.

115. van den Bosch, P. P. J., and Tjahjadi, P. I., Model updating improves MRAC performance, IEEE Trans. Auto. Contr., AC-29: 1106-1108 (1984).

116. Wong, K. Y., and Bayoumi, M. M., Self-tuning control of multi-locomotive powered long freight trains, Proc. 4th IFAC Conf. on Control in Transportation Systems, Baden-Baden, F.R.G., pp. 25-32 (1983).
117. Murthy, S. S., and Fenton, R. E., A velocity adaptive, microprocessor-based vehicle lateral controller, Proc. 4th IFAC Conf. on Control in Transportation Systems, Baden-Baden, F.R.G., pp. 365-372 (1983).
118. Sinha, P. K., and Hulme, A. J., Adaptive control of a nonlinear system, Proc. IFAC Symp. on Digital Control, New Delhi, pp. 119-125 (1982).
119. Andersson, P., and Ljung, L., A test case for adaptive control: Car steering, Proc. IFAC Symp. on Digital Control, New Delhi, pp. 71-76 (1982).
120. Fenton, R. E., Melocik, G. C., and Olson, K. W., On the steering of automated vehicles: Theory and experiment, IEEE Trans. Auto. Contr., AC-21: 306-315 (1976).
121. Irving, E., and Dang van Mien, H., Discrete-time model reference multivariable adaptive control—Applications to electric power plants, Proc. IFAC World Congress, Kyoto, Japan, pp. 3143-3257 (1981).
122. Borisson, U., Self-tuning regulators—Industrial applications and multivariable theory, Report No. 7513, Department of Auto. Contr., Lund Institute of Technology, Lund, Sweden (1975).
123. Fortuna, L., Gallo, A., and La Cava, M., Approach to the adaptive control of a solar plant, Proc. 6th IFAC Symp. Identification and System Parameter Estimation, Washington, D.C., pp. 1689-1694 (1982).
124. Oliva, D. N., Morris, E. L., and Oliva, M. T., Adaptive control of a steam turbine, Proc. IFAC/IFIP Symp. on Real Time Digital Control Appl., Guadalajara, Mexico, pp. 139-144 (1983).
125. Oliva, D. N., Application of adaptive control techniques to a steam turbine, Doctoral Thesis, Univ. of Exeter, U.K. (1981).
126. Walker, B. K., Chia, T.-L., Stern, K. S., and Katona, P. G., Parameter identification and adaptive control for blood pressure, Proc. IFAC Symp. Identification and System Parameter Estimation, Washington, D.C., pp. 1193-1198 (1982).
127. Slate, J. B., and Sheppard, L. C., A model-based adaptive blood pressure controller, Proc. IFAC Symp. Identification and System Parameter Estimation, Washington, D.C., pp. 1437-1442 (1982).
128. Kaufman, H., Roy, R., and Xu, X., Model reference adaptive control of drug infusion rate, Proc. IFAC Symp. Identification and System Parameter Estimation, Washington, D.C., pp. 277-282 (1982).
129. Kaufman, H., Roy, R., and Xu, X., Model reference adaptive control of drug infusion rate, Automatica, 20: 205-209 (1984).

130. McInnis, B. C., Everett, R. L., Wang, J. C., Vajapeyam, B., and Akutsu, T., A microcomputer-based adaptive control system for the artificial heart, Proc. 8th IFAC World Congress, Kyoto, Japan, pp. 3753-3760 (1981).
131. Kitumura, T., and Akashi, H., A design of an adaptive control system for left ventricular assist, Proc. 8th IFAC World Congress, Kyoto, Japan, pp. 3767-3772 (1981).
132. McInnis, B. C., Guo, Z.-W., Lu, P. C., and Wang, J.-C., Adaptive control of left ventricular bypass assist devices, IEEE Trans. Auto. Contr., AC-30: 322-329 (1985).
133. Goodwin, G. C., Elliott, H., and Teoh, E. K., Deterministic convergence of a self-tuning regulator with covariance resetting, IEE Proc., 130, Pt. D: 6-8 (1983).
134a. McInnis, B. C., Wang, J.-C., and Goodwin, G. C., Adaptive control systems for the artificial heart, Proc. IEEE Frontiers Eng. Health Care, pp. 121-124 (1982).
134b. McInnis, B. C., Everett, R. L., Vajapeyam, B., Lin, C. H., Iyer, M., and Akutsu, T., Digital system for P-wave detection and synchronization of the artificial heart, Int. J. Biomed. Comput., 14: 381-388 (1983).
135. Williams, D., Yousefpour, P., and Swanick, B. H., On-line adaptive control of a fermentation process, IEE Proc., 131, Pt. D: 117-124 (1984).
136. Bastin, G., Dochain, D., Haest, M., Installe, M., and Opdenacker, P., Modeling and adaptive control of a continuous anaerobic fermentation process, in Modelling and Control of Biotechnical Processes (A. Halme, ed.), Pergamon Press, Oxford, pp. 299-306, 1983.
137. Bastin, G., Dochain, D., Haest, M., Installe, M., and Opdenacker, P., Identification and adaptive control of a biomethanization process, in Modelling and Data Analysis in Biotechnology and Medical Engineering (G. C. Vansteenkiste and P. C. Young, eds.), North-Holland, Amsterdam, pp. 271-282, 1983.
138. Halme, A. (ed.), Proceedings of 1st IFAC Congress on Modelling and Control of Biotechnical Processes, Pergamon Press, Oxford, 1983.
139. Dochain, D., and Bastin, G., Adaptive identification and control algorithms for nonlinear bacterial growth systems, Automatica, 20: 621-634 (1984).
140. Ko, K.Y.-J., McInnis, B. C., and Goodwin, G. C., Adaptive control and identification of the dissolved oxygen process, Automatica, 18: 727-730 (1982).
141. Watanabe, T., and Iwai, S., Design of an adaptive control constraint system of a milling machine tool, Proc. IFAC Symp. on CAD of Multivariable Systems, West Lafayette, Ind., pp. 553-564 (1982).
142. Haber, R., Hetthessy, J., Keviczky, L., Vajk, I., Feher, A., Czeiner, N., Csaszar, Z., and Turi, A., Identification and adaptive control of a glass-furnace, Automatica, 17: 175-185 (1981).

143. Goodwin, G. C., and Sin, K. S., Adaptive Filtering, Prediction, and Control, Prentice-Hall, Englewood Cliffs, N. J. (1984).
144. Boehm, H., Adaptive control of a dry etch process by microcomputer, Automatica, 18: 665-673 (1982).

BIBLIOGRAPHY

Aguado, A., and Gomez, A., Identification and adaptive control of a sugar factory vacuum pan, Proc. IFAC/IFIP Symp. Real Time Digital Control Appl., Guadalajara, Mexico, pp. 139-144 (1983).

Allidina, A. Y., Hughes, F. M., and Tye, C., Self-tuning control of a nuclear reactor, Proc. Int. Conf. Control and its Applications, University of Warwick, U.K. (1981).

Andeen, R. E., Self-adaptive autopilots, Space/Aeronautics, pp. 46-52 (1965).

Andeen, R. E., and Shipley, P. P., Digital adaptive flight control system for aerospace vehicles, AIAA J., 1: 1105-1109 (1963).

Assilian, S., Artificial intelligence in the control of real dynamic systems, Doctoral Thesis, Queen Mary's College, London (1974).

Åström, K. J., and Källström, C. G., Identification of ship steering dynamics, Automatica, 12: 9-22 (1976).

Åström, K. J., Källström, C. G., Norrbin, N. H., and Bystrom, L., The identification of linear ship steering dynamics using maximum likelihood parameter estimation, SSPA Report No. 75, Gothenberg, Sweden (1975).

Athans, M., Castanon, D., Dunn, K.-P., Greene, C. S., Lee, W. H., Sandel, Jr., N. R., and Willsky, A. S., The stochastic control of the F-8C aircraft using a multiple model adaptive control method—Part I: Equilibrium flight, IEEE Trans. Auto. Contr., AC-22: 768-780 (1977).

Badr, O., Rey, D., and Ladet, P., Adaptive dual control for a multivariable distillation process, Proc. IFAC/IFIP Symp., The Hague, Netherlands, pp. 141-147 (1977).

Badr, O., Rey, D., and Ladet, P., On the dual-adaptive control and its practical applications, Automatica, 15: 91-96 (1979).

Balhoff, R. A., and Corripio, A. B., An adaptive feedforward control algorithm for computer control of waste-water neutralization, Proc. IFAC/IFIP Symp. Real Time Digital Control Applications, Guadalajara, Mexico, pp. 395-406 (1983).

Bell, D., Stability analysis of an adaptive aircraft control system, Control, 11 (1967).

Berezuai, G. T., and Sinha, N. K., Adaptive nuclear reactor control for integral quadratic cost functions, Automatica, 9: 522-536 (1973).

Berger, C. S., and Brown, W. A., Adaptive control of muscle relaxation, Proc. IFAC/IFIP Symp. Real Time Digital Control Appl., Guadalajara, Mexico, pp. 109-118 (1983).

Bergmann, S., and Lachmann, K. H., Digital parameter adaptive control of a pH process, Preprints JACC, San Francisco (1980).

Betz, R., Nonlinear adaptive control with application to a.c. motor drives, Doctoral Thesis, University of Newcastle, N.S.W., Australia (1983).

Biladovich, F., Sabanovic, A., Music, O., and Izosimov, D. B., Current inverter in the sliding mode for induction motor control, 3rd IFAC Symp. Control in Power Electronics and Electrical Drives, Lausanne, Switzerland, pp. 139-144 (1983).

Bittanti, S., Clarke, D. W., Romeo, F., and Scattolini, R., Self-tuning control of liquid-saturated steam exchanger, IFAC/IFIP Symp. Real-Time Digital Control Appl., Guadalajara, Mexico, pp. 631-638 (1983).

Bittanti, S., Romeo, F., and Scattolini, R., Stochastic identification and digital control of a heat exchanger: A simulation test case, J. Franklin Inst., 318: 29-57 (1984).

Bonanomi, P., and Güth, G., Adaptive regulators for the excitation of large turbogenerators, in Methods and Applications in Adaptive Control (H. Unbehauen, ed.), Springer-Verlag, Berlin, pp. 242-250, 1980.

Bonanomi, P., Güth, G., Blaser, G., and Glavitsch, H., Concept of a practical adaptive regulator for excitation control, Proc. IEEE Summer Power Meeting, Vancouver, B.C., Canada, Paper No. A-79-453-2 (1979).

Boon, L. A., Voors, A. W., and Pronk, C., Adaptive system to determine the position of an overhead crane, Proc. 4th IFAC Symp. Automation in Mining, Mineral and Metal Processing, Helsinki, pp. 303-310 (1983).

Borisson, U., and Syding, R., Self-tuning control of an ore-crusher, Automatica, 12: 1-7 (1976).

Borisson, U., and Wittenmark, B., Moisture control of paper machine—An application of a self-tuning regulator, Report No. 7737, Lund Inst. of Tech., Lund, Sweden (1973).

Borisson, U., and Wittenmark, B., An industrial application of a self-tuning regulator, Proc. IFAC Symp. Digital Computer Applications to Process Control, Zurich, Switzerland (1974).

Breddermann, R., Realization and application of a self-tuning on-off controller, in Methods and Applications in Adaptive Control (H. Unbehauen, ed.), Springer-Verlag, Berlin, pp. 74-83, 1980.

Breitholtz, C., and Hillberg, O., Development of a reference model for the drying zone of a travelling grate pelletizing plant, Proc. 3rd IFAC Symp. on Automation in Mining, Mineral and Metal Processing, Montreal, Canada (1980).

Brickwedde, A., Microprocessor-based adaptive speed and position control for electric drives, Proc. IEEE-IAS Annual meeting, pp. 411-417 (1984).

Buchholt, F., Clement, K., and Bay Jorgensen, S., Self-tuning control of a fixed bed chemical reactor, Preprints 5th IFAC Symp. Identification and System Parameter Estimation, Darmstadt, F.R.G. (1979).

Carter, G., and Rutherford, D., A heuristic adaptive controller for a sinter plant, Proc. IFAC Symposium, Johannesburg, pp. 315-324 (1976).

Cegrell, T., and Hedquist, T., Successful adaptive control of paper machines, Automatica, 11: 53-59 (1975).

Centner, R. M., and Idelsohn, J. M., Adaptive controller for a metal cutting process, JACC, Minneapolis, Minn. (1963).

Clark, D. C., The model reference self-adaptive control system as applied to flight control of a supersonic transport, JACC, Stanford, Calif. (1964).

Claussen, U., Adaptive time-optimal position control with microprocessor, in Methods and Applications in Adaptive Control (H. Unbehauen, ed.), Springer-Verlag, Berlin, pp. 270-279, 1980.

Claussen, U., Adaptive zeitoptimale Lageregelung eines linearen Stellantriebs mit Synchronem Langstatormotor, Doctoral Thesis, Tech. Univ. of Braunschweig, F.R.G. (1980).

Claussen, U., and Leonhard, W., Microprocessor-controlled linear synchronous motor as positioning drive, Proc. Int. Conf. Elec. Machines (1978).

Corliss, L. D., and Gossett, T. D., A study of quadratic performance index for a BTOL pre-filter model reference attitude control system, AIAA Guidance, Control and Flight Mechanics Conf., Princeton, N.J., Paper No. 69-884 (1969).

Cormier, W. H., and Fenton, R. E., On the steering of automated vehicles—A velocity-adaptive controller, IEEE Trans. Vehicular Tech., VT-29: 375-385 (1980).

Courtiol, B., Applying model reference adaptive techniques for the control of electromechanical systems, Proc. 6th IFAC Congress, Vol. ID, Boston, Mass., pp. 58.2-1-58.2-9 (1976).

Craven, R., and Glavitsch, H., Large disturbance performance of an adapted linear excitation control system, IEEE Summer Power Meeting, Vancouver, B.C., Canada, Paper No. A-79-452-4 (1979).

Csaki, K., Keviczky, L., Hetthessy, J., Hilger, M., and Kolostori, J., Simultaneous adaptive control of chemical composition, fineness and maximum quality of ground materials at a closed circuit ball mill, Proc. 7th IFAC Congress, Helsinki, pp. 453-460 (1978).

Dahhon, B., Nijim, K., M'Saad, M., and Youlal, B., MRAC of an industrial phosphate drying furnace, Proc. IFAC Workshop Adaptive Systems in Control and Signal Processing, San Francisco, pp. 315-321 (1983).

Davison, E. J., Taylor, P. A., and Wright, J. D., On the application of tuning regulators to control a commercial heat exchanger, IEEE Trans. Auto. Contr., AC-25: 361-375 (1980).

De Keyser, R. M. C., and van Cauwenberghe, A. R., Simulation and self-tuning control in a nuclear power plant, in Simulation of Control

Systems (I. Troch, ed.), North-Holland, Amsterdam, pp. 187-195 (1978).

De Maria, G., and Marino, P., An adaptive pulse-width modulator for control of d.c. drives, IEEE-IAS Annual Meeting, Mexico (1983).

D'Hulster, F. M., and van Cauwenberghe, A. R., An adaptive controller for slow processes with variable control period, Proc. IFAC Congress, Helsinki, pp. 461-467 (1978).

DODCO Inc., Theory and application of optimum limited information adaptive flight control techniques, WADC Tech. Report No. 59-468 (1959).

Dumont, G. A., and Belanger, P. R., Self-tuning control of a titanium dioxide kiln, IEEE Trans. Auto. Contr., AC-23: 532-538 (1978).

Dumont, G. A., Gendron, S., Belanger, P., and Rochon, L., Self-tuning control of a Kamyr digester chip level, Can. Ind. Comp. Society Conf., Ottawa, Canada (1984).

Ehlers, H. L., and Smyth, R. K., Adaptive control applications to aerospace vehicles, AIAA J. Aircraft, 7: 2 (1970).

Erschler, J., Roubellat, F., and Vernhes, J., Automation of hydroelectric power station using variable structure control systems, Automatica, 10: 31-36 (1974).

Fujii, S., Fujimoto, H., Shibata, A., Shimada, S., and Hajiri, H., Application of an adaptive control of a refrigerant compressor test, Proc. IFAC/IFIP Symp., The Hague, Netherlands, Paper No. M2-5 (1977).

Fung, P. T. K., Optimal and self-tuning control of dynamically-positioning vessel, Research Report No. EEE/64, Sheffield City Polytechnic, U.K. (1980).

Fung, P. T. K., and Grimble, M. J., Dynamic ship position using a self-tuning Kalman filter, IEEE Trans. Auto. Contr., AC-28: 339-350 (1983).

Gaylor, R., Schaeperkoetter, R. L., and Cunningham, D. C., An adaptive tracking filter for the stabilization of bending modes in flexible vehicles, AIAA/JACC Guidance and Control Conf., Seattle, pp. 441-447 (1966).

Gentil, S., Sandraz, J. P., and Foulard, C., Different methods for dynamic identification of an experimental paper machine, Proc. 3rd IFAC Symp. on Identification and System Parameter Estimation, The Hague, Netherlands, pp. 473-484 (1973).

Glattfelder, A. H., Jeanneret, F., Glauser, A., and Huser, L., CAD of adaptive hydraulic turbine governors, Proc. IFAC Symp. on CAD of Control Systems, Zurich (1979).

Glattfelder, A. H., and Schaufelberger, W., Adaptive control by self-selection: An application to hydropower control, in Methods and Applications in Adaptive Control (H. Unbehauen, ed.), Springer-Verlag, Berlin, pp. 251-260, 1980.

Green, W. L., Sauger, D. J., and Suresh, B. N., Using the self-optimizing control of an electro-hydraulic servo system to minimize the power loss, in Methods and Applications in Adaptive Control (H. Unbehauen, ed.), Springer-Verlag, Berlin, pp. 280-290, 1980.

Grimble, M. J., A combined state estimate feedback solution to the ship positioning control problem, Opt. Contr. Appl. Meth., 1: 55-67 (1980).

Haber, R., Hetthessy, F., Keviczky, L., and Kovacs, K., Some investigations on classical and adaptive load-frequency control, Proc. Workshop Applications of Adaptive Control, Yale University, New Haven, Conn., pp. 88-94 (1979).

Hagender, P., and Wittenmark, B., A self-tuning filter for fixed lag smoothing, IEEE Trans. Inf. Th., IT-23: 377-384 (1977).

Hahn, V., Unbehauen, H., and Nadolph, U., Model reference adaptive control of a multivariable non-minimum phase pilot plant, 3rd Yale Workshop on Applications of Adaptive Systems Theory, Yale University, New Haven, Conn. (1983).

Hallager, L., and Bay Jorgensen, S., Adaptive control of chemical engineering processes, Proc. IFAC Workshop on Adaptive Systems in Control and Signal Processing, San Francisco, pp. 323-330 (1983).

Halme, A., A two level realization of a self-tuning regulator in a multi-microcomputer process control system, in Analysis and Optimization of Systems (A. Bensoussan and J. L. Lions, eds.), Springer-Verlag, Berlin, 1980.

Haque, S. I., and Monopoli, R. V., Discrete adaptive control of a radio telescope, Proc. Workshop on Applications of Adaptive Control, Yale University, New Haven, Conn., pp. 170-181 (1979).

Harland, G. E., Design of model reference adaptive control for an internal combustion engine, Meas. Contr., 6: 167-173 (1973).

Harris, C. J., and Billings, S. A. (eds.), Self-Tuning and Adaptive Control: Theory and Applications, Peter Peregrinus Ltd., London, 1981.

Harris, C. J., McGregor, J. F., and Wright, J. D., An application of self-tuning regulators to catalytic reactor control, JACC, Philadelphia, (1978).

Harris, C. J., McGregor, J. F., and Wright, J. D., Self-tuning and adaptive controllers: An application to catalytic reactor control, Technometrics, 22: 153 (1980).

Harris, T. J., MacGregor, J. F., and Wright, J. D., An application of self-tuning regulators to catalytic reactor control, Proc. JACC, Philadelphia, pp. 43-58 (1978).

Heydt, G. T., A self-tuning excitation controller for maximum power transfer over a long transmission line, Elec. Machines and Power Systems (to appear).

Hodgson, A. J. F., and Clarke, D. W., Self-tuning applied to batch reactors, IEE Conf. Applications of Adaptive and Multivariable Control, Hull, U.K. (1982).

Hoffmann, U., Mueller, U., and Schuermann, B., A microprocessor-based self-tuning on-off controller for thermal processes, Proc. 5th Int. IFAC/IMEKO Conf. Instrumentation and Automation in the Paper, Rubber, Plastics and Polyerization Industries, Antwerp, Belgium, pp. 95-100 (1983).

Hollstein, R. B., and Gruber, W. H., A response regulation technique for self-adaptive control of flexible aircraft, AGARD Advanced Control System Concepts, pp. 121-138 (1970).

Howell, G. C., and Shanks, G. T., Self-adaptive maneuver demand control systems: Some practical design considerations, AGARD Advanced Control System Concepts, pp. 149-168 (1970).

Irving, E., Implicit reference model and optimal aim strategy for electric generator adaptive control, in Methods and Applications in Adaptive Control (H. Unbehauen, ed.), Springer-Verlag, Berlin, pp. 219-241, 1980.

Irving, E., Barrat, J. P., Charcossey, C., and Monville, J. P., A method to improve the power network steady-state stability and to reduce the unit stress: The multivariable adaptive control of generators, IFAC Symp. Auto. Contr. and Protection of Elec. Power Systems, Melbourne, Australia (1977).

Irving, E., and van Mien, H. D., Discrete-time model reference multivariable adaptive control: Applications to electrical power plants, Preprints IEEE Conf. Decision and Control, Fort Lauderdale, Fla. (1979).

Isermann, R., Digital control methods for power station plants based on identified models, IFAC Symp. Automatic Control in Power Generation, Distribution and Protection (1980).

Johnson, J. M., Crow, R. K., and Lutes, C. L., Study of structural bending adaptive control techniques for large launch vehicles, Tech. Report No. NAS 8-20056, Autonetics, Inc. (1966).

Jongkind, W., Model reference adaptive satellite attitude control using Euler's theorem on rotations, Master's Thesis, Lab. for Contr. Eng., Delft Univ. of Tech., Netherlands (1977).

Kanniah, J., Microprocessor-based dual-rate sampling regulator for a synchronous generator, Doctoral Thesis, Dept. of Elec. Eng., University of Calgary, Calgary, Canada (1983).

Kaufman, H., and Berry, P., Development of a digital adaptive optimal linear regulator flight controller, Proc. 6th IFAC World Congress, Boston, pp. 58.5.1-58.5.8 (1975).

Kelly, W. H., Self-tuning control strategies: Modern solutions to paper making control problems, Can. Ind. Comp. Society Conf., Hamilton, Ontario (1982).

Kezer, A., Hoffmann, L. G., and Eugel, A. G., Application of model reference adaptive control techniques to provide improved bending response of large flexible missiles, 6th Symp. Ballistic Missiles and Space Tech., New York, pp. 113-151 (1961).

Kidd, P. T., Munro, N., and Winterbone, D. E., Adaptive multivariable control of a ship propulsion system, Proc. Inst. M.C. Conf. Applns. of Multivariable Systems Theory, Plymouth, U.K. (1982).

Kochenburger, R. J., Self-adaptive method for accommodating large variations of plant gain in control systems, Proc. IFAC Congress, London (1963).

Kryukov, Yu. I., An adaptive approach to the problem of control and learning in manipulator systems, Proc. Int. Conf. Computers, Systems and Signal Processing, Bangalore, India, pp. 316-320 (1984).

Kunze, E., and Salaba, M., Praktische Erprobung Eines Adaptiven Regelungsverfaurens an Einer Zementmahlanlage, Preprints Factagung "Regelungstechnique in Zementwerken, Bielefeld, F.R.G. (1978).

Kurz, H., Digital parameter adaptive control of processes with unknown constant or varying dead time, Proc. IFAC Symp. Identification and System Parameter Estimation, Darmstadt, F.R.G. (1979).

Landau, I. D., and Courtiol, B., Adaptive model following for flight control and simulation, J. Aircraft, 9: 668-674 (1972).

Landau, I. D., Grossu, A. V., and Gavat, S., L'utilisation de la commande adaptive à l'aide d'un modèle dans la régulation des moteurs electriques, Automatisme, 13: 146-152 (1968).

Lee, J. F., A digital adaptive flight control system for flexible missiles, Proc. 7th Symp. on Ballistic Missile and Space Tech., Academic Press, New York, pp. 115-147, 1962.

Lieuson, H., Evaluation of self-tuning control of a binary distillation column, Master's Thesis, Univ. of Alberta, Canada (1980).

Lieuson, H., Morris, A. J., Nazer, Y., and Wood, R. K., Experimental evaluation of self-tuning controllers applied to pilot plant units, in Methods and Applications in Adaptive Control (H. Unbehauen, ed.), Springer-Verlag, Berlin, pp. 301-309, 1980.

Lizr, A., Lizrova, A., Hejda, J., and Dusek, F., Application of adaptive control algorithms in the Czechoslovak paper industry, Proc. 5th Int. IFAC/IMEKO Conf. Instrumentation and Automation in the Paper, Rubber, Plastics and Polymerization Industries, Antwerp, Belgium, pp. 213-218 (1983).

Mabius, L., and Kaufman, H., An adaptive flight controller for the F-8 without explicit parameter identification, IEEE Conf. Decision and Control, Florida (1976).

Malik, O. P., Hope, G. S., and Ramanujam, R., Real time MRAC of synchronous machine excitation, IEEE Winter PAS Meeting (1978).

Mehra, R., and Eterno, J., Adaptive power plant control: Problems and prospects, Proc. Workshop Application of Adaptive Control, Yale University, New Haven, Conn., pp. 69-78 (1979).

Morris, A. J., Fenton, T. P., and Nazer, Y., Application of self-tuning regulator to the control of chemical process, Proc. 5th IFAC/IFIP Conf. Digital Computer Applns. to Process Control, The Hague, Netherlands, pp. 447-455 (1977).

Morris, A. J., and Nazer, Y., Self-tuning control of a binary distillation column, Report, Dept. of Chem. Eng., Univ. of Newcastle-upon-Tyne, U.K. (1977).

Morris, A. J., Nazer, Y., Wood, R. K., and Lieuson, H., Evalution of self-tuning controllers for distillation column control, Proc. 6th IFAC/IFIP Conf. Digital Computer Applns. to Process Control, Dusseldorf, F.R.G. (1980).

Morris, E. L., and Abaza, B. A., Adaptive digital control of a steam turbine, IEE Proc., 123: 549-553 (1976).

Nader, A. D., Modelisation et commande adaptive multivariable des caisses de tête de machine à papier, Doctoral Thesis, Inst. Polytechnique, Grenoble, France (1978).

Nader, A. D., Dual model reference adaptive control on paper machine head-boxes, 4th IFAC-PRP Conf., Ghent (1980).

Nagaraja, T., and Agarwal, R. B. D., Fast acting adaptive AVR, Computers and Elec. Eng., 5: 369-371 (1978).

Najim, K., Najim, M., and Youlal, H., Self-tuning control of an industrial phosphate dry process, Opt. Contr. Appln. Meth., 3 (1982).

Narendra, K. S., and Monopoli, R. V. (eds.), Applications of Adaptive Control, Academic Press, New York, 1980.

Negoesco, S., Courtiol, B., and Francon, C., Application de la commande adaptive à la fabrication des guides d'ondes hélicoidaux, Automatisme, 23: 7-14 (1978).

Nikiforuk, P. N., and Pronovost, R. J., A three variable self-adaptive flight control system using discrete interval binary noise perturbation signals, Proc. IFAC Symp. on Identification in Automatic Control Systems, Paper 3.2 (1967).

Oldenberg, J., Experiments with a new adaptive autopilot intended for controlled turns as well as for straight course steering, Proc. 4th Ship Control Systems Symp., The Hague, Netherlands (1975).

Oster, S. S., Adaptive flight control system, Proc. Self Adaptive Flight Control System Symp., WADC Tech. Report No. 59-49, pp. 81-122 (1959).

Parks, P. C., Schaufelberger, W., Schmid, Ch., and Unbehauen, H., Applications of adaptive control systems, in Methods and Applications in Adaptive Control (H. Unbehauen, ed.), Springer-Verlag, Berlin, pp. 161-198 (1980).

Phung, V. A., and Gibbard, M. J., Practical implementation on a 5 KVA synchronous generator of adaptive excitation controller strategy for a wide range of operating conditions, IEE Proc., 125: 1009-1014 (1978).

Prince, L. T., Design, development and flight research of an experimental adaptive control technique for advanced booster systems, Report No. ASD-TDR-62-178, Wright-Patterson Airforce Base, Dayton, Ohio (1962).

Prince, L. T., A three-axis adaptive flight control system, in Adaptive Control Systems (F. P. Caruthers and H. Levenstein, eds.), Pergamon Press, Oxford, pp. 277-287, 1963.

Rabut, C., Self-adaptive numerical control of an epitaxy reactor using Kalman-Bucy filtering, Proc. IFAC Congress, Boston, Mass., Paper No. 52.4 (1975).

Roy, R., and Jenkins, K. W., Identification and control of a flexible launch vehicle, Report No. CR-551, NASA, Washington, D.C. (1966).

Rynaski, E. G., Adaptive control application to aircraft, Proc. Workshop Applications of Adaptive Control, Yale University, New Haven, Conn. (1979).

Sabonovic, A., and Izosimov, D. B., Application of sliding modes to industrial motor control, Proc. IEEE-Industrial Appln. Soc. Meeting, Cleveland, Ohio, pp. 793-801 (1979).

Sabonovic, A., Izosimov, D. B., Bilalovic, F., and Music, O., Sliding modes in controlled motor drives, Proc. 3rd IFAC Symp. Control in Power Electronics and Elec. Drives, Lausanne, Switzerland, pp. 133-138 (1983).

Sastry, V. A., Self-tuning control of Kamyr digester chip level, Pulp and Paper, 5: 160 (1977).

Sastry, V. A., Seaborg, D. E., and Wood, R. K., A self-tuning regulator applied to a binary distillation column, Automatica, 13: 417-424 (1977).

Schulze, H., Erfahrungen im Systemprogramm für Adaptation und Regelung in Zementwerken, Proc. IFAC/IFIP Symp., Zurich, pp. 436-450 (1974).

Seyfried, H. W., and Stöle, D., Application of adaptive control in rolling mill area, especially for plate mills, Proc. IFAC Congress, Boston, Paper No. 46.1 (1975).

Sheirah, M. A. H., Malik, O. P., and Hope, G. S., A self-tuning automatic voltage regulator, Elec. Power System Res., 2: 199-213 (1979).

Shipley, R. P., Ergel, A. G., and Hung, J. W., Self-adaptive flight control by multi-variable parameter identification, Report No. AFFDL-TR-65-90, Wright-Patterson Airforce Base, Dayton, Ohio (1965).

Sikora, R. F., Self-tuning control of moisture content, 3rd Int. Pulp and Paper Proc. Contr. Symp., Vancouver, Canada (1983).

Simmons, J. C., and Waymeyer, W. K., Adaptive autopilot study for a high performance interceptor missile, in Adaptive Control Systems (F. P. Caruthers and H. Levenstein, eds.), Pergamon Press, Oxford, pp. 263-276, 1963.

Sinha, N. K., and Bereznai, G. T., Adaptive nuclear reactor control for minfuel-weighted absolute error and minimax criteria, Proc. 6th IFAC Congress, Boston (1975).

Sinner, E., Etude d'un régulateur adaptif pour la commande d'une classes de procédés industriels, Doctoral Thesis, Université Scientifique et Médicale de Grenoble (1973).

Sinner, E., Régulateur adaptif à variables d'Etat pour procédés industriels, Rev. RAIRO, 1: 103-123 (1975).

Smith, G. W., Synthesis of self-adaptive autopilot for a large elastic booster, IRE Trans. Auto. Contr., AC-5: 229 (1961).

Smyth, R. K., and Davis, J. C., A self-adaptive control system for a space booster of the Saturn class, Proc. JACC (1962).

Smyth, R. K., and Ehlers, H. L., Survey of adaptive control applications to aerospace vehicles, AGARD Advanced Control System Concepts, pp. 1-21 (1971).

Smyth, R. K., Du Plessis, R. M., and Mattingly, L. K., A digital self-adaptive body bending filter for flexible airframe control, Proc. Nat. Aerospace Elec. Conf., pp. 369-376 (1960).

Speth, W., Simple methods for rapid self-adaptation of automatic controllers in drive applications, Proc. 4th IFAC Congress, Warsaw, Paper No. 50.4 (1969).

Speth, W., Self-adjusting control systems in electrical drive technology, Doctoral Thesis, Tech. University, Braunschweig, F.R.G. (1971).

Spiegel, P., Parameter control in electric drives (in German), Zu Messen. Steuern. Regeln., 12: 197-202 (1969).

Stefani, R. T., Design and simulation of a high performance, digital, adaptive, normal acceleration control system using modern parameter estimation techniques, Report No. DAC-60637, Douglas Aircraft Co., Santa Monica, Calif. (1967).

Stein, G., Adaptive flight control—A pragmatic view, in Applications of Adaptive Control (K. S. Narendra and R. V. Monopoli, eds.), Academic Press, New York, 1980.

Stoten, D. P., Discrete adaptive control of a manipulator arm, Opt. Contr. Appln. Meth., 3: No. 4 (1982).

Stute, G., and Götz, F. R., Anwendung adaptiver systeme bei spanenden Werkzengmaschinen, Deutsch-Franzosischer Aussprachetag Industrielle Anwendung adaptiver Systeme, Freiburg, F.R.G., pp. 263-278 (1973).

Tan, Chung-I., and McInnis, B. C., Adaptive digital control implemented using residue number systems, IEEE Trans. Auto. Contr., AC-27: 499-502 (1982).

Taylor, L. W., and Adkins, E. J., Adaptive control and the X-15, Proc. Conf. on Aircraft Flying Qualities, Princeton University, Princeton, N.J. (1965).

Thirurooran, C., Modeling and control of a turbo-charged diesel engine, Doctoral Thesis, UMIST, Manchester, U.K. (1979).

Thomas, R. J., Thorp, J. S., and Pottle, C., A model referenced controller for stabilizing large transient swings in power systems, IEEE Trans. Auto. Contr., AC-21: 746-750 (1976).

Tremblay, J. P., and Wright, J. D., Multivariable MRAC of a pilot scale packed bed tubular reactor, Proc. IFAC/IFIP Symp., Den Haag, pp. 513-525 (1977).

Udink ten Cate, A. J., and van de Vooren, J., Digital adaptive control of a greenhouse heating system, Proc. IFAC/IFIP Symp., The Hague, Netherlands, Paper No. M2-2 (1977).

Unbehauen, H. (ed.), Methods and Applications in Adaptive Control, Springer-Verlag, Berlin (1980).

Unbehauen, H., and Diekmann, K., Application of MIMO-identification to a blast furnace, Proc. 6th IFAC Symp. Identification and System Parameter Estimation, Washington, D.C. (1982).

Unbehauen, H., and Göhring, B., Application of the extended maximum likelihood method for the identification of a blast furnace, Auto. Contr. Theor. Appln., 4: 36-42 (1976).

Unbehauen, H., and Schmid, Ch., Adaptive Regelung eines Warmetauschers mit dem Prozeßrechner, Proc. 21st Int. Wiss. Koll., TH Ilmenan, pp. 109-117 (1976).

Unbehauen, H., and Schmid, Ch., Application of adaptive systems in process control, Proc. Workshop on Applications of Adaptive Control, Yale University, New Haven, Conn. (1979).

Vajk, I., Keviczky, L., Kovacks, K., and Hetthessy, J., Application of self-adjusting regulator in interconnected power system, Proc. 1978 Summer Computer Simulation Conf., Newport Beach, Calif. (1978).

Waymeyer, W. K., and Sporing, R. W., Closed loop adaptation applied to missile control, Proc. JACC, pp. 18-23 (1962).

Wellstead, P. E., Montagner, G., and Gale, S., A self-adaptive model for prediction of blast furnace hot metal production and quality, Report, Control System Center, UMIST, Manchester, U.K. (1978).

Wellstead, P. E., Thirurooran, C., and Winterbone, D. E., Identification of a turbo-charged diesel engine, Proc. 7th IFAC World Congress, Helsinki, Finland (1978).

Wellstead, P. E., and Zanker, P. M., Application of self-tuning to engine control, in Self-Tuning and Adaptive Control (C. J. Harris and S. A. Billings, eds.), Peter Peregrinus Ltd., London, 1981.

White, A. D., and Linkens, D. A., An adaptive Kalman filter for marine navigation and gravity measurement, IEE Proc., 125: 1311-1317 (1978).

Wilhelm, R. G., Self-tuning control strategies: Multifaceted solutions to paper machine control problems, ISA Spring Symp., Columbus, Ohio (1982).

Woo, J., and Rootenberg, J., Lyapunov redesign of MRAC system for long term ventilation of lung, ISA Trans., 14: 89-98 (1975).

Young, K.-K. D., and Kwatny, H. G., Variable structure servomechanism design and application to overspeed protection control, Automatica, 18: 385-400 (1982).

Zanker, P. M., and Wellstead, P. E., On self-tuning diesel engine governors, Report No. 42, Control Systems Center, UMIST, Manchester, U.K. (1978).

Zhivagladov, V., and Mirkin, B., Adaptation and direct digital control of cement industry technological process with application of distributed check, Proc. IFAC/IFIP Symp., Zurich, Switzerland, pp. 451-462 (1974).

Zinober, A. S. I., Self adaptive control of overhead crane operations, Proc. 5th IFAC Symp. Identification and System Parameter Estimation, Darmstadt, F.R.G., pp. 1161-1167 (1979).

10

Future of Adaptive Control

> We belong not to the past dawns,
> But to the noons of the future.
>
> Sri Aurobindo

10.1 PERSPECTIVES FOR THE FUTURE

Technological change often arises not from persistent looking at the past, but from occasional glimpses of what might be in the future. The purpose of this chapter is not to prophesy, but to identify certain trends so that we may "act that each tomorrow brings us farther than today." As a matter of fact, we cannot envisage, or even imagine, the full range of possibilities for future technological advances. In his Profiles of the Future, Arthur C. Clarke offers a rule for assessing prophecies: "When a distinguished but elderly scientist states that something is possible, he is almost certainly right. When he states that something is impossible, he is very probably wrong." We cannot envisage *all* possibilities, simply because we cannot imagine *everything*. Technology is fast becoming the art of the "impossible."

Adaptive control is one of the most challenging and active areas of intense research in recent years. Modern control systems are becoming more and more sophisticated and demand certain exacting capabilities, and complicated decision-making abilities in the presence of uncertainties and changing environment. The very word *adaptive* is indicative of human capabilities and so we can say that adaptive control has a close kinship with intelligence.

Extensive research is going on in the field of industrial automation, robotics, man-machine systems, anticipatory systems, learning theory, pattern recognition, and fuzzy logic controllers. Adaptive control is of vital interest in these areas. In all complex, large-scale problems, the control schemes must be invested with some "intelligence" in terms of memory, learning capability, anticipatory ability, multilevel decision making, and also the capacity to respond to fuzzy qualitative statements. An autonomous robot is an example of a multipurpose "intelligent" system provided with human-like capabilities, such as perception, modeling of environment, efficient locomotion, high-level decision-making, and other advanced abilities for remote manipulation in hazardous situations. The concept of adaptive control will play an important role, and adaptive controllers of the future will have to be provided with special capabilities, including the ability to improve performance.

10.2 AREAS FOR FURTHER RESEARCH

Numerous problems must still be solved to bridge the existing gap between theory and the conditions of the real world. A number of still unresolved problems have already been mentioned at relevant places in the earlier chapters. We can list the following areas for further research:

1. A unifying view of different adaptive control methodologies
2. A unified approach to self-tuning algorithms and the criteria for the choice of the best algorithm
3. A theoretical investigation to find less stringent conditions for asymptotic stability
4. Robustness redesign of adaptive control systems
5. Design of MRAC systems in the presence of stochastic disturbances
6. Design of a low-order adaptive controller for a higher-order plant which can lead to solution of other open questions in adaptive control
7. Adaptive control of strongly nonlinear processes which are also time-varying and stochastic
8. Adaptive control of distributed parameter systems, which is still an unexplored area
9. Extremal adaptive control for systems with unknown parameters and higher order dynamics
10. Hybrid adaptive control with reference to both MRAC and STR
11. Adaptive control of fast time-varying processes in which computation time poses a difficulty
12. Study of convergence rates and acceleration of convergence of existing algorithms
13. Methods to overcome some practical difficulties such as (a) unsatisfactory control behavior during the adaptation phase, (b) undesirable effects such as "bursting" of parameter estimates, (c) the necessity to provide a priori information about model order, dead time, and sample time
14. Single-chip implementation of self-tuning controllers for low cost, small size and weight, implementation in bionic systems, and consumer applications

15. Adaptive controllers for decentralized situations in which each controller has only partial information
16. Adaptive control in Markov decision processes
17. Adaptive optimal solution of a nonlinear two-point boundary value problem in which the desired end product in a process such as steelmaking is achieved over a fixed time interval with unknown initial state and unknown and variable plant parameters
18. Development of fuzzy logic adaptive controllers
19. Development of multivariable adaptive model-following systems using the theory of variable structure systems
20. Design of simple restricted complexity adaptive control systems

Adaptive control techniques will take a quantum jump if solutions are found for the preceding issues and will leap forward into a new technological era.

Artificial intelligence expert systems technology has developed dramatically in recent years. Knowledge engineering is a newly emerging technological field that has great promise in developing human-like intelligent behavior. A human expert possesses heuristic capbilities. Judgment was until recently an exclusive domain of the human brain. Now, however, the rapidly developing technology of artificial intelligence is moving in this direction and it is possible to develop ambiguity-tolerant software that can cope—like a human—with incomplete and uncertain information. An intelligent expert system can be designed to be able to take decisions in an uncertain domain of knowledge. Recent methodologies involving approximate reasoning, rule-based programming, and fuzzy logic have helped in the understanding and design of such expert systems. Prototype expert systems are applied in such areas as robotics, industrial automation, medical diagnosis, and so on. A significant research effort is underway to exploit this technology and extend it to new applications. This will be a key technology in the next decade for the development of adaptive control systems. In the coming years, researchers have many challenging tasks ahead before the full potential of adaptive control can be attained.

EPILOGUE

A million lotuses swaying
on one stem,
World after coloured and
ecstatic world,
Climbs towards some far
unseen epiphany.

Sri Aurobindo

Appendix 1

Softwear Packages

This is a listing of software packages developed by researchers around the world for adaptive control systems and related problems. This is extracted mostly from Ref. A1, except packages 1 and 5, which are the latest updated versions.

1. ADPAC

ADPAC (ADaptive control PACkage) is a package for the interactive design of SISO adaptive control systems. It allows the implementation of a wide class of STR schemes and simulation of the overall controlled system. Identification routines: recursive least squares (RLS) and extended least squares (ELS) for ARMAX models, with Givens orthogonal factorization. Control strategies: LQG type, with or without control weighting and with control horizon ranging from one to several steps (minimum variance, extended minimum variance, m-step, receding horizon, infinite horizon) and pole assignment. Implicit (i.e., closed-loop) ARX system models can be identified for ARMAX systems by means of RLS routines. The implicit LQG infinite horizon case is handled by computing the variation of the control variable, instead of the optimal control itself (Refs. A2–A4).

Details can be obtained from Franco Davoli, DIST, University of Genoa, Via Opera Pia 11 A, 16145 Genoa, Italy.

2. KEDDC

KEDDC is a versatile package which makes possible the analysis, identification, and design of a variety of adaptive control systems. Analysis and synthesis of SISO and MOMO systems: linear, nonlinear, state space, frequency domain, description using polynomial matrices, continuous, discrete time. Process identification: deterministic and stochastic methods. Controller design: classical approaches, LQ techniques, pole-zero placement method, inverse Nyquist array technique, parameter optimization, compensator controllers, observers, Kalman filters. Adaptive control: classical approaches, MRAC methods, and STR schemes (Refs. A5–A8).

Details can be obtained from Dr.-Ing. Chr. Schmid, Lehrstuhl für Electrische Steuerung und Regelung, Ruhr University, P.B. 102148, D-4630 Bochum 1, F.R.G.

3. CYPROS

CYPROS is a package that covers a wide range of problems: system identification, parameter estimation, simulation, analysis and design. Different model representation forms can be handled: multivariable state-space models in continuous or discrete time, multivariable transfer functions in continuous or discrete form or matrix polynomial form. Transformation programs are available. The control package in CYPROS consists of programs based on multivariable time-domain and frequency-domain methods for analysis and design. An adaptive program is based on a pole placement technique. A generalized minimum-variance control algorithm is also available. Classical control methods for SISO systems are also included. Parameter estimation package contains programs for SISO and MIMO linear and nonlinear models. The maximum likelihood method and extended Kalman filter are implemented. In the simulation package different methods are available for solving differential and difference equations (Ref. A9).

Details can be obtained from Arne Tysso, CAMO, Computer-Aided Modeling A/S, P.B. 2893 Elgesaeter, N-7001 Trondheim, Norway.

4. $MATRIX_X$ and SYSTEM-BUILD

$MATRIX_X$ is a package for control system design and analysis, system identification, data analysis, and simulation. System identification tools include nonparametric frequency-domain techniques, maximum likelihood algorithms in batch and recursive forms, and adaptive algorithms. Simulation of differential/algebraic systems and fast propagation of sparse systems are also available. A powerful two- and three-dimensional graphics package is included.

The SYSTEM-BUILD capability in $MATRIX_X$ allows graphic building of nonlinear differential/algebraic system models. Each subsystem may be declared to be continuous or discrete with a given sampling interval. Multirate time simulations can be performed on sampled data systems (Ref. A10).

Details can be obtained from Eleanor Vade Bon Coeur, Integrated Systems Inc., 101 University Avenue, Palo Alto CA 94301-1695.

5. DESIGN MASTER

DESIGN MASTER has five programs: SIMNON, IDPAC, MODPAC, SYNPAC, and POLPAC.

SIMNON permits simulation and optimization of nonlinear continuous-time and discrete-time systems. The program has choices of different integration routines, Runge-Kutta method with fixed and variable step length, predictor corrector, and routines for stiff equations. Systems of order 100 can be simulated.

IDPAC allows data analysis, spectral analysis, correlation analysis, parameter estimation in linear models with many inputs and many outputs, and model validation.

MODPAC incorporates transformations of models, graphical outputs, matrix operations, polynomial operations, system transformations, Kalman decompositions, controllability, observability, transformation from state space to frequency domain and polynomial representations, transformation to diagonal and Hessenberger form, and transformation to observability and canonical forms.

SYNPAC deals with the design of feedback and feedforward controllers for multivariable linear dynamic systems.

POLPAC is a polynomial-oriented design package. It can be applied for deadbeat, minimum variance, and pole placement designs.

Details can be obtained from K. J. Åström, Department of Automatic Control, Lund Institute of Technology, P.B. 725, S-220-07 Lund 7, Sweden.

6. ISER-CSD

ISER-CSD (Interactive System for Education and Research in Control System Design) is a package meant for simulation of dynamic systems in time or frequency domains, SISO and MIMO systems, continuous-time or discrete-time models, and models in transfer function form or state-space form. It is also intended for identification of systems in the multiinput/single-output case (four methods), and optimal control design based on the linear quadratic Gaussian methodology. This package has the following main characteristics: performs data structure and dynamic memory management and modular structure for further modifications.

Details can be obtained from Cevdet Suleyman, Institut Privé Control Data, 59 rue Nationale, 75013 Paris, France.

7. DPACS-F

DPACS-F is a software package for the analysis of multivariable control systems based on state-space and frequency-domain methods. Pole assignment and LQG methodologies are used for design. This package includes programs for identification of multivariable systems based on maximum likelihood and generalized least squares. It can handle 40×40 matrix and time series and 20×2100 data. Dimension will be increased on MV version (Refs. A11, A12).

Details can be obtained from K. Furuta, Department of Control Engineering, Tokyo Institute of Technology, Oh-Okayama, Meguro-ku, Tokyo, Japan.

8. ORACLS

ORACLS (Optimal Regulator Algorithms for Control of Linear Systems) is a package for the design of linear time-invariant multivariable observer-based compensators primarily through LQG approach. It has 62 special-purpose subroutines. Among other capabilities, the design algorithms can generate transient/steady-state LQG, implicit/explicit model following, single-input pole placement, and sampled data linear quadrature regulator control laws (Ref. A13).

Details can be obtained from Earnest A. Armstrong, NASA Langley Research Center, Hampton, VA 23665.

9. CTRL-C

CTRL-C is an interactive computer language for analysis and design of multivariable control systems. A powerful matrix environment enables system simulation, signal generation, matrix analysis, and graphics. Systems may be described in state-space, transfer function or discrete time forms. Transformations between representations are possible. Design primitives are provided for eigenstructure assignment, LQG regulators, LQG filters, and model following. Time analysis, frequency analysis, matrix analysis and digital signal processing primitives are also available.

Details can be obtained from John Little, Systems Control Technology, 1801 Page Mill Road, Palo Alto, CA 94303.

10. TRIP

TRIP (TRansformation and Identifcation Program) is an interactive program for the analysis and design of SISO systems. It distinguishes seven different models for linear systems. TRIP supports the transformation of one model to another model. Among a variety of problems TRIP can handle, the least squares method is one (Ref. A14).

Details can be obtained from P. P. J. van den Bosch, Delft University of Technology, P.B. 5031, 2600 GA-Delft, The Netherlands.

REFERENCES

A1. Frederick, D. K., Herget, C. J., and McFarland, F., Software summaries, in Computer-Aided Control Systems Engineering (M. Jamshidi and C. J. Herget, eds.), Elsevier Science Publishing (North-Holland), Amsterdam, 1985.

A2. Bartolini, G., Casalino, G., Davoli, F., and Minciardi, R., A package for multivariable adaptive control, Proc. 3rd IFAC Symp. on Software for Computer Control, Madrid, Spain, Pergamon Press, Oxford (1982).

A3. Bartolini, G., Casalino, G., Davoli, F., Minciardi, R., and Zappa, G., On implicit modeling theory and its application to adaptive control, Proc. IFAC World Congress, Budapest, Hungary, Pergamon Press, Oxford, pp. 949-954 (1984).

A4. Casalino, G., Davoli, F., and Minciardi, R., Finite and infinite horizon LQ adaptive controllers based on implicit model identification, Proc. IEEE-CDC, Fort Lauderdale, Fla., pp. 853-855 (1985).

A5. Schmid, Ch., CAD of adaptive systems, in Computer-Aided Design of Control Systems (M. A. Cuenod, ed.), Pergamon Press, Oxford (1980).

A6. Schmid, Ch., KEDDC—A computer-aided analysis and design package for control systems, Proc. American Contr. Conf., Arlington, Va., pp. 211-212 (1982).

A7. Schmid, Ch., and Unbehauen, H., Identification and CAD of adaptive systems using the KEDDC package, Proc. 5th IFAC Symp. on Identification and Syst. Par. Estn., Darmstadt, F.R.G., pp. 1087-1094 (1979).

A8. Schmid, Ch., and Unbehauen, H., KEDDC, A general-purpose CAD software system for application in control engineering, in Software for Computer Control (M. Novak, ed.), Pergamon Press, Oxford (1979).

A9. Tysso, A., CYPROS—Cybernetic program packages, MIC, 1 (1980).

A10. Walker, R., Shah, S., Gregory, C. Z., and Varvell, D., MATRIX_X: A model building, nonlinear simulation and control design program, in Computer-Aided Control Systems Engineering (M. Jamshidi and C. J. Herget, eds.), Elsevier Science Publishers (North-Holland), Amsterdam (1985).

A11. Furuta, K., et al., Computer-aided design program for linear control systems, Proc. IFAC Symp. on CAD, Zurich, pp. 267-272 (1980).

A12. Furuta, K., et al., Computer system design for furnace by CAD, Proc. IFAC Symp. on Digital Control, New Delhi, pp. 31-38 (1982).
A13. Armstrong, E. S., ORACLS: A Design System for Linear Multivariable Control, Marcel Dekker, New York (1980).
A14. van den Bosch, P. P. J., Interactive computer-aided control system analysis and design, in Computer-Aided Control Systems Engineering (M. Jamshidi and C. J. Herget, eds.), Elsevier Science Publishers (North-Holland), Amsterdam (1985).

ADDITIONAL REFERENCES

Cassalino, G., Davoli, F., Minciardi, R., and Zappa, G., On implicit modeling theory: Basic concepts and application to adaptive control, to appear in Automatica, March 1987.
Schmid, Ch., KEDDC—A computer-aided analysis and design package for control systems, in Computer-Aided Control Systems Engineering (M. Jamshidi and C. J. Herget, eds.), Elsevier Science Publishers (North-Holland), Amsterdam (1985).
Tuffs, P. S., and Mohtadi, C., FAUST: Users Guide, Report No. OUEL-1582/85, Oxford University, Oxford (1985).
Tuffs, P. S., and Mohtadi, C., FAUST: Programmers Reference Guide, Report No. OUEL-1583/85, Oxford University, Oxford (1985).

Appendix 2

Additional Bibliography

Allidina, A. Y., and Yin, H., Explicit pole-assignment self-tuning algorithms, Int. J. Contr., 42: 1113-1130 (1985).

Altray, B. K., On the stability of parallel MRAS with an AEF, IEEE Trans. Auto. Contr., AC-30: 889-891 (1985).

Anbumani, K., Patnaik, L. M., and Sarma, I. G., Self-tuning pole-placement in nonlinear systems of Hammerstein model, IEEE Trans. Ind. Electro., IE-32: 166-170 (1985).

Andersen, T. R., and Nielsen, S. B., An efficient single output fuzzy control algorithm for adaptive applications, Automatica, 21: 539-546 (1985).

Anderson, B. D. O., Bitmead, R. R., Johnson, C. R., Jr., Kokotovic, P. V., Kosut, R. L. Mareels, I., Praly, L., and Riedle, B. D., Stability of Adaptive Systems: Passivity and Averaging Analysis, M.I.T. Press, Cambridge, MA (1986).

Andersson, P., Adaptive forgetting in recursive identification through multiple models, Int. J. Contr., 42: 1175-1194 (1985).

Arzen, K. E., Experiments with expert system for process control, Proc. 1st Int. Conf. Appl. Artificial Intelligence Eng. Practice, Southampton, U.K. (1986).

Åström, K. J., Autotuning, adaptation and expert control, Proc. Am. Contr. Conf., Boston MA, Vol. 3, pp. 1514-1519 (1985).

Åström, K. J., Anton, J. J., and K. E. Arzen, Expert control, Automatica, 22: 277-286 (1986).

Åström, K. J., et al., The self-tuning regulators revisited, Proc. 7th IFAC/IFORS Symp. Ident. Syst. Param. Estn. (1985).

Bai, E. W., and Sastry, S. S., Persistency of excitation, sufficient richness and parameter convergence in discrete time adaptive control, Syst. Contr. Lett., 6: 153-163 (1985).

Bai, E. W., and Sastry, S. S., Parameter identification using prior information, Int. J. Contr., 44: 455-473 (1986).

Bai, E. W., and Sastry, S. S., Discrete-time adaptive control utilizing prior information, IEEE Trans. Auto. Contr., AC-31, 779-782 (1986).

Baker, R. P. D., and Morris, A. J., A generalized minimum variance self-tuning controller for systems with large dead times, Proc. Am. Contr. Conf., Boston, MA, pp. 408-409 (1985).

Bar-Kana, I., Direct multivariable adaptive control with application to large structure systems, Doctoral Thesis, Rensselaer Polytechnic Institute, Troy, NY (1983).

Bar-Kana, I., Positive realness in continuous-time adaptive control, Proc. Am. Control Conf., Seattle, WA (1986).

Bar-Kana, I., Positive realness in discrete-time adaptive control systems, Int. J. Syst. Sci., 17: 1001-1006 (1986).

Bar-Kana, I., and Kaufman, H., Global stability and performance of a simplified adaptive algorithm, Int. J. Contr., 42: 1491-1505 (1985).

Bar-Kana, I., and Kaufman, H., Robust simplified adaptive control for a class of multivariable continuous time systems, Proc. 24th IEEE-CDC, Fort Lauderdale, FL (1985).

Bartolini, G., Casalino, G., Davoli, F., Mastretta, M., Miniciardi, R., and Morten, E., Development of adaptive fuzzy controllers with application to continuous casting plants, in Industrial Applications of Fuzzy Control (M. Sugeno, ed.), North-Holland, Amsterdam, pp. 73-86 (1985).

Baruch, I. S., Square-root algorithm for real-time identification and adaptive control of multivariable systems, Proc. Int. Conf. Systems Science, Wroclaw, Poland (1986).

Berger, C. S., The use of inter-sampling in adaptive control, IEEE Trans. Auto. Contr., AC-30: 1227-1229 (1985).

Betz, R. E., and Evans, R. J., Nonlinear adaptive control of a synchronous machine, IEEE Trans. Aerosp. and Electron. Syst., AES-21: 631-643 (1985).

Bitmead, R. R., Anderson, B. D. O., and Ng, T. S., Convergence rate determination for gradient-based adaptive estimators, Automatica, 22: 185-191 (1986).

Bittanti, S., and Scattolini, R., Multivariable self-tuning controller: A model-following approach, Int. J. Contr., 42: 1035-1048 (1985).

Boel, R. K., Teo, K. L., and Moore, J. B., Convergence in mean for ELS-based adaptive control, Syst. Contr. Lett., 5: 389-396 (1985).

Bortsov, Yu. A., and Junger, I. B., Adaptive electromechanical systems with indirect discontinuous control, Electric Tech., U.S.S.R., 1: 34-44 (1985).

Brickwedde, A., Microprocessor-based adaptive speed and position control for electric drives, IEEE Trans. Ind. Appl., IA-21: 1154-1161 (1985).

Bundell, G. A., Robust decentralized MRAC in the manipulator control problem, Proc. Int. Conf. Control, Cambridge, U.K., Vol. 1, pp. 321-326 (1985).

Burton, J. A., and Zinober, A. S. I., Continuous approximation of variable structure control, Int. J. Syst. Sci., 17: 875-886 (1986).

Byrnes, J., Katebi, R., and Grimble, M. J., LQG adaptive autopilot design, Proc. IFAC Symp. Identification and Syst. Param. Estn., York, U.K. (1985).

Caines, P. E., and Meyn, S., The zero divisor problem of stochastic adaptive control, Proc. IFAC Symp. Identification and Syst. Param. Estn., York, U.K. (1985).

Chalam, V. V., Adaptive control: Recent trends, Systems Sci. (Wroclaw, Poland), 12: No. 1-2 (1986).

Chan, W. P., and Hang, C. C., Discrete MRAC using only input and output measurements, Int. J. Contr., 42: 161-176 (1985).

Chandrasekhar, J., and Vittal Rao, M. P. R., A new differentiator-free Lyapunov MRAC with increased flexibility and simple structure, Proc. 9th World Congress of IFAC, Budapest, Vol. 2, Pergamon Press, Oxford, pp. 883-888 (1985).

Chang-Huan Liu, et al., An adaptive control strategy for robotic manipulators, J. Chin. Inst. Eng. (Taiwan), 8: 251-260 (1985).

Chen, H. F., and Caines, P. E., On the adaptive control of a class of systems with random parameters and disturbances, Automatica, 21: 737-741 (1985).

Chen, H. F., and Guo, L., Optimal stochastic adaptive control with quadratic index, Int. J. Contr., 43: 869-882 (1986).

Cheng, Shi-jie, Malik, O. P., and Hope, G. S., Self-tuning stabilizer for a multimachine power system, IEE Proc., 133, Pt. C: 176-185 (1986).

Cheng, W.-M., et al., Model reference fuzzy adaptive control, Proc. 9th World Congress of IFAC, Budapest, Vol. 2, Pergamon Press, Oxford, pp. 889-892 (1985).

Chien, I. L., et al., A self-tuning controller for systems with unknown or varying time delays, Ind. J. Contr., 42: 949-964 (1985).

Chiu, T. H., Koivo, A. J., and Lewczyk, R., Experiments on manipulator gross motion using self-tuning controller and visual information, J. Robotic Syst., 3: 59-70 (1986).

Clarke, D. W., Kanjilal, P. P., and Mahfadi, C., A generalized LQG approach to STC: Part I—Aspects of design; Part II—Implementation and simulation, Int. J. Contr., 41: 1493-1508, 1525-1544 (1985).

Dasgupta, S., Adaptive identification and control, Doctoral thesis, Australian National University, Canberra (1984).

Dasgupta, S., and Johnson, C. R., Jr., Some basic questions about two popular adaptive algorithm modifications, Proc. Am. Contr. Conf., Boston, MA, Vol. 2, pp. 970-972 (1985).

Delariminat, P., Adaptive control based on identification by instrumental variables, RAIRO, 20: 99-118 (1986).
de la Sen, M., On the improvement of the adaptive transient: An adaptive sampling approach, Proc. 9th World Congress of IFAC, Budapest, Vol. 2, Pergamon Press, Oxford, pp. 829-834 (1985).
de la Sen, M., Adaptive sampling for improving the adaptation transients in hybrid adaptive control, Int. J. Contr., 61: 1189-1205 (1985).
de la Sen, M., and Paz, H. B., Discrete multivariable adaptive control, Int. J. Contr., 42: 1071-1088 (1985).
de la Sen, M., Multirate hybrid adaptive control, IEEE Trans. Auto. Contr., AC-31: 582-586 (1986).
Dexter, A. L., Self-tuning controllers applied to environmental control, IEE Vacation School on Industrial Digital Control Systems, Oxford (1985).
Dorling, C. M., The design of variable structure control systems: A manual for VASSYD package, University of Sheffield, U.K. (1985).
Dorling, C. M., and Zinober, A. S. I., Two approaches to hyperplane design in multivariable variable structure control systems, Int. J. Contr., 44: 65-82 (1986).
Drews, P., and Starke, G., Development approaches for advanced adaptive control in automated arc welding, Proc. Int. Conf. on Automation and Robotization in Welding and Allied Processes, Pergamon Press, Oxford, pp. 115-126 (1985).
Dugard, L., and Dion, J. M., Direct adaptive control for linear multivariable systems, Int. J. Contr., 42: 1251-1282 (1985).
Dumont, G. A., Application of advanced control methods in the pulp and paper industry: A survey, Automatica, 22: 143-153 (1986).
Durrant-Whyte, H., Practical adaptive control of actuated spatial mechanisms, Proc. IEEE Int. Conf. Robotics and Automation, Silver Spring, MD, pp. 650-655 (1985).
Elicabe, G. E., et al., Model reference adaptive control of a certain class of nonlinear systems, Proc. Am. Contr. Conf., Boston, MA, Vol. 1, pp. 399-400 (1985).
Fernandez del Busto, R., et al., Application of a model reference adaptive algorithm to the combustion control of a power unit boiler, Trans. ASME, J. Dyn. Syst. Meas. Contr., 107: 290-291 (1985).
Feuer, A., A parameterization for model reference adaptive pole placement, IEEE Trans. Auto. Contr., AC-31: 782-785 (1986).
Fu, L., Bodson, M., and Sastry, S. S., New stability theorems for averaging and their applications to the convergence analysis of adaptive identification and control schemes, Proc. IEEE-CDC, Ft. Lauderdale, FL (1985).
Fujii, S., et al., A discrete time multivariable MRAC algorithm with application to a real plant, Proc. 9th World Congress of IFAC, Budapest, Vol. 2, Pergamon Press, Oxford, pp. 877-882 (1985).

Garipov, E., et al., Laboratory system for self-tuning controllers analysis and implementation, AFCET Congress Automatique, Toulouse, France (1985).

Gavel, D. T., and Siljak, D. D., High gain adaptive decentralized control, Proc. Am. Contr. Conf., Boston, MA, Vol. 1, pp. 568-573 (1985).

Gawthrop, P. J., Multiloop self-tuning control: Cascade system, Proc. 9th World Congress of IFAC, Budapest, Vol. 2, Pergamon Press, Oxford, pp. 905-910 (1985).

Gawthrop, P. J., Self-tuning PID controllers: Algorithms and implementation, IEEE Trans. Auto. Contr., AC-31: 201-209 (1986).

Gomart, O., and Caines, P., On the extension of robust global adaptive control results to unstructured time-varying systems, IEEE Trans. Auto. Contr., AC-31: 370-373 (1986).

Goodwin, G. C., Hill, D. J., and Palaniswami, M., Towards an adaptive robust controller, IFAC Symp. Identification and Syst. Param. Estn., York, U.K. (1985).

Goodwin, G. C., Lozano Leal, R., Mayne, D. Q., and Middleton, R. H., Rapprochment between continuous and discrete MRAC, Automatica, 22: 199-207 (1986).

Goodwin, G. C., and Mayne, D. Q., A continuous-time adaptive controller with integral action, IEEE Trans. Auto. Contr., AC-30: 793-795 (1985).

Goodwin, G. C., and Middleton, R. H., Continuous and discrete adaptive control, in Control and Dynamic Systems, Vol. XXXIV (C. T. Leondes, ed.), Academic Press, New York (1985).

Grimble, M. J., Implicit and explicit LQG self-tuning controllers, Proc. 9th World Congress of IFAC, Budapest, Vol. 2, Pergamon Press, Oxford, pp. 941-947 (1985).

Grimble, M. J., Controllers for LQG self-tuning applications with colored measurement noise and dynamic costing, IEE Proc., 133, Pt. D: 19-29 (1986).

Grimble, M. J., and Moir, T. J., Multivariable weighted minimum variance self-tuning controllers, Int. J. Contr., 42: 1283-1308 (1985).

Gupta, M. M., et al., Some new adaptive algorithms for a class of MRAC systems, Proc. 9th World Congress of IFAC, Budapest, Vol. 2, Pergamon Press, Oxford, pp. 871-876 (1985).

Hagglund, T., Adaptive control of systems subject to large parameter changes, Proc. 9th World Congress of IFAC, Budapest, Pergamon Press, Oxford, pp. 993-998 (1985).

Hahn, V., Rock, H., Schmid, Ch., and Weimer, P., Some experiences with the application of multivariable adaptive control in chemical and electromechanical plants, Opt. Contr. Appl. Meth., 7: 225-248 (1986).

Halme, A., Sekainaho, J., and Soinien, J., Adaptive control with nonlinear filtering, Automatica, 21: 453-463 (1985).

Hamida, A. A., et al., Self-tuning regulator of an ore crusher, Conf. Rec. ASILOMAR, Pacific Grove, CA, IEEE Computer Soc. Press, pp. 312-316 (1985).

Higham, E., Expert systems in self-tuning controllers, The Chem. Engineer (London), No. 423: 41-44 (1986).

Hiram, Y., et al., Overcoming difficulties in the application of self-tuning controllers, Proc. Am. Contr. Conf., Boston, MA, Vol. 3, pp. 1507-1513 (1985).

Hiroi, M., et al., Microprocessor-based decoupled control of manipulator using modified model-following method with sliding mode, IEEE Trans. Ind. Electro., IE-33: 110-113 (1986).

Holst, J., and Poulsen, N. K., Self-tuning control of plants with abrupt changes, Proc. 9th World Congress of IFAC, Budapest, Vol. 2, Pergamon Press, Oxford, pp. 923-982 (1985).

Hong, H. D., and Zohdy, M. A., An adaptive control system scheme for oscillator servo system, IEEE Trans. Ind. Electro., IE-32: 37-40 (1985).

Hori, N., et al., A discrete adaptive observer for a multioutput aircraft control system, Proc. 9th World Congress of IFAC, Budapest, Vol. 2, Pergamon Press, Oxford, pp. 1057-1062 (1985).

Huang, H. P., et al., Adaptive design of model-based controllers, Proc. Am. Contr. Conf., Boston, MA, Vol. 3, pp. 1520-1527 (1985).

Hulliger, P., Adaptive control algorithms for machines (in French), Bull. Assoc. Suisse Electr. (Switzerland), 76: 920-923 (1985).

Ichikawa, K., Adaptive control of delay system, Int. J. Contr., 43: 1659 (1986).

Ih, C. C., Wang, S. J., and Leondes, C. T., An investigation of adaptive control techniques for space stations, Proc. Am. Contr. Conf., Boston, MA, Vol. 1, pp. 81-94 (1985).

Ioannou, P. A., Decentralized adaptive control of interconnected systems, IEEE Trans. Auto. Contr., AC-31: 291-298 (1986).

Ioannou, P. A., Robust adaptive controller with zero residual tracking errors, IEEE Trans. Auto. Contr., AC-31: 773-776 (1986).

Ioannou, P. A., and Kokotovic, P., Decentralized adaptive control of interconnected systems with reduced order models, Automatica, 21: 401-412 (1985).

Irisa, T., et al., A novel approach on a parameter self-tuning method in an a.c. servo system, Automatica, 22: 287-294 (1986).

Isermann, R., and Lachmann, K.-H., Parameter-adaptive control with configuration aids and supervision functions, Automatica, 21: 625-638 (1985).

James, D. J. G., Burnham, K. J., and Shields, D. N., Developments in self-tuning control for bilinear systems, Proc. Int. Conf. Syst. Science, Wroclaw, Poland (1986).

Janecki, D., On some generalization of Landau's theorem applicable to analysis of adaptive systems, IEEE Trans. Auto. Contr., AC-30: 1097-1098 (1985).

Johnson, B. W., and Aylor, J. H., Design of an adaptive controller microcomputer implementation, IEEE Trans. Ind. Electro., IE-83: 28-38 (1986).

Johnson, C. D., Adaptive controller design using disturbance-accommodation technique, Int. J. Contr., 42: 193-210 (1985).

Kanniah, J., et al., Development of model and a method of analysis for a dual-rate sampled STR, Proc. 9th World Congress of IFAC, Budapest, Vol. 2, Pergamon Press, Oxford, pp. 899-904 (1985).

Karastojanov, D. N., and Nachev, G. N., Adaptive control of industrial robots for arc welding, Proc. 9th World Congress of IFAC, Budapest, Vol. 5, Pergamon Press, Oxford, pp. 2405-2410 (1985).

Kawala, S., et al., Minimum variance self-tuning control of a continuous reheating furnace, Proc. 9th World Congress of IFAC, Budapest, Vol. 4, Pergamon Press, Oxford, pp. 1873-1878 (1985).

Kelly, R., and Mazet, E., A simple adaptive control of an aerotherm, Rev. Générale Therm., 25: 161-165 (1986).

Khurana, H., Ahson, S. I., and Lamba, S. S., On stabilization of large scale control systems using VSS theory, IEEE Trans. Auto. Contr., AC-31: 176-177 (1986).

Kidd, P. T., Robustness of MRAC in the presence of unmodeled dynamics and output disturbances, Int. J. Contr., 42: 1213-1226 (1985).

Kisselevski, F., and Zelenski, K., Adaptive control for welding robots, Proc. Int. Conf. on Automation and Robotization in Welding and Allied Processes, Pergamon Press, Oxford, pp. 135-142 (1985).

Koboyashi, T., Adaptive control for infinite dimensional systems, Int. J. Syst. Sci., 17: 887-896 (1986).

Koivo, A. J., Self-tuning manipulator control in Cartesian base coordinate system, Trans. ASME, J. Dyn. Syst. Meas. Contr., 107: 316-323 (1985).

Kokotovic, P., Riedle, B., and Praly, L., On a stability criterion for continuous slow adaptation, Syst. Contr. Lett., 6: 7-14 (1985).

Konishi, K., et al., An application of model reference adaptive technique to a pressure control system, Trans. Soc. Instr. Contr. Eng. (Japan), 21: 996-998 (1985).

Kosut, R. L., Input/output properties of adaptive control systems: Robustness and reduced order design, Proc. 9th World Congress of IFAC, Budapest, Vol. 2, Pergamon Press, Oxford, pp. 973-978 (1985).

Kosut, R. L., and Anderson, B. D. O., A local stability analysis for a class of adaptive systems, IEEE Trans. Auto. Contr., AC-31: 86-89 (1986).

Kosut, R. L., Anderson, B. D. O., and Mareels, I., Stability theory for adaptive systems: Methods of averaging and persistency of excitation, Proc. IEEE-CDC, Ft. Lauderdale, FL (1985).

Kosut, R. L., and Friedlander, B., Robust adaptive control: Condition for global stability, IEEE Trans. Auto. Contr., AC-30: 610-624 (1985).

Koutchoukali, M. S., et al., MRAC systems for a catalytic fluidized bed reactor, Automatica, 22: 101-104 (1986).

Kraft, S. D., and Wall, E. T., Experimental microprocessor-based adaptive control system, IEEE Control Syst. Magazine, 6: 35-39 (1986).

Kreisselmeier, G., An approach to stable indirect adaptive control, Automatica, 21: 385-399 (1985).

Kreisselmeier, G., A robust indirect adaptive control approach, Int. J. Contr., 43: 161-175 (1986).

Kreisselmeier, G., and Anderson, B. D. O., Robust model reference adaptive control, IEEE Trans. Auto. Contr., AC-31: 127-133 (1986).

Kreisselmeier, G., and Smith, M. C., Stable adaptive regulation of arbitrary nth order plants, IEEE Trans. Auto. Contr., AC-31: 299-305 (1986).

Krolikowski, A., Application of input signal design in system identification for adaptive control, Int. J. Syst. Sci., 17: 305-318 (1986).

Kurz, H., Adaptive control of a waste-water neutralization process, Proc. 9th World Congress of IFAC, Budapest, Vol. 6, Pergamon Press, Oxford, pp. 3257-3261 (1985).

Kwalik, K. M., and Schork, A. J., Adaptive control of a continuous polymerization reactor, Proc. Am. Contr. Conf., Vol. 2, Boston, MA, pp. 872-877 (1985).

Lakov, D., Adaptive robot under fuzzy control, Fuzzy Sets and Syst., 17: 1-8 (1985).

Lal, K., and Trybus, L., Adaptive PID control of robot accurate motions, Proc. Int. Conf. Systems Science, Wroclaw, Poland (1986).

Lang, S.-J., Gu, X.-Y., and Chai. T.-Y., A multivariable self-tuning controller with decoupling design, IEEE Trans. Auto. Contr., AC-31: 474-477 (1986).

Ledwich, G., et al., An adaptive power system stabilizer based on linear quadratic concept, Proc. 9th World Congress of IFAC, Budapest, Vol. 4, Pergamon Press, Oxford, pp. 2091-2095 (1985).

Lee, J. Y., and Sun, Y.-Y., Adaptive harmonic control in PWD inverter with fluctuating input voltage, IEEE Trans. Ind. Electro., IE-83: 92-98 (1986).

Lee, T.-H., and Narendra, K. S., Stable discrete adaptive control with unknown high frequency gain, IEEE Trans. Auto. Contr., AC-31: 477-479 (1986).

Lefebvre, S., et al., Adaptive control for HVDC power transmission systems, IEEE Trans. Power App. Syst., PAS-104: 2329-2335 (1985).

Lim, K.-Y., and Eslami, M., Adaptive controller designs for robot manipulator systems using Lyapunov direct method, IEEE Trans. Auto. Contr., AC-30: 1229-1233 (1985).

Lin, W., Kumar, P. R., and Seidman, T. I., Will the self-tuning approach work for general cost criterion? Syst. Contr. Lett., 6: 77-85 (1985).

Linkens, D. A., et al., Smith predictor and self-tuning control of muscle relaxant drug administration, IEE Proc., 132, Pt. D: 212-218 (1985).

Lozano, R., and Carrillo, J., Adaptive control of a fermentation process, Proc. 9th World Congress of IFAC, Budapest, Vol. 4, Pergamon Press, Oxford, pp. 1879-1883 (1985).

Lozano-Leal, R., and Goodwin, G. C., A globally stable adaptive pole-placement algorithm without a persistency of excitation requirement, IEEE Trans. Auto. Contr., AC-30: 795-798 (1985).

Martensson, B., The order of any stabilizing regulator is sufficient a priori information for adaptive stabilization, Syst. Contr. Lett., 6: 87-91 (1985).

Martin-Sanchez, J. M., Adaptive control for time in variant processes, Int. J. Contr., 44: 315-329 (1986).

McDermott, P. E., and Mellichamp, D. A., A de-coupling pole-placement STC for a class of multivariable processes, Opt. Contr. Appl. Meth., 7: 55-79 (1986).

McInnis, B. C., et al., Adaptive control system for the total artificial heart, Proc. 9th World Congress of IFAC, Budapest, Vol. 6, Pergamon Press, Oxford, pp. 3051 3055 (1985).

Meyn, S. P., and Caines, P. E., The zero divisor problem of multivariable stochastic adaptive control, Syst. Contr. Lett., 6: 235-238 (1985).

Mikles, A., et al., Adaptive control of tubular chemical reactors, Proc. 9th World Congress of IFAC, Budapest, Vol. 4, Pergamon Press, Oxford, pp. 1891-1896 (1985).

Millnert, M., Adaptive control of abruptly changing systems, Proc. 9th World Congress of IFAC, Budapest, Vol. 2, Pergamon Press, Oxford, pp. 999-1003 (1985).

Minamide, N., Nikiforuk, P. N., and Gupta, M. M., Design of a reduced order adaptive observer, IEE Proc., 133, Pt. D: 133-136 (1986).

Moore, J. B., A globally convergent recursive adaptive LQG regulator, Proc. 9th World Congress of IFAC, Budapest, Vol. 2, Pergamon Press, Oxford, pp. 955-959 (1985).

Moore, J. B., and Boel, R. K., Asymptotically optimum recursive prediction error methods in adaptive estimation and control, Automatica, 22: 237-240 (1986).

Morris, A. S., and Mahieddine, F., Microcomputer implementation of a MIMO pole-assignment STR, Electron. Lett., 22: 56-58 (1986).

Morse, A. S., A three-dimensional universal controller for the adaptive stabilization of any strictly proper minimum phase systems with relative degree not exceeding two, IEEE Trans. Auto. Contr., AC-30: 1188-1191 (1985).

Mosca, F., et al., Multistep horizon self-tuning controller: The MUSMAR approach, Proc. 9th World Congress of IFAC, Budapest, Vol. 2, Pergamon Press, Oxford, pp. 935-939 (1985).

M'Saad, M., Ortega, R., and Landau, I. D., Adaptive controllers for discrete time systems with arbitrary zeros—An overview, Automatica, 21: 413-423 (1985).

Mudgett, D. R., and Morse, A. S., Adaptive stabilization of discrete linear systems with unknown high frequency gain, IEEE Trans. Auto. Contr., AC-30: 798-799 (1985).

Murdoch, P., and Oliveras, E. V., A discrete adaptive observer for a system of known order, Int. J. Contr., 41: 477-492 (1985).

Narendra, K. S., and Annaswamy, A. M., Robust adaptive control in the presence of bounded disturbances, IEEE Trans. Auto. Contr., AC-31: 306-315 (1986).

Narendra, K. S., Annaswamy, A. M., and Singh, R. P., A general approach to the stability analysis of adaptive systems, Int. J. Contr., 41: 193-216 (1985).

Nemir, D. C., and Kashyap, R. L., Certainty equivalence and the minimax principle in self-tuning control, IEEE Trans. Auto. Contr., AC-31: 587-590 (1986).

Neumann, P., Predictive integral and STR for load-frequency simulation of interconnected power systems, Proc. 9th World Congress of IFAC, Budapest, Vol. 4, Pergamon Press, Oxford, pp. 2231-2236 (1985).

Ohkawa, F., A model reference adaptive control system for a class of discrete linear time varying system with time delay, Int. J. Contr., 42: 1227-1238 (1985).

Ohkawa, F., MRAC system for discrete linear time-varying systems with periodically varying parameters and time delay, Int. J. Contr., 44: 171-180 (1986).

Ohsuga, R., et al., Adaptive control system for incubator oxygen treatments, Proc. 9th World Congress of IFAC, Budapest, Vol. 6, Pergamon Press, Oxford, pp. 3033-3038 (1985).

Ohta, H., et al., Adaptive flight control of unstable helicopters during transition, Proc. 9th World Congress of IFAC, Budapest, Vol. 2, Pergamon Press, Oxford, pp. 1045-1050 (1985).

Omani, F. K., and Sinha, N. K., Modified approach to state-space STC with pole assignment, IEE Proc., 132, Pt. D: 257-262 (1985).

Opitz, H. P., Robustness properties of discrete variable structure controllers, Int. J. Contr., 43: 1003-1014 (1986).

Ortega, R., et al., A globally stable multivariable adaptive controller with reduced prior knowledge, Proc. Am. Contr., Conf., Boston, MA, Vol. 1, pp. 418-422 (1985).

Ortega, R., et a ., Practical requirements and theoretical results in robust adaptive control, Proc. 9th World Congress of IFAC, Budapest, Vol. 2, Pergamon Press, Oxford, pp. 979-986 (1985).

Ortega, R., et al., Robustness of discrete time direct adaptive controllers, IEEE Trans. Auto. Contr., AC-30: 1179-1187 (1985).

Park, G. T., Self-tuning control application to a nuclear power plant, Int. J. Contr., 43: 1823-1840 (1986).

Pavbov, B. V., and Solovev, I. G., Design of main circuit in direct adaptive control systems without measuring the complete state vector of the plant, Auto. Rem. Contr., 46: 629-636 (1985).

Payne, A. N., Adaptive one-step ahead control subject to an input-amplitude constraint, Int. J. Contr., 43: 1257-1270 (1986).

Platzer, D., and Kaufman, H., MRAC of thyristor driven d.c. motor systems subject to current limitations, Proc. 9th World Congress of IFAC, Budapest, Vol. 4, Pergamon Press, Oxford, pp. 1991-1995 (1985).

Puls, F. A., and Barash, M. M., An adaptive control algorithm for robotic deburring, J. Manuf. Syst., 4: 169-178 (1985).

Radke, F., and Isermann, R., A parameter adaptive PID controller with step-wise parameter optimization, Proc. 9th World Congress of IFAC, Budapest, Vol. 4, Pergamon Press, Oxford, pp. 1885-1890 (1985).

Riedle, B. D., and Kokotovic, P. V., Stability analysis of and adaptive system with unmodelled dynamics, Int. J. Contr., 41: 389-402 (1985).

Riedle, B. D., and Kokotovic, P. V., A stability-instability boundary for disturbance-free slow adaptation and unmodeled dynamics, IEEE Trans. Auto. Contr., AC-30: 1027-1030 (1985).

Riedle, B. D., and Kokotovic, P. V., Integral manifold of slow adaptation, IEEE Trans. Auto. Contr., AC-31: 316-326 (1986).

Rohrs, C. E., et al., Some design guidelines for discrete-time adaptive controllers, Proc. 9th World Congress of IFAC, Budapest, Vol. 2, Pergamon Press, Oxford, pp. 857-862 (1985).

Rohrs, C. E., et al., A practical robustness theorem for adaptive control, Proc. 9th World Congress of IFAC, Budapest, Vol. 2, Pergamon Press, Oxford, pp. 979-983 (1985).

Rohrs, C. E., et al., Robustness of continuous time adaptive control algorithms in the presence of unmodeled dynamics, IEEE Trans. Auto. Contr., AC-30: 881-889 (1985).

Romero, D. R., and Heydt, G. T., An adaptive excitation system controller in a stochastic environment, IEEE Trans. Power Syst., PWRS-1: 168-175 (1986).

Rubvo, F. R., et al., Chaotic motion in an adaptive control system, Int. J. Contr., 42: 353-360 (1985).

Saelid, S., et al., Design considerations, analysis and practical experience with an adaptive ship's autopilot, Proc. 9th World Congress of IFAC, Budapest, Vol. 5, Pergamon Press, Oxford, pp. 2895-2900 (1985).

Sano, A., and Kukuchi, M., Adaptive control of internal oxygen pressure of new born infants under incubator oxygen treatments, IEE Proc., 132, Pt. D: 205-211 (1985).

Scattolini, R., and Clarke, D. W., Multivariable model following STC with off-set rejection, Int. J. Contr., 42: 1309-1322 (1985).

Sharaf, S. M. Z., et al., Multivariable adaptive controller for a turbo-generator, IEE Proc., 133, Pt. D: 83-89 (1986).

Shri, D. H., and Kozin, F., On almost sure convergence of adaptive algorithms, IEE Trans. Auto. Contr., AC-31: 471-474 (1986).

Singh, S. N., Adaptive model following control of nonlinear robotic systems, IEEE Trans. Auto. Contr., AC-30: 1099-1100 (1985).

Sirisena, H. R., and Teng, F. C., Multivariable pole-zero placement STC, Int. J. Syst. Sci., 17: 345-352 (1986).

Slotine, J. J. E., and Coetsee, J. A., Adaptive sliding controller synthesis for nonlinear systems, Int. J. Contr., 43: 1631-1651 (1986).

Stankovic, S. S., and Radenkovic, M. S., Some practical aspects of adaptive regulators of arbitrary orders, Proc. 9th World Congress of IFAC, Budapest, Vol. 2, Pergamon Press, Oxford, pp. 917-922 (1985).

Stankovic, S. S., and Radenkovic, M. S., Restricted complexity adaptive controllers, in the case of stochastic reference, Proc. 7th IFAC Symp. Identification and Syst. Param. Estn., York, U.K., p. 1323 (1985).

Stankovic, S. S., and Radenkovic, M. S., Self-tuning servo for stochastic references, Automatica, 22: 241-244 (1986).

Stoten, D. P., A unified approach to the decoupled discrete adaptive control of manipulators, IEE Proc., 132, Pt. D: 151-157 (1985).

Sundaresan, M. K., and Koenig, M. A., Decentralized MRAC of robotic manipulators, Proc. Am. Contr. Conf., Boston, MA, Vol. 1, pp. 44-49 (1985).

Svoronos, S. A., An adaptive internal model controller, Proc. Am. Contr. Conf., Boston, MA, Vol. 3, pp. 1560-1565 (1985).

Swiniarski, R., Microcomputer application to adaptive time-optimal control, Proc. Int. Conf. Systems Science, Wroclaw, Poland (1986).

Szabados, B., and Storoschuk, O., Microprocessor adaptive control of oil temperature for robot arm valves: A practical implementation, IEEE Trans. Ind. Electro., IE-33: 21-27 (1986).

Tahmassebi, T., et al., Self-tuning control with feedforward for process with variable time delays, Int. J. Contr., 42: 1323-1336 (1985).

Tjokro, S., et al., Adaptive PID control, Proc. Am. Contr. Conf., Boston, MA, Vol. 3, pp. 1528-1534 (1985).

Tornizuka, M., and Zhang, S., Modeling and conventional/adaptive PI control of a lathe cutting process, Proc. Am. Contr. Conf., Boston, MA, Vol. 2, pp. 745-750 (1985).

Trankle, T. L., et al., An expert system for control system design, Proc. IEE Int. Conf. Control, Cambridge, U.K. (1985).

Trulsson, E., On adaptive control with prescribed robustness properties, Proc. Am. Contr. Conf., Boston, MA, Vol. 1, pp. 403-404 (1985).

Trulsson, E., and Ljung, L., Adaptive control based on explicit criterion minimization, Automatica, 21: 385-399 (1985).

Tsiligiannis, C. A., and Svoronos, S. A., Deterministic convergence of a Clarke-Gawthrop self-tuning controller, Automatica, 22: 193-197 (1986).

Tuffs, P. S., Self-tuning control: Algorithms and applications, Report OUEL—1567/85, Oxford University, Oxford (1985).

Vadali, S. R., Variable structure control of spacecraft large-angle manoevres, J. Guidance Contr. Dyn., 9: 235-239 (1986).

Vagi, F., et al., Self-tuning control of distillation columns, Proc. Am. Contr. Conf., Boston, MA, Vol. 3, pp. 1269-1274 (1985).

Van Cauwenberghe, A. R., and De Keyser, R. M. C., Self-adaptive long range predictive control, Proc. Am. Contr. Conf., Boston, MA, Vol. 2, pp. 1155-1160 (1985).

van den Bosch, P. P. J., et al., Adaptive attitude control for large-angle slew manoeuvres, Automatica, 22: 209-215 (1986).

Viviani, G. L., and Saet, Y. A., New type of multi-state detector for adaptive control: An application to network switching, Int. J. Electro., 60: 531-534 (1986).

Vukobratovic, M., and Kircanski, N., An approach to adaptive control of robotic manipulators, Automatica, 21: 639-647 (1985).

Vukobratovic, M., Stokic, D., and Kircanski, N., Non-adaptive and Adaptive Control of Manipulation Robots, Springer-Verlag, Berlin (1985).

Wang, S. J., et al., Space station dynamic modeling, disturbance accommodation, and adaptive control, Workshop on Identification and Control of Flexible Space Structures, San Diego, CA (1984).

Watanabe, E., et al., A design method for a STR for a distributed parameter system, Int. J. Contr., 43: 271-283 (1986).

Watanabe, T., A model-based approach for adaptive control optimization in milling, Trans. ASME, J. Dyn. Syst. Meas. Contr., 108: 56-64 (1986).

White, B. A., Range-space dynamics of scalar variable structure control systems, IEE Proc., 133, Pt. D: 35-40 (1986).

Ydstie, B. E., Extended horizon adaptive control, Proc. 9th World Congress of IFAC, Budapest, Vol. 2, Pergamon Press, Oxford, pp. 911-915 (1985).

Zachariah, K. J., and Warwick, K., Auto-weighted minimum variance self-tuning controller, Opt. Contr. Appl. Meth., 7: 41-54 (1986).

Zarrop, M. B., and Fischer, M., Reduced variance pole-assignment STR, Int. J. Contr., 42: 1013-1034 (1985).

Zhang, Q., and Tomizuka, M., Multivariable direct adaptive control of thermal mixing processes, Trans. ASME, J. Dyn. Syst. Meas. Contr., 107: 278-283 (1985).

Zoubeidi, M., et al., Application of learning in adaptive control, Proc. Am. Contr. Conf., Boston, MA, Vol. 3, pp. 1495-1500 (1985).

Zwartbol, T., et al., Attitude estimation and control of manoevering space craft, Automatica, 21: 513-526 (1985).

Additional Books Added in Proof Stage:

Bielecki, T. R., Adaptive control of a Markov chain over finite parameter set without continuity assumptions on the control laws, IEEE Trans. Auto. Contr., AC-31: 985-987 (1986).

Cristi, R., Adaptive control with finite time persistency of excitation, IEEE Trans. Auto. Contr., AC-31: 982-985 (1986).

Ioannou, P. A., and Tsakalis, K. S., A robust adaptive controller, IEEE Trans. Auto. Contr., AC-31: 1033-1043 (1986).

Lai, T. L., and Wei, C.-Z., Extended least squares and their applications to adaptive control and prediction in linear systems, IEEE Trans. Auto. Contr., AC-31: 898-906 (1986).

Mahmoud, M. S., et al., Reduced-order performance of adaptive control systems, IEEE Trans. Auto. Contr., AC-31: 1076-1079 (1986).

Moore, W. E., The parallel implementation of a cascade adaptive identification algorithm, IEEE Trans. Auto. Contr., AC-31: 1074-1076 (1986).

Ortega, R., Stability analysis of discrete-time adaptive controllers for systems with unknown delay, IEEE Trans. Auto. Contr., AC-31: 980-982 (1986).

Nesler, C. G., Adaptive control of thermal processes in buildings, IEEE Control Systems Magazine, 6: 9-13 (1986).

Tsiligiannis, C. A., and Svoronos, S. A., Multivariable self-tuning control via the right interactor matrix, IEEE Trans. Auto. Contr., AC-31: 987-989 (1986).

Yeo, Y. K., and Williams, D. C., An adaptation algorithm for parallel model reference adaptive bilinear systems, IEEE Trans. Auto. Contr., AC-31: 1071-1073 (1986).

Index